Advances in Predictive, Preventive and Personalised Medicine

Volume 16

What this book series is about... Current healthcare: What is behind the issue? For many acute and chronic disorders, the current healthcare outcomes are considered as being inadequate: global figures cry for preventive measures and personalised treatments. In fact, severe chronic pathologies such as cardiovascular disorders, diabetes and cancer are treated after onset of the disease, frequently at near end-stages. Pessimistic prognosis considers pandemic scenario for type 2 diabetes mellitus, neurodegenerative disorders and some types of cancer over the next 10-20 years followed by the economic disaster of healthcare systems in a global scale. Advanced healthcare tailored to the person: What is beyond the issue? Advanced healthcare promotes the paradigm change from delayed interventional to predictive medicine tailored to the person, from reactive to preventive medicine and from disease to wellness. The innovative Predictive, Preventive and Personalised Medicine (PPPM) is emerging as the focal point of efforts in healthcare aimed at curbing the prevalence of both communicable and non-communicable diseases such as diabetes, cardiovascular diseases, chronic respiratory diseases, cancer and dental pathologies. The cost-effective management of diseases and the critical role of PPPM in modernisation of healthcare have been acknowledged as priorities by global and regional organizations and health-related institutions such as the Organisation of United Nations, the European Union and The National Institutes of Health. Why integrative medical approach by PPPM as the medicine of the future? PPPM is the new integrative concept in healthcare sector that enables to predict individual predisposition before onset of the disease, to provide targeted preventive measures and create personalised treatment algorithms tailored to the person. The expected outcomes are conducive to more effective population screening, prevention early in childhood, identification of persons at-risk, stratification of patients for the optimal therapy planning, prediction and reduction of adverse drug-drug or drug-disease interactions relying on emerging technologies, such as pharmacogenetics, pathology-specific molecular patters, sub/cellular imaging, disease modelling, individual patient profiles, etc. Integrative approach by PPPM is considered as the medicine of the future. Being at the forefront of the global efforts, The European Association for Predictive, Preventive and Personalised Medicine (EPMA, http://www.epmanet.eu/) promotes the integrative concept of PPPM among healthcare stakeholders, governmental institutions, educators, funding bodies, patient organisations and in the public domain. Current Book Series, published by Springer in collaboration with EPMA, overview multidisciplinary aspects of advanced bio/medical approaches and innovative technologies. Integration of individual professional groups into the overall concept of PPPM is a particular advantage of this book series. Expert recommendations focus on the cost-effective management tailored to the person in health and disease. Innovative strategies are considered for standardisation of healthcare services. New guidelines are proposed for medical ethics, treatment of rare diseases, innovative approaches to early and predictive diagnostics, patient stratification and targeted prevention in healthy individuals, persons at-risk, individual patient groups, sub/populations, institutions, healthcare economy and marketing.

Springer and the Series Editors welcome book ideas from authors. Potential authors who wish to submit a book proposal should contact Nathalie L'horset-Poulain, Executive Editor (e-mail: nathalie.lhorset-poulain@springer.com)

Nadiya Boyko • Olga Golubnitschaja

Editors

Microbiome in 3P Medicine Strategies

The First Exploitation Guide

 Springer

Editors
Nadiya Boyko
Clinical Laboratory Diagnostics and
Pharmacology
Uzhhorod National University &
Ediens LLC
Uzhhorod, Ukraine

Olga Golubnitschaja
3P Medicine, Radiation Oncology
Universitätsklinik Bonn
Bonn, Nordrhein-Westfalen, Germany

ISSN 2211-3495   ISSN 2211-3509   (electronic)
Advances in Predictive, Preventive and Personalised Medicine
ISBN 978-3-031-19563-1   ISBN 978-3-031-19564-8   (eBook)
https://doi.org/10.1007/978-3-031-19564-8

This Springer imprint is published by the registered company Springer Nature Switzerland AG
The registered company address is: Gewerbestrasse 11, 6330 Cham, Switzerland

Preface

The human body is inhabited by trillions of diverse microorganisms collectively called "microbiome" or "microbiota" consisting of bacteria, viruses, and fungi, amongst others. Microbiota composition is highly individual, being reciprocally involved in development and progression of metabolic particularities. Multi-faceted effects of microbiome are associated with human physical and mental health status, and can act as supportive or protective factors, in favour or against a developing pathology depending on the personalised patient profile. Consequently, an individual microbiome composition provides an option to modulate modifiable risk factors and, therefore, creates a highly attractive operational area for the translational biomedical research with multi-professional expertise and healthcare-relevant output in the framework of predictive, preventive and personalised medicine (PPPM/3PM).

Microbiota composition modulates individual immunity at the local (such as skin, gut and airway) and global (systemic immune reactions distal from sites of their colonisations) levels. For example, microbiota plays a crucial role in individual vulnerability/resistance against the SARS-CoV-2 infection. On the other hand, microbiota profiles are highly relevant for vaccination efficacy, determining, therefore, individual outcomes and mortality rates, e.g. under the COVID-19 pandemic conditions.

Microbiome plays a pivotal role in the urogenital health in female and male populations. To this end, patient phenotyping and stratification are highly recommended to include individualised microbiota composition tests, e.g. in female patients diagnosed with the vulvar-vaginal dryness. Immune system and vaginal microbiota composition interplay represents the modifiable risk factor in HPV-induced carcinogenesis. Consequently, relevant bio-fluids (vaginal swabs and cervico-vaginal lavage/secretions) provide important information about novel immunological targets for advanced diagnostics and therapy algorithms. Similarly, urinary tract infections are associated with primary PCa risks: a specific urinary microbiota composition is decisive for the low-grade inflammation frequently associated with more aggressive forms of metastatic PCa. Further, the microbiome of semen is of great interest of translational medicine as strongly impacting the reproductive health of men, the health of the couple and even the health of offspring

Since individual microbiome composition is a strong contributor to potential organ damage and chronic pathologies, personalised microbiota correction and application of pre- and probiotics are at the forefront of advanced treatments tailored to the person.

Advanced 3PM strategies implementing in the microbiome area patient stratification, predictive and companion diagnostics, targeted preventive measures and personalised treatment algorithms, seem to hold a promis of therapeutic modalities without side-effects, significantly improve individual outcomes and overall cost-efficacy of healthcare. According to the accumulated research data, corresponding diagnostic and treatment approaches are applicable to primary (health risk assessment in individuals with sub-optimal health conditions and prevention of a disease development), secondary (personalised treatment of clinically manifested disorders preventing a disease progression) and tertiary (optimal management of non-curable diseases) care.

In the current book, we do highlight the implementation potential of the microbiome-relevant research in the framework of predictive diagnostics, targeted prevention and treatments tailored to the individualised patient profile.

Bonn, Germany   Olga Golubnitschaja

Uzhhorod, Ukraine   Nadiya Boyko

What This Book Series Is About...

Why currently applied reactive medicine is unsatisfactory for healthcare?

First two decades and the running third one of the twenty-first century are characterised by epidemics of both communicable (COVID-19) and non-communicable diseases. Regarding the latter, there is an unprecedented decrease in the age of the affected subpopulation including mood disorders, "young" strokes (below 50 years of age) and aggressive metastatic cancers reported for the 20+ years old patients with particularly poor outcomes. Consequent socio-economic burden is tremendous.

EPMA promotes the paradigm change from reactive to predictive, preventive and personalised medicine (PPPM / 3PM) benefitting healthcare and society as a whole

Nobody lives forever, but everyone wants to remain mentally sharp and physically fit as long as they live. Traditional reactive medicine provides medical services to the group of people who are already ill. In contrast, predictive, preventive and personalised medicine (3PM) aims to keep the whole population healthy. 3PM is thus superior, both from an ethical and a cost perspective.

The current book series, published by Springer in collaboration with EPMA, overviews advanced bio/medical approaches and innovative technologies focused on predictive diagnostics, targeted prevention and personalisation of treatments tailored to the person. Patient needs are in the focus. Integration of professional groups with complementary expertise into holistic concepts of 3PM is a particular advantage of this book series. Highly innovative technological tools are presented and steadily updated for health risk assessment in individuals with sub-optimal health conditions and prevention of a disease development, secondary care (personalised treatment of clinically manifested disorders preventing a disease progression), and tertiary care (making palliation to an optimal management of non-curable diseases). Expert recommendations are focused on the cost-effective management tailored to the person in health and disease. Innovative strategies are considered for translational research and standardisation of healthcare services. New guidelines are proposed for health policy, medical ethics and improved healthcare economy.

Bonn, Germany   Olga Golubnitschaja

Book Series Editor

Prof. Dr. Olga Golubnitschaja is the head of the world's first predictive, preventive and personalised (3P) medicine unit in the Department of Radiation Oncology, University Hospital Bonn, Rheinische Friedrich-Wilhelms-Universität Bonn, Germany. Dr. Golubnitschaja is educated in journalism, biotechnology and medicine and has been awarded research fellowships in Austria, Russia, the UK, Germany, the Netherlands, and Switzerland (early and predictive diagnostics in paediatrics, neurosciences and cancers).

Dr. Golubnitschaja is the author of more than 400 international publications (research and review articles, position papers, books, book and congress contributions) in the innovative field of predictive, preventive and personalised medicine (3PM) with the main research focusing on sub-optimal health conditions, pre- and perinatal diagnostics, diagnostics of cardiovascular disease and neurodegenerative pathologies, and predictive diagnostics in cancer and diabetes. Please note, in years 1990–2002, she published as *Olga Labudova*. Her cumulative Google Scholar *h*-index is 60.

Awards: National & International Fellowship of the Alexander von Humboldt-Foundation; Highest Prize in Medicine and Eiselsberg-Prize in Austria; Springer-Nature Award; EMA Award.

In years 2009–2021, Dr. Golubnitschaja was the secretary general, and since September 2021, she is the president of the "European Association for Predictive, Preventive & Personalised Medicine" (EPMA, Brussels), networking over 50 countries worldwide, www.epmanet.eu. She is editor-in-chief of the *EPMA J.* (actual Clarivate Analytics IF 8.836, Scopus CiteScore 11.3) and editor-in-chief of the book series Advances in Predictive, Preventive & Personalised Medicine, Springer Nature.

Dr. Golubnitschaja is the European Representative in the EDR-Network at the National Institutes of Health USA, http://edrn.nci.nih.gov/.

Dr. Golubnitschaja acts as a regular reviewer in over 50 clinical and scientific journals and as a grant reviewer of international and national funding bodies in European and other countries.

Dr. Golubnitschaja is an evaluation expert at the European Commission. In years 2010–2013, she was involved in creating the PPPM-related contents of the European Programme "Horizon 2020".

Currently, Dr. Golubnitschaja is vice-chair of the Habilitation Committee (responsible for all medical specialisations) at the Medical Faculty, University of Bonn, Germany.

Dr. Golubnitschaja is vice-chair of the Evaluation Panel for Marie Curie Mobility Actions at the European Commission in Brussels.

Contents

About the Editors

Nadiya Boyko, Professor, Doctor of Sciences of Microbiology, Head of the Department of Clinical Laboratory Diagnostics and Pharmacology, Director of Research Development and Educational Centre of Molecular Microbiology and Mucosal Immunology of Uzhhorod National University, Ukraine. The research field priorities particularly are personalised nutrition and individual pharmabiotics in regulation of human [gut] microbiota for prevention of noncommunicable diseases, precise diagnostics, P4 medicine.

Job-related skills: 35 years' experience in research and leading research in bioeconomy, food and health relevant fields, mainly dealing with food safety, human nutrition and human microbiome, individual nutrition needs, FCDB, impact of food on human host, mucosal immune response etc.

She is a lecturer for native and foreign (English) speaking students, supervising PhD students, regularly acting as invited lecturer for the variety of international and national meetings, and master classes. She is an intensive participant with oral presentations, discussions, debates, etc. in health and food relevant EU and other countries leader associations' workshops, congresses, committees, ambassador for Global Harmonization Initiative, president for SOMED (2022–2024).

She is dealing with a lot of national and international groups of contacts on regular bases and participating as p.i. in around 50 projects. She is also the author of about

350 scientific works, including 100 papers in professional scientific journals, more than 50 publications in the peer-reviewed journals, and Chapter in Elsevier press, index H=12. She is specialist with expertise in life sciences, participated in international projects – BaSeFood, JSO-ERA EU – FP7, CAPINFOOD (SEE), BacFoodNet (COST), ODiN (FP7), FoodWARD (Erasmus) and H2020 (SKIN) projects.

Olga Golubnitschaja is the head of the world's first predictive, preventive and personalised (3P) medicine unit in the Department of Radiation Oncology, University Hospital Bonn, Rheinische Friedrich-Wilhelms-Universität Bonn, Germany. Dr. Golubnitschaja is educated in journalism, biotechnology and medicine and has been awarded research fellowships in Austria, Russia, the UK, Germany, the Netherlands, and Switzerland (early and predictive diagnostics in paediatrics, neurosciences and cancers).

Dr. Golubnitschaja is the author of more than 400 international publications (research and review articles, position papers, books, book and congress contributions) in the innovative field of predictive, preventive and personalised medicine (3PM) with the main research focusing on sub-optimal health conditions, pre- and perinatal diagnostics, diagnostics of cardiovascular disease and neurodegenerative pathologies, and predictive diagnostics in cancer and diabetes. Please note, in years 1990–2002, she published as *Olga Labudova*. Her cumulative Google Scholar *h*-index is 60.

Awards: National & International Fellowship of the Alexander von Humboldt-Foundation; Highest Prize in Medicine and Eiselsberg-Prize in Austria; Springer-Nature Award; EMA Award.

In years 2009–2021, Dr. Golubnitschaja was the secretary general, and since September 2021, she is the president of the "European Association for Predictive, Preventive & Personalised Medicine" (EPMA, Brussels), networking over 50 countries worldwide, www.epmanet.eu. She is editor-in-chief of the *EPMA J.* (actual Clarivate Analytics IF 8.836, Scopus CiteScore 11.3) and editor-in-chief of the book series Advances in Predictive, Preventive & Personalised Medicine, Springer Nature.

Dr. Golubnitschaja is the European Representative in the EDR-Network at the National Institutes of Health USA, http://edrn.nci.nih.gov/.

Dr. Golubnitschaja acts as a regular reviewer in over 50 clinical and scientific journals and as a grant reviewer of international and national funding bodies in European and other countries.

Dr. Golubnitschaja is an evaluation expert at the European Commission. In years 2010–2013, she was involved in creating the PPPM-related contents of the European Programme "Horizon 2020".

Currently, Dr. Golubnitschaja is vice-chair of the Habilitation Committee (responsible for all medical specialisations) at the Medical Faculty, University of Bonn, Germany.

Dr. Golubnitschaja is vice-chair of the Evaluation Panel for Marie Curie Mobility Actions at the European Commission in Brussels.

Chapter 1
Microbiome in the Framework of Predictive, Preventive and Personalised Medicine

Nadiya Boyko, Vincenzo Costigliola, and Olga Golubnitschaja

Abstract The human body is inhabited by trillions of diverse microorganisms collectively called "microbiome" or "microbiota". Microbiota consists of bacteria, viruses, fungi, protozoa, and archaea. Microbiome demonstrates multi-faceted effects on human physical and mental health. Per evidence there is a multi-functional interplay between the whole-body microbiome composition on the epithelial surfaces including skin, nasal and oral cavities, airway, gastro-intestinal and urogenital tracts on one hand and on the other hand, the individual health status. Microbiota composition as well as an option to modulate it – together create a highly attractive operation area for the translational bio/medical research with multi-professional expertise and healthcare-relevant output in the framework of predictive, preventive and personalised medicine (PPPM/3 PM). Advanced PPPM strategies implemented in the microbiome area are expected to significantly improve individual outcomes

N. Boyko
RDE Center of Molecular Microbiology and Mucosal Immunology, Uzhhorod National University, Uzhhorod, Ukraine

Department of Clinical Laboratory Diagnostics and Pharmacology, Uzhhorod National University, Uzhhorod, Ukraine

Ediens LLC, Uzhhorod, Ukraine
e-mail: nadiya.boyko@uzhnu.edu.ua

V. Costigliola
European Association for Predictive, Preventive and Personalised Medicine, EPMA, Brussels, Belgium

European Medical Association, EMA, Brussels, Belgium
e-mail: vincenzo@emanet.org

O. Golubnitschaja (✉)
Predictive, Preventive and Personalised (3P) Medicine, Department of Radiation Oncology, University Hospital Bonn, Rheinische Friedrich-Wilhelms-Universität Bonn, Bonn, Germany

3PMedicon, Taufkirchen an der Pram, Austria
e-mail: olga.golubnitschaja@ukbonn.de

© The Author(s), under exclusive license to Springer Nature
Switzerland AG 2023
N. Boyko, O. Golubnitschaja (eds.), *Microbiome in 3P Medicine Strategies*,
Advances in Predictive, Preventive and Personalised Medicine 16,
https://doi.org/10.1007/978-3-031-19564-8_1

and overall cost-efficacy of healthcare. According to the accumulated research data, corresponding diagnostic and treatment approaches are applicable to primary care (health risk assessment in individuals with sub-optimal health conditions and prevention of a disease development), secondary care (personalised treatment of clinically manifested disorders preventing a disease progression) and tertiary care (making palliation to an optimal management of non-curable diseases). In the current book, we do highlight the implementation potential of the microbiome-relevant research in the framework of predictive diagnostics, targeted prevention and treatments tailored to the individualised patient profile.

Keywords Predictive Preventive Personalised Medicine (PPPM/3PM),
Microbiome · Microbiota composition · Suboptimal health · Primary, secondary, tertiary care · Male and female health · Bacteria, viruses, fungi, protozoa, archaea · Immunity · Gut · Skin · Airway · Urogenital tract · Gastro-intestinal tract · Prostate · Vagina · Microbiota-host interaction · Inflammation · Wound healing · Phenotyping · BMI · Overweight · Underweight · COVID-19 · Organ damage · Chronic diseases · Cancers · Muti-level diagnostics · Treatment · Individual outcomes

1.1   Microbiota Composition Modulates Individual Immunity with Far-Reaching Consequences

The microbiota colonisation of the human body starts at birth establishing a symbiotic lifelong relationship with the host by performing a great number of functions such as preventing pathogen colonisation and modulating individual immunity [1] As recently analysed in detail by de Jong SE et al. [2] there are two levels of the immune response modulation by microbiota, namely the local and global ones. The local immune response modulation can be well exemplified by skin, gut and airway microbiota. The gut immune system is capable to response towards specific metabolite patterns produced and regulated by the intestinal microbiota. Microbiota-host interaction at the skin can activate the interleukin-1 (IL-1) signalling pathway to trigger the recruitment of innate lymphoid cells as demonstrated immune-related skin disorders. Although being less diverse than in the gut, microbiota is also present in the lung locally modulating the immune response. To this end, tissue-resident memory T cells in the lung can be reactivated by metabolites produces by microflora present in the lung.

Besides the local interplay, microbiota modulates systemic immune reactions distal from sites of their colonisations. The evidence-based mechanisms are the systemic circulation of bacterial products and dissemination of the microbiota-derived metabolites. Consequently, low microbiota diversity early in life is associated with a development of corresponding immune phenotypes, which in turn might be decisive for a lifelong disease predisposition that should be taken into account considering individual patient profiles and predictive approach for treating acute and chronic pathologies. To this end, microbiota profile has been demonstrated as being highly relevant for individual vaccination outcomes [2] such as under COVID-19 pandemic conditions (see below subchapter).

1.2   Microbiome and Individual COVID-19 Outcomes

Per evidence, microbiota plays a crucial role in individual COVID-19 outcomes. To this end, over 50% of deaths in COVID-19-infected patients are associated with bacterial superinfections [3]. For the patients with poor COVID-19 outcomes specifically high levels of Prevotella, Staphylococcus and Fusobacterium representing periodontopathic bacteria were demonstrated as the specific feature of their microbiota profiles. For 80% of patients treated at intensive care units, a particularly high oral bacterial load has been recorded. To this end, altered oral microbiome profiles, systemic inflammatory processes and poor oral health are well acknowledged risk factors specific for some vulnerable patient cohorts with poor COVID-19 outcomes, such as elderly, diabetes mellitus, hypertension and cardiovascular disease [4]. The disease-progression specific interaction between the viral particles and the systemic host microbiota is the proposed pathomechanism of the COVID-19 infection [5]. Since an aspiration of periodontopathic bacteria induces the expression of angiotensin-converting enzyme 2—the receptor for SARS-CoV-2—and production of inflammatory cytokines in the lower respiratory tract, poor oral hygiene and periodontal disease have been proposed as leading to the COVID-19 aggravation [6]. Consequently, the issue-dedicated expert recommendations are focused on the optimal oral hygiene [7] as being crucial for improved individual outcomes and reduced morbidity under the COVID-19 pandemic conditions [3, 8]. For an effective prevention, an application of oral probiotics has been proposed connecting the gut-lung axis with the viral and microbial pathogenesis, inflammation, secondary infections and severe complications linked to COVID-19 [9, 10].

1.3   Microbiome in Female Health

Vulvar-vaginal dryness (VVD) affects both pre- and postmenopausal women at any age and may lead to severe concomitant pathologies including female genital cancers. VVD carries a multi-factorial character, and there are several modifiable risk factors considered as preventable within the primary care. Consequently, treatment algorithms tailored to the individualised patient profile have a potential to milder or even reverse the VVD. To this end, VVD-relevant individualised patient profiling is highly recommended to involve phenotyping (e.g. Flammer Syndrome and Sicca Syndrome) and follow-up examinations including biomarker panels reflecting abnormal BMI, stress overload, disturb microcirculation and microbiota composition, amongst others [11].

Further, the immune system and vaginal microbiota composition interplay represents the modifiable and important risk factors in HPV-induced carcinogenesis. In turn, HPV infection leads to a gradually increasing colonisation by anaerobic bacteria significantly contributing to the severity of cervical dysplasia. This evidence is of great clinical utility for prediction of the malignant potential. Identifying microbiota profiles specific for high- versus low-grade dysplasia is helpful to identify patients at high risk of the disease progression. To this end, any unnecessary surgical treatment of cervical dysplasia negatively affects obstetrical outcomes and life quality of the treated patients. Therefore, microbiome-based therapies tailored to the individualised patient profiles are pivotal for the clinically relevant decision, conservative disease management of HPV-associated pre-cancer conditions and improved individual outcomes. From the practical point of view, a detailed evaluation of HPV capabilities to evade immune mechanisms from various bio-fluids (vaginal swabs, cervico-vaginal lavage/secretions, or blood) could promote the identification of new immunological targets for novel individualized diagnostics and therapy. Associated health risk assessment may represent the crucial tool for predictive diagnostics and personalised mitigating and preventive measures essential for improving state-of-the-art primary, secondary and tertiary care in individuals predisposed to the disease, patients with cervical pre-cancer conditions and cervical malignancies [12].

1.4   Microbiome in Male Health

Human microbiome plays an important and nuanced role in controlling immunity and cancer development in male subpopulations. With about 1.41 million new cases annually registered, prostate cancer (PCa) is the world leading male cancer. The corresponding socio-economic burden is enormous: the costs of treating PCa are increasing more rapidly than those of any other cancer [13]. Anti-cancer mRNA-based therapy is a promising approach but pragmatically considered, it will take years or even decades to make mRNA therapy working for any type of cancers, and if possible, for individual malignancy sub-types which are many specifically for the PCa. Multi-professional expertise is pivotal to create and implement anti-PCa programmes in the population considering primary (health risk assessment), secondary (prediction and prevention of metastatic disease in PCa) and tertiary (making palliative care to the management of chronic disease) care in the framework of predictive, preventive and personalised medicine [13].

To this end, urinary tract infections are associated with a primary PCa risk linked to chronic inflammation, in which a specific urinary microbiota composition plays a crucial role. Corresponding pathomechanisms consider the chronic inflammation initiated by microbial species persisting in the urinary tract as the trigger which promotes prostate inflammatory atrophy increasing the risk of PCa development.

Moreover, specifically the low-grade inflammation could be associated with the presence of more aggressive forms of PCa on one hand and, on the other hand, with specific microbiota composition of the genital-urinary tract [14]. The prostate microbiota includes viral, bacterial, fungal, and parasitic contributions. Based on the urine samples, a unique microbiota signature has been demonstrated for higher Gleason score cancers with a marked number of viral genomic insertions into host DNA; microbiota profiles are distinguishable between the prostate malignancy and benignancy [15] The gut microbiota composition is also associated with the prostate cancer development and progression, apparently by an impaired balance between inflammatory and anti-inflammatory bacterial lipopolysaccharides, production of bile salts, and metabolism of dietary fiber to short chain fatty acids. Collectively gut and urogenital microbiomes impact both – the PCa development and individual treatment outcomes [15].

Finally, the microbiome of semen is of increasing scientific interest: per evidence, the seminal microbiota set-up strongly impacts the reproductive health of men, the health of the couple and even the health of offspring, owing to transfer of microorganisms to the partner and offspring [16]. This field is currently under development with highly promising clinically relevant output.

1.5 Outlook in the Framework of 3P Medicine: Microbiome-Relevant Fields Exemplified

1.5.1 Microbiome Is Instrumental for Primary Healthcare Towards Suboptimal Health Conditions: Focus on Reversing Epidemic Trends of Non-communicable Diseases

First two decades of the twenty-first century are characterised by epidemics of non-communicable diseases. There is an unprecedented decrease in the age of the affected subpopulation including mood disorders, "young" strokes (below 50 years of age), aggressive metastatic cancers reported for 20 + years old patients with particularly poor outcomes. Consequent socio-economic burden is tremendous. To reverse the trend, the paradigm shift from reactive medical services to predictive approach, targeted prevention and personalization of treatments has been proposed by the European Association for Predictive, Preventive and Personalised Medicine (EPMA, Brussels, www.epmanet.eu). Particularly modifiable risk factors are instrumental for the cost-effective prevention of illnesses in the population. To this end, many persons complain poor healthy perception but are not in diagnosable disorders. This intermediate state between health and a diagnosable disease is known as suboptimal health status, which is a reversible borderline condition between optimal health and clinically manifested pathology. Consequently, suboptimal health conditions are considered the operational timeframe for implementing predictive diagnostics and cost-effective targeted prevention in the population. For advanced health risk assessment, specifically liquid biopsy (body fluids) and microbiome analysis are strongly recommended as essential pillars of the comprehensive individualised patient profiling at the level of suboptimal health (primary health-care) [17].

1.5.2   Microbiome of Overweight Versus Underweight Individuals: A Disease-Specific Phenotyping

Microbiotic profiles differ significantly between overweight and underweight individuals: whereas 75% of the obesity-enriched genes originate from Actinobacteria (compared with 0% of lean-enriched genes; the other 25% are from Firmicutes), 42% of the lean enriched genes originate from Bacteroidetes (compared with 0% of the obesity-enriched genes) [18]. To this end, microbiota composition specific for individuals with abnormally high BMI is in focus of many studies dedicated to metabolic syndrome with complications. Although being highly clinically relevant, only little evidence is currently available about the microbiome set-up characteristic for individuals with a low BMI. In extreme cases, underweight patients are, further diagnosed with anorexia nervosa (AN) which is one of the prominent eating disorders linked to cascading pathologies such as mood disorders, amongst others, which in turn, are associated with a shifted enteric microbiota composition. To this end, diseased eating significantly limits gut flora diversity that, per evidence, results in intestinal dysbiosis (ID) typical for AN. ID of a different severity grade, presents a disruption to the microbiota homeostasis caused by an imbalanced microflora. Compared to healthy individuals with normal BMI, *Phylum Bacteroidetes* is decreased, whereas *Phylum Firmicutes* is increased in AN. Consequently, intestinal microbiota profiling may be of pivotal importance to predict persistence, recovery from and/or relapse of the eating disorders and evidence-based personalised diet prescription.

In conclusion, being amongst others linked to abnormal BMI, a disease-specific phenotyping is pivotal for individualised prediction and targeted prevention as schematically presented in Fig. 1.1. Currently available evidence demonstrates microbiota composition as reciprocally involved in development and progression of above listed metabolic particularities and associated pathologies. Advanced PPPM strategies implemented in the microbiome area are expected to significantly improve individual outcomes and overall cost-efficacy of healthcare. According to the accumulated research data, corresponding diagnostic and treatment approaches are applicable to primary care (health risk assessment in individuals with sub-optimal health conditions and prevention of a disease development), secondary care (personalised treatment of clinically manifested disorders preventing a disease progression) and tertiary care (making palliation to an optimal management of

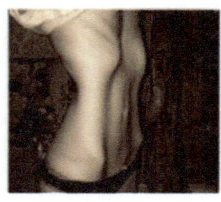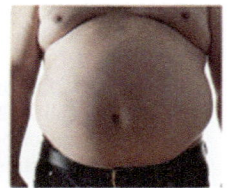

Low BMI   High BMI
High physical activity Low physical activity
Low blood pressure   Dyslipidemia

The general population

Anorexic Phenotype; increased risk for
- Primary vascular dysregulation
- Flammer syndrome
- Systemic hypoxia / ischemic lesions
- Eye disorders (normal-tension glaucoma)
- Mental disorders (depression)
- Neurological disorder (mulitple-sclerosis)
- Reproductive dysfunction
- Xerostomia / Siccca syndrome
- Chronic inflammation
- Impaired wound healing
- Predisposition to cancer
with particularly poor outcomes

Obese Phenotype; increased risk for
Arteriosclerosis -
Hypertension -
Stroke -
Metabolic syndrome -
Eye disorders -
Mental and neurological disorders -
Reproductive dysfunction -
Xerostomia / Sicca syndrome -
Chronic inflammation -
Impaired wound healing -
Predisposition to cancer -
with particularly poor outcomes

Fig. 1.1 Phenotyping is instrumental for the health risk assessment: Anorexic versus obese phenotype [18]

non-curable diseases). In the current book, we do highlight the implementation potential of the microbiome-relevant research in the framework of predictive diagnostics, targeted prevention and treatments tailored to the individualsed patient profile.

References

1. Yu JC, Khodadadi H, Baban B (2019) Innate immunity and oral microbiome: a personalized, predictive, and preventive approach to the management of oral diseases. EPMA J 10(1):43–50. https://doi.org/10.1007/s13167-019-00163-4
2. de Jong SE, Clin A, Pulendran B (2020) The impact of the microbiome on immunity to vaccination in humans. Cell Host Microbe 28(2):169–179. https://doi.org/10.1016/j.chom.2020.06.014
3. Sampson V, Kamona N, Sampson A (2020) Could there be a link between oral hygiene and the severity of SARS-CoV-2 infections? Br Dent J 228(12):971–975. https://doi.org/10.1038/s41415-020-1747-8
4. Pitones-Rubio V, Chávez-Cortez EG, Hurtado-Camarena A, González-Rascón A, Serafín-Higuera N (2020) Is periodontal disease a risk factor for severe COVID-19 illness? Med Hypotheses 144:109969. https://doi.org/10.1016/j.mehy.2020.109969
5. Xiang Z, Koo H, Chen Q, Zhou X, Liu Y, Simon-Soro A (2020) Potential implications of SARS-CoV-2 oral infection in the host microbiota. J Oral Microbiol 13(1):1853451. https://doi.org/10.1080/20002297.2020.1853451

6. Takahashi Y, Watanabe N, Kamio N, Kobayashi R, Iinuma T, Imai K (2020) Aspiration of periodontopathic bacteria due to poor oral hygiene potentially contributes to the aggravation of COVID-19. J Oral Sci 63(1):1–3. https://doi.org/10.2334/josnusd.20-0388
7. Tachalov VV, Orekhova LY, Kudryavtseva TV, Loboda ES, Pachkoriia MG, Berezkina IV, Golubnitschaja O (2021) Making a complex dental care tailored to the person: population health in focus of predictive, preventive and personalised (3P) medical approach. EPMA J 12(2):129–140. https://doi.org/10.1007/s13167-021-00240-7
8. Botros N, Iyer P, Ojcius DM (2020) Is there an association between oral health and severity of COVID-19 complications? Biom J 43(4):325–327. https://doi.org/10.1016/j.bj.2020.05.016
9. Abdelhamid AG, El-Masry SS, El-Dougdoug NK (2019) Probiotic Lactobacillus and Bifidobacterium strains possess safety characteristics, antiviral activities and host adherence factors revealed by genome mining. EPMA J 10(4):337–350. https://doi.org/10.1007/s13167-019-00184-z
10. Baindara P, Chakraborty R, Holliday ZM, Mandal SM, Schrum AG (2021) Oral probiotics in coronavirus disease 2019: connecting the gut-lung axis to viral pathogenesis, inflammation, secondary infection and clinical trials. New Microbes New Infect 40:100837. https://doi.org/10.1016/j.nmni.2021.100837
11. Goncharenko V, Bubnov R, Polivka J Jr, Zubor P, Biringer K, Bielik T, Kuhn W, Golubnitschaja O (2019) Vaginal dryness: individualised patient profiles, risks and mitigating measures. EPMA J 10(1):73–79. https://doi.org/10.1007/s13167-019-00164-3
12. Kudela E, Liskova A, Samec M, Koklesova L, Holubekova V, Rokos T, Kozubik E, Pribulova T, Zhai K, Busselberg D, Kubatka P, Biringer K (2021) The interplay between the vaginal microbiome and innate immunity in the focus of predictive, preventive, and personalized medical approach to combat HPV-induced cervical cancer. EPMA J 12(2):199–220. https://doi.org/10.1007/s13167-021-00244-3
13. Ellinger J, Alajati A, Kubatka P, Giordano FA, Ritter M, Costigliola V, Golubnitschaja O (2022) Prostate cancer treatment costs increase more rapidly than for any other cancer-how to reverse the trend? EPMA J 13(1):1–7. https://doi.org/10.1007/s13167-022-00276-3
14. Kucera R, Pecen L, Topolcan O, Dahal AR, Costigliola V, Frank A, Giordano FA, Golubnitschaja O (2020) Prostate cancer management: long-term beliefs, epidemic developments in the early 21st century and 3PM dimensional solutions. EPMA J 11(3):399–418. https://doi.org/10.1007/s13167-020-00214-1
15. Wheeler KM, Liss MA (2019) The microbiome and prostate cancer risk. Curr Urol Rep 20(10):66. https://doi.org/10.1007/s11934-019-0922-4
16. Altmäe S, Franasiak JM, Mändar R (2019) The seminal microbiome in health and disease. Nat Rev Urol 16(12):703–721. https://doi.org/10.1038/s41585-019-0250-y
17. Wang W, Yan Y, Guo Z, Hou H, Garcia M, Tan X, Anto EO, Mahara G, Zheng Y, Li B, Kang T, Zhong Z, Wang Y, Guo X, Golubnitschaja O (2021) Suboptimal health study consortium and european association for predictive, preventive and personalised medicine. All around suboptimal health – a joint position paper of the suboptimal health study consortium and European Association for Predictive, Preventive and Personalised Medicine. EPMA J 12(4):403–433. https://doi.org/10.1007/s13167-021-00253-2
18. Golubnitschaja O, Liskova A, Koklesova L, Samec M, Biringer K, Büsselberg D, Podbielska H, Kunin AA, Evsevyeva ME, Shapira N, Paul F, Erb C, Dietrich DE, Felbel D, Karabatsiakis A, Bubnov R, Polivka J, Polivka J Jr, Birkenbihl C, Fröhlich H, Hofmann-Apitius M, Kubatka P (2021) Caution, "normal" BMI: health risks associated with potentially masked individual underweight-EPMA position paper 2021. EPMA J 12(3):243–264. https://doi.org/10.1007/s13167-021-00251-4

Chapter 2
Artificial Intelligence-Based Predictive, Preventive, and Personalised Medicine Applied to Bacteraemia Diagnosis

Oscar Garnica ⓘ, José M. Ruiz-Giardín ⓘ, and J. Ignacio Hidalgo ⓘ

Abstract One of the most promising aspects of applying artificial intelligence is in the field of predictive, preventive, and personalised medicine (PPPM). PPPM principles can be applied to all medical domains. To this aim, PPPM can use methodologies, techniques, and tools provided by any other discipline, such as mathematics or engineering. Among them, one of the most promising is computer science, particularly artificial intelligence. Artificial intelligence is the new paradigm that will change how many of the tasks currently done by humans will be done shortly. This potential has been envisioned by World Health Organization that has deposited great expectations on how artificial intelligence will improve health care provisioning.

This chapter illustrates the application of PPPM principles to early bacteraemia prediction using machine learning techniques.

Bacteraemia can produce severe sepsis, one of the most common causes of morbidity and mortality, and its prognosis depends on a rapid diagnosis and an appropriate antibiotic treatment. Currently, the bacteraemia diagnosis is based on laboratory tests that can take up to 6 days. An early prediction based on machine learning techniques using electronic health records of hospital patients along with the physician analysis of patient variables would shorten up to 6 days the start of early administration of personalised antibiotics and additional treatments that would prevent subsequent complications.

The chapter summarises the current approach to treating bacteraemia, its deficits and its clinical, economic, and structural consequences. Afterwards, it presents all the steps to apply machine learning of hospital records: cleaning the data, detecting bias, handling missing data, and applying machine learning techniques. Data have

O. Garnica (✉) · J. I. Hidalgo
Department of Computer Architecture, Universidad Complutense de Madrid, Madrid, Spain
e-mail: ogarnica@ucm.es; hidalgo@ucm.es

J. M. Ruiz-Giardín
Department of Internal Medicine, Hospital Universitario de Fuenlabrada, Madrid, Spain
e-mail: josemanuel.ruiz@salud.madrid.org

© The Author(s), under exclusive license to Springer Nature Switzerland AG 2023
N. Boyko, O. Golubnitschaja (eds.), *Microbiome in 3P Medicine Strategies*, Advances in Predictive, Preventive and Personalised Medicine 16, https://doi.org/10.1007/978-3-031-19564-8_2

9

been analysed using three machine learning techniques: Support Vector Machine, Random Forest, and K-Nearest Neighbours.

The machine learning techniques provide state-of-the-art results in metrics of interest in predictive medical models with values that exceed the medical practice threshold and previous results in the literature using classical modelling techniques in specific types of bacteraemia.

Keywords Predictive preventive personalised medicine (PPPM/3 PM) · Artificial intelligence · Bacteraemia · Modelling · Machine learning · Bid data analysis · Expert recommendations

2.1   Predictive, Preventive and Personalised Medicine and Bacteraemias

The predictive, Preventive and Personalised Medicine (PPPM/3PM) paradigm is being promoted to improve healthcare practices. A core element in 3P medicine is the need for a predictive patient-tailored medicine focused on predicting disease development and customising the physician practices to the patient's variables [20]. In addition, preventing the disease also avoids the severe complications accompanying many of them.

Sepsis with bacteraemia is one of the medical complications with higher morbidity and mortality [87]. Hence, sepsis provokes around 19 million cases, 5 million deaths annually worldwide [17], and according to [58], the case-fatality bacteraemia rate is 12%.

Many diseases targeted by 3PM are originated by bacteraemia or have bacteraemia among their complications. For such a reason, 3PM principles can be used to predict the risk of complications in bacteraemia, prevent them, and customise the medical practice [21, 77].

The application of preventive actions has had several successes, as for example, the Michigan-keystone project that was focused in reducing the catheter-related bloodstream infections in children [54]. A personalised patient-tailored antibiotic treatment follows the prediction of bacteraemia and its source; that is, each patient needs a specific antibiotic treatment according to the bacteraemia's focus and his/her clinical situation: i.e., age, vaccination coverage, fever, hemodynamic situation, source of infection, type of bacterial infection, laboratory markers, if he has suffered previous hospital incomes, received antibiotics before, or been exposed to invasive procedures, or if a multiresistant microorganism has colonised him, among others. The correct antibiotic election is intimately related to the morbidity and mortality of the patient [19, 57, 76].

2.2 Relevance of Bacteraemia Prediction

Bacteraemia prognosis primarily depends on a rapid diagnosis. A patient-tailored antibiotic treatment should be promptly administered whenever a serious bacterial infection is suspected, but, if possible, after taking blood cultures [23, 68]. For such a reason, the bacteraemia prediction will bring an improvement in patient healthcare.

Bacteraemia is detected via blood cultures that can take up to 6 days to provide a definitive diagnosis. So, the prediction of bacteraemia would reduce this diagnosis lag, shortening up to 6 days the diagnosis. In conjunction with patient variables, this prediction should be considered to initiate the early administration of personalised antibiotic treatment and medical services, select specific diagnostic techniques, and determine the need for additional treatments that would prevent or reduce the subsequent bacteraemia complications.

Therefore, the bacteraemia prediction would decrease its morbidity and mortality by starting an early, appropriate, and specific antibiotic treatment.

2.3 Artificial Intelligence as a Key Technology for PPPM

Predicting bacteriemia and its complications applying artificial intelligence (AI) techniques on the patient's data in the electronic hospital records is a case of using multidisciplinary PPPM/3PM strategies to improve healthcare. Hence, AI techniques will improve the prevention of bacteraemia by identifying patients with bacteraemia and their specific bacteraemia source earlier. The bacteraemia's source is key knowledge because it determines the diagnostic techniques to search its origin, the specific and most appropriate antibiotic treatment, and it helps determine additional treatments that sometimes must be combined with the antibiotic treatment [85].

Bacteriemia prediction is just one of many examples where AI techniques can contribute to predicting some diagnosis and provide physicians with relevant information that help them make informed decisions. In this way, ML techniques will contribute an important added value to the three pillars of 3P medicine.

2.4 World Health Organisation's Recommendations About Artificial Intelligence

AI is transforming our society thanks to its ability to perform tasks that previously seemed reserved for human beings. AI techniques are widely known in our day-to-day in multiple domains, such as internet searches, automatic recommendation systems, conversational bots, natural language processing, or computer programs capable of beating human champions in games like chess or go.

ML is a type of AI that focuses on techniques that allow computers to generate models for decision-making in predictions from observed behaviour on previously collected data sets. To do this, the algorithm analyses large volumes of data, identifies and synthesises patterns and, based on them, generates predictions about the behaviour of new data not used during model training. For all these models, it is critical to have a significant volume of data: the larger the volume of input data, the more relevant patterns it will be able to identify, the more information can be extracted from the patterns, and the more accurate the model's predictions on data will be -a.k.a. the generalisation of the model-.

Currently, AI is being applied in many different sectors. In [52], the World Health Organisation (WHO) exposes the great expectations that AI offers, and therefore machine learning, to improve the provision of health care and medicine around the world. In this report, WHO proposes that the governance of AI be based on six principles. Especially relevant from a computational point-of-view are two of those six guiding principles:

- Ensuring transparency, clarity, and intelligibility of AI, and
- promoting responsibility and accountability.

In order to satisfy transparency and clarity, AI applications on health projects will document with sufficient information the conception of AI technology to be used in the project, establishing regular meetings with all stakeholders to facilitate knowledge and debate on the design and possible use to be made of this technology.

2.5 Interpretability Is a Key Issue in Artificial Intelligence for Health

On the other hand, ML's intelligibility and accountability are related to the interpretability of the generated models. Interpretability can be assessed as the degree to which a human can consistently predict the model's result [31]. Not only do we want to know the model's prediction, but we also want to know why the prediction was taken; that is, the model should explain the outputs [15].

2.6 Machine Learning Model's Taxonomy According to Its Interpretability

ML models can be classified into black-box models and white-box models regarding interpretability. The former are created directly from data and are inherently non-interpretable; even the designers who created them cannot understand how variables are combined to provide the outputs. On the other hand, white-box models are intrinsically interpretable; it is possible to know the variables' importance to generate the output. Unfortunately, the most successful ML classifiers in the literature are black-box models (neural networks, random forest (RF), support vector machine (SVM)). Among the latter, the advantage of interpretability is opposed by a low performance in terms of precision. However, there are post-hoc interpretation techniques of the models that serve to interpret the predictions based on the values of the variables: for example, permutation variable importance or Shapley additive explanations, to name two very frequently used. These techniques can be used on black-box models (for example, RF or SVM) to provide a post-hoc interpretation of the predictions of the ML model that can serve for the intelligibility and accountability of the generated model.

2.7   Examples of the Usage of Artificial Intelligence in Medicine

ML techniques have had a successful history in their applications to many areas of medicine [62]: neural networks [74] and K-Nearest Neighbours (KNN) [44] for the cancer diagnosis, neural networks for blood glucose levels prediction [80], bladder cancer [9] or colorectal cancer [46], ensemble classifiers in bioinformatics [83], deep residual networks for carcinoma subtype identification [18], Tree-Lasso logistic regression [28], Bayesian networks for the prediction of the causal pathogen in children with osteomyelitis [86], decision trees [70], Markov chains and bioinspired algorithms [27], drug identification using Support Vector Machine (SVM) [38], or predicting risk of disease using Random Forest (RF) [29] to cite some illustrative examples.

2.8   Bacteraemia's Definition

Bacteraemia is the presence of bacteria in the bloodstream. The blood does not contain bacteria in healthy patients, so its presence is associated with infections that can impact the patient's life [75].

2.8.1   Bacteraemia's Origin

The most frequent origin for bacteraemia is an infection in a specific location in the body that facilities the bacteria's movement into the blood. The infections that most frequently produce bacteraemia are urinary (*prostatitis* or *pyelonephritis*), respiratory (*pneumonia*), and vascular (infected catheters). Approximately 70% of hospitalised patients receive some type of venous catheter, and between 15% and 30% of all nosocomial bacteraemias are associated with intravascular devices [61, 65]. In Spain, 49% of nosocomial bacteraemias were related to venous catheters [66, 82], according to the 2016 National Study of Nosocomial Infections [59]. Other origins for bacteraemia are bones (*osteomyelitis*), skin and soft tissues (*cellulitis* or *myositis*), or digestive (*cholecystitis* or *cholangitis*).

Some medical procedures also facilitate bacteria's movement from sites usually colonised by bacteria into the blood, i.e., endoscopies of the digestive tract (colonoscopies) or urinary catheters in the bladder. Similarly, habits such as the use of intravenous drugs can facilitate the movement of bacteria from the skin to the blood [69].

The bacteria in the blood can spread the infection to other places in the body, producing *osteomyelitis, arthritis, endocarditis, meningitis*, or brain abscesses, among others. There is a connection between the type of bacteraemia microorganism, the acquisition site, and the associated mortality. Hence, bacteraemia-related mortality varies between 11% and 37% depending on the type and place of microorganism [10].

A high mortality rate is associated with bacteraemias, and blood cultures are the gold standard for diagnosing bloodstream infections. Due to the high morbidity and mortality associated with bacteraemia, it is mandatory to initiate effective antibiotic treatment as soon as possible to reduce the death rate [33, 34].

2.8.2   Bacteraemia and Acquisition

Depending on the bacteraemia way of acquisition, bacteraemias are classified as nosocomial-acquired bacteraemias which are suffered after 48 h of hospital income, healthcare-associated bacteraemias which are related to diagnosis or treatment procedures but without hospital incomes, and community-acquired bacteraemias which are acquired in the community and unrelated to hospital income or health care.

2.8.3   Diagnosis of Bacteraemia and Blood Cultures

A high mortality rate is associated with bacteraemia, and the mean of detecting them is via blood cultures in vials that contain growth media of two types: aerobic and anaerobic. The diagnosis of bacteraemia is probably one of the most critical functions of microbiology laboratories. There are two types of blood culture methods: conventional methods with manual systems (such as biphasic blood culture, manometer methods, lysis filtration–centrifugation) and automatic systems. Both provide results within 2–5 days.

The blood cultures proceed as follows. The patient's blood—between 20 and 40 ml – is drawn and introduced into vials. The blood volume extracted for each blood culture is the most important variable when harvesting bacteria and fungi from patients with bacteraemia [11, 81]. Next, the vials are placed within a system that maintains the optimal environmental conditions (humidity, temperature, light) for the microorganism's growth. The microorganism's growth produces CO_2, and the system detects its production. This process can take between hours and 5 days. If the system does not detect CO_2 during this time frame, it reports a negative culture (no bacteraemia), whereas if it does detect CO_2 production, then it reports a positive culture. Nevertheless, a positive culture does not imply bacteraemia. Therefore, it is also important to determine if this growth is true bacteraemia or a contaminant (negative bacteraemia) [50, 53].

2.8.4 Contamination and Blood Cultures

Blood culture contamination is common, very costly for the health system, and often confuses physicians. Blood culture contamination rate should be less than 3%, and contamination rates above this value should be studied and corrected with educational programs [64].

To discard contamination once the positive culture appears, it is needed to identify the bacteria species in the vials. The complete process of identifying the microorganism can take up to three extra days. In many cases, the species identified came from the skin or was introduced in the blood sample during the blood extraction or culture. In these cases, the culture is considered contaminated and has no bacteraemia. Therefore, proper skin preparation for venepuncture minimises the risk of blood culture contamination with normal skin flora.

2.8.5 Timing and Diagnosis of Bacteraemia

The prediction of true bacteraemia has two important milestones. First, when the physician decides to extract blood from the patient for the blood culture. Blood cultures should be obtained under severe sepsis, suspected infection with organ dysfunction, high blood lactate levels, infectious processes associated with bacteraemia (for example, severe *pneumonia, meningitis, pyelonephritis, cholangitis*, suspected *endocarditis*, or endovascular infections), or patients with fever and at least another sign or symptom of infection in the absence of a known alternative diagnosis. Nevertheless, blood cultures should not be decided indiscriminately because the number of contaminated blood cultures would lead to unneeded antibiotic therapy and increase economic costs.

The second milestone is when the blood culture is positive (i.e., the system detects CO_2). Nevertheless, the conclusive identification of the microorganism can take two or three extra days. Once identified, only those cultures where bacteria species come from an infection are declared true bacteraemia, whereas those with contaminants species will be considered negative bacteraemia.

2.8.6 Bacteraemia and Physicians

It would be extremely useful for the physician to predict bacteraemia before deciding to obtain blood cultures. Unfortunately, physicians have difficulties in predicting which patients have bacteraemia. These difficulties are appreciable in the low rate of true positive blood cultures; reports rates between 5% and 8% and reports values as low as 3.6% per analysis [3, 39, 45].

The type of positive blood culture (aerobic or anaerobic) and the time-lapse to detect CO_2 could be useful variables so that the physician could predict whether the bacteria is true or negative in the middle of the procedure to obtain a conclusive identification of the microorganism.

2.8.7 Bacteraemia and Treatment

The association between mortality and the origin of bacteraemia seems to depend on the use and timing of treatment with adequate empirical antibiotics [2, 43, 67].

Regarding the timing, the start times for adequate antibiotic therapy are critical for the prognosis of patients with sepsis and bacteraemia. For this reason, physicians should be notified whenever a positive blood culture is detected because the microorganism can often represent an infection that may lead to death.

Regarding the treatment of bacteraemia, one of the most important issues is the interpretation of positive blood cultures. When identified in blood cultures, some organisms should never be considered contaminants, such as *Staphylococcus aureus*, gram-negative roads, or *Candida* spp. On the other hand, organisms such as *Corynebacterium* sp. or coagulase-negative *Staphylococcus* spp. are common skin contaminants, and they usually do not need antibiotic treatment. However, sometimes this last group, usually contaminants, could produce bacteraemia mostly related to catheters or prosthetic valves. These considerations influence the decision regarding antibiotic treatment and how long a patient should be treated.

An ML model capable of predicting true bacteraemia could help the physician start earlier the treatment and make the appropriate decision about the antibiotic treatment. In this sense, PPPM/3PM has a very important point of intervention in suspected bacteraemia and its treatment.

2.9 Clinical, Economic and Structural Consequences of Positive Blood Cultures

The usefulness of blood cultures in predicting bacteraemia varies between 4.1% and 7% [3, 55], whereas false-positive results due to contamination vary between 0.6% and over 8.0% [63]. The low usefulness of blood culture has an important economic effect, increasing a 20% hospital costs for patients with false-positive blood cultures [25, 60]. In this sense, economic analyses estimate the costs of a single false-positive blood culture between $6878 and $7502 per case [1].

To reduce the medical waste and the overuse of blood culture, in 2012, the American Board of Internal Medicine introduced the Choosing Wisely campaign that set clear guidelines for the use of blood cultures. Studies assessing risk factors for bacteraemia have led to multiple stratification systems without consensus [84].

2.10 Bacteraemia and Prediction Models

Specialised prediction models aim to provide patient risk stratification to support a tailored clinical decision. They use variables thought to be associated, either negatively or positively, with the outcome of interest [73]. Risk prediction models can be used to estimate the probability of either having (diagnostic model) or developing a particular disease or outcome (prognostic model) [26].

Regarding prediction models for bacteraemia, the physician's bacteraemia suspicion is not a good predictor since it lacks specificity, sensitivity, or predictive values to be clinically useful. Due to this, bacteraemia clinical prediction models have been developed for specific infections (pneumonia [30, 35], skin infections [40], and community-acquired bacteraemias [42]) using classical methodologies, from stepwise logistic regression, multivariable analysis, or multiple mutually exclusive stepwise logistic regression [30]. Unlike theirs, our model applies to any intra- or extra-hospital bacteraemia source and uses ML techniques.

2.11 Applying Machine Learning to Predict Bacteraemias

It would be a major improvement to predict which patients suffer from true bacteraemia before the two milestones mentioned above: before deciding on blood sample extraction or when the blood culture is positive. The prediction would reduce the diagnosis lag and avoid waiting for up to 6 days for the definitive results.

To the best of our knowledge, there are no useful clinical, analytical, or epidemiological studies that allow physicians to predict bacteraemia at the patient's initial assessment nor the application of ML techniques to create diagnostic bacteraemia models. Hence, our work's main objective is to implement ML techniques on a set of patient data from electronic hospital records to predict the appearance of bacter-aemia, thus eliminating the lag for the results of blood cultures and anticipating the application of therapeutic treatments.

This work explores the application of three ML techniques: Support Vector Machine (SVM), Random Forest (RF), and K-Nearest Neighbours (KNN). The potential of these models in terms of PPPM/3PM is that they can be useful in the decision-making process regarding blood culture collection, clinical monitoring, and empirical antimicrobial therapy when used in conjunction with clinical judge-ment. Therefore, the benefits of this work are twofold: first, the possibility of start-ing the personalised patient's treatment earlier; second, the number of blood cultures would be reduced since they would only be prescribed in cases where the tech-niques' predictions did not have high reliability.

2.12   Databases Have to Represent the Patient's Casuistry

ML techniques must be provided with datasets that faithfully and fairly represent the distribution values in the patient's health and demographic parameters so that their prediction is unbiased.

The Hospital Universitario de Fuenlabrada, Madrid, Spain, has provided our database. It is a 350-bed hospital with the following services: general surgery, urol-ogy, orthopaedic surgery, gynaecology and obstetrics, paediatrics, intensive care units (ICUs), haematology-oncology, internal medicine, and cardiology. The data-base was gathered from 2005 to 2015, and it consists of 4357 anonymous patient records containing 117 variables per patient, 43.9% female with age 65.1 ± 19.7 year, and 56.1% male with age 62.7 ± 20.2 year. Each record contains demographic and medical data (medical history, clinical analysis, comorbidities, etc.) and the blood culture result that take one of two values: bacteraemia and no bacteraemia. The database contains 2123 bacteraemia (51.3%), including aerobic, strictly anaerobic, and facultative anaerobic bacteria, and 2234 no bacteraemia (48.7%), including 1844 contaminations. The final classification of true bacteraemia was done in pro-spective time by an infectious disease physician, using all the previous data, includ-ing microbiological, clinical, and analytical data.

47 out of the 117 variables were discarded from the database because they were derived from other variables, irrelevant to the study, or assessed after the blood culture.

2.13   Two Datasets for Predictions at Two Different Times

Two datasets were created as of the hospital database. The first dataset contains the variables known previously to the blood culture, accounting for 65 variables. We call it the pre_culture dataset. The second dataset, called mid_culture, contains the data at the second milestone stated in Sect. 2.8.5; that is, data available when the concentration of CO_2 rises, accounting for 69 variables; the variables in pre_culture plus four new ones: the time to CO_2 detection, the type of media with bacterial growth, either aerobic or anaerobic, and the first vial where the growth is detected (Refer to Appendix in [19] for an enumeration of the variables understudy). As stated in that section, this rise of CO_2 does not necessarily mean bacteraemia since it could be due to true bacteraemia or contamination (a.k.a. no bacteraemia) of the blood sample during extraction.

2.14   Data Preprocessing Is Mandatory Previous to Applying Machine Learning Techniques

The dataset can contain numerous artefacts that bias the ML outcomes. They should be detected and fixed, if possible, previous to applying ML techniques. These artefacts are mostly related to bias in data gathering, the handling of categorical variables and missing data, and data values in different scales.

2.15   Detection of Data Bias During Information Gathering

The distribution of values in the dataset should be analysed. Hence, datasets should contain a balanced percentage of values, and missing data should not be statistically related to the predicted variable. This analysis attempts to identify if the gathering process was accomplished under some biased assumption about the behaviour of the data. The biased variable should be removed from the dataset in case of such a bias.

2.16   Handling Categorical Variables

Many ML techniques, for example, SVM and KNN, require a definition of distance to measure the separation between instances. This requirement constraints how the categorical variables can be mapped into numerical values: when a notion of distance does not exist among the categorical values, the mapping into numerical values should avoid creating it. The one-hot encoder is the most used codification technique to avoid this problem. It loops through the dataset, and for each class of a categorical variable creates a new variable with only two values: true or false. Once the categorical variables have been removed, the Euclidean distance is chosen.

2.17   Handling Missing Data

Hospital records do not contain all the values for all the variables. These values that have not been recorded are called missing data, and they can reduce the statistical power of the analysis and bias the ML techniques. The best method to handle missing data depends on the nature of the data missingness that can be classified into three categories [41]:

- Missing completely at random (MCAR) in which the missingness is random, unrelated to the outcomes, and does not contain valid information for analysis,
- Missing at random (MAR) when the missingness depends on the outcomes observed, and
- Missing not at random (MNAR) when missingness depends on unobserved measurements.

Therefore, the first step is to determine the missingness of the data. To this aim, for each variable at a time, two classes are defined, missing and non-missing, and an RF classifier is built upon only this variable. If the RF classification accuracy is high, then a MAR behaviour is concluded for the variable and discarded from the dataset [14].

Once the nature of the data missingness is determined, several approaches have been proposed to handle the high number of missing data [24].

The complete case data approach removes the instances with missing data to obtain a new dataset without misses; all instances have valid data in all variables. This approach suffers from two limitations: (i) it would not allow evaluating a new record with missing data once the ML model is deployed, and (ii) it significantly reduces the number of records in the training dataset.

A different approach attempts to keep the largest possible ratio of complete records in the dataset sorting the variables in the dataset according to the decreasing ratio of missing data and sequentially removing the variables following the ranking [4]. The total quantity of data in the complete instances, i.e., the number of complete instances times the number of instances, is calculated in each iteration. As the number of variables decreases, the total amount of non-missing data in the complete instances increases to a maximum, beyond which the quantity of non-missing data instances decreases. This maximum determines the number of variables that most contribute to complete case instances, and it is the best option.

Finally, the separate class approach [14] defines a new category for each variable to represent the missing data of such a variable. In the case of numerical variables, the missing data receive a value outside the range of the variable's values. In this way, the required separation between the missing data and the correct values is created.

Each approach creates a different dataset so that the best one must be explicitly determined by experimentation with the dataset under study.

2.18   Data Values' Renormalisation

All the numerical variables have to be renormalised so that the range of values is approximately the same for all of them, which is especially relevant for those techniques such as SVM or KNN that use the notion of distance in a metric space. Hence, all numerical variables are rescaled to values in [0, 1]. The separate classes associated with the missing data are assigned with the value −0.5 since there are no negative values in any dataset.

2.19   Machine Learning Techniques
for Bacteraemia Prediction

The next three sections are devoted to introducing the ML techniques used in this study. WHO's requirements have been satisfied by choosing ML techniques that pose post-hoc interpretation techniques of the models.

2.19.1   Support Vector Machine

SVM is a supervised ML technique [7, 12] that can be used in binary classification problems. It classifies the instances of the dataset by defining a plane that separates the two classes; those instances in a side of the plane pertain to one class, whereas the instances on the other side pertain to another class. To this aim, this technique finds the equations of the plane that separates the two different classes, maximising the margin; that is, the distance of the closest instances in the dataset – called support vectors – to the plane. Consequently, SVM requires a definition of distance on the dataset's variables to evaluate the separation between the instances and the plane, and all the categorical variables should be translated into numerical values.

There are two types of SVM classifiers: linear and nonlinear. In the former, SVM operates on the raw data under the supposition that the data are linearly separable. In contrast, the latter transforms the original instances by adding extra similarity variables, trying to create a linearly separable dataset under the supposition that the original was not. The most used similarity function is the Gaussian Radial Basis Function [71].

SVM models can also be classified into two types depending on whether a few instances of one class are allowed to be located within the margin region or even in the region assigned to the other class. A hard margin classification is defined if no instance of one class can be within the margin region or in the region assigned to the other class. On the contrary, a soft margin classification allows the miss-classification of some instances but provides higher margins in the classification and better generalisation capabilities, lower overfitting, whereas hard margin classification typically provides a clean but narrower margin. Typically, SVM implementations provide a hyperparameter to control the softness of the margin, C. The highest the C, the strictest the classification.

2.19.2 Random Forest

RF is a supervised ML technique used in classification and regression [8]. In binary classification problems, RF creates multiple binary decision trees – trees with binary outputs –, each providing its classification output, and combines the results of all the trees using an aggregation function, i.e., the majority vote of individual tree predictions, to provide the classification of the given instance. Nevertheless, high accuracy requires the technique to satisfy certain requirements, the first of which is the independence of the individual trees which is achieved by using different subsets of instances to train every individual binary classification tree. The subsets can be sampled using two different schemas: sampling with replacement, called bagging, or without replacement, called pasting. Nevertheless, even though using the same subset, each tree could be different if the variables are chosen in a different order during the training. One of the most used algorithms to train decision trees is the classification and regression tree (CART). For each tree's node, CART splits the training subset into two subsets using a single variable and a threshold for such variable, searching for the tuple variable/threshold that provides the two purest subsets in that node. The purity of the node classification is evaluated as a weighted mean of the purity of the two subsets provided by the node [47]. The purity of the subsets can be evaluated using any of two common metrics: the Gini impurity and the entropy-based impurity.

Finally, the decision tree can be regularised with the following hyperparameter [6]: the maximum depth of the trees, the minimum number of samples in a node to be split, the minimum number of samples of a leaf node, the maximum number of leaf nodes, and the maximum number of variables to be tested in order to split a node.

RF achieves high-accuracy predictions thanks to the aggregation of weak learners. Thus, each tree has a larger bias than trained using the complete training set, but the aggregation of trees provides a lower bias-aggregated classification.

2.19.3 K-Nearest Neighbours

KNN can be used either as a supervised or unsupervised classification method. In the supervised flavour of this simple nonparametric ML technique to classify the binary-class instances [56], the output class of a given new is assigned by finding the k of nearest records in the dataset and aggregating their classifications (i.e., averaging or voting), so that the technique returns the output aggregation of the k nearest neighbours. Like SVM, this technique requires a definition of distance.

However, this technique does not need a training phase, and it achieves a very high capacity: the larger the training set, the higher the capacity.

The value for k should be selected according to these rules:

- The value should be a prime number to avoid ties.
- It should be less than the number of reference instances in an instance class.
- Its value should be large enough to avoid false classification caused by outliers.

The fine-tuning of the value for k usually requires a heuristic guided search on a range of reasonable values.

2.20 Model Validation

A k-fold cross-validation method is applied to validate the obtained models. A total of k validation subsets of data are generated from the original dataset by dividing it into k parts. In particular, 10 is established as the value for k. The process uses one of these parts as a validation subset while the other nine compound a unique training subset. Consequently, the training process is repeated 10 times, and the assessment of the ML models will be measured as the average of the 10 models on each corresponding validation subset.

Model validation also serves the purpose of detecting under and overfitting of the model. A model is underfitted when it is unable to learn accurately enough the data patterns during the training so that it mainly learns the noise. On the contrary, a model is overfitted when it learns the data pattern so accurately the training data that also learns the noise in the data. When overfitted, the model fails in the prediction because it is unable to filter the noise in the new unseen data.

2.21 Data Analysis Tools

Many different tools can be applied to this goal. Among the most frequently used are R, Matlab, or Python, and all the tasks that comprise the creation and analysis of an ML model can be performed with any of them. In this work, the analysis was performed in Python 3.7 using sklearn 0.23 for model inference and ELI5 0.10.1 for the permutation importance method.

2.22 Identification of Data Bias in Hospital Records

Fortunately, the dataset provided by the Hospital is balanced in terms of the samples of both no bacteraemia and bacteraemia (51/49% approx.). It should be considered that the no bacteraemia class includes actual negative bacteraemia results and those samples corresponding to cultures that were contaminated during the analysis process. In addition, the correlation between missing data and the predicted variable was checked to assure that the MAR assumption holds for the input data. Figure 2.1 presents the classification accuracy for all the variables in the dataset, one variable at a time.

In Fig. 2.1, a variable stands out for the rest that corresponds to the missing class of the suspected source (with 40% of missing values, see Fig. 2.2), whose value in the histogram is 82.6%, indicating being a good predictor for the negative class. Conversely, the other variables present a slight bias in the prediction. A ratio of missing data so high indicates that the importance of the variable should be questioned, indicating a correlation between the predicted variable and the missing-data class. Hence, 7.2% of the missing suspected sources are bacteraemia, and 72.4% of the samples with a suspected source are classified as bacteraemia.

These figures state a missing at random (MAR) [41] behaviour for this variable. A plausible explanation for this fact could be that the physician includes the suspected source in the database only when the bacteria is already detected, deciding whether the source of infection is interesting for no bacteraemia classification or not. Consequently, the suspected source was removed from the datasets.

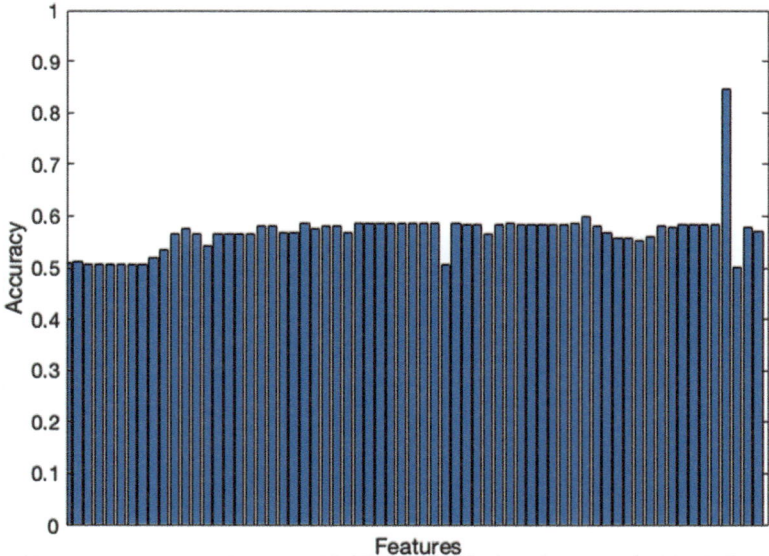

Fig. 2.1 Accuracy of the individual variables when only two classes (missing and non-missing) are used to predict bacteraemia

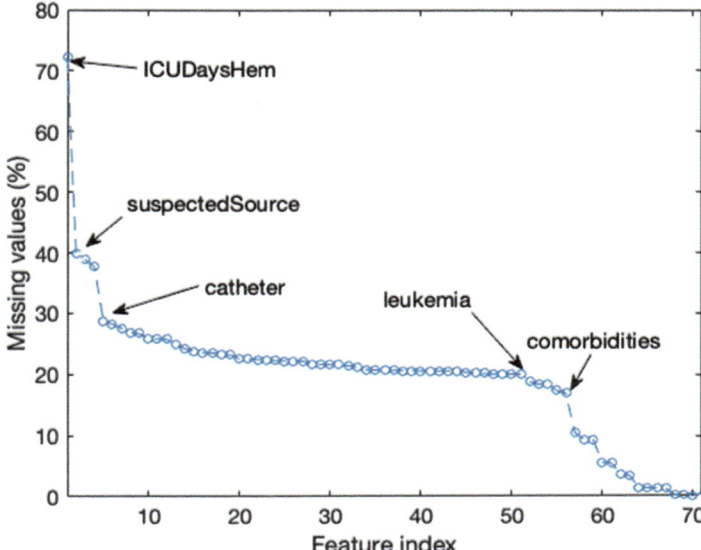

Fig. 2.2 Percentage of missing values for all the variables. (See [19] for a detailed description)

2.23   Ratio of Missing Data in Datasets

In this section, the distribution and number of missing data per variable are presented. Four variables stand out from the rest: the suspected origin of the bacteraemia previous to culture, the results of CPR testing, the source of bacteraemia in the last hospital department, or the number of days in ICU previous to culture, which is the worst variable with around 70% of missing data. For the other three, the percentage is close to 40%. After them, we find other 50 variables with a ratio between 20% and 30% of missing data.

If a complete case strategy would be applied to the data, i.e., removing instances with missing values, a dataset of 476 instances would be obtained from the original dataset of 4357 instances. As the reader can see, this option loses a large volume of data, being inappropriate. Nevertheless, we evaluated its achievements to classify bacteraemia accurately.

A second approach could be to remove only the variables with a high number of missing values. Figure 2.3 illustrates the evolution of the total volume of data in all complete instances versus the number of complete instances. In our case, the optimal number is 51 variables with 2760 instances, totalling 140,760 non-missing values in the dataset. As in the previous approach, we think this is also inappropriate due to the same reasons. First, some critical variables could be eliminated, as it would be the case of the mentioned suspected medical source of the patient's infection. Second, we would eliminate up to 33.8% of the variables and 44.6% of

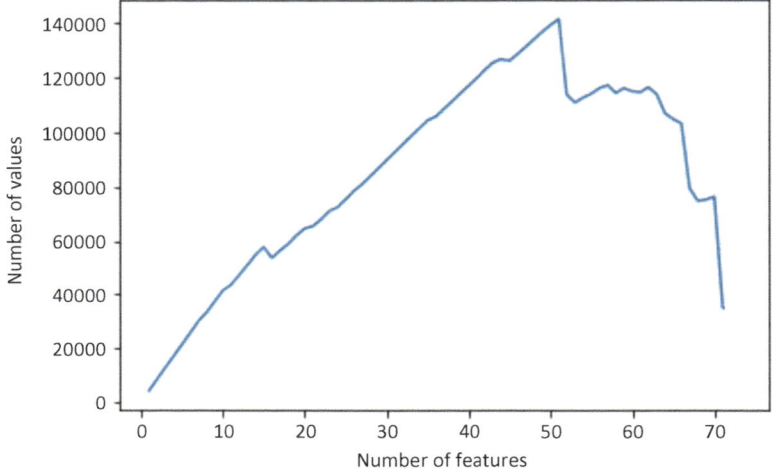

Fig. 2.3 Number of variables versus the number of non-missing values in the dataset

instances. As in the complete case strategy, its achievements to classify bacteraemia accurately were also evaluated.

The separate class method [14] was also evaluated to handle missing data. After comparing the performance of the three missing-data methods using RF as test-bench, the renormalised separate class method was selected since it obtains the best performance.

2.24 Method to Parameterise and Validate Machine Learning Techniques

A three steps procedure was used for evaluating the three ML techniques applied in this work:

1. The dataset is divided into two sets corresponding to 80/20 percentages of the data for training and testing sets, respectively.
2. Then a grid-search 10-fold cross-validation is run on training data to find their best hyperparameters of the ML techniques.
3. The testing phase is performed using the corresponding dataset with the best hyperparameters found.

2.25   SVM's Hyperparameters and Performance Metrics for Prediction Previous to Culture

SVM models have two hyperparameters, C and γ. In this work, they are combined swiping the ranges C = {0.1, 0.2, …, 1, 2, …, 10, 20, …, 100} and γ = {1/L, 1/Lσ, 0.1, 0.2, …, 1} respectively, being σ the variance and L the number of variables in dataset, by using the Gaussian Radial Basis Function. C is the hyperparameter that controls the softness of the SVM margin.

Experimental results show that the best pre_culture SVM model is with γ = 1/L, and C = 9, implying that the instances are separable. With those hyperparameters, the averaged values for accuracy are 76.9 ± 1.7% in the training phase and 75.9% in the testing phase. Those values indicate a good generalisation capability of the model since the value of the accuracy in the testing phase is very close to the value obtained during the training phase. Table 2.1 summarises the key metrics usually applied for evaluating the predictive capacity of ML models for the four techniques. In particular, the SVM model has a sensitivity of 80.7% with a specificity of 71.4%, positive predictive value (PPV) of 72.8%, and negative predictive value (NPV) of 79.6%. The values for the other ML models are commented on in the following subsections.

2.26   SVM's Hyperparameters and Performance Metrics for Prediction During Culture

Following the same procedure, mid_culture models were also obtained by SVM, obtaining the best model using γ = 1 and C = 8 as hyperparameters, which implies that the instances are slightly more separable than in the pre_culture dataset. In this case, the accuracies are 83.0 ± 1.4% and 80.5% for the training and testing phases, respectively. Although the difference is higher than in the pre_culture model, a good generalisation capability is shown. The values of the other metrics are sensitivity of 81.3%, specificity of 79.7%, PPV of 80.5%, and NPV of 80.5%. As can be observed,

Table 2.1 Accuracy, specificity, sensitivity, positive predictive value (PPV), negative predictive value (NPV), and area under the curve (AUC) of the models

		Accuracy (%)		Sensitivity	Specificity	PPV	NPV	AUC
ML	Model	Training	Test	(%)	(%)	(%)	(%)	
SVM	pre_culture	76.9 ± 1.7	75.9	80.7	71.4	72.8	79.6	0.85
	mid_culture	83.0 ± 1.4	80.5	81.3	79.7	80.5	80.5	0.88
RF	pre_culture	79.5 ± 1.4	78.2	86.1	70.7	73.6	84.3	0.86
	mid_culture	85.6 ± 1.4	85.9	87.4	84.4	85.2	86.6	0.93
KNN	pre_culture	72.8 ± 2.3	76.5	89.6	65.2	69.0	87.9	0.85
	mid_culture	78.0 ± 2.7	78.4	87.4	69.6	73.6	85.2	0.88

the usage of intermediate results of the blood culture improves the values in all the metrics between 5.0% and 8.0%.

2.27   Interpretation of SVM's Results

In order to interpret the SVM's results, a variables' importance analysis has been deployed using importance sampling. Table 2.2 presents the top 10 most important variables for both SVM models. In the case of the pre_culture model, the top three to predict bacteraemia are chronic respiratory disease, the number of days in ICU before blood extraction, and the presence of catheters. According to this table, for the mid_culture model, three out of the four new variables rank in the top five most relevant variables: growth in anaerobic and aerobic vials and the number of days until CO_2 detection.

2.28   RF's Hyperparameters and Performances Metrics for Prediction Previous to Culture

In the case of RF models, we use the Gini impurity metric, while the maximum depth, the minimum number of samples in a node, or any other hyperparameters were not constrained. However, the number of trees were evaluated in the range {1, 2, ..., 90}. In this case, the best pre_culture RF model averages an accuracy of

Table 2.2 Variable importance for SVM

pre_culture		mid_culture	
Importance	Variable	Variable	Importance
0.0408 ± 0.0254	ChrRes	VialAnae	0.1495 ± 0.0206
0.0381 ± 0.0228	IcuDay	VialAer	0.0931 ± 0.0202
0.0367 ± 0.0462	CatTyp	CO_2	0.0289 ± 0.0185
0.0229 ± 0.0050	PolMic	UriSed	0.0234 ± 0.0155
0.0220 ± 0.0273	Dept	Fever	0.0211 ± 0.0168
0.0220 ± 0.0335	FevSym	Consc	0.0165 ± 0.0079
0.0216 ± 0.0287	UriSed	LocSyn	0.0147 ± 0.0160
0.0179 ± 0.0204	Anaero	ResMani	0.0133 ± 0.0045
0.0119 ± 0.0191	Fever	CatTyp	0.0128 ± 0.0085
0.0106 ± 0.0155	ParDrug	ChrRes	0.0102 ± 0.0122

The left-hand side of the table ranks the top 10 variables for the pre_culture model, whereas the right-hand side ranks the top 10 variables for the mid_culture model. In blue, the new variables included in the mid_culture model. The meaning of the variable's names is explained in Table 2.3

Table 2.3 Description of the important variables for the three ML classifiers

Variable	Description
1erVial	First blood culture vial with growth
Age	Age
Anaero	Anaerobic bacteraemia versus other bacteraemia
CatTyp	Catheter type
ChrRes	Chronic respiratory disease
CO_2	Days to CO_2 detection
Coagul	Altered coagulation values
Consc	Consciousness level at the moment of bacteraemia
CPR	C-reactive protein level
Day	Day of blood extraction
DayHosp	Days in Hospital before blood extraction
Dept	Speciality where bacteraemia is suspected
Fever	Fever. Armpit temperature>38.0 °C at the time of blood extraction
FevSym	Symptoms related to the source of fever
IcuDay	Days in Intensive Care Unit before blood culture extraction
Leuko	Leukocytes (μl^{-1})
LocSyn	Syndromes related to the source of fever
Month	The month of blood extraction
OthCom	Other comorbidities
Platelets	Platelets (μl^{-1})
PolMic	Polymicrobial bacteraemia microorganisms
ResMani	Respiratory manipulations
Steroi	Steroids
UriSed	Urine sediment
Vasopre	Use of vasopressor agents at the time of bacteraemia
VialAer	Growth at least in aerobic environments
VialAnae	Growth at least in anaerobic environments

The variable names are sorted in alphabetical order. In blue, those variables included in the mid_ culture models

79.5 ± 1.4% using 10-fold cross-validation with 86 trees and an accuracy of 78.2% during the testing phase. The small differences in accuracy on training and test data-sets refutes the overfitting of the model. Table 2.1 summarises the key metrics that clinical practitioners use to evaluate the models' predictive capacity. The variables' importance has been evaluated using the permutation importance algorithm, and Table 2.4 presents the most critical variables of the model.

2.29   RF's Hyperparameters and Performances Metrics for Prediction During Culture

Regarding the model with RF for prediction during the culture (mid_culture), we obtain that the best result (training accuracy of 85.6 ± 1.4%) is obtained with an ensemble model of 68 trees. The mid_culture RF model improves all the metrics for predicting compared with the pre_culture model. Moreover, mid_culture RF models are much better not only in training (+ 6.1%) but also in test (+7.7%). All the other metrics were also improved in relation to the pre_culture model (sensitivity by 1.3%, PPV by 11.6%, specificity by 13.7% and NPV by 2.3%.) (Table 2.1).

An analysis of the importance of the variables to predict bacteraemia for RF models (see Table 2.4) shows that the most critical variables are the number of days at CO_2 detection, the positive in anaerobic vials, the first blood culture vial with growth, and the positive in aerobic vials in order of importance. Contrary to SVM models, the two RF rankings are more unbalanced than the former, although we can observe the dominance of an outstanding variable in both cases, which doubles the importance of the second variable in the pre_culture model and which is 8× for the mid_culture model.

2.30   KNN's Hyperparameters and Performances Metrics for Both Models

KNN classifier only uses a hyperparameter, k. This study explored an integer value in the range [1, 20]. We obtained 15 and 9 as the best values of k for the pre_culture and mid_culture models, respectively. In Table 2.1, we can find the values of the

Table 2.4 Variable importance for RF

pre_culture		mid_culture	
Importance	Variable	Variable	Importance
0.0434 ± 0.0214	Dept	CO_2	0.1530 ± 0.0035
0.0253 ± 0.0169	CatTyp	VialAnae	0.0197 ± 0.0013
0.0148 ± 0.0011	IcuDay	1erVial	0.0109 ± 0.0017
0.0094 ± 0.0011	FevSym	VialAer	0.0061 ± 0.0010
0.0074 ± 0.0008	LocSyn	Age	0.0028 ± 0.0006
0.0051 ± 0.0007	Month	CPR	0.0026 ± 0.0005
0.0043 ± 0.0005	Platalets	LocSyn	0.0024 ± 0.0005
0.0041 ± 0.0007	Fever	Leuko	0.0024 ± 0.0005
0.0040 ± 0.0008	UriSed	Fever	0.0024 ± 0.0005
0.0037 ± 0.0006	DayHosp	Day	0.0020 ± 0.0005

The left-hand side of the table ranks the top 10 variables for the pre_culture model, whereas the right-hand side ranks the top 10 variables for the mid_culture model. Table 2.3 presents the meaning of the variable's names

Table 2.5 Variable importance for KNN

pre_culture		mid_culture	
Importance	Variable	Variable	Importance
0.0239 ± 0.0136	Fever	VialAnae	0.0186 ± 0.0061
0.0227 ± 0.0122	UriSed	VialAer	0.0135 ± 0.0084
0.0222 ± 0.0069	LocSyn	1erVial	0.0122 ± 0.0061
0.0213 ± 0.0059	Vasopre	UriSed	0.0080 ± 0.0025
0.0211 ± 0.0069	FevSym	Anaero	0.0078 ± 0.0034
0.0183 ± 0.0099	CatTyp	Vasopre	0.0069 ± 0.0119
0.0161 ± 0.0029	Steroi	Fever	0.0067 ± 0.0084
0.0147 ± 0.0102	Month	CatTyp	0.0064 ± 0.0043
0.0147 ± 0.0108	OthCom	CO_2	0.0044 ± 0.0080
0.0144 ± 0.0097	Coagul	IcuDay	0.0041 ± 0.0064

The left-hand side of the table ranks the top ten variables for the pre_culture model, whereas the right-hand side ranks the top ten variables for the mid_culture model. In blue, the new variables included in the mid_culture model. Table 2.3 presents the meaning of the variable's names

main predictive capacity metrics of the KNN models. In the testing phase, the best pre_culture KNN model averages an accuracy of 76.5%, and as in other models explained above, the performance of the models is improved when we include mid_culture features, although in this case only a 1.9% of increment in testing accuracy; a slight decrease of 2.2% in sensitivity and of 2.7% in NPV are observed. As in the case of RF models, new features reduce the size of models, i.e., the number of relevant neighbours.

Table 2.5 presents the top 10 most important variables in the KNN model according to importance sampling criteria.

2.31   ROC Comparison of the Three Machine Learning Models

The set of ML techniques evaluated for the two datasets shows very good figures for ROC, as the reader can observe in Fig. 2.4. The area under the curve (AUC) values are 0.88 for the mid_culture SVM model and 0.85 for the pre_culture SVM model. As shown, RF models are better with values of 0.93 for the mid_culture RF and 0.86 in the case of the pre_culture RF model. In the case of the KNN models, the AUC values indicate a lower predictive power, 0.85 and 0.88, respectively.

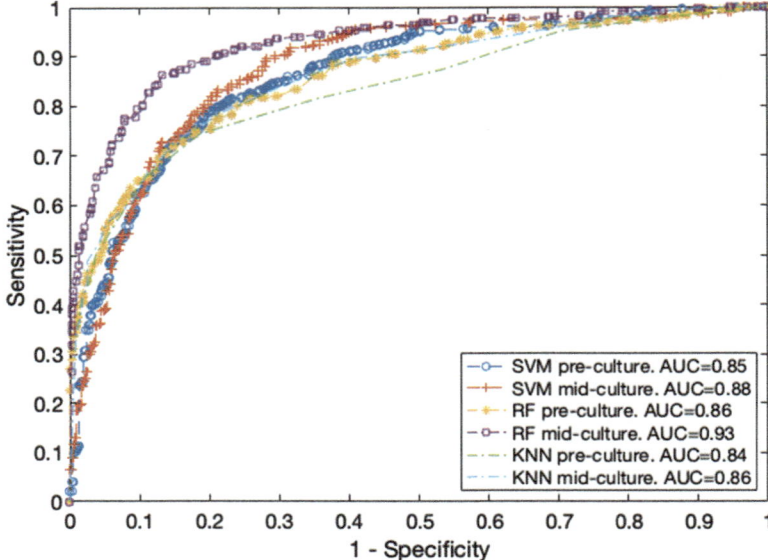

Fig. 2.4 ROC for the best models of SVM, RF, and KNN

2.32   Data Interpretation

It is well known that when obtaining ML models from medical records, attention must be put to the nature and amount of missing data in order to avoid biassing in the conclusions of the ML models. Following this idea, we did not evaluate imputation methods based on ML algorithms in this study. We applied the separate class method to handle the missing data because it preserves the number of patients in the study and provides good classifier metrics. We did not consider KNN or missForest [78] because KNN can infer relationships among the features that could distort the data structure [5]. MissForest, although considered a more efficient imputation method, must be run with every new patient, which would increase the computational cost of every new prediction when the system is in production.

The evaluation of the importance rankings of variables for all the three ML methods considered in this study indicated a significant ratio of common top features for both datasets. In this sense, the number of days in ICU before blood culture extraction, the presence of catheters, fever and the presence of symptoms related to the source of fever, and the presence of urine sediments are critical features of major importance for the pre_culture models. KNN and RF features analysis detect seasonality in bacteraemia since the month of the blood culture appears for both techniques, although it has low importance in both technique.

Results are different for the mid_culture dataset. In this case, features added to the dataset, and not used in the pre_culture model, are the most important for accurately predicting bacteraemia. The mid_culture RF model ranks the four new features among the top, whereas SVM and KNNs only include three out of the four new features. The importance of the mid_culture features in the model exceeds the importance of all the features in the pre_culture model. This consistency highlights that prediction capability is a characteristic intrinsically related to the data already available in most hospital health records.

The feature importance for the pre_culture SVM and RF models is balanced. The ranking of important features can be divided into two parts. The three first features are within a range of 10.0% of the most important ones. Then the importance of the remaining seven features is gradually reduced generating accurate predictions. In particular, the two KNN rankings are the most balanced of the three ML techniques. The first five features in the pre_culture model and the first three features in the mid_culture model have very similar values, although the dispersion of accuracy in the training stage doubles the dispersion values of the other ML techniques, which justifies why the KNN technique produces less predictable accuracy for the model.

The feature importance of the RF model is less balanced than the KNN one since there is a critical feature followed by two less relevant variables, being irrelevant the remainder features. This behaviour is more evident in the mid_culture model, where the mid_culture features dominate the classification. Hence, the physician may make a prediction based on the presence of these features. Nevertheless, the features coincide in almost half of the cases.

Another important conclusion of the ML models is that the high number of features that appear in the analysis of the models as important in the predictions justifies physicians' difficulty in generating accurate predictions.

If we use accuracy as the most important metric, the best model is the mid_culture RF model, which obtains a value of 85.9% in accuracy, while the pre_culture models range between 75.9% for SVM and 78.2% for RF. Hence, the accuracy of ML techniques is 8x human accuracy (from 3.6% to 10.0% according to [39]).

The key metrics to evaluate the predictive capacity of the model range from 80.7% to 89.6% for sensitivity, 65.2% to 84.4% in specificity, 69.0% to 85.2% for PPV, and 79.6% to 86.6% for NPV, with the mid_culture RF model outperforming the other models and achieving an average accuracy of 85.9 ± 1.4%, sensitivity 87.4%, specificity 84.4%, PPV 85.2%, NPV 86.6%, and an outstanding AUC of 0.93 with improvements of 6.7% with regards to the accuracy of second best technique, SVM, 6.1% in sensitivity, 4.7% in specificity, 4.7% in PPV, and 6.1% in NPV. So, we can state that the ML values of sensitivity, specificity, predictive positive, and negative values exceed the results described in the literature. This is also true for AUC, whose value is higher than 0.85 in all the ML models. The threshold AUC value of a model in the medical practice for being considered a valid model is 0.7. Models should exceed this value to be considered, and a good predictive model should have at least an AUC value of 0.8. The previous results in the literature using classical modelling techniques in specific types of bacteraemia are as follows: pneumonia [35] with AUC 0.79, skin-related [40] with AUC 0.71, or any type [42] with AUC 0.77. ML models clearly outperform those values.

The results presented in this study indicate that good bacteraemia predictions can be achieved using already available hospital records with better figures of merit than the physicians' predictions. These predictions can help physicians make as soon as possible correct diagnosis and prevent complications with the most specific and personalised antibiotics and treatment for each patient.

2.33   Interplay Between COVID-19 and Bacteraemia

In the times of COVID-19 pandemia, it would be necessary to associate it with bacteraemia and how ML models can help COVID-19 patients. It is worth noting that bacteraemia is rare in those patients, which supports the judicious use of blood cultures in the absence of compelling evidence for bacterial co-infection [72]. Although there are a reduced number of reports, bacteraemia with *S. aureus* in patients hospitalised with COVID-19 is associated with high mortality rates. The infection with *S. aureus* was also reported as a complication of other viral pandemics, such as the Spanish flu in 1918–1919 and the H1N1 influenza pandemic in 2009–2010. This indicates that the interaction of *S. aureus* with SARS CoV-2 is similar to influenza [48].

The proposed mechanisms of viral-induced bacterial co-infections include the viral modification of airway structures, as well as the initiation of immune-suppressive responses [22]. A similar mechanism has been described in another report of oral infections where the authors suggest that poor oral hygiene and periodontal disease could produce the aggravation of COVID-19 [79].

Bacteraemia as secondary to pneumonia or typical hospital-acquired infection has not been defined, although secondary bacteraemia was also reported in 37% (27/73) of patients with acute respiratory distress syndrome [88]. ML techniques could help physicians predict bacteraemia as a secondary infection in critical COVID-19 patients, who suffer these secondary infections more frequently [32].

2.34   Conclusions and Recommendations in the Framework of 3P Medicine

In this study, three ML supervised techniques were applied, obtaining accurate predictive models of the blood culture outcome using hospital electronic health records. Records include data previous to blood extraction and data measured in the first hours/days of the blood culture. The coincidence of the three classifiers in terms of results reinforces the power of the conclusions and confirms not only the viability of ML techniques as a key technology for applying the PPPM/3PM principles to improving patients' survival rates significantly but also providing more cost-effective management of bacteraemia.

Early diagnosis of bacteraemia is critical and does so an appropriate early antibiotic treatment due to its high morbidity and mortality. ML techniques can play an important role in helping physicians in diagnosing, early treating and managing bacteraemia. In addition, ML models can be applied to determine preventive actions and hence reduce the medical costs associated with bacteraemia. As an example, an early prediction of bacteraemia by physicians could avoid the intervention to obtain blood samples, could reduce the number of bottles for blood culture per patient (from six to four), the time-lapse devoted to the culture, and the procedures to identify possible contaminant microorganisms with theirs costs in time and money.

Another important application of bacteraemia prediction is related to the selection of antibiotic treatment and its duration, which can change depending on whether the patient is suffering bacteraemia or not since bacteraemia related diseases need longer antibiotic treatments. Using ML models, physicians could optimise the duration of the treatments in the presence or absence of bacteraemia infection. The shortest the duration of the antibiotic treatment, the lower the cost of the treatment per patient. Reducing the length in time of antibiotics implies also avoiding secondary effects associated, such as antibiotic resistance [13].

Automatic and continuous data extraction from electronic medical records combined with models could help physicians reduce mortality and morbidity [36, 49] by identifying early the appearance of bacteraemia and avoiding the progression to a severe disease and providing timely appropriate antibiotic treatments.

Patient databases play a central role in 3P medicine [20] since ML technologies in the framework of 3P medicine depend entirely on the accuracy of their models, which is related to the availability of datasets with low missing value rates and structured information. Another crucial aspect is to avoid the physician's *a priori* interpretation of the data, which can lead to biases in the missing values. This requirement should be included in future database design specifications and the design of database user interfaces.

As mentioned, the success of ML techniques also depends on the availability of structured datasets. Unfortunately, most health information that is stored in hospitals, according to the European Commission's Recommendation on Electronic Health Records [16], suffers from a lack of structured format impeding automatic manipulation of the features. An additional effort must be made to construct protocols that permit the record of information in good conditions to extract automatic models and not use natural language that hinders the extraction of structured information.

Predictive models play a key role in bolstering decision systems, and ML techniques have outstanding potential to create models with an excellent level of accuracy [37]. They have been used to identify useful correlations between biometric, genetic, and environmental data with the potential risks and benefits of certain therapeutic choices [51]. They also have great potential to exceed the performance of physicians' heuristics, reducing lags in diagnosis and treatment costs when their application is extended from the genomic and biometric data to the clinical and demographic data in the patient's records.

Although we strongly recommend avoiding non-structured features (medical texts described in natural language), future work should also include the treatment of this kind of information since a lot of data could be in the database expressed in natural language. Having more information could improve the accuracy of ML models. Automatic translation of natural language is also of interest to read this information. The development of an app for mobile devices that provide a prediction to the physician at the bedside based on the latest available patient records should also be considered in the future.

These ideas are directed to improve predictive and personalised treatment in a disease as bacteraemia that currently continues producing a high level of mortality.

References

1. Alahmadi YM, Aldeyab MA, McElnay JC, Scott MG, Darwish Elhajji FW, Magee FA, Dowds M, Edwards C, Fullerton L, Tate A, Kearney MP (2011) Clinical and economic impact of contaminated blood cultures within the hospital setting. Hosp Infect 77(3):233–236. https://doi.org/10.1016/j.jhin.2010.09.033
2. Arpi M, Renneberg J, Andersen HK, Nielsen B, Larsen SO (1995) Bacteraemia at a danish university hospital during a twenty-five-year period (1968–1892). Scand J Infect Dis 27(3):245–251. https://doi.org/10.3109/00365549509019017
3. Bates DW, Cook EF, Goldman L, Lee TH (1990) Predicting bacteraemia in hospitalised patients: a prospectively validated model. Ann Intern Med 113(7):495–500. https://doi.org/10.7326/0003-4819-113-7-495
4. Batista GEAPA, Monard MC (2003) An analysis of four missing data treatment methods for supervised learning. Appl Artif Intell 17(5–6):519–533. https://doi.org/10.1080/713827181
5. Beretta L, Santaniello A (2016) Nearest neighbor imputation algorithms: a critical evaluation. BMC Med Inform Decis Mak 16(3):74. https://doi.org/10.1186/s12911-016-0318-z
6. Bernard S, Heutte L, Adam S (2009) Influence of hyperparameters on random forest accuracy. In: Proceedings of the 8th international workshop on multiple classifier systems, MCS '09. Springer, Berlin/Heidelberg, pp 171–180. https://doi.org/10.1007/978-3-642-02326-2_18
7. Boser BE, Guyon IM, Vapnik VN (1992) A training algorithm for optimal margin classifiers. In: COLT '92: proceedings of the fifth annual workshop on computational learning theory. Association for Computing Machinery, New York, pp 144–152. https://doi.org/10.1145/130385.130401
8. Breiman L (2001) Random forests. Mach Learn 45(1):5–32
9. Catto JW, Linkens DA, Abbod MF, Chen M, Burton JL, Feeley KM, Hamdy FC (2003) Artificial intelligence in predicting bladder cancer outcome: a comparison of neuro-fuzzy modeling and artificial neural networks. Cancer Res 9(11):4172–4177
10. Cisneros-Herreros JM, Cobo-Reinoso J, Pujol-Rojo M, Rodríguez-Baño J, Salavert-Lletí M (2007) Guía para el diagnóstico y tratamiento del paciente con bacteriemia. guías de la sociedad española de enfermedades infecciosas y microbiología clínica (seimc). Enferm Infecc Microbiol Clín 25(2):111–130. https://doi.org/10.1016/S0213-005X(07)74242-8
11. Cockerill FRI, Wilson JW, Vetter EA, Goodman KM, Torgerson CA, Harmsen WS, Schleck CD, Ilstrup DM, Washington JAI, Wilson WR (2004) Optimal testing parameters for blood cultures. Clin Infect Dis 38(12):1724–1730. https://doi.org/10.1086/421087
12. Cristianini N, Shawe-Taylor J (2000) Support vector machines, Chap. 6. Cambridge University Press, 93–124. https://doi.org/10.1017/CBO9780511801389.008
13. Davey P, Marwick CA, Scott CL, Charani E, McNeil K, Brown E, Gould IM, Ramsay CR, Michie S (2017) Interventions to improve antibiotic prescribing practices for hospital inpatients. Cochrane Database Syst Rev 2. https://doi.org/10.1002/14651858.D003543.pub4
14. Ding Y, Simonoff JS (2010) An investigation of missing data methods for classification trees applied to binary response data. J Mach Learn Res 11(6):131–170

15. Doshi-Velez F, Kim B (2017) Towards a rigorous science of interpretable machine learning. https://doi.org/10.48550/arXiv.1702.08608

16. European Commission: Commission Recommendation (EU) 2019/243 of 6 February 2019 on a European electronic health record exchange format. Tech. ep. MSU-CSE-06-2, European Commission (2019). URL: https://eur-lex.europa.eu/legal-content/EN/TXT/?uri=uriserv:O J.L_.2019.039.01.0018.01.ENG

17. Fleischmann C, Scherag A, Adhikari NKJ, Hartog CS, Tsaganos T, Schlattmann P, Angus DC, Reinhart K (2016) Assessment of global incidence and mortality of hospital-treated sepsis. Current estimates and limitations. Am J Respir Crit Care Med 193(3):259–272. https://doi.org/10.1164/rccm.201504-0781OC

18. Gandomkar Z, Brennan PC, Mello-Thoms C (2018) MuDeRn: multi-category classification of breast histopathological image using deep residual networks. Artif Intell Med 88:14–24. https://doi.org/10.1016/j.artmed.2018.04.005

19. Garnica O, Gómez D, Ramos V, Hidalgo JI, Ruiz-Giardín JM (2021) Diagnosing hospital bacteraemia in the framework of predictive, preventive and personalised medicine using electronic health records and machine learning classifiers. PMA Journal 12(3):365–381. https://doi.org/10.1007/s13167-021-00252-3

20. Golubnitschaja O, Kinkorova J, Costigliola V (2014) Predictive, preventive and personalised medicine as the hardcore of 'horizon 2020': EPMA position paper. PMA J 5(1):6–6. https://doi.org/10.1186/1878-5085-5-6

21. Golubnitschaja O, Topolcan O, Kucera R, Costigliola V, Akopyan M et al (2020) 10th anniversary of the European Association for Predictive, Preventive and Personalised (3p) Medicine – EPMA world congress supplement 2020. PMA J 11(1):1–133. https://doi.org/10.1007/s13167-020-00206-1

22. Goncheva MI, Conceicao C, Tuffs SW, Lee HM, Quigg-Nicol M, Bennet I, Sargison F, Pickering AC, Hussain S, Gill AC, Dutia BM, Digard P, Fitzgerald JR, Palese P (2020) Staphylococcus aureus lipase 1 enhances influenza a virus replication. mBio 11(4):e00975–e00920. https://doi.org/10.1128/mBio.00975-20

23. Gudiol F, Aguado JM, Almirante B, Bouza E, Cercenado E, Ángeles Domínguez M, Gasch O, Lora-Tamayo J, Miró JM, Palomar M, Pascual A, Pericas JM, Pujol M, Rodríguez-Baño J, Shaw E, Soriano A, Vallés J (2015) Diagnosis and treatment of bacteraemia and endocarditis due to staphylococcus aureus. a clinical guideline from the Spanish Society of Clinical Microbiology and Infectious Diseases (SEIMC). Enferm Infecc Microbiol Clín 33(9):625.e1–625.e23. https://doi.org/10.1016/j.eimc.2015.03.015

24. Guyon IM, Elisseeff A (2003) An introduction to variable and variable selection. J Mach Learn Res 3(Mar):1157–1182

25. van der Heijden YF, Miller G, Wright PW, Shepherd BE, Daniels TL, Talbot TR (2011) Clinical impact of blood cultures contaminated with coagulase-negative staphylococci at an Academic Medical Center. Infect Control Hosp Epidemiol 32(6):623–625. https://doi.org/10.1086/660096

26. Hendriksen JMT, Geersing GJ, Moons KGM, de Groot JAH (2013) Diagnostic and prognostic prediction models. J Thromb Haemost 11(s1):129–141. https://doi.org/10.1111/jth.12262

27. Hidalgo JI, Botella M, Velasco JM, Garnica O, Cervigón C, Martínez R, Aramendi A, Maqueda E, Lanchares J (2020) Glucose forecasting combining Markov chain based enrichment of data, random grammatical evolution and bagging. Appl Soft Comput 88:105923. https://doi.org/10.1016/j.asoc.2019.105923

28. Jovanovic M, Radovanovic S, Vukicevic M, Van Poucke S, Delibasic B (2016) Building interpretable predictive models for pediatric hospital readmission using tree-lasso logistic regression. Artif Intell Med 72:12–21. https://doi.org/10.1016/j.artmed.2016.07.003

29. Khalilia M, Chakraborty S, Popescu M (2011) Predicting disease risks from highly imbalanced data using random forest. BMC Med Inform Decis Mak 11(1):51. https://doi.org/10.1186/1472-6947-11-51

30. Kim B, Choi J, Kim K, Jang S, Shin TG, Kim WY, Kim JY, Park YS, Kim SH, Lee HJ, Shin J, You JS, Kim KS, Chung SP (2017) Bacteraemia prediction model for community-acquired pneumonia: external validation in a multicenter retrospective cohort. Acad Emerg Med 24(10):1226–1234. https://doi.org/10.1111/acem.13255
31. Kim B, Khanna R, Koyejo OO (2016) Examples are not enough, learn to criticise! Criticism for interpretability. In: Lee D, Sugiyama M, Luxburg U, Guyon I, Garnett R (eds) Advances in neural information processing systems, vol 29. Curran Associates, Inc. URL https://proceedings.neurips.cc/paper/2016/file/5680522b8e2bb01943234bce7bf84534-Paper.pdf
32. Lai CC, Wang CY, Hsueh PR (2020) Co-infections among patients with Covid-19: the need for combination therapy with non-anti-SARS-CoV-2 agents? J Microbiol Immunol Infect 53(4):505–512. https://doi.org/10.1016/j.jmii.2020.05.013
33. Laupland KB, Church DL (2014) Population-based epidemiology and microbiology of community-onset bloodstream infections. Clin Microbiol Rev 27(4):647–664. https://doi.org/10.1128/CMR.00002-14
34. Lee CC, Lee CH, Hong MY, Tang HJ, Ko WC (2017) Timing of appropriate empirical anti-microbial administration and outcome of adults with community-onset bacteraemia. Crit Care (Lond Eng) 21(1):119. https://doi.org/10.1186/s13054-017-1696-z
35. Lee J, Hwang SS, Kim K, Jo YH, Lee JH, Kim J, Rhee JE, Park C, Chung H, Jung JY (2014) Bacteraemia prediction model using a common clinical test in patients with community-acquired pneumonia. The American Journal of Emergency Medicine 32(7):700–704. https://doi.org/10.1016/j.ajem.2014.04.010
36. Lee KH, Dong JJ, Jeong SJ, Chae MH, Lee BS, Kim HJ, Ko SH, Song YG (2019) Early detection of bacteraemia using ten clinical variables with an artificial neural network approach. J Clin Med 8(10). https://doi.org/10.3390/jcm8101592
37. Lella L, Licata I, Minati G, Pristipino C, Belvis AGD, Pastorino R (2019) Predictive AI models for the personalised medicine. In: Moucek R, Fred ALN, Gamboa H (eds) Proceedings of the 12th international joint conference on biomedical engineering systems and technologies (BIOSTEC 2019) – volume 5: HEALTHINF, Prague, Czech Republic, February 22–24, 2019. ciTePress, pp 396–401. https://doi.org/10.5220/0007472203960401
38. Lin J, Chen H, Li S, Liu Y, Li X, Yu B (2019) Accurate prediction of potential druggable proteins based on genetic algorithm and Bagging-SVM ensemble classifier. Artif Intell Med 98:35–47. https://doi.org/10.1016/j.artmed.2019.07.005
39. Linsenmeyer K, Gupta K, Strymish JM, Dhanani M, Brecher SM, Breu AC (2016) Culture if spikes? Indications and yield of blood cultures in hospitalised medical patients. J Hosp Med 11(5):336–340. https://doi.org/10.1002/jhm.2541
40. Lipsky BA, Kollef MH, Miller LG, Sun X, Johannes RS, Tabak YP (2010) Predicting bacteraemia among patients hospitalised for skin and skin-structure infections: derivation and validation of a risk score. Infect Control Hosp Epidemiol 31(8):828–837. https://doi.org/10.1086/654007
41. Little RJ, Rubin DB (2002) Statistical analysis with missing data. Wiley, New York. https://doi.org/10.1002/9781119013563
42. Lizarralde Palacios E, Gutiérrez Macías A, Martínez Odriozola P, Franco Vicario R, García Jiménez N, Miguel de la Villa F (2004) Bacteriemia adquirida en la comunidad: elaboración de un modelo de predicción clínica en pacientes ingresados en un servicio de medicina interna. Med Clín 123(7):241–246. https://doi.org/10.1016/S0025-7753(04)74477-2
43. Lombardi DP, Engleberg N (1992) Anaerobic bacteraemia: incidence, patient characteristics, and clinical significance. Am J Med 92, 53(1):–60. https://doi.org/10.1016/0002-9343(92)90015-4. URL: https://www.sciencedirect.com/science/article/pii/0002934392900154
44. Mahfouz MA, Shoukry A, Ismail MA (2020) EKNN: ensemble classifier incorporating connectivity and density into KNN with application to cancer diagnosis. Artif Intell Med:101985. https://doi.org/10.1016/j.artmed.2020.101985
45. Makadon HJ, Bor D, Friedland G, Dasse P, Komaroff AL, Aronson MD (1987) Febrile inpatients. J Gen Intern Med 2(5):293–297. https://doi.org/10.1007/BF02596161

46. Martínez-Romero M, Vázquez-Naya JM, Rabuñal JR, Pita-Fernández S, Macenlle R, Castro-Alvariño J, López-Roses L, Ulla JL, Martínez-Calvo AV, Vázquez S, Pereira J, Porto-Pazos AB, Dorado J, Pazos A, Munteanu CR (2010) Artificial intelligence techniques for colorectal cancer drug metabolism: ontology and complex network. Curr Drug Metab 11(4):347–368. https://doi.org/10.2174/138920010791514289
47. Menze BH, Kelm BM, Masuch R, Himmelreich U, Bachert P, Petrich W, Hamprecht FA (2009) A comparison of random forest and its Gini importance with standard chemometric methods for the variable selection and classification of spectral data. BMC Bioinformatics 10(1):213. https://doi.org/10.1186/1471-2105-10-213
48. Morens DM, Taubenberger JK, Fauci AS (2008) Predominant role of bacterial pneumonia as a cause of death in pandemic influenza: implications for pandemic influenza preparedness. J Infect Dis 198(7):962–970. https://doi.org/10.1086/591708
49. Murdoch TB, Detsky AS (2013) The inevitable application of big data to health care. AMA 309(13):1351–1352. https://doi.org/10.1001/jama.2013.393
50. Mylotte JM, Tayara A (2000) Blood cultures: clinical aspects and controversies. Eur J Clin Microbiol Infect Dis 19(3):157–163. https://doi.org/10.1007/s100960050453
51. Nardini C, Osmani V, Cormio PG, Frosini A, Turrini M, Lionis C, Neumuth T, Ballensiefen W, Borgonovi E, D'Errico G (2021) The evolution of personalised healthcare and the pivotal role of European regions in its implementation. Pers Med 18(3):283–294. https://doi.org/10.2217/pme-2020-0115
52. World Health Organisation (2021) Ethics and governance of artificial intelligence for health: WHO guidance. World Health Organization, Geneva
53. Ortiz E, Sande M (2000) Routine use of anaerobic blood cultures: are they still indicated? Am J Med 108(6):445–447. https://doi.org/10.1016/s0002-9343(99)00410-6
54. Pai S, Enoch DA, Aliyu SH (2015) Bacteraemia in children: epidemiology, clinical diagnosis and antibiotic treatment. Expert Rev Anti-Infect Ther 13(9):1073–1088. https://doi.org/10.1586/14787210.2015.1063418
55. Perl B, Gottehrer NP, Raveh D, Schlesinger Y, Rudensky B, Yinnon AM (1999) Cost-effectiveness of blood cultures for adult patients with cellulitis. Clin Infect Dis 29(6):1483–1488. https://doi.org/10.1086/313525
56. Peterson LE (2009) K-nearest neighbor. Cholarpedia 4(2):1883
57. Phua AIH, Hon KY, Holt A, O'Callaghan M, Bihari S (2019) Candida catheter-related blood-stream infection in patients on home parenteral nutrition – rates, risk factors, outcomes, and management. Clin Nutr ESPEN 31:1–9. https://doi.org/10.1016/j.clnesp.2019.03.007
58. Pien BC, Sundaram P, Raoof N, Costa SF, Mirrett S, Woods CW, Reller LB, Weinstein MP (2010) The clinical and prognostic importance of positive blood cultures in adults. Am J Med 123(9):819–828. https://doi.org/10.1016/j.amjmed.2010.03.021
59. Sociedad Española de medicina preventiva, s.p.e.h.: Estudio epine-epps 2016. Protocolo. Tech. rep., European Centre for Disease Prevention and Control (2016). URL: http://www.sempsph.om/media/com_jnews/upload/EPINE-EPPS%202016%20Protocolo%20(v9.0).pdf
60. Qamruddin A, Khanna N, Orr D (2008) Peripheral blood culture contamination in adults and venepuncture technique: prospective cohort study. J Clin Pathol 61(4):509–513. https://doi.org/10.1136/jcp.2007.047647
61. Raad I, Hanna H, Maki D (2007) Intravascular catheter-related infections: advances in diagnosis, prevention, and management. The Lancet Infectious Diseases 7(10):645–657. https://doi.org/10.1016/S1473-3099(07)70235-9. URL: https://www.sciencedirect
62. Ramesh AN, Kambhampati C, Monson JRT, Drew PJ (2004) Artificial intelligence in medicine. Ann R Coll Surg Engl 86(5):334–338. https://doi.org/10.1308/147870804290
63. Ratzinger F, Dedeyan M, Rammerstorfer M, Perkmann T, Burgmann H, Makristathis A, Dorffner G, Lötsch F, Blacky A, Ramharter M (2014) A risk prediction model for screening bacteremic patients: a cross sectional study. PLoS One 9(9):1–10. https://doi.org/10.1371/journal.pone.0106765

64. Richter SS, Beekmann SE, Croco JL, Diekema DJ, Koontz FP, Pfaller MA, Doern GV (2002) Minimising the workup of blood culture contaminants: implementation and evaluation of a laboratory-based algorithm. Journal of Clinical Microbiology 40(7):2437–2444. https://doi. org/10.1128/JCM.40.7.2437-2444.2002.   URL: https://journals.asm.org/doi/abs/10.1128/ JCM.40.7.2437-2444.2002

65. Rodríguez-Baño J, López-Prieto M, Portillo M, Retamar P, Natera C, Nuño E, Herrero M, del Arco A, Muñoz A, Téllez F, Torres-Tortosa M, Martín-Aspas A, Arroyo A, Ruiz A, Moya R, Corzo J, León L, Pérez-López J (2010) Epidemiology and clinical variables of community-acquired, healthcare-associated and nosocomial bloodstream infections in tertiary-care and community hospitals. Clinical Microbiology and Infection 16(9):1408–1413. https://doi. org/10.1111/j.1469-0691.2010.03089.x. URL: https://www.sciencedirect.com/science/article/ pii/S1198743X14606856

66. Ruiz-Giardin JM, Ochoa Chamorro I, Velázquez Ríos L, Jaqueti Aroca J, García Arata MI, SanMartín López JV, Guerrero Santillán M (2019) Blood stream infections associated with central and peripheral venous catheters. BMC Infect Dis 19(1):841. https://doi.org/10.1186/ s12879-019-4505-2

67. Ruiz-Giardín JM, Martin-Díaz RM, Jaqueti-Aroca J, Garcia-Arata I, San Martín-López JV, Sáiz-Sánchez Buitrago M (2015) Diagnosis of bacteraemia and growth times. International Journal of Infectious Diseases 41:6–10. https://doi.org/10.1016/j.ijid.2015.10.008. URL: https://www.sciencedirect.com/science/article/pii/S1201971215002428

68. Sakarikou C, Altieri A, Bossa MC, Minelli S, Dolfa C, Piperno M, Favalli C (2018) Rapid and cost-effective identification and antimicrobial susceptibility testing in patients with gram-negative bacteraemia directly from blood-culture fluid. J Microbiol Methods 146:7–12. https:// doi.org/10.1016/j.mimet.2018.01.012

69. Schaefer G, Campbell W, Jenks J, Beesley C, Katsivas T, Hoffmaster A, Mehta SR, Reed S (2016) Persistent bacillus cereus bacteraemia in 3 persons who inject drugs, San Diego, California, USA. Merging Infect Dis 22(9):1621–1623. https://doi.org/10.3201/eid2209.50647

70. Schetinin V, Jakaite L, Krzanowski W (2018) Bayesian averaging over decision tree models for trauma severity scoring. Artif Intell Med 84:139–145. https://doi.org/10.1016/j. artmed.2017.12.003

71. Scholkopf B, Sung K-K, Burges CJC, Girosi F, Niyogi P, Poggio T, Vapnik V (1997) Comparing support vector machines with Gaussian kernels to radial basis function classifiers. IEEE Trans Signal Process 45(11):2758–2765. https://doi.org/10.1109/78.650102

72. Sepulveda J, Westblade LF, Whittier S, Satlin MJ, Greendyke WG, Aaron JG, Zucker J, Dietz D, Sobieszczyk M, Choi JJ, Liu D, Russell S, Connelly C, Green DA, Carroll KC (2020) Bacteraemia and blood culture utilisation during covid-19 surge in New York city. J Clin Microbiol 58(8):e00875-20. https://doi.org/10.1128/JCM.00875-20

73. Shipe ME, Deppen SA, Farjah F, Grogan EL (2019) Developing prediction models for clinical use using logistic regression: an overview. J Thorac Dis 11(4). https://doi.org/10.21037/ jtd.2019.01.25

74. Sidey-Gibbons JAM, Sidey-Gibbons CJ (2019) Machine learning in medicine: a practical introduction. BMC Med Res Methodol 19(1):64. https://doi.org/10.1186/s12874-019-0681-4

75. Smith DA, Nehring SM (2019) Bacteraemia. StatPearls Publishing, Treasure Island. URL: http://europepmc.org/books/NBK441979

76. Song Y, Himmel B, Öhrmalm L, Gyarmati P (2020) The microbiota in hematologic malignancies. Curr Treat Options in Oncol 21(1):2. https://doi.org/10.1007/s11864-019-0693-7

77. Stanski NL, Wong HR (2020) Prognostic and predictive enrichment in sepsis. Nat Rev Nephrol 16(1):20–31. https://doi.org/10.1038/s41581-019-0199-3

78. Stekhoven DJ, Bühlmann P (2011) MissForest—non-parametric missing value imputation for mixed-type data. Bioinformatics 28(1):112–118. https://doi.org/10.1093/bioinformatics/btr597

79. Tachalov VV, Orekhova LY, Kudryavtseva TV, Loboda ES, Pachkoriia MG, Berezkina IV, Golubnitschaja O (2021) Making a complex dental care tailored to the person: population health in focus of predictive, preventive and personalised (3p) medical approach. PMA J 12(2):129–140. https://doi.org/10.1007/s13167-021-00240-7

80. Tena F, Garnica O, Lanchares J, Hidalgo JI (2021) Ensemble models of cutting-edge deep neural networks for blood glucose prediction in patients with diabetes. Sensors 21(21). https://doi.org/10.3390/s21217090. URL: https://www.mdpi.com/1424-8220/21/21/7090

81. Towns ML, Jarvis WR, Hsueh PR (2010) Guidelines on blood cultures. Journal of Microbiology, Immunology and Infection 43(4):347–349. https://doi.org/10.1016/S1684-1182(10)60054-0. URL: https://www.sciencedirect.com/science/article/pii/S1684118210600540

82. Vaqué J, Rosselló J, Arribas J (1999) Prevalence of nosocomial infections in Spain: EPINE study 1990–1997. Journal of Hospital Infection 43:S105–S111. https://doi.org/10.1016/S0195-6701(99)90073-7. URL: https://www.sciencedirect.com/science/article/pii/S01956701 99900737. The 4th International Conference of the Hospital Infection Society

83. Wei L, Wan S, Guo J, Wong KK (2017) A novel hierarchical selective ensemble classifier with bioinformatics application. Artif Intell Med 83:82–90. https://doi.org/10.1016/j.artmed.2017.02.005

84. Wildi K, Tschudin-Sutter S, Dell-Kuster S, Frei R, Bucher HC, Nüesch R (2011) Factors associated with positive blood cultures in outpatients with suspected bacteraemia. Eur J Clin Microbiol Infect Dis 30(12):1615–1619. https://doi.org/10.1007/s10096-011-1268-0

85. Wilson M (2020) Critical factors in the recovery of pathogenic microorganisms in blood. Clin Microbiol Infect 26(2):174–179. https://doi.org/10.1016/j.cmi.2019.07.023

86. Wu Y, McLeod C, Blyth C, Bowen A, Martin A, Nicholson A, Mascaro S, Snelling T (2020) Predicting the causative pathogen among children with osteomyelitis using Bayesian networks – improving antibiotic selection in clinical practice. Artif Intell Med 107:101895. https://doi.org/10.1016/j.artmed.2020.101895

87. Yu JC, Khodadadi H, Baban B (2019) Innate immunity and oral microbiome: a personalised, predictive, and preventive approach to the management of oral diseases. PMA J 10(1):43–50. https://doi.org/10.1007/s13167-019-00163-4

88. Zangrillo A, Beretta L, Scandroglio AM, Monti G, Fominskiy E, Colombo S, Morselli F, Belletti A, Silvani P, Crivellari M, Monaco F, Azzolini ML, Reineke R, Nardelli P, Sartorelli M, Votta CD, Ruggeri A, Ciceri F, De Cobelli F, Tresoldi M, Dagna L, Rovere-Querini P, Serpa Neto A, Bellomo R, Landoni G, COVID-BioB Study Group (2020) Characteristics, treatment, outcomes and cause of death of invasively ventilated patients with covid-19 ARDS in Milan, Italy. Crit Care Resusc 22(3):200–211

Chapter 3
Vaginal Microbiome and Its Role in HPV Induced Cervical Carcinogenesis

Erik Kudela, Veronika Holubekova, Zuzana Kolkova, Ivana Kasubova, Marek Samec, Alena Mazurakova, and Lenka Koklesova

Abstract HPVs representing the most common sexually transmitted disease are a group of carcinogenic viruses with diferent oncogenic potential. Vaginal microbiome represent the modifiable and important risk factor in HPV-induced carcinogenesis. HPV infection significantly increases vaginal microbiome diversity and induces local inflammation, leading to gradual increases in the abundance of anaerobic bacteria and consequently the severity of cervical dysplasia. Anaerobic bacteria produce pro-inflammatory mediators and induce oxidative stress with subsequent epigenetic alterations resulting in formation of tumor microenvironment. Delineation of the exact composition of the vaginal microbiome, epigenetic state of cervical epithelium and immune environment before HPV acquisition, during persistent/ progressive infections and after clearance, provides insights into the complex mechanisms of cervical carcinogenesis. It gives hints regarding the prediction of malignant potential. Relative high HPV prevalence in the general population is a challenge for modern and personalized diagnostics and therapeutic guidelines. Identifying the dominant microbial as well as epigenetic and immune response biomarkers of high-grade and low-grade dysplasia could help us to triage the patients with marked chances of lesion regression or progression. Any unnecessary surgical treatment of cervical dysplasia could negatively affect obstetrical outcomes and sexual life. Therefore, understanding the effect and role of microbiome-based therapies is a breaking point in the conservative management of HPV-associated precancerous lesions. Sequentially, the immune response and epigenetic rearrangement of cervical epithelium could help to control the therapy outcome. Qualitative and quantitative assessment of local microbial environment and associated risk factors constitutes

E. Kudela (✉) · A. Mazurakova · L. Koklesova
Clinic of Gynecology and Obstetrics, Jessenius Faculty of Medicine,
Comenius University in Bratislava, Martin, Slovakia
e-mail: erik.kudela@uniba.sk

V. Holubekova · Z. Kolkova · I. Kasubova · M. Samec
Biomedical Center Martin, Jessenius Faculty of Medicine in Martin,
Comenius University in Bratislava, Martin, Slovakia

© The Author(s), under exclusive license to Springer Nature
Switzerland AG 2023
N. Boyko, O. Golubnitschaja (eds.), *Microbiome in 3P Medicine Strategies*,
Advances in Predictive, Preventive and Personalised Medicine 16,
https://doi.org/10.1007/978-3-031-19564-8_3

the critical background for preventive, predictive, and personalized medicine that is essential for improving state-of-the-art medical care in patients with cervical precancerous lesions and cervical cancer. This chapter focuses on the influence and potential diagnostic and therapeutic applications of the microbial markers in HPV-related cancers in the context of 3P medicine.

Keywords Predictive preventive personalized medicine (PPPM/3 PM) · Vaginal microbiome · HPV · Cervical malignancy · Female health · Gynecology · Oncology · Modifiable risks · Molecular pathways · Biomarker patterns · Prebiotics · Probiotics

3.1   Global Burden of HPV-Associated Cervical Cancer

Predictive, preventive, and personalized (3P) medicine has a cardinal priority in women's health. 3P medicine is a cornerstone in the fight against gynecologic cancers and represents an innovative approach in individualized treatment strategies and precision medicine [1, 2].

Cervical cancer (CC) is the fourth most common malignancy in women, with an estimated incidence of 13.1 per 100,000 women in age-standardized rate and mortality of 6.9 per 100,000 women in age-specific rate globally [3]. In 2018, around 570,000 new cases of CC were identified and more than 311,000women die each year. Incidence and mortality are higher in low and middle-income countries where organized screening and vaccination programms are lacking [4, 5]. Globally, 4.5% of cancers are due to the human papillomavirus infection [6]. Moreover, almost 95% of CC biopsies contain high-risk human papillomavirus (HPV) genotypes [7]. The incidence of HPV infection poses a significant burden on individuals and the broader healthcare system. The global burden of HPV is high, with around 630,000 new cases of HPV-associated cancer occurring each year [8]. It is necessary to say that infection with high-risk human papillomavirus alone does not cause cervical cancer.

HPV infection is one of the causes of pre-invasive and invasive cervical disease [7]. The most HPV infections do not cause symptoms and resolve spontaneously, but persistent infection can cause a wide range of diseases, including benign lesions, precancerous lesions and cancers [9]. Persistent HPV infection has a slow progression to invasive cervical cancer (average 7–10 years), and during this period, precancerous lesions can be detected by cytological screening [10]. More than 200 HPV genotypes belong to the *Papillomaviridae* family of DNA viruses, and about 30 HPV genotypes were found in he anogenital tract. Based on their oncogenic potential, HPV types are classified as high-risk (HPV 16, 18, 31, 33, 35, 39, 45, 51, 52, 56, 58, and 59) and probably high-risk (HPV 26, 53, 66, 67, 68, 70, 73, and 82) with an increased affinity for mucosa [11]. Low-risk HPV genotypes (HPV 6, 11, 42, 43, and 44) have an affinity to skin and evoke a benign lesions, including genital warts, papillomas and many other skin lesions [12]. Mucosal HPV genotypes are associated with cervical, penile, vaginal, vulvar, anal, and oropharyngeal pre-cancers and cancers [13]. Among women with normal cytology, the most common HPV genotype is HPV16, but the prevalence of other genotypes varies in different geographic regions [14]. HPV types 16 and 18 are clear, powerful carcinogens and cause approximately 70% of invasive cervical cancers [9].

HPV is one of the aetiological causes of cervical cancer and belongs to the most common sexually transmitted diseases (STDs). Almost 80% of women are affected by HPV infection during the lifetime [15]. Therefore, the researchers explore the interaction between HPV virus and the host organism that seems to be much more complex as expected. Fortunately, most HPV infections are cleared spontaneously (79% of infections in 24 months) by the immune system and do not lead to cervical dysplasia [16]. In some cases, permanent HPV infection may develop low- and high-grade lesions [17]. Besides, 90% of screening results in US population is represented by negative cytological finding and HPV negativity. The rest iss character-ized mainly by mild cytological abnormalities, including ASCUS (atypical squamous cells of undetermined significance) and LSIL (low-grade squamous intraepithelial lesion) either linked with presence or absence of hrHPV (high-risk HPV). HSILs (high-grade squamous intraepithelial lesions) represent only a small fraction of cytological results. Currently, management of women with mild cyto-logical abnormalities is based on a conservative approach, as in non-suspicious col-poscopic findings [18, 19]. HPV virus has the highest incidence in nulliparous women and therefore, the postponed surgical treatment of cervical lesion could have an auspicious effect on future pregnancies [20].

3.2   Risk Factors Associated with HPV Infection

Behavioral modifications can be suggested to help reduce the prevalence of HPV infection among women. Geographical disparities in the cervical cancer disease burden are stark and reflect the availability, coverage, and quality of preventive strategies and the prevalence of risk factors [21]. It remains unclarified why the HPV virus can persist or form the dysplasia in cervical epithelium. Biologically based non-modifiable and behaviorally based modifiable risk factors likely play an important in preventing the progression of the viral infection and prediction of its course. Obviously, given the serious socio-economic burden on society, overall can-cer management requires a shift from a reactive to 3P medicine in order to imple-ment cost-effective and individualized healthcare that will benefit the whole society [22–25]. The vaginal microbiome and the innate immune system are directly linked to the pathogenesis of HPV-induced CC [7]; therefore, the individualized patient profiles and targeted preventive, early predictive, or therapeutic strategies should be implemented in CC management as fundamental pillars of 3P medicine in order to achieve better outcomes for the individual and society as a whole. In this regard, improved CC management requires the identification of new liquid biopsy biomark-ers that should be obtained from specific body fluids. In summary, elucidating the malignant potential of specific HPV infections and related cervical lesions with respect to the local microenvironment is pivotal to the preventive and personalized approach to state-of-the-art medicine.

3.2.1 Non-modifiable Risk Factors

Risk factors for the development of cervical malignancy such as HPV infections, patient age, ethnic factors, host genetic factors, and the family history cannot be influenced in any way [26]. It is difficult to prevent their impact on the transformation of primary infections into persistent infections and persistent infections into pre-cancerous lesions. In addition, HPV infections, viral factors, viral load, and co-infections with multiple HPV genotypes and / or other sexually transmitted diseases (STDs) are major factors in the development of malignancy [27]. Human papillomavirus is a well-known carcinogenic agent [28]. In study Ciccarese et al. [28] the persistence of genital HPV infection was statistically associated with female gender, HR-HPV infection, smoking, and Ureaplasma parvum infection.

The age of the virus host itself is also associated with an increased risk of developing HPV infection. HPV is most prevalent among adolescents and young adults between 15 and 25 years of age. It is supposed that 75% of young individuals acquire HPV in this age range [29–31]. On the other hand, adolescent and younger girls have a greater risk of HPV infection due to a lack of immune responses in cervical epithelium and the presence of squamous metaplasia undergoing an endocervical reconfiguration to the ectocervix in response to an acidic environment. Therefore, basal cells in the cervical epithelium are more susceptible to HPV infection during the metaplastic transformation that may lead to cell proliferation and the development of cervical dysplasia or squamous cell cancer [32, 33]. It is also interesting that the prevalence of HPV infection increases in postmenopausal women over 50 years of age. This may be due to impaired immunity and subsequent reactivation of latent infections, and also the number of births and sexual partners over a lifetime represents a cumulative risk of worsening cervical cytology [34].

3.2.2 Modifiable Risk Factors

Influencing factors in HPV infection include sexual behavior, lifestyle, as well as socioeconomic status. Moreover, cultural and geographic variations influence the sexual behavior of women and their partners. An estimated 33–37% of incident cancers in Canada are attributable to modifiable risk factors [35]. There are some established cofactors for progression of cervical HPV infection to cancer in the long run, such as early age of first sexual intercourse, multiple sexual partners, tobacco use, immune suppression such as co-infection with human immunodeficiency virus (HIV), high parity, long-term hormonal contraceptive use, and poor nutritional status [36]. These factors include certain aspects of a woman's sexual history: the age of first sexual intercourse, the age at first birth, parity [37], and the intake of oral contraceptives [38]. Also, increased HPV risk is associated with numerous sexual partners. According to study of Kitamura et al. [39] number of lifetime sex partners ($\geq$6) and present history of sexually transmitted infection were the common significant predictors of high-risk and low-risk HPV infection .The use of condoms and hormonal contraceptives are recommended to reduce the risk of HPV infection. Condoms moreover, have some protective effects against the transmission of HPV and other STDs, including HIV [40]. If the parity may influence CC risk remains still unclear. HPV-positive women with 7 or more reported full-term pregnancies have a fourfold increased risk of CC compared to HPV-positive nulliparous women [37]. Hormonal factors related to pregnancy and cervical trauma associated with delivery may increase the risk of cervical carcinogenesis [41]. Also, the question of the increased risk of CC in the use of oral contraceptives is still unanswered. Many studies show that there is an increased risk of cervical adenocarcinoma due to the excess of progesterone-free estrogen caused by oral tablets; endometrial cells respond to this hormonal imbalance through endometrial hyperplasia [42]. On the other hand, women using hormonal contraceptives have a lower risk of developing the disease because they undergo regular screening tests and are also under the supervision of a doctor [43].

Known risk factors for the disease are, in particular, unhealthy lifestyles, smoking, alcohol and drug use. Negative life events and impacts pose a higher risk, especially for women from lower socio-economic groups. Cigarette smokers are at an increased risk for SIL and CC compared to HPV-positive non-smokers [44]. Cervical tissue turnover may be caused by the action of carcinogens in tobacco smoke, which cause immunosuppression. The persistence of HPV virus has also been confirmed in cervical mucus. Progression of HPV infection to malignancy is possible due to the transfer of viral DNA into the host genome. [45]. Local immunosuppression can result from nicotine and its metabolite cotinine that are present in cervical mucus and from the reduced numbers of Langerhans cells in the cervical epithelium of smokers [28].Substance abuse (e.g. alcohol consumption [46] and illegal drugs [47]) may reduce immune function and thus affect the cervical squamous epithelial microenvironment and support persistent HPV infection. Studies reported that smoking habit and HR-HPV genotypes influence virus persistence [28].

The risk of health problems is higher for women from low socio-economic backgrounds because they have limited incomes and restricted access to health care. These women are often poorly informed about CC risks and suffer from nutritional deficiencies [48]. DNA damage can be prevented by the intake of vitamins C and E, carotenoids and lycopene contained in the amount of fresh fruits and vegetables [49]. Accordingly, fresh vegetable consumption reduced the risk of HPV persistence by more than 50% [50]. A healthy lifestyle with moderate sexual behavior may reduce the risk of long-term HPV infection and enhance the immune response. In the United States women from Hispanic background experiencing a higher burden of disease compared to women of Caucasian origin [51]. There are considerable racial and ethnic disparities in terms of burden of HPV infection and cervical cancer. Interventions targeting modifiable cancer risk factors could prevent a substantial number of incident cancer cases. HPV vaccines helps to eradicate most types of HPV condition [52]. HPV vaccination is safe and efficacious to prevent persistent HPV infection, precancerous anogenital lesions and cervical cancer [53]. The vaccine is recommended initially for women who have not been vulnerable with HPV during sexual intercourse. Acceptance of HPV vaccine varies across racial and ethnic groups. Addition of HPV vaccination to the existing immunization programs calls for logical discussion and consideration to preserve the highest ethical standard in administering this vaccine to a sensitive age group of adolescence [36]. Before HPV vaccination, periodic Pap testing in women was the main way to prevent cervical cancer. Effective primary prevention and secondary prevention approaches can prevent most cervical cancer. Cervical cancer screening followed by treatment once abnormal lesions are identified can help reduce cervical cancer incidence and mortality [10].

3.3   The Vaginal Microbiome: Its Composition and Interactions

A complex system comprising of the mucosal epithelial barrier, the immune system, and a healthy VM generating lactic acid, hydrogen peroxide, halides, and antimicrobial peptides protects the female genital tract from infections. Furthermore, the VM influences local inflammatory immunological responses, such as cytokine production [54]. Maintaining or enhancing the VM is a novel and successful treatment method for HPV infections and associated precancerous lesions [55].

Estrogen levels have a major impact on the diversity of the vaginal microbiota. Estrogen affects the volume and viscosity of vaginal secretions, the glycogen content, and the amounts of oxygen and carbon dioxide in the vagina [56]. Regular vaginal lubrication, an acidic vaginal pH, and maintaining a healthy vaginal microbiome are all vital and efficient defensive mechanisms against alien microbial infections. Vaginal dryness as a part of Sicca syndrome implies a low Lactobacillus status with a high proclivity for vaginal infections and even lichen sclerosus of the vulva [57]. Reduced estrogen levels in prepubertal and postmenopausal women are related to low *Lactobacillus* levels and presence of mixture of anaerobic bacteria [58]. The vaginal microbiota of pregnant women, on the other hand, is more stable and often dominated by *L. crispatus* or *L. iners* [59]. Two longitudinal studies found that the phase of the menstrual cycle and sexual activity influence changes in the vaginal microbiota diversity [60, 61].

Lactobacillus species can colonize both the urinary system and the rectum. Through direct or estrogen-mediated pathways, the female reproductive tract microbiome interacts with the gut (vagina–gut axis), the urinary tract (vagina–bladder axis), and other places such as the oral cavity. The rectum is an important lactobacilli reservoir that helps to maintain a healthy vaginal microbiota and local immune system, resulting in a decreased infection rate [62]. Enteric bacteria are capable of

Table 3.1 Factors influencing the composition of vaginal microbiome [62]

Genetics/host	Environmental	Socioeconomic	Behavioural	STI
Aging	Geography	Education	Sexual behavior	Bacterial
Genomics	Early life factors	Income	Contraception	Viral
Epigenetics	Toxins and carcinogens	Race, ethnicity	Hygiene practices	Fungal
Pregnancy	ATB, prebiotics, xenobiotics	Access to healthcare	Smoking	Parasitic
Hormonal status	Stress	Social policy	Alcohol consumption	
Comorbidities	HPV vaccine		Diet/nutrition	
Altered immunity			Obesity	
Obesity			Physical activity	

STI sexually transmitted infections, *ATB* antibiotics

deconjugating estrogens and facilitate their reabsorption into the circulatory system [63]. This causes increased glycogen and mucus production, as well as thickening of the lower genital tract epithelium. As a result, a decrease in estrogen-metabolizing bacteria may impact *Lactobacillus* dominance in vaginal flora [64]. Moreover, vaginal and fecal microbiota transplantation (VMT, FMT) is a unique, intriguing therapy option being researched for women with BV or vaginal diseases [65].

Tangled interactions between the microbiome and host that increase the risk of gynecological malignancies can be influenced by behavioral, social, genetic, environmental, and host factors, including early life factors such as gestation, type of delivery, and infancy. Table 3.1 summarizes these factors.

Recent studies have investigated the association between the composition of the VM and HPV infections that contribute to carcinogenesis. Ravel et al. [66] studied the VM and vaginal pH of 396 asymptomatic, sexually active women and discovered five primary VM community state types (CSTs). *L. crispatus* dominated CST I, which occurred in 26.2% of the women, whereas *L. gasseri, L. iners,* and *L. jensenii* dominated CST II (6.3%), CST III (34.1%), and CST V (5.3%), respectively. The remaining CST IV detected in 27% of the women was heterogeneous, with larger proportions of strictly anaerobic bacteria such as *Gardnerella, Prevotella, Megasphaera,* and *Sneathia* species [66]. CST I had the lowest median pH (4.0 0.3), suggesting that other CSTs may create less lactic acid than group I or have different buffering properties [3].

3.4 Two Faces of *Lactobacillus* Species

Lactobacillus spp. protect the vaginal epithelium by preventing harmful bacteria from adhering to the epithelial tissue via a number of barrier (self-aggregation, adhesion) and interference (receptor binding interference, coaggregation with potential pathogens) mechanisms [67]. *Lactobacillus* creates organic acids by degrading glycogen to keep the vaginal environment acidic [68], which can prevent pathogenic bacteria from invading. *Lactobacilli* that produce H2O2 enhance the release of antimicrobial compounds by epithelial cells and boost the antibacterial activity of preexisting protective factors (muramidase and lactoferrin) [69]. *Lactobacilli* release a variety of metabolites and surfactants, including exopolysaccharides, phosphorylated polysaccharides, and peptidoglycans, which have the ability to prevent pathogenic microbes and carcinogenesis [70–72]. *Lactobacillus* spp. also stimulates the immune system's cellular and humoral components [73].

In comparison, *L. crispatus* is more successful than *L. iners* in avoiding bacterial dysbiosis, which is missing a protective role in vaginal health [74]. Because *L. iners* can only run the synthesis of L-lactic acid and cannot create H2O2, it has neither antibacterial or antiviral action [75]. *L. iners*, on the other hand, generates inerolysin, a cytotoxin comparable to *Gardnerella vaginalis*. Inerolysin causes pores to develop in the vaginal epithelium, increasing the risk of infection [76].

HPV clearance may be influenced by microbiome modulation using *Bifidobacteria* and *Lactobacillus* [77]. *Lactobacillus* spp. are linked to a lower detection of high-risk HPV subtypes (OR 0.64), cervical dysplastic lesions (OR 0.53), and invasive malignancies (OR 0.12). *L. crispatus* (CST I) exhibits even superior qualities in terms of the occurrence of high-risk HPV (hrHPV) infections (OR 0.49) and neoplastic alterations (OR 0.50) [78]. Furthermore, a *Lactobacillus gasseri* (CST II) dominated microbiome promotes the quick clearance of HPV infections (aTRR = 4.43) [79, 80]. Moreover, a subset of less prevalent *Lactobacilli*, including *L. agilis* and *L. sanfranciscensis*, is significantly reduced in HPV-positive women and may have a role in cervical carcinogenesis [81]. Overexpression of H2O2-producing *L. jensenii* and *L. coleohominis*, according to Mitra et al., inhibits the development of low-grade cervical dysplasia [82].

A *L. iners*-dominated microbiome, on the other hand, is frequently related with HPV positivity. Despite having a greater relative abundance in hrHPV-negative women (mean 19.3) than in hrHPV-positive women (11.9), *L. iners* remains the dominating species in CST IV [83]. The most common transition identified in the vagina is from CST III to CST IV, suggesting that *L. iners* is less capable of inhibiting anaerobic bacterial colonization than other *Lactobacillus* spp. [84]. *L. iners* was even shown to be more prevalent just before to the commencement of HPV16 infections [85]. *Lactobacillus* depleted CST IV and *L. iners* are responsible for HPV persistence and development of preinvasive and invasive lesions [80]. *Gardnerella vaginalis* combined with *L. iners* or other unclassified lactobacilli enhances the likelihood of high-grade cervical lesions [86, 87]. This CST type increases the incidence of cervical dysplasia by a factor of six [88]. Another study evaluated L. types and discovered that *L. iners* had a greater likelihood of prevalent hrHPV than *L. crispatus* (OR – 1.31). When compared to *L. crispatus*-dominated CST, *L. iners* is linked with nearly double the overall risk for LSIL and malignant lesions (OR 1.95) [87].

3.5   Prognostic Role of Vaginal Microbiome Composition

Bacterial vaginosis (BV) is quite common (around 9% in the UK [88] and up to 29% in the US [89]). In women with BV, invading pathogens such as *Gardnerella vaginalis, Prevotella*, and *Mobiluncus* species displace native vaginal flora [90]. The substitution of *G. vaginalis* for *Lactobacilli* results in a more basic pH, which promotes BV. *G. vaginalis* creates a biofilm that serves as a matrix for other pathogenic bacteria to attach to; this biofilm also lowers antibiotic potency [84]. CST IV in the cervicovaginal niche has been linked to an increased likelihood of acquiring persistent HPV infections and, as a result, developing cervical lesions [91].

The nature of the VM changes between HPV-positive and HPV-negative people. Furthermore, bacterial vaginosis is more common in persistent high-risk HPV infections than in cases with HPV elimination [80]. *L. acidophilus, L. crispatus, L. jensenii, L. psittaci, L. ultunensis*, and *L. vaginalis* typically colonize HPV negative women. HPV-positive women, on the other hand, have a more diversified microbiome that includes bacteria such as *A. vaginae, G. vaginalis, D. micraerophilus, G. vaginalis, S. amnii, S. sanguinegens*, and *Prevotella* species (*P. amnii, P. buccalis*, and *P. timonensis*) [83]. Shannon et al. found that HPV-positive individuals are more likely than HPV-negative women to develop cervico-VMs consistent with CST-IV (58.8% vs. 29.4%) [92]. The dominance of some pathogens during the early stages of HPV infection may contribute to the development of cervical dysplasia later on [93]. A CST IV subgroup, defined by the predominance of *Gardnerella, Prevotella, Megasphaera*, and *Atopobium* species, was found in 43% of women with persistent HPV infections but only 7.4% of women with HPV clearance. *G. vaginalis* is a significant risk factor for the development of cervical lesions (OR 10.19) [74]. Its prevalence may come from a change in immune responses from anti-microbial to anti-viral, with a loss of bacterial control induced by HPV itself [85]. Following the production of bacterial sialidase by *G. vaginalis*, a biofilm is formed that entraps anaerobic bacteria such as *Prevotella* and *Atopobium*, resulting in their proliferation and HPV persistence [94]. *G. vaginalis* may be a significant indicator of HPV infection progression to HSIL lesions by inducing higher microbiome diversity [95], which has an immunosuppressive impact [96]. This vicious loop alters mucosal metabolism and immunological responses generally [97, 98]. A pro-inflammatory milieu therefore promotes viral DNA integration, which is critical in cervical carcinogenesis, viral persistence, and disease progression [82].

HPV infection enhances VM heterogeneity and richness, resulting in a steady increase in the CST IV fraction and, as a result, a worsening of cervical dysplasia severity [99]. The mycobiome correlates with HPV infection and CIN severity in a similar way, with more fungal diversity in hrHPV infections (*Malassezia*) and ASCUS cytology findings (*Sporidiobolaceae, Sacharomyces*) [100].

3.5.1   Microbial Markers of LSIL

Chen et al. identified certain microorganisms and vaginal bacterial structure as being linked to the advancement of CINs. For example, HPV infection without CINs or malignant lesions was closely related with *Megasphaera*, whereas *Prevotella* amnii was the most prevalent bacteria in the low-grade squamous intraepithelial lesion group [99]. When compared to HPV-negative women with low-risk microbial scores, HPV-positive individuals with high-risk microbial patterns have an OR of 34.1 for LSIL development [80].

Snaethia is another common species associated with HPV infection because it causes severe inflammation in the vaginal microenvironment. Some research suggests that it may be a sign of low-risk HPV (lrHPV) genital infection [101]. Lr HPV infection is also linked to an increase in *Actinobacteria* and *Atopobium*, both of which alter epithelial barriers [102]. *Gardnerella, Bifidobacterium, Hydrogenophylus, Burkholderia,* and *Fusobacterium* are also prevalent bacterial species in the lrHPV group, and all have carcinogenic potential [82, 103, 104].

CST IV incidence rise twofold in LSIL, thrice in HSIL, and fourfold in CC groups [81]. Higher amounts of *Peptostreptococcus anaerobius, Anaerococcus tetradius,* and *Snaethia sanguinensis* are related with HSIL lesions; these species may serve as biomarkers for HSIL lesions [80, 82, 105]. *Mycoplasma* is more common in HSIL lesions than in LSIL lesions, however it is not seen in malignancy. *Mycoplasma* spp. may be involved in the early stages of HPV infection and may aid in HPV persistence [93].

3.5.2   Microbial Markers of HSIL

So et al. calculated the total risks for HSIL and CC development related to certain bacteria, finding the highest values for *Atopobium* (OR 4.33), *Finegoldia* (6.00), *Prevotella timonensis* (6.00), G. vaginalis (7.33), and *Prevotella buccalis* (11.00) [74, 81]. In situations of chronic hrHPV infection, *Prevotella* spp. might be predictive of CIN2+ lesion progression [81]. CIN2 is a complex condition with a high rate of remission. CST IV species, such as *Megasphera, Prevotella,* and G. *vaginalis,* had an OR of 3.85 for 12 month persistence and an OR of 4.25 for 24 month persistence [106]. CIN3 lesions had considerably different VM composition than CIN2 (*Lactobacillus* spp., A. *vaginae,* G. *vaginalis,* and U. *parvum*), with *Aerococcus, Leptotrichia, M. hominis, Prevotella, Snaethia,* and L. *crispatus* prevalence declining from 70% to 47% [107, 108].

Another article by Chao X. et al. [109] evaluated the potential for the vaginal microbiome to be used as a marker for HSIL. The study revealed an increasing abundance of *Stenotrophomonas, Streptococcus, Pseudomonas,* and a paucity of *Bifidobacterium* and other genera like *Prevotellaceae* and *Faecalibacterium* in patients with HSIL. These findings implicated these genera as potential biomarkers of HSIL and their possible use in cervical screening [109].

Mitra et al. [110] confirmed the depletion of *Lactobacillus* species, high diversity of the vaginal microbiome, and high levels of pro-inflammatory cytokines in women with CIN. The study also evaluated the vaginal microbiome of women after excisional treatment for CIN at their 6-month follow-up visit; they suggested that failure to re-establish a *Lactobacillus*-enriched CST may lead to higher risk of development and persistence of cervical precancerous lesions, even after surgical treatment [110].

According to these research, significant decreases in lactobacilli are detected in the third grade of dysplastic alterations of the cervical epithelium. *Bacillus, Snaethia, Acidovirus, Oceanobacillus profundus, Fusobacteria, Veillonellaceae, Anaerococcus*, and *Porphyromonas* are all related with invasive cancer [99]. Figure 3.1 depicts an overview of bacterial compositions in HPV infection, various degrees of dysplasia, and invasive malignancy.

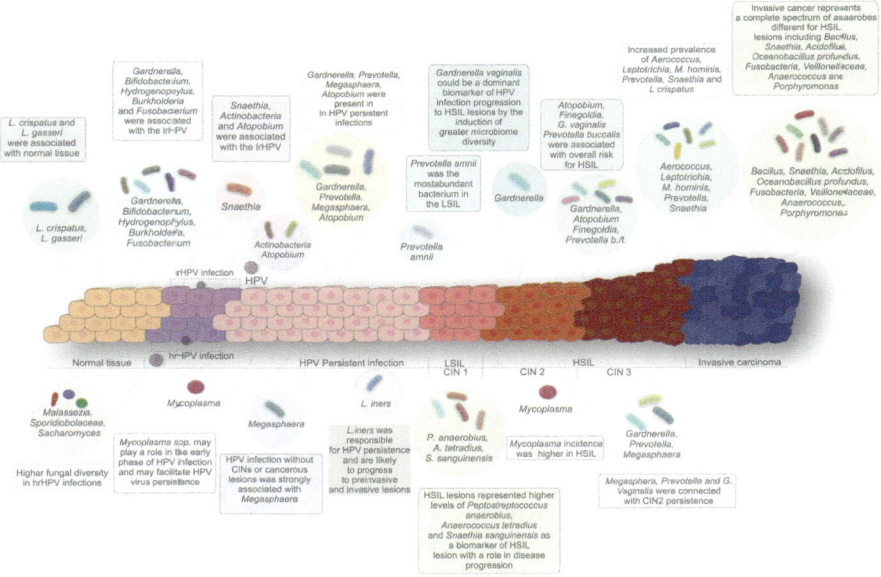

Fig. 3.1 Vaginal microbiota composition during cervical carcinogenesis. Abbreviations: *HPV* human papillomavirus, *hrHPV* high-risk HPV, *lrHPV* low-risk HPV, *LSIL* low-grade squamous intraepithelial lesion, *HSIL* high-grade squamous intraepithelial lesion, *CIN* cervical intraepithelial neoplasia, *L. crispatus* Lactobacillus crispatus, *L. gasseri* Lactobacillus gasseri, *G. vaginalis* Gardnerella vaginalis, *Prevotella b.* Prevotella buccalis, *Prevotella t.* Prevotella timonensis, *M. hominis* Mycoplasma hominis, *L. iners* Lactobacillus iners, *P. anaerobius* Peptostreptococcus anaerobius, *A. tetradius* Anaerococcus tetradius, *S. sanguinensis* Snaethia sanguinensis

3.6 The Relationship Between the Microbiome and the Epigenome of Cervical Cells

An epigenome is composed of an enrollment of the chemical changes to the DNA and histone proteins of an organism. Epigenetic changes are heritable alterations that do not modify the DNA sequence. The most relevant regulators of gene expression are posttranslational histone modification and related chromatin remodelling, DNA methylation occurring in 5′-C-phosphate-G-3′ (CpG) islands, and non-coding RNAs [111]. Epigenetic changes can modify the cellular microenvironment and therefore are well studied in tumor formation, metastasis and also in cervical lesion progression and cervical carcinogenesis. There is a growing interest in finding biomarkers that would be responsible for the disease progression and/or would be used in targeted therapy in personalized medicine.

It is assumed that also the specific microbiota influence the epigenome of cervical squamous epithelium but exact mechanism is still unclear. It is known, that approximately 10% of host genes is regulated by the intestinal microbiota, including genes regulating immunity, cell proliferation, and metabolism. The host defence against invading pathogens or pathogen persistence is regulated through gene expression changes in different types of immune cells. [112]. A recent investigation has shown that the microbiome can induce changes in gene expression and alteration of chromatin structure via epigenetic modification. Moreover, it has been shown that gut or intestinal microbiota can also influence the cervical microbiota [113].

The relationship between microflora and intestinal homeostasis has been studied more extensively. One of mechanism by which the microbiome influences the homeostasis is histone modification and chromatin remodelling in intestinal immune cells [114]. In a study on colonic epithelial cells, the treatment of a cell culture with live microbiota from healthy individuals confirmed changes in gene expression of more than 5000 host genes via changes in host chromatin accessibility and binding of transcription factor. For instance, the chromatin opening resulted in an increased expression of tropomyosin 4 (TPM4) that is responsible for binding of actin filament that may have a role in regulation of cytoskeleton and affects the host cell response to the microbiome [115]. Probiotic bacteria, such as *L. rhamnosus* GG and *B. breve*, inhibit LPS-induced expression of interleukins IL-17 and IL23 by inhibiting the acetylation of histones [116]. Lactic acid and/or short-chain fatty acids (SCFAs) produced by *Bifidobacterium* and *Lactobacillus* species act as histone deacetylation inhibitors and mediate the beneficial effect of probiotics [117]. Peripheral tissues beyond the gut and the gut-brain axis are directly affected by the production of SCFAs through the regulation of neuroimmune and neuroendocrine functions [118].

There are also proof that the intestinal DNA methylation pattern is affected by the inhabiting microbiota [119–121]. DNA methylation is an essential epigenetic mechanism for differentiation of intestinal stem cells and maintaining intestinal homeostasis [119, 122, 123]. A study looking at the effect of the commensal microbiota on DNA methylation pattern and host gene expression in murine intestinal epithelial cells revealed that transcription regulated by the microbiota was performed through induction of DNA methylation changes in regulatory regions. In case of acute inflammation, the microbiota induced chromatin remodelling and ten-eleven translocation protein 2/3 (TET2/3)-dependent enhancer demethylation [121]. Moreover, DNA methylation also plays an important role in the induction and regulation of the immune response to bacteria [124]. The association of cervicovaginal microbiota with DNA methylation status has not been extensively studied. Nené et al. (2020) procured an evidence of a strong local interaction between the host epigenome and the cervicovaginal microbiome, The methylation pattern determined the specific composition of microbial communities and on the basis of the methylation status in 819 CpGs, it was possible to discriminate between communities associated with hypomethylation with dominance of non-lactobacilli species and cervicovaginal samples with lactobacilli-dominant strains [125].

Unfortunately, there is a lack of studies focusing on the interaction of the cervicovaginal microbiota and DNA methylation. It was found that the pathogenic bacteria are able to induce alterations in cellular DNA methylation pattern directly via the regulation of TET and DNA methyltransferase (DNMT) enzymes or indirectly via inflammatory mediator production in response to bacterial infection [126, 127]. *In vitro* experiments on human dendritic cells (professional APCs) infected with *Mycobacterium tuberculosis* detected DNA demethylation at distal enhancer elements, mainly in genes expressing transcription factors of the immune system. The gene regions with the hypomethylation were characterized by elevated levels of 5-hydroxymethylcytosine (5-hmC) that demonstrated the activation of TET enzymes in the process of demethylation induced by mycobacteria [126]. Immunohistochemical analysis of cervical tissues, showed upregulated expression of TET1 dioxygenase in precancerous lesions and downregulated in invasive cancers. TET1 interaction with chromatin-modifying suppressors (lysine-specific histone demethylase [LSD1] and enhancer of zeste homolog 2 [EZH2]) subsequently led to silencing of epithelial-mesenchymal transition (EMT) and a reduction of zinc finger E-box binding homeobox 1 (ZEB1) and vimentin (VIM) expression [128]. Cervical cancer is frequently compared to head and neck squamous cell cancer (HNSCC) because of the similarities of the squamous epithelium and its susceptibility to HPV infection. A recent study reported *Fusobacterium* and *Peptostreptococcus* as the most frequent pathogenic bacteria in HNSCC cancer tissues and in oral rinse samples. *Fusobacterium nucleatum* has also been linked tohypermethylation and hypomethylation of the host gene promoter, presumably through gene dysregulation associated with the inflammatory response and cell proliferation through epigenetic silencing [129]. *F. nucleatum* was also present in welll differentiated cervical cancers containing cells with cancer stem cells characteristics [130].

A cervicovaginal microbiome with dominance of *Lactobacillus* spp. was found as more protective against *Chlamydia trachomatis* infection due to production of lactic acid D(−) and L(+) isoforms in a pH-dependent manner [131]. A cervicovaginal environment with presence of non-lactobacilli (O-type) communities was described to be less supportive for *Lactobacillus* colonisation, especially due to

CpG island methylation mirroring gastrointestinal differentiation [125]. *L. crispatus* and *L. jensenii* modulate genes responsible for cell proliferation by decreasing the histone deacetylase (HDAC4) and increasing the histone acetylase (HAT) EP300. Lower proliferation in epithelium is mediated through an elevation in expression of cyclin-dependent kinase inhibitor 1A (CDKN1A) and following inhibition of cyclin-dependent kinase 4 (CDK4) activity [131]. HAT EP300 is also regulated by oestrogen receptor 1 (ESR1) that mediates vaginal cell proliferation through ligand binding to the receptor. Cervical epithelial cells exposed to *L. crispatus* and *L. jensenii* but not to *L. iners* or *G. vaginalis* entailed in lower expression of ESR1 induced by EP300 that could control cell homeostasis [131].

Cervical epithelium colonised with *Lactobacillus* spp. communities is also protected by microbial modulation of human miRNAs. Small RNA transcriptome sequencing (small RNA-seq) of cervical specimens and *In vivo* experiments on cervical epithelial cells showed abnormal miR-183, miR-193b, and miR-223 expression in dominance of *Lactobacillus* spp. These miRNAs are linked with transcription, cell cycle, cellular signalling, development, and hypoxia. Other investigated miRNAs (miR-203b and miR-320b-1) had no experimentally validated targets [131]. On the other side, non-*Lactobacillus*-dominant microbial communities have been associated with to elevated levels of miR-23a-3p and miR-130a-3p in young women [132]. Other *in vivo* study revealed significantly increased expression of miR-143, miR-145, miR-146, miR-148b, miR-193b, miR-223, and miR-15a in ectocervical cells exposed to bacteria-free supernatants from *L. iners* or *G. vaginalis*. *L. crispatus* free bacterial supernatants did not alter miRNA expression. Interestingly, the significantly increased miRNAs were prominently reduced after exposure of ectocervical cells to *L. crispatus* bacteria-free supernatant [133]. The downstream targets of these miRNAs remain unknown.

As we mentioned before, epigenetic biomarkers are well studied due to their potential use in early diagnosis, disease prognosis and targeted therapy. When introducing diagnostics focused on specific epigenetic biomarkers, it is necessary to verify the presence of the biomarker first in a small group of patients and to analyze the whole genome and assess all epigenetic changes. When small epigenetic changes are verified without the context of whole genome sequencing, it is usually impossible to verify these results in larger clinical trials [134].

3.7 Inflammation in Response to Cervical Microbiome

Resolving the dilemma of whether the changes in microbial composition in the distinct tissues lead to a dysregulation of physiological functions promoting cancer, or whether the accrue of cancer or a pro-inflammatory process in a specific tissue can modify the environment and promote the growth of some bacterial species before others, could subsequently be used in the therapeutic administration of bacteria or their metabolites to exploit their immunosuppressive and unique properties that can shape the tumor microenvironment and support a favorable local microbiome [135].

Pathogenic or commensal bacteria are a part of common microbiome that help to keep the homeostasis in the tissues where grow. The cervical epithelial barrier holds the microbiota in a mucosal layer and avoids the penetration of viruses and pathogenic bacteria into the cervical stroma. Excreted biosurfactants produced by *Lactobacillus* form a biofilm on the cervical surface [136]. Bacterial metabolites and secreted factors, such as lipopolysaccharide (LPS), can induce an inflammatory response and disruption of the epithelial barrier [133, 137]. Epithelial cells recognize pathogens by bacterial LPS through Toll-like receptors (TLRs), NOD-like receptors, RIG-I-like receptors, and C-type lectin receptors [138, 139]. The detection of bacteria triggers signalling pathways in innate immune cells what initiates the production of chemokines, cytokines and antimicrobial peptides that elicit defensive and inflammatory reactions. Adaptive immunity contributes to the immune response mediated by T and B cells that are activated by the presentation of bacterial antigens by antigen-presenting cells (APCs) [124]. The effect of cervicovaginal microbiome on epithelial barrier permeability and function of cervix has been studied *in vitro* by Anton et al. (2018). They revealed that bacterial supernatants derived from *L. crispatus*, *L. iners,* and *G. vaginalis* alter the integrity of the epithelial barrier by various mechanisms. Both *L. iners* and *G. vaginalis* induce changes in miRNA expression and inflammatory mediator profile (IL-6 and IL-8) leading to increased epithelial barrier permeability. In addition, *L. iners* is able to disrupt of adherent junctions through a cleavage of E-cadherin. Conversely, *L. crispatus* protect the epithelial barrier in the presence of LPS that induced increased cell permeability and inflammation [133]. A detailed examination of metabolites during vaginal dysbiosis and inflammation of the cervical epithelium may help to identify inflammatory-induced biomarkers that have also been described in cervical cancer. Such markers could aid in preventive medicine in predicting the development of cervical pre-canceroses.

Metabolic and immunological changes in cervical epithel induced by composition of vaginal microbiota were described by Łaniewski and Herbst-Kralovetz (2021). They examined the complex impact of cervicovaginal bacteria (*L. crispatus* as a commensal bacteria connected with healthy microflora and *G. vaginalis, P. bivia, A. vaginae,* and *S. amnii* representing microflora in BV) on cervical epithelial cells in a 3D cervical cell model. The immune–metabolic analysis showed that *S. amnii* and *A. vaginae* had the highest pro-inflammatory potential because of their triggering of pro-inflammatory chemokine and cytokine production, iNOS, and oxidative stress. Their results also confirmed that *G. vaginalis, P. bivia*, and *S. amnii* negatively influence the integrity of the epithelial barrier by reduced levels of mucins and matrix metalloproteinases. Contrary, phenyllactic acid produced by *L. crispatus* has an antimicrobial effect and helps to create a protective microenvironment [140]. Two-dimensional (2D) and three-dimensional (3D) cell models can be used in personalized medicine to understand the interaction of host and microbial cells and their response to external factors. There is a little information on the mechanism of STD-permissive microbiota interaction and increment of infection risk. These mechanisms can be studied partially using three-dimensional (3D) models of the cervical epithelium, which facilitate the study of host epithelial interactions as well as sexually transmitted infectious agents, and bacteria commonly present in the vaginal flora. In this way, it is possible to get closer to the conditions in the infected human host, to better understand the complexity of human physiology that can help to develop preventive therapies [141].

Microbial imbalance can adjust the inflammatory response through the production of pro-inflammatory chemokines, cytokines, and induction of oxidative stress [142]. A study by Anahtar et al. (2015) revealed that local immune response and production of pro-inflammatory cytokines is species-specific and coheres mainly with very diverse bacterial communities in CST-IV. Authors performed a transcriptional analysis and recognized differentially expressed genes in cervical APCs in women with *Prevotella*-dominant communities compared to women with *Lactobacillus*-dominant communities. Elevated genes are involved in NF-κB, TLR, NOD-like receptor, and TNF-α signalling pathways. The summaries of the study suggest that endocervical APCs percept microbial LPS and trigger immune response via TLR-4 signalling and NF-κB pathways, resulting in secretion of cytokines and chemokines that initiate inflammation [143]. Moreover, the production of TNF-α and IFN-γ cytokines has negative effect on function of endocervical epithelial barrier by disruption of tight junction leading to viral and bacterial translocation [144, 145]. The presence of *Fusobacterium* spp. probably contributes to an immunosuppressive microenvironment [104] based on induced elevated expression of anti-inflammatory IL-4 and transforming growth factor β1 (TGFβ1) in cervical cells. On the other hand, *Fusobacterium* spp. can also promote cervical carcinogenesis by inhibition of immune cytotoxicity [146]. A schematic difference between the epithelial homeostasis and bacterial vaginosis is displayed on Fig. 3.2.

A study focusing on classification of CST and profiling of the local immune response assessed that a different type of inflammation is triggered and controlled by both the dominant non-*Lactobacillus* and by *Lactobacillus* spp. bacterial colonies. One mechanism is through balance between the pro-inflammatory (IL-1α and IL-18) and anti-inflammatory (IL-1ra) response. The second, independent immune mechanism is specific to *L. iners* in CST-IV, with elevated inflammation and production of pro-inflammatory factors, for example TNF-α and macrophage migration inhibitory factor (MIF) [147].

Prolonged inflammation and aggregated metabolites produced by immune cells also entail DNA damage and activate epigenetic alterations that can trigger process of tumorigenesis. There are parallelism in the pattern of DNA methylation in cancer

Fig. 3.2 (continued) low levels of mucins and MMPs, and trigger TLR, NOD-like receptor, NF-κB, and TNF-α signalling pathways. *Fusobacterium* spp. creat the immunosuppressive increased expression of anti-inflammatory TGFβ1 and IL-4 in cervical cells was associated with the presence of microenvironment. **Abbreviations**: *A. vaginae Atopobium vaginae, G. vaginalis Gardnerella vaginalis, L. crispatus. Lactobacillus crispatus, L. iners Lactobacillus iners, P. bivia Prevotella bivia, S. amnii Sneathia amnii, APC* antigen presenting cells, *CLR* C-type lectin receptors, *IL-4* interleukin 4, *IL-6* interleukin 6, *IL-8* interleukin 8, *iNOS* inducible nitric oxide synthase, *LPS* lipopolysaccharide, *miRNA* microRNA, *MMPs* matrix metalloproteinases, *mRNA* messenger RNA, *NK* natural killer, *NLR* NOD-like receptors, *RIG-I* Retinoic acid-inducible gene I, *RISC* RNA-induced silencing complex, *TGFB1* transforming growth factor beta 1, *TLRs* toll-like receptors

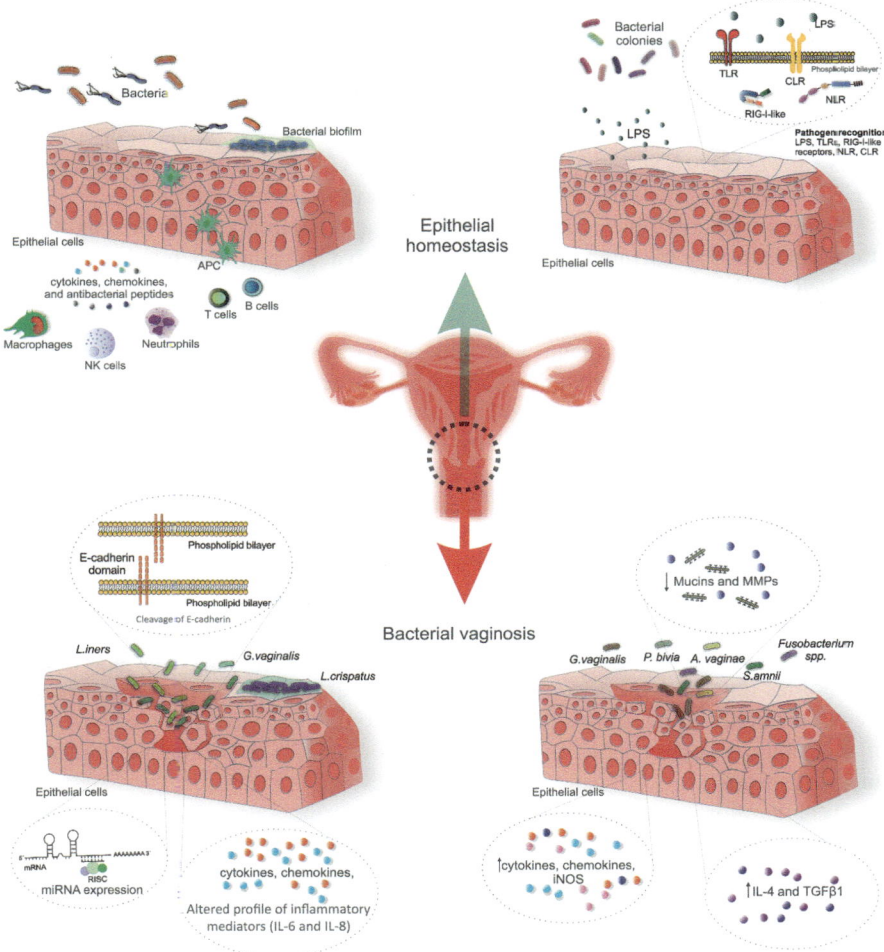

Fig. 3.2 The changes in composition of cervical epithelium in epithelial homeostasis and bacterial vaginosis (BV); Bacterial colonies influence the epithelial cells with their metabolites and secreted factors, as LPS. Epithelial cells recognize pathogen by interaction with bacterial LPS through TLRs, NLRs, CLRs and RIG-1-like receptors,. **Epithelial homeostasis** is specific with presence of Lactobacillus spp. communities producing biosurfactants in a biofilm that protects the epithelium against adhesion of pathogenic bacteria. The innate immune cells (macrophages, NK cells and neutrophils) as a part of local epithelial immune response produce cytokines, chemokines, and antimicrobial peptides that trigger defensive inflammatory responses. Adaptive immunity with activated T and B cells contributes to the immune response through the presentation of bacterial antigens by APCs. **Bacterial vaginosis (BV)** is known with strictly anaerobic non-lactobacilli species, as L. iners, and higher pH. *L. iners* and *G. vaginalis* elevate the permeability of the epithelial barrier by altered inflammatory mediator profile (IL-6 and IL-8), abnormal miRNA expression profile, and disruption of adherent junctions through cleavage of E-cadherin. A protective effect of *L. crispatus* produce phenyllactic acid with protective effect on the epithelial barrier is distributed. BV associated microflora in CST-IV group is represented by *G. vaginalis, P. bivia, A. vaginae,* and *S. amnii* species and produce pro-inflammatory cytokines and chemokines, iNOS, and oxidative stress;

cells and epithelial cells influenced by chronic inflammation [148]. In gastric cancer, there is direct proof that infection with *Helicobacter pylori* and chronic inflammation cause *CDH1* promoter methylation and differnet methylation in other CpG islands of genes encoding components of inflammation (*TFF-2* and *COX-2*), DNA repair (*hMHL-1* and *MGMT*), cell cycle regulation (*CDKN2A*), transcription factors (*RUNX3, FOXD3, USF1* and *USF2, GATA-4* and *GATA-5*), intracellular junctions (*VEZT, Cx32*, and *Cx43*), and other tumour suppressor genes [149].

Metabolic changes during bacterial infection and inflammation induce oxidative stress followed by elevated production of reactive oxygen and nitrogen species (ROS/NOS) [150–152]. Oxidative stress can be thought to be a mediator of epigenetic changes, especially in the state of DNA methylation. Formation of 8-OHdG, a typical DNA oxidative damage with mutagenic effect, can also adversely affect nearby cytosine DNA methylation [153, 154]. In addition, ROS are also able to activate DNA demethylation by oxidation of 5-mC to 5-hmC [155]. Furthermore, DNA methylation can be catalysed by oxidative stress through upregulation of DNMT expression by the formation of superoxide anions or by the formation of a new complex containing DNMT [156, 157]. The exact role of oxidative stress as a mediator in the interaction between the epigenome of the cervical epithelium, microbiome and inflammation remains to be elucidated.

Previous studies brought strong evidence that the microbiome plays a pivotal role in estimating cervicovaginal homeostasis and inflammatory response, but the exact mechanisms have not been fully described. The antitumor effect of microbial metabolites together with the activation of the local immune system may serve in preventive medicine to regress precancerous lesions. The intervention of new types of microbiome supports the expansion of favourable over pathogenic genera of microorganisms.

The microflora, whether intestinal or vaginal, forms a complex ecosystem. An imbalance develops dysbiosis, in which various exogenous and endogenous factors are involved. In modern precision medicine, the task will be to reconstruct and shape such a microflora that will restore a state of healthy balance and reverse the process of disease outbreak. However, this is an extremely challenging task because the microbiome is affected by many factors. Currently, precision medicine is practiced in various ways including the diet and uptake of prebiotics, probiotics and synbiotics or phage therapy and antibiotics as well as microbiota transplantation [158].

3.8 Molecular Pathways Contributing to Inflammatory Response in the Epithelium of the Cervix

Currently, many studies identify a link between the cervical microbiome, inflammation processes, and pathological conditions such as HPV infection and carcinogenesis. Microbial imbalance leads to increased secretion of cytokines and chemokines related to inflammation that subsequently accelerate immune dysregulation in female reproductive organs, promoting suitable conditions for cancer development

[159]. Inflammation processes are associated with tissue remodeling and functional changes in different cells (i. e., epithelial, vascular, and immune) influenced by alterations in various molecular pathways involving specific signaling molecules such as growth factors, cytokines, or chemokines. The mechanism of immune evasion is critical for persistent infection of HPV, which is necessary for the development of cervical cancer [160]. Therefore, chronic inflammation triggers cellular events connected to the development of a tumor associated environment promoting initiation, promotion, and progression of cervical cancer. Activated immune cells produce pro-inflammatory signaling molecules connecting inflammation, infection, immunity, and cancer [161]. Recent evidence revealed an increased level of circulating markers associated with inflammation, such as TNFα, IFNβ, or interleukins (IL-6, IL-1β) observed in HPV+ women with cervical cancer [162]. TNF regulates cellular mechanisms including apoptosis, inflammation, proliferation and differentiation [163]. Based on current data, TNF-α is defined as a proinflammatory cytokine characterized by a broad spectrum of functions such as the first line of immune defense against pathogens, including HPV infection. On the other hand, a systemic elevated level of TNF-α can influence steps leading to the initiation or promotion of cancer [164]. Another signaling molecule, proinflammatory IL-17 produced by different cells occurring in the tumor microenvironment, activates various molecular pathways such as JAK2/STAT3, PI3K/Akt, and NF-κB to promote the tumorigenesis of cervical cancer [165].

In addition, the transcription factor the nuclear factor kappa B (NF-kB) is involved in regulating various cellular processes such as inflammatory and immune responses, viral replication, and cancer development [159]. HPV infection downregulates NF-kB to allow its replication activated by the immune system resulting in HPV persistence. On the other hand, NF-kB activity is restored during the progression to high-grade intraepithelial neoplasia and cervical cancer. Furthermore, an elevated level of NF-kB signaling is connected to mutations in upstream signaling molecules [166]. Principally, NF-kB play a central role connecting cancer and inflammation. NF-kB activation is initiated by underlying inflammation or due to the inflammatory microenvironment associated with cancer development. Moreover, NF-kB induce expression of proinflammatory as well as tumor promoting signaling molecules including IL-6 or TNF-α, and survival genes, such as Bcl-XL [167]. Additionally, the induction of IL-6 by E6 protein via NF-κB, IL-6 stimulate activation of STAT3 that induce IL-6, IL-10 and TGF-β and support the aggregation of regulatory T cells and induce maintaining immunosuppressive state of tumor microenvironment [159].

Among others, Wnt/β-catenin signaling regulates inflammatory processes of the cell at least partly via suppressing or stimulating NF-κB pathway, while NF-κB pathway also modulates Wnt/β-catenin in two manners; positively or negatively [168]. In fact, aberrantly activated Wnt/β-catenin signaling components are connected with various tumor types including cervical cancer [169]. Moreover, recent evidence reveals an important role of Notch signaling in the modulation of inflammation. In addition, it has been observed that Notch is associated with NF-κB pathway. Therefore, Notch NF-κB pathway cross-link demonstrates the ability to modulate inflammatory cytokines, including TNF -α and IL-1β and TLR agonists [170].

Also, TLR play a crucial role by recognizing pathogens (endogenous and exogenous), thus mobilizing immune responses. For example, TLR-4 regulate immune-inflammatory responses in reaction to the stimulation by exogenous pathogens (bacterial LPS). Stimulation of MAPKs and NF-κB signal pathways and upregulation of genes associated with proinflammatory cytokines are induced by TLR4 activation. Indeed, TLR-4 play a crucial role in cancer development. TLR-4 promotes an immunosuppressive state of the tumor microenvironment and facilitates HPV+ cervical cancer growth [171]. Increase in TLR4 expression was observed in cervical cancer cells [172].

It is well known that oxidative stress, as a consequence of the disbalance of free radicals and antioxidants, contributes to cell and tissue damage. Numerous clinical and experimental data document crosslink between uncontrolled oxidative stress that can lead to chronic inflammation with the potential for cancer development [173]. Oxidative stress through increased production of ROS accelerates oxidative damage of DNA, leading to inflammation, mutation of DNA and genomic instability, which promote carcinogenic transformation [174]. An increase levels in oxidative stress biomarkers (i.e., 8-oxo-7,8-dihydro-2'-deoxyguanosine (8-oxodG) and 8-nitroguanine) were identified in cancer-associated inflammation. Indeed, these biomarkers were found in the tissues of cancer and precancerous lesions as o consequence of the infection, such as a human papillomavirus-related cervical cancer [175]. Moreover, ROS affects DNA integrity and influences the viral life cycle via anti-apoptotic and pro-survival effects. Furthermore, uncontrolled oxidative stress increase E6 and E7 gene expression resulting in oncogene deregulation and expression of miRNAs [176].

Cyclooxygenase-2 (COX-2), the enzyme contributing to the synthesis of prostanoids (prostaglandins and thromboxanes), is also an essential mediator of inflammation and carcinogenesis. COX-2 overexpression was documented in various epithelial cancer cells, including cervical cancer. Notably, Cox-2 is involved in various deregulated inflammatory processes that can accelerate cancer development, mainly in a milieu of viral inflammation typical for cervical carcinogenesis [177]. In addition, it has been suggested that HPV E5, E6, and E7 could increase COX-2 and PG E2 expressions leading to stimulation of COX-PG pathway. The activated COX-PG cascade is considered as one of the major causes of inflammation induced by HPV [178]. Last but not least, a tumor microenvironment can be thought of as a complex ecosystem, while inflammatory cells represent important components participating in carcinogenesis. Tumor-associated macrophages (TAMs) orchestrate cancer-associated inflammation by releasing different mediators such as proangiogenic VEGF or proinflammatory PDGF. Cyclin-dependent kinases (CDKs) are important regulators that coordinate cell cycle progression; for example, CDK12 is significantly upregulated in cervical cancer, while its knockdown accelerates macrophages infiltration and modulates the immune microenvironment in cervical cancer cells [179].

Above discussed associations between cervical microbiome, inflammation, and cancer can be regulated by epigenetic mechanisms. Thus, a brief overview of study results to exemplify the possible regulation of selected pathways that can be also associated with inflammation and cervical cancer by epigenetic mechanisms are provided in Table 3.2. In addition, above-mentioned signaling pathways and associated epigenetic changes related cervicovaginal imbalance and inflammation are shown in Fig. 3.3.

Table 3.2 Study results to exemplify the role of epigenetic regulation in cervical microbiome, inflammation, and tumorigenesis

Targeted pathway	Study design (e.g. cell line)	Effect	References
miRNA			
TNF-α	HeLa cells	miRNA-21 regulates apoptosis and proliferation of cervical cancer cells (via TNF-α)	[180]
TNF-α, NF-κB	C33A and HeLa cells	NF-κB activation by TNF-α leads to reduced miR-130a expression (miR-130a downregulates TNF-α)	[181]
NF-κB	C33A and HeLa cells and cervical cancer tissue	Involvement of miR-429 NF-κB pathway regulation (via IKKβ targeting); functions as a tumor suppressor	[182]
Notch	HeLa, SiHa, C33A, Caski, ME-180, and SW756 cells	MiRNA-873-5p supports cervical cancer progression (ZEB1 regulation via notch signaling)	[183]
Wnt / β-catenin	HeLa cells	miR-142-5p supports cervical cancer progression (LMX1A targeting via Wnt/β--catenin pathway)	[184]
COX-2	HeLa cells	miRNA-101 suppresses proliferation and invasion and promotes apoptosis (via COX-2 regulation)	[185]
TNF	Cervical cancer tissue specimens and cell lines	miR-484/MMP14 could inhibit proliferation and invasion of cervical cancer cells (via inhibition of TNF signaling); DNMT1-mediated silencing of miR-484	[163]
Oxidative stress pathway	Cervical cancer cells	E6 disrupts the expression of miR-23b, miR-218, and miR-34a; the increased expression of miR-15a/miR-16-1 induces the inhibition of cell proliferation, survival, and invasion	[176]
Histone modifications			
Wnt / β-catenin	HeLa, SiHa, C33A, MS751, and End1/E6E7 cells	HDAC10 exerts antitumor activity in cervical cancer (modulated via microRNA-223/TXNIP/Wnt/β-catenin pathway)	[186]
NF-κB	In silico prediction and *in vitro* assays	The interaction of HDAC1 with p50 NF-κB subunit (via nuclear localization sequence) leading to limited expression of inflammatory genes	[187]
DNA methylation			
Notch	HeLa, SiHa, C33A, CASKI, C4-1, and Ect1/E6E7	*MSX1* gene function as tumor supressor in cervical cancer (via notch signaling); *MSX1* promoter methylation demonstrated in 42% of primary cervical tumor tissues	[188]
Notch	Cervical cancer patients (n = 110)	Alterations in promoter methylation (Notch1 and Notch3 receptor genes) detected in cervical cancer	[189]

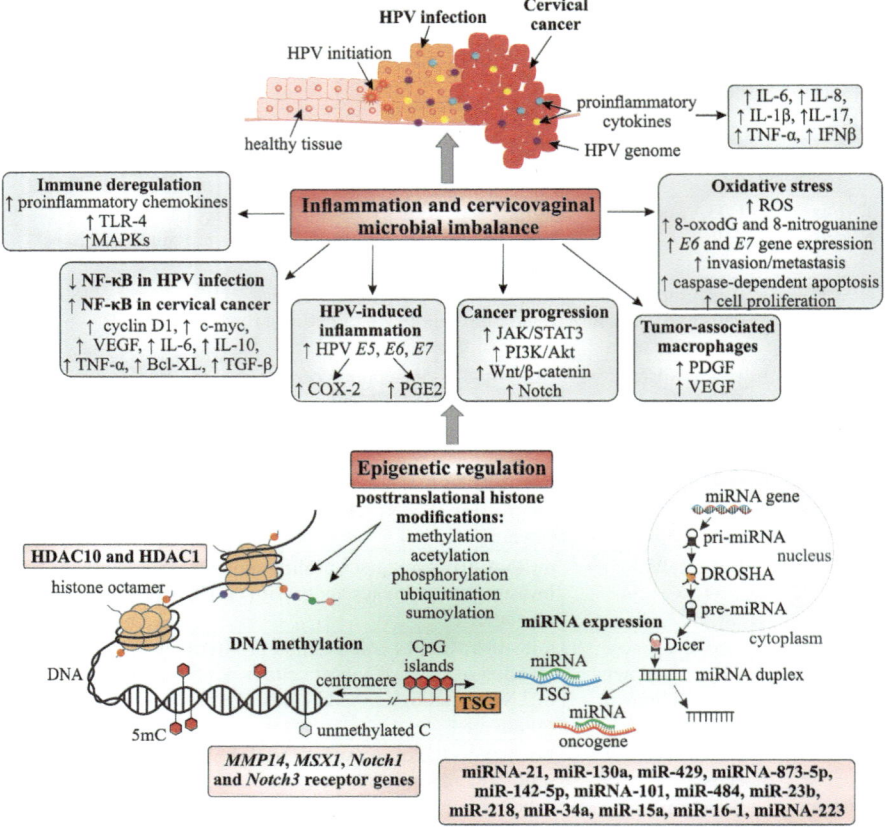

Fig. 3.3 Signaling pathways and associated epigenetic changes related to cervicovaginal imbalance and inflammation. **Abbreviations:** *5mC* 5-methylcytosine, *8-oxodG* 8-hydroxy-2′ -deoxy-guanosine, *Akt* protein kinase B, *COX-2* cyclooxygenase-2, *HPV* human papillomavirus, *IFNβ* interferon β, *IL* interleukin, *JAK* the Janus kinase, *MAPKs* mitogen-activated protein kinases, *miRNA* microRNA, *MMP14* matrix metallopeptidase 14, *MSX1* Msh homeobox 1, *PDGF* platelet derived growth factor, *PGE2* prostaglandin E2, *PI3K* phosphoinositide 3-kinase, *ROS* reactive oxygen species, *STAT3* signal transducer and activator of transcription 3, *TFG-β* transforming growth factor-β, *TLR-4* Toll-like receptor 4, *TNF-α* tumor necrosis factor alpha, *TSG* tumor suppressor gene, *VEGF* vascular endothelial growth factor

3.9 Personalized Diagnostics and Treatment of HPV-Associated Cervical Disease Based on Vaginal Microbiome Composition

3.9.1 State-of-the-Art Diagnostics of VM Composition

The vaginal microbiome is constantly evolving and is heavily impacted by age-ing, hormones, lifestyle, diet, alcohol intake, smoking, the menstrual cycle, sex-ual activity, and contraceptive usage. These parameters should be considered when evaluating the association between vaginal dysbiosis and HPV infec-tion [190].

We distinguish between culture-dependent and culture-independent approaches (metagenomics) for analyzing microbial composition. The precision of sequence-based microbial profiling depends on the characteristics of the sequencing data. The sequencing errors lead to incorrect taxonomic classification (limited number of dif-ferentiated microbiome genera; mostly on family or genus level only) and false estimates of microbial diversity [190, 191]. The correction of sequencing errors is a unique challenge in 16S rRNA gene sequencing output. Pyrosequencing, capillary Sanger sequencing, and Mi Seq or Hi Seq (Illumina) are examples of current sequencing technologies. Each of them has a distinct read length, accuracy, speed, and cost. Bio (Pacific Biosciences), Ion Torrent (Thermo Fisher Scientific), and MinION (Oxford Nanopore Technologies) platforms are examples of new sophisti-cated sequencing systems [190, 192].

The evolution of bioinformatic tools based on AI-models (machine learning and deep learning approaches) show a promising potential in disease prediction bringing considerable clinical benefits [190, 193]. Moreover, Andralojc et al. recently introduced a unique circular probe-based RNA sequencing (CiRNAseq) technology able to profile and quantify the cervicovaginal microbiome with improved sequencing sensitivity and specificity. CiRNAseq is a highly promising tool for studying the interplay between the cervico-vaginal microbiome and hrHPV in cervical carcinogenesis prompting the discovery of new prognostic and predictive biomarkers. CiRNAseq is also able to detect viral RNAs and host gene expression products, which may allow investigations of host-microbiome interac-tions in a single test [83].

3.9.2 Individualized Therapy

The complete lack of effective medical treatments for HPV infection currently con-stitutes a serious challenge. Conization therapy is still the gold standard in situations with histologically proven high-grade cervical lesions with potential long-term complications [80]. Immunohistochemical markers such as p16 and Ki-67 improve the sensitivity of cytological or histological analyses. Nonetheless, no diagnostic marker-based therapies are included in international guidelines. As a result, research priorities should shift away from the ocean of molecular markers with limited therapeutic utility toward the potential interactions inside and between human cells and their bacterial symbionts.

Bubnov et al. conducted a thorough assessment of probiotic-based regimens in the treatment of immunological and atopic disorders, metabolic and inflammatory illnesses, and malignancies [194]. There is strong evidence that the gut microbiome is connected to proper brain, heart, skin, pulmonary, and urogenital system functioning, which is leading to innovative ways to preserving health, disease prevention, and treatment of numerous chronic illnesses. As a result, the use of probiotics and prebiotics has the potential to have a huge impact on patient treatment [195].

3.9.3 *Probiotics*

Bacterial therapeutics, probiotics, prebiotics, antibiotics, and microbiota transplantation are all examples of ways to modify microbiome (sometimes known as 'bugs as drugs') [62]. World Health Organization and the United Nations' Food and Agriculture Organization define probiotics as "live microorganisms, which when administered in adequate amounts confer a health benefit on the host" [196]. Probiotics, such as *Bifidobacteria* species, *Lactobacilli*, and *Streptococci*, that promote health, immune system, inflammatory state, favourable epigenetic pattern and micronutrient absorption, provide a potential therapy method for HPV-induced cervical pre-cancerous conditions.

Lactobacillus is a genus with over 170 species that cannot be distinguished phenotypically using standard methods [197]. We can now identify microorganisms and their products using culture-independent molecular detection methods thanks to recent technical advancements. *Lactobacillus* spp. supernatants contain antimicrobials (mevalonolactone, benzoic acid, and 5-methyl-hydantoin), surfactants (distearin, dipalmitin, and 1,5-monolinolein), anesthetics (barbituric acid derivatives), and autoinducer type 2 precursors, lactic acid, phenolics, hydrogen peroxide, sodium, calcium, potassium, magnesium, and DNAases [198]. This rich antimicrobial composition emphasizes the influence of *Lactobacilli* on vaginal health.

Prebiotics and probiotics that promote a lactobacillus-dominant environment in the VM represent possible low-cost therapy options with little negative effects [199]. Furthermore, probiotics are harmless and have been shown to have antiviral action [200]. Probiotics such as *Lactobacillus paracasei* and *Lactobacillus rhamnosus* acidify the vaginal milieu, inhibit bacterial adhesion, and work in tandem with the host immune system [201]. This straightforward treatment, using oral or vaginal regimens, might enhance HPV clearance or possibly reverse carcinogenesis, reducing eventual morbidity [199, 202]. HPV clearance was significantly higher in probiotic-using patients compared to controls (25% vs. 7.7%) [203]. The same results were observed in individuals with cytological abnormalities, with a twofold increase in clearance and normalization [204].

Orally administered lactobacilli can enter the vagina, repair its microbiota in the presence of microbial imbalance, and eliminate or minimize the prevalence of urogenital infections [205, 206]. Probiotics that have been ingested orally climb to the vaginal tract after being expelled from the rectum. Vaginal administration, on the other hand, provides for the direct replacement of probiotics for damaged vaginal flora [207]. The use of probiotics including *L. casei* on a daily basis increased HPV clearance in LSIL lesions [203]. Bifidobacteria may improve cancer immunity and immunotherapy effectiveness [208]. Clarification of the precise relationships between certain bacterial kinds, the immune system, and HPV will aid in the creation of highly effective probiotics.

Recent research based on new culturomics technologies has shown a plethora of possible next-generation probiotics with enhanced ameliorative benefits. *Prevotella copri, Christensenella minuta, Parabacteroides goldsteinii, Akkermansia muciriphila, Bacteroides thetaiotaomicron*, and *Bacteroides fragilis* are only a few examples [209, 210]. Innovative ways for monitoring and modifying the vaginal microbiota can result in the creation of novel biomarkers and cutting-edge customized therapy solutions [201].

Metabolites of probiotics may affect the epigenetics of the host organism. The authors Morovic and Budinoff (2021) propose a new branch of epigenetics called 'para-epigenetics', which means the influence of the organism on the epigenetics of various organisms. Understanding the mode of action of probiotics is essential in an epigenetic and para-epigenetic context and is possible through meta-epigenomic studies using new methods and sound science. The results should be verified by appropriate technologies, such as CRISPR-Cas targeted demethylation, base editing and nuclease-deactivated cas9. Applying and validating of these strategies to probiotics will reduce the cost of new therapies and therapeutic products to improve human health [211].

3.9.4 Prebiotics

There is also considerable interest in using prebiotics to influence other microbiomes inside the host, such as the female urogenital tract, where they may operate as bacterial growth inhibitors and virus attenuators. We now have a limited selection of prebiotics, which includes galactans and fructans. New compounds such as carbohydrate-based substances generated from plants, yeast-based substances, and numerous non-carbohydrate substances such as polyphenolics, fatty acids, and other micronutrients are anticipated to be provided [212, 213]. Furthermore, synbiotics, which combine the effects of fermentable chemicals with living microorganisms, may be the future method of customized therapy [212].

Prebiotics also substantialy decrease HPV positivity and the incidence of low-grade cervical lesions. Selected prebiotics improve the ectopic structure of the cervical mucosa and promote maturation of the metaplastic epithelium, resulting in negative colposcopic findings. Furthermore, prebiotics generate a mucoadhesive coating on the cervical surface that protects it from pathogenic microbial agents [55].

3.9.5   Penile Microbiota

Surprisingly, research into the penile microbiota offers promising implications. Given the sexually transmitted nature of the pathogenic agents in question, it is wise to consider the male microbiome. *Prevotella*, a penile anaerobic bacterium, may increase the risk of genital diseases such as bacterial vaginosis. Penile *Corynebacteria* and *Staphylococcus*, on the other hand, are linked to a healthy cervicovaginal microbiota [214]. In conclusion, the penis can serve as a reservoir for BV-associated bacteria, and numerous interactions occur in both the male and female genital biomes [215, 216]. Based on the arguments presented above, topical microbicides may be effective in combating HPV infection in men [214].

3.10   Signature of Immune System and the Composition of Cervicovaginal Microbiome in Preventive, Predictive, and Personalized Medicine: Association with Cervical Cancer and HPV Clearance

Due to the association between CC and persistent HPV, effective cancer prevention requires the precise understanding of the mechanisms resulting in HPV clearance for cancer prevention. The human microbiome is of a great importance health and proper immune functions [200, 217]. Therefore, the significance of microbiome in the concept of 3P medicine should be evaluated in terms of evidence-based knowledge [194]. *G. vaginallis* contribute in immune responses, probably via a shift from anti-microbial to anti-viral responding. In addition, these bacteria are associated with a certain level of inflammatory surveillance in the maintenance of HPV-negative state. Therefore, probiotics or pro-inflammatory agents could be beneficial in the treatment of persistent HPV [85]. Furthermore, the analysis of cervical secretions revealed cervicovaginal microbiota dysbiosis to be closely related to persistent HPV infection; indeed, identified biomarkers of dysbiosis (supported by blood markers) could provide new concepts for CC prevention [218–220]. A detailed overview of liquid biopsy biomarkers potentially utilizable in cervical pathogenesis associated with HPV infection is shown in Table 3.3.

Indeed, HPV could use several mechanisms to evade immune responses to ensure the progression from infection to chronic dysplasia and eventually to cancer [223]. Invariant natural killer T (iNKT) cells are present in cervical tissue during the progression from HPV infection to CIN; therefore, this observation suggest the important role of iNKT cells in immunosuppression, while the hindering of their

Table 3.3 Signatures of the immune system and the cervicovaginal microbiota used in cervical carcinogenesis and HPV infection studies

Aim of study	Study design	Results	References
Understanding the connection between host, virus, and vaginal microbiota by determining the association between locally secreted cancer biomarkers and characteristics of the cervicovaginal microenvironment	Evaluation of cervicovaginal lavage in Ctrl HPV+ (n = 11), Ctrl HPV− (n = 18), LSIL (n = 12), ICC (n = 10) and HSIL (n = 27)	Increased cancer biomarkers in ICC compared to ctrl HPV−: Apoptosis-related proteins (sFas, sFasL, TRAIL), growth and angiogenic factors (HGF, SCF, VEGF), proinflammatory cytokines (TNF-α), multi-functional proteins (OPN, CYFRA 21-1, AFP) and hormones (leptin, prolactin)	[221]
		biomarkers for differentiation of ICC and healthy ctrl HPV−: CYFRA 21-1, TNF-α, prolactin, MIF, and SCF (measured in the CVL)	
		The expression of cancer biomarkers associated with VM composition and genital inflammation:	
		Significant positive correlations with inflammatory scores in 19 out of the 23 cancer biomarkers; negative correlations with *Lactobacillus* abundance in 9 cancer biomarkers: TNFα, MIF, TRAIL, sFasL, prolactin, OPN, FGF2, and SCF; positive correlations with *Lactobacillus* abundance and negative correlation with vaginal pH in vaginal pH. HE4 (an antimicrobial peptide)	
Identification of the relationship between vaginal pH, HPV, genital immune mediators, vaginal microbiota, and the severity of cervical neoplasms	Evaluation of cervicovaginal lavage and vaginal swabs in Ctrl HPV+ (n = 31), Ctrl HPV− (n = 20), ICC (n = 10), LGD (n = 12), and HGD (n = 27)	Association of abnormal pH with cancer: elevated vaginal pH at different stages of cervical cancer	[222]
		Underrepresented *Lactobacillus* spp. in LGD, ICC, Ctrl HPV+, and HGD; enriched *Sneathia* spp.	
		In LGD and HGD: BV-associated with enriched *Parvimonas* and *Atopobium* Only in HGD: BV-associated with enriched *Prevotella, Gardnerella, Shuttleworthia, and Megasphaera*	
		In patients of Hispanic ethnicity or abnormal pH: underrepresented *Lactobacillus* spp.; enriched *Atopobium* and *Sneathia*	
		In invasive carcinoma only: Increased genital inflammatory scores and genital immune mediators	
		Positive correlation of TLR2 and negative correlations of LAG-3 and PD-L1 with health-associated *Lactobacillus* dominance	
		Positive correlation of CD28, CD40, and TLR2 with genital inflammation	

(continued)

Table 3.3 (continued)

Aim of study	Study design	Results	References
Characterization of the relations between the inflammatory milieu and microbiome during HPV-16 pre-acquisition, persistence, and clearance	Evaluation of cervical wash repository samples in patients (women) with acquired HPV-16, persistence at least 8 months and cleared (n = 14) and WNHPV (n = 8)	In HPV-16 patients: increased abundance of *L. Iners* only before infection; increased *G. vaginalis* instantly before clearance, significantly increased after clearance, and returned to baseline (months later)	[85]
		In WNHPV patients: increased *L. iners* when compared with samples from the immediate post-clearance samples of HPV-infected patients.	
		Decreased *G. vaginalis* when compared with immediate post-clearance samples but similar to pre- and post-infection levels	
		Increased cytokine levels of MIP-1α, IL-4, -5, -10, -12, and 13, TNF-α, IFNγ, and IFN-α2 at the immediate post-clearance samples when compared to the pre-acquisition and second post-clearance samples	
		Correlation between increased *G. vaginalis* and increased cytokines (IFN-γ, TNF-α, IL-4, IL-5, IL-10, IL-12, IL-13, MIP-1α) in the post-clearance visits	
		Final clearance was preceded by increased *G. vaginalis* and peaked at the time of the observed cytokine peak	
Association of HPV persistence and cervicovaginal microbiota dysbiosis	Evaluations of cervical secretions and blood in healthy women and persistent and transient HPV infected patients	Increased Actinobacteria, Proteobacteria, Fusobacteria, and Bacteroidetes in HPV-persistent infection	[218]
		Association of *L. iners* with transient HPV infection	
		Association of Prevotella, Anaerococcus, and Sphingomonas, with persistent HPV infection	
		Altered cervicovaginal microbiota dysbiosis and immune microenvironment through upregulated TNF-α and IL-6 in cervical secretions from persistent HPV infection when compared with healthy women or transient infections	
		In patients with persistent HPV infection: Significantly increased peripheral blood (myeloid-derived suppressor cells and regulatory T cells)	

Abbreviations: WNHPV women with no history of HPV, *VEGF*, vascular endothelial growth factor, *TRAIL* TNF-related apoptosis-inducing ligand, *TNF-α* tumor necrosis factor, *PD-L1* programmed cell death ligand 1, *TLR2* toll-like receptor 2, *TIM-3* T-cell immunoglobulin and mucin domain-containing 3, *sFasL* soluble Fas ligand, *SCF* stem cell factor, *OPN* osteopontin, *MIP-1α* macrophage inflammatory protein-1α, *MIF* macrophage migration inhibitory factor, *LSIL* low-grade intraepithelial lesions, *LGD* low-grade dysplasia, *IL* interleukin, *IFN-γ* interferon gamma, *ICC* invasive cervical carcinoma, *HSIL* high-grade intraepithelial lesions, *HGF* hepatocyte growth factor, *HGD* high-grade dysplasia, *CYFRA 21-1* cytokeratin fragment 21–1, *Ctrl HPV+* HPV-positive controls, *Ctrl HPV−* HPV-negative controls, *CD40* cluster of differentiation 40, *CD28* cluster of differentiation 28, *CD27* cluster of differentiation 27, *BV*, bacterial vaginosis, *AFP* α-fetoprotein

accumulation could result in the suppression of CIN development [224]. Furthermore, recent study by Gutiérrez-Hoya described that CC cells express markers that are associated with activation and inhibition of the immune responses, and also natural killer (NK) cells receptors. Therefore, the described observations suggest that these molecules could facilitate mimicry of the immune cells and evasion of immune responses [225]. Still, the processes enabling HPV to evade immune responses could promote the identification of new immunological targets applicable in HPV-related cancer management, including the use of cytokines to create tumor milieu that favours damaging of transformed cells, DC vaccines activating Th1 and cytotoxic T lymphocyte responses, and NK cells activation through allogenic or autologous transplants to induce tumor cell lysis [226].

3.11   Conclusions and Recommendations in the Framework of 3P Medicine

Virus infections are responsible for around 15% of cancer cases worldwide [227]. These oncogenic viruses exploit a number of carcinogenic strategies, including direct effects or immune responses, resulting in persistent inflammation. A common theory of virus–bacteria–host interaction involves altered host gene expression, which promotes cancer [228]. In contrast, a microbiome composed of different bacterial, viral, and fungal species can protect the host from viral infections. This defense system is strongly reliant on the microbiome's exact structure and its many interactions.

3.11.1   Predictive Diagnostics

A disturbed VM balance is one of the most important risk factors in HPV infection. It might be an appealing and useful collection of indicators for stratifying dysplastic cervical lesions and determining the development of LSIL and HSIL lesions to invasive disease. Immunoscores, for example, are novel diagnostic approaches that can help differentiate between non-advancing and progressing precursor lesions. Furthermore, a thorough examination of the local vaginal environment may reveal the malignant potential in the puzzling histological subgroup of cervical intraepithelial neoplasia grade 2 (CIN2), which has a high likelihood of regression. A similar method is required in the event of chronic LSIL lesions, where the guidelines are still uncertain.

The development of new markers is crucial for tracking and identifying microbial compositions linked with HPV infection and cervical dysplasia. Modern molecular methods for VM assessment may enable better and more personalized therapeutic techniques to reduce CC occurrence (in conjunction with HPV vaccination) [229]. As we have assessed, some epigenetic and inflammatory bio-markers can also reflect the presence of specific bacteria that can serve as predictive biomarkers in calculating the risk of developing cervical dysplasia in personalized medicine. Rapid bedside diagnostics (microchip arrays and metabolomic technolo-gies) for detecting high-risk patients have the potential to help in triage and patient selection for intense monitoring and therapy [80]. *Snaethia sanguinensis, Anaerococcus tetradius*, and *Peptostreptococcus anaerobius*, for example, have been proposed as vaginal biomarkers for CIN in Caucasian, Black, and Asian women by Mitra et al. [82]. Similar microbiological markers might be beneficial in the personalized surveillance and prognosis of women with cervical lesions [106].

In the future, metabolomic studies using nuclear magnetic resonance and mass spectroscopy are highly likely to increase our understanding of the link between vaginal microbiota composition and health and disease states. It would be ideal to have a better understanding of the effects of certain bacterial species on epithelial and immunological cell function, such as *L. iners, A. vaginae*, and *Sneathia* species. Indeed, if the function of a certain bacterial species in HPV infection is discovered, microchip array or metabolomic technologies can be utilized to identify persons who are at a higher risk of developing high-grade cervical lesions. Furthermore, innovative therapeutic strategies based on pre/probiotics will be possible to alter the makeup of the vaginal microbiota.

3.11.2   Targeted Prevention

The maintenance of the cervical epithelial barrier and a healthy vaginal microbiome has a substantial influence on women's health, including safe pregnancy and resis-tance to sexually transmitted diseases. In the treatment of HPV-related infections, prevention is equally critical. Several behavioral, environmental, and host factors impact vaginal microbiota and can be successfully and easily altered. Obesity, smoking, and a high-fat diet with a high glycemic load and nutritional density have all been linked to an increased risk of bacterial vaginosis [230]. Even hygiene prac-tices, such as regular douching and the use of vaginal lubricants, might have an impact on the local vaginal flora [231]. Concerns regarding the functional links between genital microbiota and the host, as well as the interactions between sperm and the VM, are raised by gender-specific health concepts [232]. Several investiga-tions [233, 234] discovered vaginal microbiome members on male penile skin, ure-thral, urine, and sperm specimens, demonstrating that sexual partners share and trade microorganisms inhabiting the urogenital tracts. This conclusion underscores the need of gender-neutral, universal HPV prevention initiatives that focus on modi-fiable risk factors.

3.11.3   Personalization of Medical Services

Targeting the research on the particular *lactobacillus* composition may provide insight into how to avoid persistent HPV infections and subsequent disease progression [235]. In the context of 3P medicine, the imprints of these various ecosystems, which are intricately tied to the immune system, may provide important insights for diagnosis and therapy. 3P medicine's multidisciplinary character is crucial for breaking down barriers between clinical and research areas [236].

The increasing amount of information connecting HPV-related gynecological cancers and dysbiosis leads to the microbiota as a possible target for cancer prevention and therapy. In the future, vaginal probiotics, prebiotics, novel antimicrobials, biofilm disruptors, and microbiome transplantation may be used alone or in combination to modulate the vaginal microbiome by restoring a healthy local microenvironment for cervical cancer prevention and/or the reduction of vaginal toxic effects associated with cancer therapies [62].

The cervicovaginal microbiota should be addressed in tailored treatment for each woman when examining the risk of acquiring cervical precancerous lesions and progression to cervical cancer. To help in lifestyle adjustment, information on their effects on genetic and epigenetic backgrounds should be included into specialized medical treatments. Biomarkers of cervical carcinogenesis derived from various biofluids (vaginal swabs, cervicovaginal lavage/secretions, or blood) provide important evidence for personalized, preventive, and predictive medicine, which is critical for improving medical care in the context of twenty-first-century medicine due to the need for an individualized approach and reduced economic burden of traditional health care. The inclusion of the VM's particular composition allows us to create a detailed ecosystem map for targeted prevention and the design of personalized therapeutic regimens. This may reduce medicine toxicity, undesirable side effects, and morbidity. [237]. Furthermore, high-throughput, low-cost multi-omics technologies play an essential therapeutic role in cancer treatment and research, notably in the discovery of biomarkers necessary for the effective implementation of 3P therapy [238, 239].

To recapitulate, the principles of early predictive diagnosis, targeted prevention, and personalized patient care, together with significant developments in biomedicine, give significant advantages over reactive medical therapies. The analysis of VM disbalance in HPV-induced CC based on the characterization of novel biomarkers indicates the unique approach in the work-frame of 3P medicine, and only further in-depth research might provide effective therapeutical options for CC patients.

This chapter includes concepts to the topic developed by the authors within previously performed studies [240, 241].

References

1. Golubnitschaja O, Costigliola V, EPMA (2012) General report & recommendations in predictive, preventive and personalised medicine 2012: white paper of the European Association for Predictive, preventive and personalised medicine. EPMA J 3:14
2. Golubnitschaja O, Yeghiazaryan K, Costigliola V, Trog D, Braun M, Debald M et al (2013) Risk assessment, disease prevention and personalised treatments in breast cancer: is clinically qualified integrative approach in the horizon? EPMA J 4:6
3. Arbyn M, Weiderpass E, Bruni L, de Sanjosé S, Saraiya M, Ferlay J et al (2020) Estimates of incidence and mortality of cervical cancer in 2018: a worldwide analysis. Lancet Glob Health 8:e191–e203
4. Stelzle D, Tanaka LF, Lee KK, Khalil AI, Baussano I, Shah ASV et al (2021) Estimates of the global burden of cervical cancer associated with HIV. Lancet Glob Health 9:e161–e169
5. Cohen PA, Jhingran A, Oaknin A, Denny L (2019) Cervical cancer. Lancet 393:169–182
6. Roman BR, Aragones A (2021) Epidemiology and incidence of HPV-related cancers of the head and neck. J Surg Oncol 124:920–922
7. zur Hausen H (2009) Papillomaviruses in the causation of human cancers – a brief historical account. Virology 384:260–265
8. Farmer E, Cheng MA, Hung C-F, Wu T-C (2021) Vaccination strategies for the control and treatment of HPV infection and HPV-associated cancer. In: Wu T-C, Chang M-H, Jeang K-T (eds) Viruses and human cancer: from basic science to clinical prevention [Internet]. Springer, Cham, pp 157–195. Cited 2 Feb 2022. Available from: https://doi.org/10.1007/978-3-030-57362-1_8
9. Fappani C, Bianchi S, Panatto D, Petrelli F, Colzani D, Scuri S et al (2021) HPV type-specific prevalence a decade after the implementation of the vaccination program: results from a pilot study. Vaccines (Basel) 9:336
10. Akwaowo. A 10-year retrospective study on a population-based cervical cancer screening programme in Nigeria [Internet]. Cited 2 Feb 2022. Available from: http://www.jmwan.org/article.asp?issn=1596-896X;year=2021;volume=6;issue=1;spage=21;epage=30;aulast=Akwaowo
11. Bouvard V, Baan R, Straif K, Grosse Y, Secretan B, Ghissassi FE et al (2009) A review of human carcinogens – part B: biological agents. Lancet Oncol 10:321–322
12. Egawa N, Doorbar J (2017) The low-risk papillomaviruses. Virus Res 231:119–127
13. Halec G, Alemany L, Lloveras B, Schmitt M, Alejo M, Bosch FX et al (2014) Pathogenic role of the eight probably/possibly carcinogenic HPV types 26, 53, 66, 67, 68, 70, 73 and 82 in cervical cancer. J Pathol 234:441–451
14. Bergqvist L, Kalliala I, Aro K, Auvinen E, Jakobsson M, Kiviharju M et al (2021) Distribution of HPV genotypes differs depending on behavioural factors among young women. Microorganisms 9:750
15. Bosch FX, de Sanjosé S (2003) Chapter 1: human papillomavirus and cervical cancer--burden and assessment of causality. J Natl Cancer Inst Monogr:3–13
16. Jaisamrarn U, Castellsagué X, Garland SM, Naud P, Palmroth J, Del Rosario-Raymundo MR et al (2013) Natural history of progression of HPV infection to cervical lesion or clearance: analysis of the control arm of the large, randomised PATRICIA study. PLoS One 8:e79260
17. Burchell AN, Winer RL, de Sanjosé S, Franco EL (2006) Chapter 6: epidemiology and transmission dynamics of genital HPV infection. Vaccine 24(Suppl 3):S3/52–S3/61
18. Demarco M, Lorey TS, Fetterman B, Cheung LC, Guido RS, Wentzensen N et al (2017) Risks of CIN 2+, CIN 3+, and cancer by cytology and human papillomavirus status: the foundation of risk-based cervical screening guidelines. J Low Genit Tract Dis 21:261–267
19. Demarco M, Egemen D, Raine-Bennett TR, Cheung LC, Befano B, Poitras NE et al (2020) A study of partial human papillomavirus genotyping in support of the 2019 ASCCP risk-based management consensus guidelines. J Low Genit Tract Dis 24:144–147

20. Sasieni P, Castanon A, Landy R, Kyrgiou M, Kitchener H, Quigley M et al (2016) Risk of preterm birth following surgical treatment for cervical disease: executive summary of a recent symposium. BJOG 123:1426–1429
21. Stelzle D, Tanaka LF, Lee KK, Ibrahim Khalil A, Baussano I, Shah ASV et al (2020) Estimates of the global burden of cervical cancer associated with HIV. Lancet Glob Health 9:e161–e169
22. Alena L, Marek S, Lenka K, Erik K, Peter K, Olga G (2021) Mitochondriopathies as a clue to systemic disorders: "Vicious circle" of mitochondrial injury, analytical tools and mitigating measures in context of predictive, preventive, and personalized (3P) medicine
23. Crigna AT, Samec M, Koklesova L, Liskova A, Giordano FA, Kubatka P et al (2020) Cell-free nucleic acid patterns in disease prediction and monitoring-hype or hope? EPMA J:1–25
24. Liskova A, Samec M, Koklesova L, Giordano FA, Kubatka P, Golubnitschaja O (2020) Liquid biopsy is instrumental for 3PM dimensional solutions in cancer management. J Clin Med 9:2749
25. Gerner C, Costigliola V, Golubnitschaja O (2020) Multiomic patterns in body fluids: techno-logical challenge with a great potential to implement the advances paradigm of 3P medicine. Mass Spectrom Rev 39(5–6):442–451
26. Zoodsma M, Sijmons RH, de Vries EG, van der Zee AG (2004) Familial cervical cancer: case reports, review and clinical implications. Hered Cancer Clin Pract 2:99
27. Oyervides-Muñoz MA, Pérez-Maya AA, Sánchez-Domínguez CN, Berlanga-Garza A, Antonio-Macedo M, Valdéz-Chapa LD et al (2020) Multiple HPV infections and viral load association in persistent cervical lesions in Mexican women. Viruses [Internet] 12. Cited 21 Feb 2021. Available from: https://www.ncbi.nlm.nih.gov/pmc/articles/PMC7232502/
28. Ciccarese G, Herzum A, Pastorino A, Dezzana M, Casazza S, Mavilia MG et al (2021) Prevalence of genital HPV infection in STI and healthy populations and risk factors for viral persistence. Eur J Clin Microbiol Infect Dis 40:885–888
29. Franceschi S, Herrero R, Clifford GM, Snijders PJF, Arslan A, Anh PTH et al (2006) Variations in the age-specific curves of human papillomavirus prevalence in women world-wide. Int J Cancer 119:2677–2684
30. Wendland EM, Villa LL, Unger ER, Domingues CM, Benzaken AS (2020) Prevalence of HPV infection among sexually active adolescents and young adults in Brazil: the POP-Brazil study. Sci Rep 10:4920
31. Río-Ospina LD, León SCS-D, Camargo M, Sánchez R, Mancilla CL, Patarroyo ME et al (2016) The prevalence of high-risk HPV types and factors determining infection in female Colombian adolescents. PLoS One 11:e0166502
32. Giroux V, Rustgi AK (2017) Metaplasia: tissue injury adaptation and a precursor to the dys-plasia–cancer sequence. Nat Rev Cancer 17:594–604
33. Bedin R, Gasparin VA, Pitilin ÉDB (2017) Fatores associados às alterações cérvico-uterinas de mulheres atendidas em um município polo do oeste catarinense Factors associated to uterine-cervix changes in women assisted in a pole town in western Santa Catarina. R Pesq Cuid Fundam Online 9:167
34. Kang L-N, Castle PE, Zhao F-H, Jeronimo J, Chen F, Bansil P et al (2014) A prospective study of age trends of high-risk human papillomavirus infection in rural China. BMC Infect Dis 14:96
35. Ruan Y, Poirier AE, Pader J, Asakawa K, Lu C, Memon S et al (2021) Estimating the future cancer management costs attributable to modifiable risk factors in Canada. Can J Public Health 112:1083–1092
36. Salwa M, Al-Munim TA (2021) HPV vaccination in Bangladesh: ethical views. In: Bauer AW, Hofheinz R-D, Utikal JS (eds) Ethical challenges in cancer diagnosis and ther-apy [Internet]. Springer, Cham, pp 31–37. Cited 2 Feb 2022. Available from: https://doi.org/10.1007/978-3-030-63749-1_3
37. Muñoz N, Franceschi S, Bosetti C, Moreno V, Herrero R, Smith JS et al (2002) Role of par-ity and human papillomavirus in cervical cancer: the IARC multicentric case-control study. Lancet 359:1093–1101

38. Moreno V, Bosch FX, Muñoz N, Meijer CJLM, Shah KV, Walboomers JMM et al (2002) Effect of oral contraceptives on risk of cervical cancer in women with human papillomavirus infection: the IARC multicentric case-control study. Lancet 359:1085–1092

39. Kitamura T, Suzuki M, Shigehara K, Fukuda K (2021) Prevalence and risk factors of human papillomavirus infection among Japanese female people: a nationwide epidemiological survey by self-sampling. Asian Pac J Cancer Prev 22:1843–1849

40. Orlando G, Tanzi A, Rizzardini G (2016) Modifiable and non-modifiable factors related to HPV infection and cervical abnormalities in women at high risk: a cross-sectional analysis from the VALHIDATE study 21. Ann Virol Res 2(2):1013

41. Roura E, Travier N, Waterboer T, de Sanjosé S, Bosch FX, Pawlita M et al (2016) The influence of hormonal factors on the risk of developing cervical cancer and pre-cancer: results from the EPIC cohort. PLoS One [Internet] 11. Cited 3 Mar 2021. Available from: https://www.ncbi.nlm.nih.gov/pmc/articles/PMC4726518/

42. Asthana S, Busa V, Labani S (2020) Oral contraceptives use and risk of cervical cancer – a systematic review & meta-analysis. Eur J Obstet Gynecol Reprod Biol 247:163–175

43. Mignot S, Ringa V, Vigoureux S, Zins M, Panjo H, Saulnier P-J et al (2019) Pap tests for cervical cancer screening test and contraception: analysis of data from the CONSTANCES cohort study. BMC Cancer 19:317

44. Torres-Poveda K, Ruiz-Fraga I, Madrid-Marina V, Chavez M, Richardson V (2019) High risk HPV infection prevalence and associated cofactors: a population-based study in female ISSSTE beneficiaries attending the HPV screening and early detection of cervical cancer program. BMC Cancer 19:1205

45. Fonseca-Moutinho JA (2011) Smoking and cervical cancer. ISRN Obstet Gynecol [Internet] 2011. Cited 3 Mar 2021. Available from: https://www.ncbi.nlm.nih.gov/pmc/articles/PMC3140050/

46. Oh HY, Kim MK, Seo S, Lee DO, Chung YK, Lim MC et al (2015) Alcohol consumption and persistent infection of high-risk human papillomavirus. Epidemiol Infect 143:1442–1450

47. Minkoff H, Zhong Y, Strickler HD, Watts DH, Palefsky JM, Levine AM et al (2008) The relationship between cocaine use and human papillomavirus infections in HIV-seropositive and HIV-seronegative women. Infect Dis Obstet Gynecol 2008:587082

48. Shah SC, Kayamba V, Peek RM, Heimburger D (2019) Cancer control in low- and middle-income countries: is it time to consider screening? J Glob Oncol 5:1–8

49. Liu Z, Ren Z, Zhang J, Chuang C-C, Kandaswamy E, Zhou T et al (2018) Role of ROS and nutritional antioxidants in human diseases. Front Physiol [Internet] 9. Cited 3 Mar 2021. Available from: https://www.ncbi.nlm.nih.gov/pmc/articles/PMC5966868/

50. Koshiyama M (2019) The effects of the dietary and nutrient intake on gynecologic cancers. Healthcare (Basel) [Internet] 7. Cited 3 Mar 2021. Available from: https://www.ncbi.nlm.nih.gov/pmc/articles/PMC6787610/

51. Koulova A, Tsui J, Irwin K, Van Damme P, Biellik R, Aguado MT (2008) Country recommendations on the inclusion of HPV vaccines in national immunization programmes among high-income countries, June 2006-January 2008. Vaccine 26:6529–6541

52. Ramakrishnappa DS, Siddharthan S, Naing NN, Priyadarshini K, Neeraja T, Ramesh K et al (2021) Acceptingfactors of HPV vaccination. Eur J Mol Clin Med 8:1297–1315

53. Escriva-Boulley G, Mandrik O, Préau M, Herrero R, Villain P (2021) Cognitions and behaviours of general practitioners in France regarding HPV vaccination: a theory-based systematic review. Prev Med 143:106323

54. Castanheira CP, Sallas ML, Nunes RAL, Lorenzi NPC, Termini L (2021) Microbiome and cervical cancer. Pathobiology 88(2):187–197

55. Lavitola G, Della Corte L, De Rosa N, Nappi C, Bifulco G (2020) Effects on vaginal microbiota restoration and cervical epithelialization in positive HPV patients undergoing vaginal treatment with carboxy-methyl-beta-glucan. Biomed Res Int 2020:1–8

56. Godha K, Tucker KM, Biehl C, Archer DF, Mirkin S (2018) Human vaginal pH and microbiota: an update. Gynecol Endocrinol 34:451–455

57. Goncharenko V, Bubnov R, Polivka J, Zubor P, Biringer K, Bielik T et al (2019) Vaginal dryness: individualised patient profiles, risks and mitigating measures. EPMA J 10:73–79
58. Muhleisen AL, Herbst-Kralovetz MM (2016) Menopause and the vaginal microbiome. Maturitas 91:42–50
59. Romero R, Hassan SS, Gajer P, Tarca AL, Fadrosh DW, Nikita L et al (2014) The composition and stability of the vaginal microbiota of normal pregnant women is different from that of non-pregnant women. Microbiome 2:4
60. van de Wijgert JHHM, Borgdorff H, Verhelst R, Crucitti T, Francis S, Verstraelen H et al (2014) The vaginal microbiota: what have we learned after a decade of molecular characterization? Fredricks DN, editor. PLoS One 9:e105998
61. Gajer P, Brotman RM, Bai G, Sakamoto J, Schutte UME, Zhong X et al (2012) Temporal dynamics of the human vaginal microbiota. Sci Transl Med 4:132ra52
62. Laniewski P, Ilhan ZE, Herbst-Kralovetz MM (2020) The microbiome and gynaecological cancer development, prevention and therapy. Nat Rev Urol 17:232–250
63. Flores R, Shi J, Fuhrman B, Xu X, Veenstra TD, Gail MH et al (2012) Fecal microbial determinants of fecal and systemic estrogens and estrogen metabolites: a cross-sectional study. J Transl Med 10:253
64. Plottel CS, Blaser MJ (2011) Microbiome and malignancy. Cell Host Microbe 10:324–335
65. Lev-Sagie A, Goldman-Wohl D, Cohen Y, Dori-Bachash M, Leshem A, Mor U et al (2019) Vaginal microbiome transplantation in women with intractable bacterial vaginosis. Nat Med 25:1500–1504
66. Ravel J, Gajer P, Abdo Z, Schneider GM, Koenig SSK, McCulle SL et al (2011) Vaginal microbiome of reproductive-age women. Proc Natl Acad Sci 108:4680–4687
67. Boris S, Suárez JE, Vázquez F, Barbés C (1998) Adherence of human vaginal lactobacilli to vaginal epithelial cells and interaction with uropathogens. Infect Immun 66:1985–1989
68. Medina-Colorado AA, Vincent KL, Miller AL, Maxwell CA, Dawson LN, Olive T et al (2017) Vaginal ecosystem modeling of growth patterns of anaerobic bacteria in microaerophilic conditions. Anaerobe 45:10–18
69. Sgibnev AV, Kremleva EA (2015) Vaginal protection by H2O2-producing lactobacilli. Jundishapur J Microbiol [Internet] 8. Cited 7 Mar 2021. Available from: https://sites.kowsarpub.com/jjm/articles/56511.html
70. Zadravec P, Štrukelj B, Berlec A (2015) Improvement of LysM-mediated surface display of designed ankyrin repeat proteins (DARPins) in recombinant and nonrecombinant strains of lactococcus lactis and lactobacillus species. Björkroth J, editor. Appl Environ Microbiol 81:2098–2106
71. Homburg C, Bommer M, Wuttge S, Hobe C, Beck S, Dobbek H et al (2017) Inducer exclusion in Firmicutes: insights into the regulation of a carbohydrate ATP binding cassette transporter from *lactobacillus casei* BL23 by the signal transducing protein P-Ser46-HPr. Mol Microbiol 105:25–45
72. Wang S, Wang Q, Yang E, Yan L, Li T, Zhuang H (2017) Antimicrobial compounds produced by vaginal lactobacillus crispatus are able to strongly inhibit candida albicans growth, hyphal formation and regulate virulence-related gene expressions. Front Microbiol [Internet] 8. Cited 7 Mar 2021. Available from: http://journal.frontiersin.org/article/10.3389/fmicb.2017.00564/full
73. Yang X, Da M, Zhang W, Qi Q, Zhang C, Han S (2018) Role of lactobacillus in cervical cancer. CMAR 10:1219–1229
74. So KA, Yang EJ, Kim NR, Hong SR, Lee J-H, Hwang C-S et al (2020) Changes of vaginal microbiota during cervical carcinogenesis in women with human papillomavirus infection. Consolaro MEL, editor. PLoS One 15:e0238705
75. Amabebe E, Anumba DOC (2018) The vaginal microenvironment: the physiologic role of lactobacilli. Front Med 5:181
76. O'Hanlon DE, Moench TR, Cone RA (2013) Vaginal pH and Microbicidal lactic acid when lactobacilli dominate the microbiota. Landay A, editor. PLoS One 8:e80074

77. Mei L, Wang T, Chen Y, Wei D, Zhang Y, Cui T et al (2022) Dysbiosis of vaginal microbiota associated with persistent high-risk human papilloma virus infection. J Transl Med 20:12
78. Wang H, Ma Y, Li R, Chen X, Wan L, Zhao W (2019) Associations of Cervicovaginal lactobacilli with high-risk human papillomavirus infection, cervical intraepithelial neoplasia, and cancer: a systematic review and meta-analysis. J Infect Dis 220:1243–1254
79. Brotman RM, Shardell MD, Gajer P, Fadrosh D, Chang K, Silver MI et al (2014) Association between the vaginal microbiota, menopause status, and signs of vulvovaginal atrophy. Menopause 21:450–458
80. Kyrgiou M, Mitra A, Moscicki A-B (2017) Does the vaginal microbiota play a role in the development of cervical cancer? Transl Res 179:168–182
81. Chao X-P, Sun T-T, Wang S, Fan Q-B, Shi H-H, Zhu L et al (2019) Correlation between the diversity of vaginal microbiota and the risk of high-risk human papillomavirus infection. Int J Gynecol Cancer 29:28–34
82. Mitra A, MacIntyre DA, Lee YS, Smith A, Marchesi JR, Lehne B et al (2015) Cervical intraepithelial neoplasia disease progression is associated with increased vaginal microbiome diversity. Sci Rep 5:16865
83. Andralojc KM, Molina MA, Qiu M, Spruijtenburg B, Rasing M, Pater B et al (2021) Novel high-resolution targeted sequencing of the cervicovaginal microbiome. BMC Biol 19:267
84. Schwebke JR, Muzny CA, Josey WE (2014) Role of Gardnerella vaginalis in the pathogenesis of bacterial vaginosis: a conceptual model. J Infect Dis 210:338–343
85. Moscicki A-B, Shi B, Huang H, Barnard E, Li H (2020) Cervical-vaginal microbiome and associated cytokine profiles in a prospective study of HPV 16 acquisition, persistence, and clearance. Front Cell Infect Microbiol 10:569022
86. Piyathilake CJ, Ollberding NJ, Kumar R, Macaluso M, Alvarez RD, Morrow CD (2016) Cervical microbiota associated with higher grade cervical intraepithelial neoplasia in women infected with high-risk human papillomaviruses. Cancer Prev Res (Phila) 9:357–366
87. Norenhag J, Du J, Olovsson M, Verstraelen H, Engstrand L, Brusselaers N (2020) The vaginal microbiota, human papillomavirus and cervical dysplasia: a systematic review and network meta-analysis. BJOG 127:171–180
88. Oh HY, Kim B-S, Seo S-S, Kong J-S, Lee J-K, Park S-Y et al (2015) The association of uterine cervical microbiota with an increased risk for cervical intraepithelial neoplasia in Korea. Clin Microbiol Infect 21(674):e1–e9
89. Kenyon C, Colebunders R, Crucitti T (2013) The global epidemiology of bacterial vaginosis: a systematic review. Am J Obstet Gynecol 209:505–523
90. Workowski KA, Bolan GA (2015) Centers for Disease Control and Prevention. Sexually transmitted diseases treatment guidelines, 2015. MMWR Recomm Rep 64:1–137
91. Curty G, de Carvalho PS, Soares MA (2019) The role of the cervicovaginal microbiome on the genesis and as a biomarker of premalignant cervical intraepithelial neoplasia and invasive cervical cancer. Int J Mol Sci [Internet] 21. Cited 27 Jan 2021. Available from: https://www.ncbi.nlm.nih.gov/pmc/articles/PMC6981542/
92. Shannon B, Yi TJ, Perusini S, Gajer P, Ma B, Humphrys MS et al (2017) Association of HPV infection and clearance with cervicovaginal immunology and the vaginal microbiota. Mucosal Immunol 10:1310–1319
93. Wei Z-T, Chen H-L, Wang C-F, Yang G-L, Han S-M, Zhang S-L (2021) Depiction of vaginal microbiota in women with high-risk human papillomavirus infection. Front Public Health 8:587298
94. Machado A, Cerca N (2015) Influence of biofilm formation by Gardnerella vaginalis and other anaerobes on bacterial vaginosis. J Infect Dis 212:1856–1861
95. Usyk M, Zolnik CP, Castle PE, Porras C, Herrero R, Gradissimo A et al (2020) Cervicovaginal microbiome and natural history of HPV in a longitudinal study. Silvestri G, editor. PLoS Pathog 16:e1008376

96. Murphy K, Mitchell CM (2016) The interplay of host immunity, environment and the risk of bacterial vaginosis and associated reproductive health outcomes. J Infect Dis 214(Suppl 1):S29–S35
97. Meng LT, Xue Y, Yue T, Yang L, Gao L, An RF (2016) Relationship of HPV infection and BV, VVC, TV: a clinical study based on 1 261 cases of gynecologic outpatients. Zhonghua Fu Chan Ke Za Zhi 51:730–733
98. Zhou D, Cui Y, Wu FL, Deng WH (2016) The change of vaginal lactobacillus in patients with high-risk human papillomavirus infection. Zhonghua Yi Xue Za Zhi 96:2006–2008
99. Chen Y, Qiu X, Wang W, Li D, Wu A, Hong Z et al (2020) Human papillomavirus infection and cervical intraepithelial neoplasia progression are associated with increased vaginal microbiome diversity in a Chinese cohort. BMC Infect Dis 20:629
100. Godoy-Vitorino F, Romaguera J, Zhao C, Vargas-Robles D, Ortiz-Morales G, Vázquez-Sánchez F et al (2018) Cervicovaginal fungi and bacteria associated with cervical intraepithelial neoplasia and high-risk human papillomavirus infections in a Hispanic population. Front Microbiol 9:2533
101. Larsen AK, Nymo IH, Briquemont B, Sørensen KK, Godfroid J (2013) Entrance and survival of Brucella pinnipedialis hooded seal strain in human macrophages and epithelial cells. PLoS One 8:e84861
102. Doerflinger SY, Throop AL, Herbst-Kralovetz MM (2014) Bacteria in the vaginal microbiome alter the innate immune response and barrier properties of the human vaginal epithelia in a species-specific manner. J Infect Dis 209:1989–1999
103. Zhou Y, Wang L, Pei F, Ji M, Zhang F, Sun Y et al (2019) Patients with LR-HPV infection have a distinct vaginal microbiota in comparison with healthy controls. Front Cell Infect Microbiol 9:294
104. Audirac-Chalifour A, Torres-Poveda K, Bahena-Román M, Téllez-Sosa J, Martínez-Barnetche J, Cortina-Ceballos B et al (2016) Cervical microbiome and cytokine profile at various stages of cervical cancer: a pilot study. PLoS One 11:e0153274
105. Lee JE, Lee S, Lee H, Song Y-M, Lee K, Han MJ et al (2013) Association of the vaginal microbiota with human papillomavirus infection in a Korean twin cohort. PLoS One 8:e63514
106. Mitra A, MacIntyre DA, Ntritsos G, Smith A, Tsilidis KK, Marchesi JR et al (2020) The vaginal microbiota associates with the regression of untreated cervical intraepithelial neoplasia 2 lesions. Nat Commun 11:1999
107. Caselli E, D'Accolti M, Santi E, Soffritti I, Conzadori S, Mazzacane S et al (2020) Vaginal microbiota and cytokine microenvironment in HPV clearance/persistence in women surgically treated for cervical intraepithelial neoplasia: An observational prospective study. Front Cell Infect Microbiol 10:540900
108. Srinivasan S, Hoffman NG, Morgan MT, Matsen FA, Fiedler TL, Hall RW et al (2012) Bacterial communities in women with bacterial vaginosis: high resolution phylogenetic analyses reveal relationships of microbiota to clinical criteria. PLoS One 7:e37818
109. Chao X, Wang L, Wang S, Lang J, Tan X, Fan Q et al (2021) Research of the potential vaginal microbiome biomarkers for high-grade squamous intraepithelial lesion. Front Med 8:1446
110. Mitra A, MacIntyre DA, Paraskevaidi M, Moscicki A-B, Mahajan V, Smith A et al (2021) The vaginal microbiota and innate immunity after local excisional treatment for cervical intraepithelial neoplasia. Genome Med 13:176
111. Takahashi K (2014) Influence of bacteria on epigenetic gene control. Cell Mol Life Sci 71:1045–1054
112. Sommer F, Nookaew I, Sommer N, Fogelstrand P, Bäckhed F (2015) Site-specific programming of the host epithelial transcriptome by the gut microbiota. Genome Biol 16:62
113. Amabebe E, Anumba DOC (2020) Female gut and genital tract microbiota-induced crosstalk and differential effects of short-chain fatty acids on immune sequelae. Front Immunol 11:2184

114. Gury-BenAri M, Thaiss CA, Serafini N, Winter DR, Giladi A, Lara-Astiaso D et al (2016) The Spectrum and regulatory landscape of intestinal innate lymphoid cells are shaped by the microbiome. Cell 166:1231–1246.e13
115. Gut Microbiota has a widespread and modifiable effect on host gene regulation – ProQuest [Internet]. Cited 13 Dec 2021. Available from: https://www.proquest.com/docvie w/2299497147?parentSessionId=x%2F6lQpphZ94ezZ5C0t3GXAQJqcckPOIG%2B9Yckc Exssw%3D&accountid=17229
116. Ghadimi D, Helwig U, Schrezenmeir J, Heller KJ, de Vrese M (2012) Epigenetic imprinting by commensal probiotics inhibits the IL-23/IL-17 axis in an in vitro model of the intestinal mucosal immune system. J Leukoc Biol 92:895–911
117. Markowiak-Kopeć P, Śliżewska K (2020) The effect of probiotics on the production of short-chain fatty acids by human intestinal microbiome. Nutrients [Internet] 12. Cited 8 Dec 2021. Available from: https://www.ncbi.nlm.nih.gov/pmc/articles/PMC7230973/
118. van der Hee B, Wells JM (2021) Microbial regulation of host physiology by short-chain fatty acids. Trends Microbiol 29:700–712
119. Yu D-H, Gadkari M, Zhou Q, Yu S, Gao N, Guan Y et al (2015) Postnatal epigenetic regulation of intestinal stem cells requires DNA methylation and is guided by the microbiome. Genome Biol 16:211
120. Pan W-H, Sommer F, Falk-Paulsen M, Ulas T, Best P, Fazio A et al (2018) Exposure to the gut microbiota drives distinct methylome and transcriptome changes in intestinal epithelial cells during postnatal development. Genome Med 10:27
121. Ansari I, Raddatz G, Gutekunst J, Ridnik M, Cohen D, Abu-Remaileh M et al (2020) The microbiota programs DNA methylation to control intestinal homeostasis and inflammation. Nat Microbiol 5:610–619
122. Elliott EN, Sheaffer KL, Schug J, Stappenbeck TS, Kaestner KH (2015) Dnmt1 is essential to maintain progenitors in the perinatal intestinal epithelium. Development 142:2163–2172
123. Sheaffer KL, Kim R, Aoki R, Elliott EN, Schug J, Burger L et al (2014) DNA methylation is required for the control of stem cell differentiation in the small intestine. Genes Dev 28:652–664
124. Qin W, Scicluna BP, van der Poll T (2021) The role of host cell DNA methylation in the immune response to bacterial infection. Front Immunol 12:696280
125. Nené NR, Barrett J, Jones A, Evans I, Reisel D, Timms JF et al (2020) DNA methylation signatures to predict the cervicovaginal microbiome status. Clin Epigenetics 12:180
126. Pacis A, Tailleux L, Morin AM, Lambourne J, MacIsaac JL, Yotova V et al (2015) Bacterial infection remodels the DNA methylation landscape of human dendritic cells. Genome Res 25:1801–1811
127. Hur K, Niwa T, Toyoda T, Tsukamoto T, Tatematsu M, Yang H-K et al (2011) Insufficient role of cell proliferation in aberrant DNA methylation induction and involvement of specific types of inflammation. Carcinogenesis 32:35–41
128. Su P-H, Hsu Y-W, Huang R-L, Chen L-Y, Chao T-K, Liao C-C et al (2019) TET1 promotes 5hmC-dependent stemness, and inhibits a 5hmC-independent epithelial-mesenchymal transition, in cervical precancerous lesions. Cancer Lett 450:53–62
129. Chen Z, Wong PY, Ng CWK, Lan L, Fung S, Li JW et al (2020) The intersection between oral microbiota, host gene methylation and patient outcomes in head and neck squamous cell carcinoma. Cancers 12:3425
130. Huang S-T, Chen J, Lian L-Y, Cai H-H, Zeng H-S, Zheng M et al (2020) Intratumoral levels and prognostic significance of fusobacterium nucleatum in cervical carcinoma. Aging (Albany NY) 12:23337–23350
131. Edwards VL, Smith SB, McComb EJ, Tamarelle J, Ma B, Humphrys MS et al (2019) The cervicovaginal microbiota-host interaction modulates chlamydia trachomatis infection. Clemente JC, editor. mBio [Internet] 10. Cited 13 Dec 2021. Available from: https://journals.asm.org/doi/10.1128/mBio.01548-19

132. Cheng L, Kaźmierczak D, Norenhag J, Hamsten M, Fransson E, Schuppe-Koistinen I et al (2021) A MicroRNA gene panel predicts the vaginal microbiota composition. mSystems [Internet]. American Society for Microbiology. Cited 30 Dec 2021. Available from: https://journals.asm.org/doi/abs/10.1128/mSystems.00175-21

133. Anton L, Sierra L-J, DeVine A, Barila G, Heiser L, Brown AG et al (2018) Common cervicovaginal microbial supernatants alter cervical epithelial function: mechanisms by which lactobacillus crispatus contributes to cervical health. Front Microbiol 9:2181

134. Locke WJ, Guanzon D, Ma C, Liew YJ, Duesing KR, Fung KYC et al (2019) DNA methylation cancer biomarkers: translation to the clinic. Front Genet 10:1150

135. González-Sánchez P, DeNicola GM (2021) The microbiome(s) and cancer: know thy neighbor(s). J Pathol 254:332–343

136. Reid G, Heinemann C, Velraeds M, van der Mei HC, Busscher HJ (1999) [31] Biosurfactants produced by Lactobacillus. Methods Enzymol [Internet] 426–433. Cited 15 Dec 2021. Available from: https://www.sciencedirect.com/science/article/pii/S0076687999100338

137. Nold C, Anton L, Brown A, Elovitz M (2012) Inflammation promotes a cytokine response and disrupts the cervical epithelial barrier: a possible mechanism of premature cervical remodeling and preterm birth. Am J Obstet Gynecol 206(208):e1–e7

138. Kumar H, Kawai T, Akira S (2011) Pathogen recognition by the innate immune system. Int Rev Immunol 30:16–34

139. Brubaker SW, Bonham KS, Zanoni I, Kagan JC (2015) Innate immune pattern recognition: a cell biological perspective. Annu Rev Immunol 33:257–290

140. Łaniewski P, Herbst-Kralovetz MM (2021) Bacterial vaginosis and health-associated bacteria modulate the immunometabolic landscape in 3D model of human cervix. NPJ Biofilms Microbiomes 7:88

141. Edwards VL, McComb E, Gleghorn JP, Forney L, Bavoil PM, Ravel J (2021) Three-dimensional models of the cervicovaginal epithelia to study host-microbiome interactions and sexually transmitted infections [Internet]. bioRxiv 2021.11.04.467382. Cited 24 Jan 2022. Available from: https://www.biorxiv.org/content/10.1101/2021.11.04.467382v1

142. Torcia MG (2019) Interplay among vaginal microbiome, immune response and sexually transmitted viral infections. Int J Mol Sci 20:E266

143. Anahtar MN, Byrne EH, Doherty KE, Bowman BA, Yamamoto HS, Soumillon M et al (2015) Cervicovaginal bacteria are a major modulator of host inflammatory responses in the female genital tract. Immunity 42:965–976

144. Madara JL, Stafford J (1989) Interferon-gamma directly affects barrier function of cultured intestinal epithelial monolayers. J Clin Invest 83:724–727

145. McGee ZA, Clemens CM, Jensen RL, Klein JJ, Barley LR, Gorby GL (1992) Local induction of tumor necrosis factor as a molecular mechanism of mucosal damage by gonococci. Microb Pathog 12:333–341

146. Gur C, Ibrahim Y, Isaacson B, Yamin R, Abed J, Gamliel M et al (2015) Binding of the Fap2 protein of fusobacterium nucleatum to human inhibitory receptor TIGIT protects tumors from immune cell attack. Immunity 42:344–355

147. De Seta F, Campisciano G, Zanotta N, Ricci G, Comar M (2019) The vaginal community state types microbiome-immune network as key factor for bacterial vaginosis and aerobic vaginitis. Front Microbiol 10:2451

148. Maiuri AR, O'Hagan HM (2016) Chapter three – interplay between inflammation and epigenetic changes in cancer. In: Pruitt K (ed) Progress in molecular biology and translational science [Internet]. Academic Press, pp 69–117. Cited 27 Dec 2021. Available from: https://www.sciencedirect.com/science/article/pii/S1877117316300606

149. Valenzuela MA, Canales J, Corvalán AH, Quest AF (2015) Helicobacter pylori-induced inflammation and epigenetic changes during gastric carcinogenesis. World J Gastroenterol 21:12742–12756

150. Ivanov AV, Bartosch B, Isaguliants MG (2017) Oxidative stress in infection and consequent disease. Oxidative Med Cell Longev 2017:3496043

151. Khansari N, Shakiba Y, Mahmoudi M (2009) Chronic inflammation and oxidative stress as a major cause of age-related diseases and cancer. Recent Patents Inflamm Allergy Drug Discov 3:73–80
152. Hardbower DM, de Sablet T, Chaturvedi R, Wilson KT (2013) Chronic inflammation and oxidative stress: the smoking gun for helicobacter pylori-induced gastric cancer? Gut Microbes 4:475–481
153. Weitzman SA, Turk PW, Milkowski DH, Kozlowski K (1994) Free radical adducts induce alterations in DNA cytosine methylation. Proc Natl Acad Sci USA 91:1261–1264
154. Turk PW, Laayoun A, Smith SS, Weitzman SA (1995) DNA adduct 8-hydroxyl-2′-deoxyguanosine (8-hydroxyguanine) affects function of human DNA methyltransferase. Carcinogenesis 16:1253–1255
155. Valinluck V, Sowers LC (2007) Endogenous cytosine damage products alter the site selectivity of human DNA maintenance methyltransferase DNMT1. Cancer Res 67:946–950
156. Campos ACE, Molognoni F, Melo FHM, Galdieri LC, Carneiro CRW, D'Almeida V et al (2007) Oxidative stress modulates DNA methylation during melanocyte anchorage blockade associated with malignant transformation. Neoplasia 9:1111–1121
157. Wu Q, Ni X (2015) ROS-mediated DNA methylation pattern alterations in carcinogenesis. Curr Drug Targets 16:13–19
158. Amedei A, Capasso C, Nannini G, Supuran CT (2021) Microbiota, bacterial carbonic anhydrases, and modulators of their activity: links to human diseases? Mediat Inflamm 2021:e6926082
159. Zhou Z-W, Long H-Z, Cheng Y, Luo H-Y, Wen D-D, Gao L-C (2021) From microbiome to inflammation: the key drivers of cervical cancer. Front Microbiol 12:3473
160. Sales KJ, Katz AA (2012) Inflammatory pathways in cervical cancer – the UCT contribution. S Afr Med J 102:493–496
161. Osiagwu DD, Azenabor AE, Osijirin AA, Awopetu PI, Oyegbami FR (2019) Evaluation of interleukin 8 and interleukin 10 cytokines in liquid based cervical cytology samples. Pan Afr Med J 32:148
162. Vitkauskaite A, Urboniene D, Celiesiute J, Jariene K, Skrodeniene E, Nadisauskiene RJ et al (2020) Circulating inflammatory markers in cervical cancer patients and healthy controls. J Immunotoxicol 17:105–109
163. Hu Y, Wu F, Liu Y, Zhao Q, Tang H (2019) DNMT1 recruited by EZH2-mediated silencing of miR-484 contributes to the malignancy of cervical cancer cells through MMP14 and HNF1A. Clin Epigenetics 11:186
164. Duvlis S, Dabeski D, Cvetkovski A, Mladenovska K, Plaseska-Karanfilska D (2020) Association of TNF-a (rs361525 and rs1800629) with susceptibility to cervical intraepithelial lesion and cervical carcinoma in women from republic of North Macedonia. Int J Immunogenet 47:522–528
165. Bai Y, Li H, Lv R (2021) Interleukin-17 activates JAK2/STAT3, PI3K/Akt and nuclear factor-κB signaling pathway to promote the tumorigenesis of cervical cancer. Exp Ther Med 22:1291
166. Tilborghs S, Corthouts J, Verhoeven Y, Arias D, Rolfo C, Trinh XB et al (2017) The role of nuclear factor-kappa B signaling in human cervical cancer. Crit Rev Oncol Hematol 120:141–150
167. Karin M (2009) NF-κB as a critical link between inflammation and cancer. Cold Spring Harb Perspect Biol 1:a000141
168. Ma B, Hottiger MO (2016) Crosstalk between Wnt/β-catenin and NF-κB Signaling pathway during inflammation. Front Immunol 7:378
169. Yang M, Wang M, Li X, Xie Y, Xia X, Tian J et al (2018) Wnt signaling in cervical cancer? J Cancer 9:1277–1286
170. Poulsen LLAC, Edelmann RJ, Krüger S, Diéguez-Hurtado R, Shah A, Stav-Noraas TE et al (2018) Inhibition of endothelial NOTCH1 signaling attenuates inflammation by reducing

cytokine-mediated histone acetylation at inflammatory enhancers. Arterioscler Thromb Vasc Biol 38:854–869

171. Jiang N, Xie F, Chen L, Chen F, Sui L (2020) The effect of TLR4 on the growth and local inflammatory microenvironment of HPV-related cervical cancer in vivo. Infect Agents Cancer 15:12

172. Zhang H, Zhang S (2017) The expression of Foxp3 and TLR4 in cervical cancer: association with immune escape and clinical pathology. Arch Gynecol Obstet 295:705–712

173. Hussain T, Tan B, Yin Y, Blachier F, Tossou MCB, Rahu N (2016) Oxidative stress and inflammation: what polyphenols can do for us? Oxidative Med Cell Longev 2016:e7432797

174. Wiseman H, Halliwell B (1996) Damage to DNA by reactive oxygen and nitrogen species: role in inflammatory disease and progression to cancer. Biochem J 313(Pt 1):17–29

175. Murata M (2018) Inflammation and cancer. Environ Health Prev Med 23:50

176. Wang X, Huang X, Zhang Y (2018) Involvement of human papillomaviruses in cervical cancer. Front Microbiol 9:2896

177. Hoellen F, Waldmann A, Banz-Jansen C, Rody A, Heide M, Köster F et al (2016) Expression of cyclooxygenase-2 in cervical cancer is associated with lymphovascular invasion. Oncol Lett 12:2351–2356

178. Hemmat N, Bannazadeh Baghi H (2019) Association of human papillomavirus infection and inflammation in cervical cancer. Pathog Dis 77:ftz048

179. Yang B, Chen J, Teng Y (2021) CDK12 promotes cervical cancer progression through enhancing macrophage infiltration. J Immunol Res 2021:6645885

180. Xu L, Xu Q, Li X, Zhang X (2017) MicroRNA-21 regulates the proliferation and apoptosis of cervical cancer cells via tumor necrosis factor-α. Mol Med Rep 16:4659–4663

181. Zhang J, Wu H, Li P, Zhao Y, Liu M, Tang H (2014) NF-κB-modulated miR-130a targets TNF-α in cervical cancer cells. J Transl Med 12:155

182. Fan J-Y, Fan Y-J, Wang X-L, Xie H, Gao H-J, Zhang Y et al (2017) miR-429 is involved in regulation of NF-κBactivity by targeting IKKβ and suppresses oncogenic activity in cervical cancer cells. FEBS Lett 591:118–128

183. Wen C-X, Tian H-L, Chen E, Liu J-F, Liu X-X (2021) MiRNA-873-5p acts as a potential novel biomarker and promotes cervical cancer progression by regulating ZEB1 via notch signaling pathway. Dose-Response 19:15593258211001256

184. Ke L, Chen Y, Li Y, Chen Z, He Y, Liu J et al (2021) miR-142-5p promotes cervical cancer progression by targeting LMX1A through Wnt/β-catenin pathway. Open Med (Wars) 16:224–236

185. Huang F, Lin C, Shi Y-H, Kuerban G (2013) MicroRNA-101 inhibits cell proliferation, invasion, and promotes apoptosis by regulating cyclooxygenase-2 in Hela cervical carcinoma cells. Asian Pac J Cancer Prev 14:5915–5920

186. Zhu J, Han S (2021) Histone deacetylase 10 exerts antitumor effects on cervical cancer via a novel microRNA-223/TXNIP/Wnt/β-catenin pathway. IUBMB Life 73:690–704

187. Cartwright TN, Worrell JC, Marchetti L, Dowling CM, Knox A, Kiely P et al (2018) HDAC1 interacts with the p50 NF-?B subunit via its nuclear localization sequence to constrain inflammatory gene expression. Biochim Biophys Acta Gene Regul Mech 1861:962–970

188. Yue Y, Zhou K, Li J, Jiang S, Li C, Men H (2018) MSX1 induces G0/G1 arrest and apoptosis by suppressing notch signaling and is frequently methylated in cervical cancer. Onco Targets Ther 11:4769–4780

189. Kadian LK, Gulshan G, Ahuja P, Singhal G, Sharma S, Nanda S et al (2020) Aberrant promoter methylation of NOTCH1 and NOTCH3 and its association with cervical cancer risk factors in north Indian population. Am J Transl Res 12:2814–2826

190. Sharma M, Chopra C, Mehta M, Sharma V, Mallubhotla S, Sistla S et al (2021) An insight into vaginal microbiome techniques. Life (Basel) 11:1229

191. Serrano MG, Parikh HI, Brooks JP, Edwards DJ, Arodz TJ, Edupuganti L et al (2019) Racioethnic diversity in the dynamics of the vaginal microbiome during pregnancy. Nat Med 25:1001–1011

192. Yang Q, Wang Y, Wei X, Zhu J, Wang X, Xie X et al (2020) The alterations of vaginal micro-biome in HPV16 infection as identified by shotgun metagenomic sequencing. Front Cell Infect Microbiol 10:286
193. Seneviratne CJ, Balan P, Suriyanarayanan T, Lakshmanan M, Lee D-Y, Rho M et al (2020) Oral microbiome-systemic link studies: perspectives on current limitations and future artificial intelligence-based approaches. Crit Rev Microbiol 46:288–299
194. Bubnov RV, Spivak MY, Lazarenko LM, Bomba A, Boyko NV (2015) Probiotics and immunity: provisional role for personalized diets and disease prevention. EPMA J 6:14
195. Reid G, Abrahamsson T, Bailey M, Bindels LB, Bubnov R, Ganguli K et al (2017) How do probiotics and prebiotics function at distant sites? Benef Microbes 8:521–533
196. Ebner S, Smug LN, Kneifel W, Salminen SJ, Sanders ME (2014) Probiotics in dietary guidelines and clinical recommendations outside the European Union. World J Gastroenterol 20:16095–16100
197. Goldstein EJC, Tyrrell KL, Citron DM (2015) Lactobacillus species: taxonomic complexity and controversial susceptibilities. Clin Infect Dis 60(Suppl 2):S98–S107
198. Ramos AN, Sesto Cabral ME, Arena ME, Arrighi CF, Arroyo Aguilar AA, Valdéz JC (2015) Compounds from lactobacillus plantarum culture supernatants with potential pro-healing and anti-pathogenic properties in skin chronic wounds. Pharm Biol 53:350–358
199. Vujic G, Jajac Knez A, Despot Stefanovic V, Kuzmic Vrbanovic V (2013) Efficacy of orally applied probiotic capsules for bacterial vaginosis and other vaginal infections: a double-blind, randomized, placebo-controlled study. Eur J Obstet Gynecol Reprod Biol 168:75–79
200. Abdelhamid AG, El-Masry SS, El-Dougdoug NK (2019) Probiotic lactobacillus and Bifidobacterium strains possess safety characteristics, antiviral activities and host adherence factors revealed by genome mining. EPMA J 10:337–350
201. Li Y, Yu T, Yan H, Li D, Yu T, Yuan T et al (2020) Vaginal microbiota and HPV infection: novel mechanistic insights and therapeutic strategies. Infect Drug Resist 13:1213–1220
202. Kyrgiou M, Koliopoulos G, Martin-Hirsch P, Arbyn M, Prendiville W, Paraskevaidis E (2006) Obstetric outcomes after conservative treatment for intraepithelial or early invasive cervical lesions: systematic review and meta-analysis. Lancet 367:489–498
203. Verhoeven V, Renard N, Makar A, Van Royen P, Bogers J-P, Lardon F et al (2013) Probiotics enhance the clearance of human papillomavirus-related cervical lesions: a prospective controlled pilot study. Eur J Cancer Prev 22:46–51
204. Palma E, Recine N, Domenici L, Giorgini M, Pierangeli A, Panici PB (2018) Long-term lactobacillus rhamnosus BMX 54 application to restore a balanced vaginal ecosystem: a promising solution against HPV-infection. BMC Infect Dis 18:13
205. Reid G, Beuerman D, Heinemann C, Bruce AW (2001) Probiotic lactobacillus dose required to restore and maintain a normal vaginal flora. FEMS Immunol Med Microbiol 32:37–41
206. Morelli L, Zonenenschain D, Del Piano M, Cognein P (2004) Utilization of the intestinal tract as a delivery system for urogenital probiotics. J Clin Gastroenterol 38:S107–S110
207. Homayouni A, Bastani P, Ziyadi S, Mohammad-Alizadeh-Charandabi S, Ghalibaf M, Mortazavian AM et al (2014) Effects of probiotics on the recurrence of bacterial vaginosis: a review. J Low Genit Tract Dis 18:79–86
208. Liwen Z, Yu W, Liang M, Kaihong X, Baojin C (2018) A low abundance of Bifidobacterium but not Lactobacillius in the feces of Chinese children with wheezing diseases. Medicine (Baltimore) 97:e12745
209. Bilen M, Dufour J-C, Lagier J-C, Cadoret F, Daoud Z, Dubourg G et al (2018) The contribution of culturomics to the repertoire of isolated human bacterial and archaeal species. Microbiome 6:94
210. Chang C-J, Lin T-L, Tsai Y-L, Wu T-R, Lai W-F, Lu C-C et al (2019) Next generation probiotics in disease amelioration. J Food Drug Anal 27:615–622
211. Morovic W, Budinoff CR (2021) Epigenetics: a new frontier in probiotic research. Trends Microbiol 29:117–126

212. Cunningham M, Azcarate-Peril MA, Barnard A, Benoit V, Grimaldi R, Guyonnet D et al (2021) Shaping the future of probiotics and prebiotics. Trends Microbiol 29:667–685
213. Morozov V, Hansman G, Hanisch F-G, Schroten H, Kunz C (2018) Human milk oligosaccharides as promising antivirals. Mol Nutr Food Res 62:e1700679
214. Onywera H, Williamson A-L, Ponomarenko J, Meiring TL (2020) The penile microbiota in uncircumcised and circumcised men: relationships with HIV and human papillomavirus infections and cervicovaginal microbiota. Front Med 7:383
215. Price LB, Liu CM, Johnson KE, Aziz M, Lau MK, Bowers J et al (2010) The effects of circumcision on the penis microbiome. PLoS One 5:e8422
216. Zozaya M, Ferris MJ, Siren JD, Lillis R, Myers L, Nsuami MJ et al (2016) Bacterial communities in penile skin, male urethra, and vaginas of heterosexual couples with and without bacterial vaginosis. Microbiome 4:16
217. Bubnov RV, Babenko LP, Lazarenko LM, Mokrozub VV, Spivak MY (2018) Specific properties of probiotic strains: relevance and benefits for the host. EPMA J 9:205–223
218. Qingqing B, Jie Z, Songben Q, Juan C, Lei Z, Mu X (2020) Cervicovaginal microbiota dysbiosis correlates with HPV persistent infection. Microb Pathog 152:104617
219. Lazarenko LM, Nikitina OE, Nikitin EV, Demchenko OM, Kovtonyuk GV, Ganova LO et al (2014) Development of biomarker panel to predict, prevent and create treatments tailored to the persons with human papillomavirus-induced cervical precancerous lesions. EPMA J 5:1
220. Qian S, Golubnitschaja O, Zhan X (2019) Chronic inflammation: key player and biomarker-set to predict and prevent cancer development and progression based on individualized patient profiles. EPMA J 10:365–381
221. Łaniewski P, Cui H, Roe DJ, Barnes D, Goulder A, Monk BJ et al (2019) Features of the cervicovaginal microenvironment drive cancer biomarker signatures in patients across cervical carcinogenesis. Sci Rep [Internet] 9. Cited 27 Jan 2021. Available from: https://www.ncbi.nlm.nih.gov/pmc/articles/PMC6517407/
222. Łaniewski P, Barnes D, Goulder A, Cui H, Roe DJ, Chase DM et al (2018) Linking cervicovaginal immune signatures, HPV and microbiota composition in cervical carcinogenesis in non-Hispanic and Hispanic women. Sci Rep [Internet] 8. Cited 27 Jan 2021. Available from: https://www.ncbi.nlm.nih.gov/pmc/articles/PMC5954126/
223. Steinbach A, Riemer AB (2018) Immune evasion mechanisms of human papillomavirus: an update. Int J Cancer 142:224–229
224. Hu T, Yang P, Zhu H, Chen X, Xie X, Yang M et al (2015) Accumulation of invariant NKT cells with increased IFN-γ production in persistent high-risk HPV-infected high-grade cervical intraepithelial neoplasia. Diagn Pathol 10:20
225. Gutiérrez-Hoya A, Zerecero-Carreón O, Valle-Mendiola A, Moreno-Lafont M, López-Santiago R, Weiss-Steider B et al (2019) Cervical cancer cells express markers associated with immunosurveillance. J Immunol Res 2019:1242979
226. Barros MR, de Melo CML, Barros MLCMGR, de Cássia Pereira de Lima R, de Freitas AC, Venuti A (2018) Activities of stromal and immune cells in HPV-related cancers. J Exp Clin Cancer Res 37:137
227. Plummer M, de Martel C, Vignat J, Ferlay J, Bray F, Franceschi S (2016) Global burden of cancers attributable to infections in 2012: a synthetic analysis. Lancet Glob Health 4:e609–e616
228. Vyshenska D, Lam KC, Shulzhenko N, Morgun A (2017) Interplay between viruses and bacterial microbiota in cancer development. Semin Immunol 32:14–24
229. Brusselaers N, Shrestha S, van de Wijgert J, Verstraelen H (2019) Vaginal dysbiosis and the risk of human papillomavirus and cervical cancer: systematic review and meta-analysis. Am J Obstet Gynecol 221:9–18.e8
230. Lewis FMT, Bernstein KT, Aral SO (2017) Vaginal microbiome and its relationship to behavior, sexual health, and sexually transmitted diseases. Obstet Gynecol 129:643–654

231. Wilkinson EM, Łaniewski P, Herbst-Kralovetz MM, Brotman RM (2019) Personal and clinical vaginal lubricants: impact on local vaginal microenvironment and implications for epithelial cell host response and barrier function. J Infect Dis 220:2009–2018

232. Mändar R, Punab M, Borovkova N, Lapp E, Kiiker R, Korrovits P et al (2015) Complementary seminovaginal microbiome in couples. Res Microbiol 166:440–447

233. Nelson DE, Van Der Pol B, Dong Q, Revanna KV, Fan B, Easwaran S et al (2010) Characteristic male urine microbiomes associate with asymptomatic sexually transmitted infection. PLoS One 5:e14116

234. Nelson DE, Dong Q, Van der Pol B, Toh E, Fan B, Katz BP et al (2012) Bacterial communities of the coronal sulcus and distal urethra of adolescent males. PLoS One 7:e36298

235. Golubnitschaja O, Baban B, Boniolo G, Wang W, Bubnov R, Kapalla M et al (2016) Medicine in the early twenty-first century: paradigm and anticipation – EPMA position paper 2016. EPMA J 7:23

236. Golubnitschaja O, Kinkorova J, Costigliola V (2014) Predictive, preventive and personalised medicine as the hardcore of 'horizon 2020': EPMA position paper. EPMA J 5:6

237. Hu R, Wang X, Zhan X (2013) Multi-parameter systematic strategies for predictive, preventive and personalised medicine in cancer. EPMA J 4:2

238. Lu M, Zhan X (2018) The crucial role of multiomic approach in cancer research and clinically relevant outcomes. EPMA J 9:77–102

239. Yu JC, Khodadadi H, Baban B (2019) Innate immunity and oral microbiome: a personalized, predictive, and preventive approach to the management of oral diseases. EPMA J 10:43–50

240. Kudela E, Liskova A, Samec M, Koklesova L, Holubekova V, Rokos T, Kozubik E, Pribulova T, Zhai K, Busselberg D, Kubatka P, Biringer K (2021) The interplay between the vaginal microbiome and innate immunity in the focus of predictive, preventive, and personalized medical approach to combat HPV-induced cervical cancer. EPMA J 12(2):199–220. https://doi.org/10.1007/s13167-021-00244-3

241. Holubekova V, Kolkova Z, Kasubova I, Samec M, Mazurakova A, Koklesova L, Kubatka P, Rokos T, Kozubik E, Biringer K, Kudela E (2022) Interaction of cervical microbiome with epigenome of epithelial cells: significance of inflammation to primary healthcare. Biomol Concepts 13(1):61–80. https://doi.org/10.1515/bmc-2022-0005

Chapter 4
Microbiome in Lean Individuals: Phenotype-Specific Risks and Outcomes

Olga Golubnitschaja and Rostyslav Bubnov

Abstract Many studies provided strong evidence that thinness as a potential individual underweight is an overlooked phenomenon in terms of health risks causality and associated pathologies which are much less explored compared to those associated with overweight and obesity. Disease predisposition linked to low body mass index has been demonstrated for individuals with Flammer Syndrome phenotype. This phenotype is manifested early in life and, therefore, is of great clinical relevance for predictive, preventive and personalised approach in primary healthcare. The transition phase between sub-optimal health condition and clinical manifestation of the disease is the operational timeframe of the reversible damage to health protecting the affected person against disease development and progression (secondary and tertiary care). Association of the FS phenotype and shifted microbiome profiles has been suggested for oral and vaginal microbiome as being linked, in particular, to the stress overload, disturbed microcirculation, Sicca syndrome and delayed/impaired healing. Anorexia nervosa (AN) is considered an extreme case of the FS phenotype. The best acknowledged risk factors predisposing individuals with FS phenotype to AN and AN-associated pathologies are specific behavioural patterns, altered psychosocial functioning, pronounced perfectionism, and thin-ideal internalisation, amongst others. It has been suggested that eating disorders may result from stress-induced maladaptive alterations in neural circuits which regulate feeding as demonstrated in animal models of eating disorders. The chapter provide in-depth analysis between the lean phenotype and microbiota profiles and

O. Golubnitschaja (✉)
Predictive, Preventive and Personalised (3P) Medicine, Department of Radiation Oncology, University Hospital Bonn, Rheinische Friedrich-Wilhelms-Universität Bonn, Bonn, Germany

3PMedicon, Taufkirchen an der Pram, Austria
e-mail: olga.golubnitschaja@ukbonn.de

R. Bubnov
Zabolotny Institute of Microbiology and Virology, National Academy of Sciences of Ukraine, Kyiv, Ukraine

Clinical Hospital 'Pheophania' of State Affairs Department, Kyiv, Ukraine

© The Author(s), under exclusive license to Springer Nature Switzerland AG 2023
N. Boyko, O. Golubnitschaja (eds.), *Microbiome in 3P Medicine Strategies*, Advances in Predictive, Preventive and Personalised Medicine 16, https://doi.org/10.1007/978-3-031-19564-8_4

suggest innovative strategies based on the principles of predictive, preventive and personalised medicine.

Keywords Disordered eating · Eating disorders · Body mass index · Underweight · Anorexia nervosa · Microbiome composition · Skin · Acne · Gingivitis · Wound healing · Individualised patient profile · Shifted metabolism · Irritable bowel disease · Flammer syndrome · Vasospasm · Thermoregulation · Behavioural patterns · Stress response · Epigenetic regulation · Predictive diagnostics · Probiotics · Prebiotics · Starvation · Preventable risk factors · Hypoxia · Mood disorders · Cancer

4.1 Suboptimal Body Weight as a Fundamental Health Risk Factor

Extensive evidence has been accumulated for overweight and obesity as a strong contributor to severe pathologies linked to metabolic disorders and shifted microbiome profiles. Contextually, an increasing interest in a physical and mental health raises questions about optimal body weight. It should be differentiated between the standardised "normal" body weight and individually optimal body weight. To this end, a borderline "normal" body mass index (BMI) might be optimal for one person but suboptimal for another one being strongly dependent a number of parameters including non-modifiable factors such as genetic predisposition, geographic origin and cultural habits as well as modifiable factors such as nutritional and lifestyle habits – all relevant for the comprehensive individual patient profile which is instrumental for predictive diagnostics, targeted prevention and personalisation of treatment algorithms [1, 2]. If deviant from the individual optimum, both overweight and underweight are well acknowledged risk factors for a shifted metabolism which in turn may strongly contribute to the development and progression of severe pathologies – see Fig. 4.1 [1].

Many studies provided strong evidence that thinness as a potential individual underweight is an overlooked phenomenon in terms of health risks causality and associated pathologies which are much less explored compared to those associated with overweight and obesity [3, 4]. Increased risks of a disease predisposition, known syndromes and severe pathologies linked to both obese and anorexic phenotypes, are summarised in Fig. 4.1. Multi-professional expertise is required to apply advanced concepts of Predictive, Preventive and Personalised Medicine (PPPM / 3 PM) to primary, secondary and tertiary care, in order to cost-effectively operate within the transitional phase between sub-optimal health condition and clinically manifested pathology as well as for optimal disease management [5].

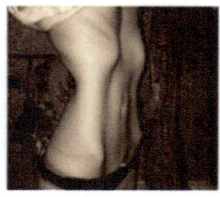
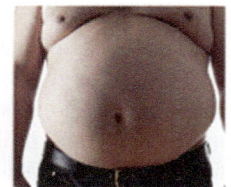

Low BMI	High BMI
High physical activity	Low physical activity
Low blood pressure	Dyslipidemia

The general population

Anorexic Phenotype; increased risk for
- Primary vascular dysregulation
- Flammer syndrome
- Systemic hypoxia / ischemic lesions
- Eye disorders (normal-tension glaucoma)
- Mental disorders (depression)
- Neurological disorder (mulitple-sclerosis)
- Reproductive dysfunction
- Xerostomia / Siccca syndrome
- Chronic inflammation
- Impaired wound healing
- Predisposition to cancer
with particularly poor outcomes

Obese Phenotype; increased risk for
- Arteriosclerosis
- Hypertension
- Stroke
- Metabolic syndrome
- Eye disorders
- Mental and neurological disorders
- Reproductive dysfunction
- Xerostomia / Sicca syndrome
- Chronic inflammation
- Impaired wound healing
- Predisposition to cancer
with particularly poor outcomes

Fig. 4.1 Overweight versus underweight in application of PPPM strategies in the population: Overweight parameters are as follows: for females, body mass index (BMI) 25–30 and for males, BMI 26–30; Obesity parameters are as follows: class I – BMI 30–35; class II – BMI 35–40; class III – BMI > 40; Underweight parameters are as follows: grade I: females, BMI < 19, males, BMI < 20; grade 2: BMI ≤ 17; grade 3: BMI ≤ 16 [1]

4.2 Eating Disorders: Epidemic Trends in Europe, Pathologies and Targeted Therapies Associated with Altered Microbiome Compositions

As early as at the begin of the millennium, scientists presented evidence about alarming trends in Europe, namely that both – disordered eating and eating disorders are becoming an epidemic; in particularly, young female subpopulations are affected [6]. Corresponding medical complications are detected in all human organ systems, including cardio-vascular, neurological, gastroenterological, gynaecological, urological, and immune disorders as well as the skin pathologies. To this end, skin signs were examined specifically in anorexia nervosa (AN) as one of the most frequently observed eating disorders in the European population. Amongst many clinically manifested skin pathologies, acne, gingivitis and poor wound healing were demonstrated [6] – per evidence, all associated with shifted metabolic profiles, altered microbiome composition, and chronic inflammation which, if not treated at early stages, may lead to cascading pathologies and systemic disorders [7–10]. Disease- and stage-specific molecular panels in body fluids (blood, saliva, urine, tears) have been demonstrated as being instrumental for predictive diagnostics followed by the targeted prevention and creation of treatment algorithms tailored to the individualised patient profiles [11].

Being strongly associated with chronic inflammation, abnormal healing and scar formation, pathophysiological pathways of acne include disease-specific patterns of interleukin (IL)-1β, IL-17, IL-23, and tumour necrosis factor alpha (TNFα); transforming growth factor-β (TGFβ), IL-6, matrix metalloproteinase (MMP), IGF-1, and B cells are found in keloid or hypertrophic scar tissues Biological antibodies targeting IL-1β, IL-17, IL-23, and TNFα could provide novel approaches for treating severe acne and related disorders [12]. Specific risk factors include shifted microbiome profiles. To this end, the understanding the acne pathophysiology currently undergoes a paradigm shift: acne development results from the loss of balance between the different Cutibacterium acnes phylotypes, together with a dysbiosis of the skin microbiome. The loss of diversity of C. acnes phylotypes acts as a trigger for innate immune system activation, leading to cutaneous inflammation. Through interactions with the skin microbiome, the gut microbiome is also involved in the acne pathophysiology. New knowledge prompts development of individualised acne therapies and will allow targeting pathogenic strains on one hand, and, on the other hand, leaving the commensal strains intact. Such alternative treatments, involving modifications of the microbiome, will form the advanced next generation of 'ecobiological' anti-inflammatory treatments [13] Indeed, oral and topical probiotics appear to be effective for the treatment of certain inflammatory skin diseases and demonstrate their promising role in the improved wound healing and novel skin cancer therapeutic approaches. However, more studies are needed to confirm the working hypothesis and preliminary results [14].

4.3   A Paradox of Non-alcoholic Fatty Liver Disease in Lean Individuals: The Clue by Altered Microbiome

A non-alcoholic fatty liver disease (NAFLD) affects a quarter of the adult population. Paradoxically, a significant subset of patients is lean. Lean patients demonstrate an altered gut microbiota profile implicated in the pathophysiology of the disease and considered as the target for the treatment options to improve steatohepatitis specifically in lean NAFLD patients [15].

The intestinal microbiota is considered an important target for innovative therapeutic modalities focused on extra-intestinal inflammatory disorders. Specifically a disruption of the intestinal-mucosal macrophage interface was hypothesised as a key factor in intestinal-liver axis disturbances. Intestinal immune responses implicated in NAFLD reflect the cross-communication between the intestinal microbiome on one hand and, on the other hand, the epithelia and mucosal immunity. Hepatixic effects result from the pro-inflammatory activity and immune imbalances implicated in NAFLD pathophysiology caused by dysbiosis of the intestinal epithelia. Intestinal dysbiosis and microbiota-mediated inflammation of the local mucosa synergistically lead to the mucosal immune dysfunction contributing the the NAFLD pathogenesis. If so, the administration of probiotics and prebiotics as a cure-all remedy is not recommended; rather the incorporation of probiotic/prebiotic formulations as adjunctive modalities based on the individualised patient profiling may be supportive to enhance lifestyle modification management efficacy to ameliorate NAFLD [16].

4.4 Phenotyping of Underweight Individuals Is Crucial for PPPM Approach

4.4.1 Flammer Syndrome Phenotype Is Linked to Low BMI and Identifiable Early in Life

Disease predisposition linked to low body mass index has been demonstrated for individuals with Flammer Syndrome phenotype [17]. This phenotype is manifested early in life and, therefore, is of great clinical relevance for predictive, preventive and personalised approach in primary healthcare. The transition phase between sub-optimal health condition and clinical manifestation of the disease is the operational timeframe of the reversible damage to health protecting the affected person against disease development and progression (secondary and tertiary care) [5]. A series of recently published studies provide strong evidence that compared to the general population, individuals with the FS phenotype may be strongly predisposed to

- Phenotype-specific behavioural patterns, disturbed microcirculation and thermo-regulation, shifted sleep patterns [17];
- stress overload and abnormal stress reactions potentially damaging mitochondrial health [18–21];
- altered metabolic pathways [5, 18, 21, 22];
- stress-associated multi-factorial disorders at young age [20, 23, 25–27, 34];
- complications in pregnancy [22, 28];
- delayed and impaired wound healing [9, 29, 30];
- chronic inflammation and compromised immune response [30];
- severe organ damage and systemic disorders such as cardio-vascular and neuro-degenerative diseases, as well as metastatic cancer subtypes [18, 31].

amongst others.

Association of the FS phenotype and shifted microbiome profiles has been suggested for oral and vaginal microbiome as being linked, in particular, to the stress overload, disturbed microcirculation, Sicca syndrome and delayed/impaired healing [24, 26, 27].

4.4.2 Anorexia Nervosa as an Extreme Case of the Flammer Syndrome Phenotype

Anorexia nervosa (AN) is considered an extreme case of the FS phenotype. The best acknowledged risk factors predisposing individuals with FS phenotype to AN and AN-associated pathologies are specific behavioural patterns, altered psychosocial functioning, pronounced perfectionism, and thin-ideal internalisation, amongst others [32]. Functional magnetic resonance imaging (fMRI) studies demonstrated disease-specific sensitivity for sensory-interoceptive-reward processes towards food consumption in AN-patients that may override homeostatic needs [33]. It has been suggested that eating disorders may result from stress-induced maladaptive alterations in neural circuits which regulate feeding as demonstrated in animal models of eating disorders [34]. Follow-up studies confirmed issue-specific alterations in the brain structure and functionality at the development of eating disorders such as AN [32, 35–37].

4.5 Microbiome Profile Is Altered in Patients with Anorexia Nervosa

Long-term dieting as well as malnutrition – both demonstrated as having sustainable effects on the systemic microbiome composition and organ-specific impacts such as brain activity patterns [38]. To this end, the adult gut microbiota is influenced by both – non-modifiable factors such as host genetics and modifiable factors, namely dietary habits, xenobiotics (e.g. antibiotics), drugs intake, lifestyle, body activity (sport/exercise), and circadian rhythms, amongst others [39–47].

Microbiotic profiles differ significantly between overweight and underweight: 75% of the obesity-enriched genes originate from Actinobacteria (compared with 0% of lean-enriched genes; the other 25% are from Firmicutes), whereas 42% of the lean enriched genes originate from Bacteroidetes (compared with 0% of the obesity-enriched genes) [46]. Whereas many studies are focused on microbiome alterations specific for individuals with abnormally high BMI, much less is known about microbiome alterations characteristic for lean, and in extreme case, underweight persons such as patients diagnosed with anorexia nervosa. Perhaps the best explored is the "lean" microbiome linked to the impaired wound healing in low BMI individuals [48]. To this end, diseased eating significantly limits gut flora diversity that per evidence leads to the intestinal dysbiosis in AN, defined as a disruption to the microbiota homeostasis caused by an imbalance in the microflora. This patient cohort suffers from mood disorders such as depression, and anxiety, peculiarity of which has been associated with the enteric microbiota composition [49]. A competitive environment evidently shapes intestinal bacterial profiles to benefits of species which are better suited to a low-energy environment characteristic for AN, where they are more likely to survive and to dominate. Indeed, compared to healthy controls with normal BMI, *Phylum Bacteroidetes* is decreased, whereas *Phylum Firmicutes* is increased in AN [50, 51] that facilitated comprehensive investigations of the intestinal microbiota in both – disordered eating and eating disorders focused on the clinically highly relevant question, whether intestinal dysbiosis may contribute to persistence, recovery from and/or relapse of the eating disorders [50]. If yes, individual microbial profiling can predict metabolic changes in organism and direct personalised diet prescription [52–62].

4.6 Starvation Strongly Impacts the Gut Microbiome Composition and Pathophysiology of AN

Bacterial species isolated from children with kwashiorkor (an acute malnutrition linked to the protein-deficient diet which is often observed in population of developing countries) after transplantation into GF mice caused significant weight loss in the experimental mice [63]. In an acute and chronic starvation animal model of anorexia (ABA), starvation-induced changes in the gut microbiome profiles have been demonstrated [64]. These alterations suggest that intestinal barrier dysfunction provoked by starvation may act as a strong contributor to the pathophysiology of AN. Considering corresponding patho-mechanisms observed in the ABA mice losing weight: a histological investigation of the colon revealed significantly decreased the muscularis layer thickness, while permeability of the colon became increased [65]. An increase in intestinal permeability was demonstrated also as being linked to physical exercises in both AN patients and ABA model [66]. A disturbed gut barrier function has been demonstrated by several studies also for other malnourishment associated disorders as well as for fasting volunteers [67]. Finally, molecular patterns specific for acute AN demonstrates a significantly increased in pro-inflammatory cytokines, the level of which was normalised after the nutritional rehabilitation [63, 69].

4.7 Irritable Bowel Disease Is Associated with AN and Specific Microflora

A strong association between irritable bowel disease (IBD) and eating disorders including AN has been suggested [70]. IBD is highly prevalent gastrointestinal pathology reported in Western countries: approximately 11% of adults in the population are affected with strongly decreased quality of life and social function, work productivity that altogether creates a significant socio-economic burden to the society [71, 72]. Specific IBD symptoms are abdominal pain, heartburn, bloating, and significant changes in the bowel functional patterns. IBD patients frequently suffer from headache, dyspepsia, and fibromyalgia, amongst others [73]. Further, up to

50% of IBD patients with extensive gastrointestinal symptoms demonstrate pronounced psychiatric symptoms, such as anxiety, depression, somatisation; their perception of negative psychosocial stressors (e.g. unemployment, bereavement) is stronger compared to individuals without IBD history [74]. A vicious circle has been demonstrated when excessive stress conditions negatively influence the microbiota composition; in turn, altered microbiota further negatively affects the stress response and behavioural patterns with concomitant visceral pain [75].

In IBD patients with particularly severe disease course, low density of microbiota and exhaled methane as well as a reduced prevalence of Methanobacteriales and Prevotella enterotype in the micro-flora profiles have been demonstrated. Prevotella enterotype might be further suppressed by increased prevalence of Bacteroides enterotype. Noteworthy, therapies based on application of laxatives, bulking agents, acid suppressants (mainly proton pump inhibitors), anti-diarrheals, and antidepressants drugs, are not dissociated with the microbial signature relevant for the IBS severity [76, 77]. Diet adapted to individualised patient profiles and application of probiotics, are recommended as the most appropriate treatment. In conclusions, a holistic approach is essential to guide clinicians towards optimal disease management [78, 79].

References

1. Golubnitschaja O, Liskova A, Koklesova L, Samec M, Biringer K, Büsselberg D, Podbielska H, Kunin AA, Evsevyeva ME, Shapira N, Paul F, Erb C, Dietrich DE, Felbel D, Karabatsiakis A, Bubnov R, Polivka J, Polivka J Jr, Birkenbihl C, Fröhlich H, Hofmann-Apitius M, Kubatka P (2021) Caution, "normal" BMI: health risks associated with potentially masked individual underweight EPMA position paper 2021. EPMA J 12(3):1–22. https://doi.org/10.1007/s13167-021-00251-4
2. Wang W, Yan Y, Guo Z, Hou H, Garcia M, Tan X, Anto EO, Mahara G, Zheng Y, Li B, Wang Y, Guo X, Golubnitschaja O (2021) All around suboptimal health. A joint position paper of the suboptimal health study consortium and European Association for predictive, preventive and personalised medicine. EPMA J. https://doi.org/10.1007/s13167-021-00253-2
3. Lazzeri G, Rossi S, Kelly C, Vereecken C, Ahluwalia N, Giacchi MV (2014) Trends in thinness prevalence among adolescents in ten European countries and the USA (1998–2006): a cross-sectional survey. Public Health Nutr 17:2207–2215. https://doi.org/10.1017/S1368-98001-30025-41
4. Garrido-Miguel M, Cavero-Redondo I, Alvarez-Bueno C, Rodriguez-Artalejo F, Aznar LM, Ruiz JR, Martinez-Vizcaino V (2017) Prevalence and trends of thinness, overweight and obesity among children and adolescents aged 3–18 years across Europe: a protocol for a systematic review and meta-analysis. BMJ Open 7:e018241. https://doi.org/10.1136/bmjopen-2017-018241
5. Flammer syndrome: from phenotype to associated pathologies, prediction, prevention and personalisation. In: Golubnitschaja O (ed) Advances in predictive, preventive and personalised medicine. Springer. 2019. ISBN 978-3-030-13549-2
6. Strumìa R, Varotti E, Manzato E, Gualandi M (2001) Skin signs in anorexia nervosa. Dermatology 203(4):314–317. https://doi.org/10.1159/000051779

7. Pei J, Li F, Xie Y, Liu J, Yu T, Feng X (2020) Microbial and metabolomic analysis of gingival crevicular fluid in general chronic periodontitis patients: lessons for a predictive, preventive, and personalized medical approach. EPMA J 11(2):197–215. https://doi.org/10.1007/s13167-020-00202-5

8. Ma X, Wang Y, Wu H, Li F, Feng X, Xie Y, Xie D, Wang W, Lo ECM, Lu H (2021) Periodontal health related-inflammatory and metabolic profiles of patients with end-stage renal disease: potential strategy for predictive, preventive, and personalized medicine. EPMA J 12(2):117–128. https://doi.org/10.1007/s13167-021-00239-0

9. Avishai E, Yeghiazaryan K, Golubnitschaja O (2017) Impaired wound healing: facts and hypotheses for multi-professional considerations in predictive, preventive and personalised medicine. EPMA J 8(1):23–33. https://doi.org/10.1007/s13167-017-0081-y

10. Maturo MG, Soligo M, Gibson G, Manni L, Nardini C (2019) The greater inflammatory pathway-high clinical potential by innovative predictive, preventive, and personalized medical approach. EPMA J 11(1):1–16. https://doi.org/10.1007/s13167-019-00195-w

11. Gerner C, Costigliola V, Golubnitschaja O (2020) Multiomic patterns in body fluids: technological challenge with a great potential to implement the advanced paradigm of 3P medicine. Mass Spectrom Rev 39(5–6):442–451. https://doi.org/10.1002/mas.21612

12. Ichiro Kurokawa I, Layton AM, Ogawa R (2021) Updated treatment for acne: targeted therapy based on pathogenesis. Dermatol Ther (Heidelb) 11(4):1129–1139. https://doi.org/10.1007/s13555-021-00552-6

13. Dréno B, Dagnelie MA, Khammari A, Corvec S (2020) The skin microbiome: a new actor in inflammatory acne. Am J Clin Dermatol 21(Suppl 1):18–24. https://doi.org/10.1007/s40257-020-00531-1

14. Yu Y, Dunaway S, Champer J, Kim J, Alikhan A (2020) Changing our microbiome: probiotics in dermatology. Br J Dermatol 182(1):39–46. https://doi.org/10.1111/bjd.18088

15. Chen F, Esmaili S, Rogers GB, Bugianesi E, Petta S, Marchesini G, Bayoumi A, Metwally M, Azardaryany MK, Coulter S, Choo JM, Younes R, Rosso C, Liddle C, Adams LA, Craxì A, George J, Eslam M (2020) Lean NAFLD: a distinct entity shaped by differential metabolic adaptation. Hepatology 71(4):1213–1227. https://doi.org/10.1002/hep.30908

16. Saltzman ET, Talia Palacios T, Thomsen M, Vitetta L (2018) Intestinal microbiome shifts, dysbiosis, inflammation, and non-alcoholic fatty liver disease. Front Microbiol 30(9):61. https://doi.org/10.3389/fmicb.2018.00061

17. Konieczka K, Ritch R, Traverso CE, Kim DM, Kook MS, Gallino A, Golubnitschaja O, Erb C, Reitsamer HA, Kida T, Kurysheva N, Yao K (2014) Syndrome. EPMA J 5(1):11. https://doi.org/10.1186/1878-5085-5-11

18. Golubnitschaja-Labudova O, Liu R, Decker C, Zhu P, Haefliger IO, Flammer J (2000) Altered gene expression in lymphocytes of patients with normal-tension glaucoma. Curr Eye Res 21:867–876

19. Moenkemann H, Flammer J, Wunderlich K, Breipohl W, Schild HH, Golubnitschaja O (2005) Increased DNA breaks and up-regulation of both G(1) and G(2) checkpoint genes p21(WAF1/CIP1) and 14-3-3 sigma in circulating leukocytes of glaucoma patients and vasospastic individuals. Amino Acids 28(2):199–205. https://doi.org/10.1007/s00726-005-0169-x

20. Golubnitschaja O (2017) Feeling cold and other underestimated symptoms in breast cancer: anecdotes or individual profiles for advanced patient stratification? EPMA J 8(1):17–22. https://doi.org/10.1007/s13167-017-0086-6

21. Yeghiazaryan K, Flammer J, Golubnitschaja O (2010) Predictive molecular profiling in blood of healthy vasospastic individuals: clue to targeted prevention as personalised medicine to effective costs. EPMA J 1(2):263–272. https://doi.org/10.1007/s13167-010-0032-3

22. Torres Crigna A, Link B, Samec M, Giordano FA, Kubatka P, Golubnitschaja O (2021) Endothelin-1 axes in the framework of predictive, preventive and personalised (3P) medicine. EPMA J 12(3):1–41. https://doi.org/10.1007/s13167-021-00248-z

23. Golubnitschaja O (2018) The keyrole of multiomics in the predictive, preventive and personalised medical approach towards glaucoma management. Klin Monatsbl Augenheilkd 235:1–5. https://doi.org/10.1055/s-0044-101164
24. Kunin A, Polivka J Jr, Moiseeva N, Golubnitschaja O (2018) "Dry mouth" and "flammer" syndromes – neglected risks in adolescents and new concepts by predictive, preventive and personalised approach. EPMA J 9(3):307–317. https://doi.org/10.1007/s13167-018-0145-7
25. Polivka J Jr, Polivka J, Pesta M, Rohan V, Celedova L, Mahajani S, Topolcan O, Golubnitschaja O (2019) Risks associated with the stroke predisposition at young age: facts and hypotheses in light of individualized predictive and preventive approach. EPMA J 10(1):81–99. https://doi.org/10.1007/s13167-019-00162-5
26. Goncharenko V, Bubnov R, Polivka J Jr, Zubor P, Biringer K, Bielik T, Kuhn W, Golubnitschaja O (2019) Vaginal dryness: individualised patient profiles, risks and mitigating measures. EPMA J 10(1):73–79. https://doi.org/10.1007/s13167-019-00164-3
27. Kunin A, Sargheini N, Birkenbihl C, Moiseeva N, Fröhlich H, Golubnitschaja O (2020) Voice perturbations under the stress overload in young individuals: phenotyping and suboptimal health as predictors for cascading pathologies. EPMA J 11(4):1–11. https://doi.org/10.1007/s13167-020-00229-8
28. Golubnitschaja O, Flammer J (2018) Individualised patient profile: clinical utility of Flammer syndrome phenotype and general lessons for predictive, preventive and personalised medicine. EPMA J 9(1):15–20. https://doi.org/10.1007/s13167-018-0127-9
29. Golubnitschaja O, Stolzenburg-Veeser L, Avishai E, Costigliola V (2018) Wound healing: proof-of-principle model for the modern hospital – patient stratification, prediction, prevention and personalisation of treatment. In Latifi R (ed) The modern hospital: patients centered, disease based, research oriented, technology driven. Springer. ISBN 978-3-030-01393-6
30. Babak B, Golubnitschaja O (2017) The potential relationship between Flammer and Sjögren syndromes: the chime of dysfunction. EPMA J 8(4):333–338. https://doi.org/10.1007/s13167-017-0107-5
31. Bubnov R, Polivka J Jr, Zubor P, Konieczka K, Golubnitschaja O (2017) "Pre-metastatic" niches in breast cancer: are they created by or prior to the tumour onset? "Flammer syndrome" relevance to address the question. EPMA J 8(2):141–157. https://doi.org/10.1007/s13167-017-0092-8
32. Stice E (2016) Interactive and mediational etiologic models of eating disorder onset: evidence from prospective studies. Annu Rev Clin Psychol 12:359–381. https://doi.org/10.1146/annurev-clinpsy-021815-093317
33. Kaye WH, Wierenga CE, Bailer UF, Simmons AN, Bischoff-Grethe A (2013) Nothing tastes as good as skinny feels: the neurobiology of anorexia nervosa. Trends Neurosci 36(2):110–120
34. Hardaway JA, Crowley NA, Bulik CM, Kash TL (2015) Integrated circuits and molecular components for stress and feeding: implications for eating disorders. Genes Brain Behav 14(1):85–97
35. Frank GK (2013) Altered brain reward circuits in eating disorders: chicken or egg? Curr Psychiatry Rep 15(10):396. https://doi.org/10.1007/s11920-013-0396-x
36. Stice E, Gau JM, Rohde P, Shaw H (2017) Risk factors that predict future onset of each DSM-5 eating disorder: predictive specificity in high-risk adolescent females. J Abnorm Psychol 126(1):38–51. https://doi.org/10.1037/abn0000219
37. Bulik CM (2016) Towards a science of eating disorders: replacing myths with realities: the fourth Birgit Olsson lecture. Nord J Psychiatry 70(3):224–230. https://doi.org/10.3109/08039488.2015.1074284
38. Herpertz-Dahlmann B, Seitz J, Baines J (2017) Food matters: how the microbiome and gut–brain interaction might impact the development and course of anorexia nervosa. Eur Child Adolesc Psychiatry 26(9):1031–1041. https://doi.org/10.1007/s00787-017-0945-7
39. Calvani R, Picca A, Lo Monaco MR, Landi F, Bernabei R, Marzetti E (2018) Of microbes and minds: a narrative review on the second brain aging. Front Med 5:53. https://doi.org/10.3389/fmed.2018.00053

40. Goodrich JK, Davenport ER, Waters JL, Clark AG, Ley RE (2016) Cross-species comparisons of host genetic associations with the microbiome. Science 352:532–535. https://doi.org/10.1126/science.aad9379

41. Singh RK, Chang H-W, Yan D, Lee KM, Ucmak D, Wong K et al (2017) Influence of diet on the gut microbiome and implications for human health. J Transl Med 15:73. https://doi.org/10.1186/s12967-017-1175-y

42. Kang SS, Jeraldo PR, Kurti A, Miller MEB, Cook MD, Whitlock K et al (2014) Diet and exercise orthogonally alter the gut microbiome and reveal independent associations with anxiety and cognition. Mol Neurodegener 9:36. https://doi.org/10.1186/1750-1326-9-36

43. Wilson ID, Nicholson JK (2017) Gut microbiome interactions with drug metabolism, efficacy, and toxicity. Transl Res 179:204–222. https://doi.org/10.1016/j.trsl.2016.08.002

44. Clarke SF, Murphy EF, O'Sullivan O, Lucey AJ, Humphreys M, Hogan A et al (2014) Exercise and associated dietary extremes impact on gut microbial diversity. Gut 63:1913–1920. https://doi.org/10.1136/gutjnl-2013-306541

45. Clark A, Mach N (2017) The crosstalk between the gut microbiota and mitochondria during exercise. Front Physiol 8:319. https://doi.org/10.3389/fphys.2017.00319

46. Thaiss CA, Levy M, Korem T, Dohnalová L, Shapiro H, Jaitin DA et al (2016) Microbiota diurnal rhythmicity programs host transcriptome oscillations. Cell 167:1495–1510. e12. https://doi.org/10.1016/j.cell.2016.11.003

47. Bubnov RV, Babenko LP, Lazarenko LM, Mokrozub VV, Spivak MY (2018) Specific properties of probiotic strains: relevance and benefits for the host. EPMA J. https://doi.org/10.1007/s13167-018-0132-z

48. Stolzenburg-Veeser L, Golubnitschaja O (2018) Mini-encyclopaedia of the wound healing – opportunities for integrating multi-omic approaches into medical practice. J Proteome 188:71–84. https://doi.org/10.1016/j.jprot.2017.07.017

49. Kleiman SC (2014) The intestinal microbiota in acute anorexia nervosa and during renourishment: relationship to depression, anxiety, and eating disorder psychopathology; Aguirre M, Jonkers DM, Troost FJ, Roeselers G, Venema K. In vitro characterization of the impact of different substrates on metabolite production, energy extraction and composition of gut microbiota from lean and obese subjects. PLoS One 9(11):e113864. https://doi.org/10.1371/journal.pone.0113864

50. Arboleya S, Watkins C, Stanton C, Ross RP (2016) Gut bifidobacteria populations in human health and aging. Front Microbiol 7:1204. https://doi.org/10.3389/fmicb.2016.01204

51. Schwensen HF. Kan C, Treasure J, Høiby N, Sjögren M (2018) A systematic review of studies on the faecal microbiota in anorexia nervosa: future research may need to include microbiota from the small intestine. Eat Weight Disord 23(4):399–418. https://doi.org/10.1007/s40519-018-0499-9

52. Lam YY, Maguire S, Palacios T, Caterson ID (2017) Are the gut bacteria telling us to eat or not to eat? Reviewing the role of gut microbiota in the etiology, disease progression and treatment of eating disorders. Nutrients 9(6):E602. https://doi.org/10.3390/nu9060602

53. Morita C, Tsuji H, Hata T, Gondo M, Takakura S, Kawai K, Yoshihara K, Ogata K, Nomoto K, Miyazaki K. Sudo N (2015) Gut Dysbiosis in patients with anorexia nervosa. PLoS One 10(12):e0145274. https://doi.org/10.1371/journal.pone.0145274

54. Kleiman SC, Watson HJ, Bulik-Sullivan EC, Huh EY, Tarantino LM, Bulik CM, Carroll IM (2015) The intestinal microbiota in acute anorexia nervosa and during renourishment: relationship to depression, anxiety, and eating disorder psychopathology. Psychosom Med 77(9):969–981. https://doi.org/10.1097/PSY.0000000000000247

55. Armougom F, Henry M, Vialettes B, Raccah D, Raoult D (2009) Monitoring bacterial community of human gut microbiota reveals an increase in lactobacillus in obese patients and methanogens in anorexic patients. PLoS One 4:e7125

56. Zheng X, Zhou K, Zhang Y, Han X, Zhao A, Liu J, Qu C, Ge K, Huang F, Hernandez B, Yu H, Panee J, Chen T, Jia W, Jia W (2018) Food withdrawal alters the gut microbiota and metabolome in mice. FASEB J 5:fj201700614R. https://doi.org/10.1096/fj.201700614R

57. Smits LP, Kootte RS, Levin E, Prodan A, Fuentes S, Zoetendal EG, Wang Z, Levison BS, Cleophas MCP, Kemper EM, Dallinga-Thie GM, Groen AK, Joosten LAB, Netea MG, Stroes ESG, de Vos WM, Hazen SL (2018) Effect of vegan fecal microbiota transplantation on carnitine-and choline-derived trimethylamine-N-oxide production and vascular inflammation in patients with metabolic syndrome. Nieuwdorp M J Am Heart Assoc 7(7):e008342. https://doi.org/10.1161/JAHA.117.008342
58. Duarte SMB, Stefano JT, Miele L, Ponziani FR, Souza-Basqueira M, Okada LSRR, de Barros Costa FG, Toda K, Mazo DFC, Sabino EC, Carrilho FJ, Gasbarrini A, Oliveira CP (2018) Gut microbiome composition in lean patients with NASH is associated with liver damage independent of caloric intake: a prospective pilot study. Nutr Metab Cardiovasc Dis 28(4):369–384. https://doi.org/10.1016/j.numecd.2017.10.014
59. Allen JM, Mailing LJ, Niemiro GM, Moore R, Cook MD, White BA, Holscher HD, Woods JA (2018) Exercise alters gut microbiota composition and function in lean and obese humans. Med Sci Sports Exerc 50(4):747–757. https://doi.org/10.1249/MSS.0000000000001495
60. Kootte RS, Levin E, Salojärvi J, Smits LP, Hartstra AV, Udayappan SD, Hermes G et al (2017) Improvement of insulin sensitivity after lean donor feces in metabolic syndrome is driven by baseline intestinal microbiota composition. Cell Metab 26(4):611–619.e6. https://doi.org/10.1016/j.cmet.2017.09.008
61. Karakuła-Juchnowicz H, Pankowicz H, Juchnowicz D, Valverde Piedra JL, Małecka-Massalska T (2017) Intestinal microbiota – a key to understanding the pathophysiology of anorexia nervosa? Psychiatr Pol 51(5):859–870. https://doi.org/10.12740/PP/65308
62. Małecka-Massalska T, Popiołek J, Teter M, Homa-Mlak I, Dec M, Makarewicz A, Karakuła-Juchnowicz H (2017) Application of phase angle for evaluation of the nutrition status of patients with anorexia nervosa. Psychiatr Pol 51(6):1121–1131. https://doi.org/10.12740/PP/67500
63. Smith MI, Yatsunenko T, Manary MJ, Trehan I, Mkakosya R, Cheng J, Kau AL, Rich SS, Concannon P, Mychaleckyj JC, Liu J, Houpt E, Li JV, Holmes E, Nicholson J, Knights D, Ursell LK, Knight R, Gordon JI (2013) Gut microbiomes of Malawian twin pairs discordant for kwashiorkor. Science 339(6119):548–554
64. Jésus P, Ouelaa W, François M, Riachy L, Guérin C, Aziz M, Do Rego JC, Déchelotte P, Fetissov SO, Coëffier M (2014) Alteration of intestinal barrier function during activity-based anorexia in mice. Clin Nutr 33(6):1046–1053
65. Jesus P, Ouelaa W, Francois M, Riachy L, Guerin C, Aziz M, Do Rego JC, Dechelotte P, Fetissov SO, Coeffier M (2014) Alteration of intestinal barrier function during activity-based anorexia in mice. Clin Nutr 33:1046–1053. https://doi.org/10.1016/j.clnu.2013.11.006
66. Uil JJ, van Elburg RM, van Overbeek FM, Mulder CJ, VanBerge-Henegouwen GP, Heymans HS (1997) Clinical implications of the sugar absorption test: intestinal permeability test to assess mucosal barrier function. Scand J Gastroenterol Suppl 223:70–78
67. Elia M, Behrens R, Northrop C, Wraight P, Neale G (1987) Evaluation of mannitol, lactulose and 51Cr-labelled ethylenediaminetetra-acetate as markers of intestinal permeability in man. Clin Sci (Lond) 73:197–204. https://doi.org/10.1042/cs0730197
68. Solmi M, Veronese N, Favaro A, Santonastaso P, Manzato E, Sergi G, Correll CU (2015) Inflammatory cytokines and anorexia nervosa: a meta-analysis of cross-sectional and longitudinal studies. Psychoneuroendocrinology 51:237–252. https://doi.org/10.1016/j.psyneuen.2014.09.031
69. Corcos M, Guilbaud O, Paterniti S, Moussa M, Chambry J, Chaouat G, Consoli SM, Jeammet P (2003) Involvement of cytokines in eating disorders: a critical review of the human literature. Psychoneuroendocrinology 28:229–249. https://doi.org/10.1016/S0306-4530(02)00021-5
70. Ilzarbe L, Fàbrega M, Quintero R, Bastidas A, Pintor L, García-Campayo J, Gomollón F, Ilzarbe D (2017) Inflammatory bowel disease and eating disorders: a systematized review of comorbidity. J Psychosom Res 102:47–53. https://doi.org/10.1016/j.jpsychores.2017.09.006

71. Lovell RM, Ford AC (2012) Global prevalence of, and risk factors for, irritable bowel syndrome: a meta-analysis. Clin Gastroenterol Hepatol 10(7):712–721.e4. https://doi.org/10.1016/j.cgh.2012.02.029

72. Suares NC, Ford AC (2011) Prevalence of, and risk factors for, chronic idiopathic constipation in the community: systematic review and meta-analysis. Am J Gastroenterol 106:1582–1591

73. Locke GR 3rd, Zinsmeister AR, Fett SL, Melton LJ 3rd, Talley NJ (2005) Overlap of gastrointestinal symptom complexes in a US community. Neurogastroenterol Motil 17:29–34

74. Drossman DA, Whitehead WE, Toner BB, Diamant N, Hu YJ, Bangdiwala SI, Jia H (2000) What determines severity among patients with painful functional bowel disorders? Am J Gastroenterol 95:974–980

75. Moloney RD, Johnson AC, O'Mahony SM, Dinan TG, Greenwood-Van Meerveld B, Cryan JF (2016) Stress and the microbiota-gut-brain axis in visceral pain: relevance to irritable bowel syndrome. CNS Neurosci Ther 22(2):102–117. https://doi.org/10.1111/cns.12490

76. Tap J, Derrien M, Törnblom H et al (2017) Identification of an intestinal microbiota signature associated with severity of irritable bowel syndrome. Gastroenterology 152:111–123

77. Hadizadeh F, Bonfiglio F, Belheouane M et al (2018) Faecal microbiota composition associates with abdominal pain in the general population. Gut 67:778–779. https://doi.org/10.1136/gutjnl-2017-314792

78. Harper A, Naghibi MM, Garcha D (2018) The role of bacteria, probiotics and diet in irritable bowel syndrome. Foods 7(2):E13. https://doi.org/10.3390/foods7020013

79. Vicari E, Salerni M, Sidoti G, Malaguarnera M, Castiglione R (2017) Symptom severity following rifaximin and the probiotic VSL#3 in patients with chronic pelvic pain syndrome (due to inflammatory prostatitis) plus irritable bowel syndrome. Nutrients 9(11):E1208. https://doi.org/10.3390/nu9111208

Chapter 5
Microbiome and Obesity

**Tetyana Falalyeyeva, Nazarii Kobyliak, Oleksandr Korotkyi,
Tamara Meleshko, Oksana Sulaieva, Iryna Hryshchenko,
Liudmyla Domylivska, and Nadiya Boyko**

Abstract Obesity is a complex metabolic disease disturbing both children and adults. Childhood obesity is a consequence of developing this pathology and associated conditions in adulthood. The disbalance/imbalance between energy expenditure and intake is the most frequent cause of overweight. Genetic susceptibility, inflammation and environmental and/or lifestyle factors may be involved in this complicated process. Modern research suggests that the microbiome plays a significant role in the pathophysiology of obesity. As a result, the gut microbiota is gaining meaningful scientific interest concerning obesity and different associated metabolic

T. Falalyeyeva (✉)
Educational and Scientific Centre "Institute of Biology and Medicine", Taras Shevchenko National University of Kyiv, Kyiv, Ukraine

Medical Laboratory CSD, Kyiv, Ukraine

N. Kobyliak
Medical Laboratory CSD, Kyiv, Ukraine

Department of Endocrinology, Bogomolets National Medical University, Kyiv, Ukraine

O. Korotkyi · I. Hryshchenko · L. Domylivska
Educational and Scientific Centre "Institute of Biology and Medicine", Taras Shevchenko National University of Kyiv, Kyiv, Ukraine

T. Meleshko
RDE Center of Molecular Microbiology and Mucosal Immunology, Uzhhorod National University, Uzhhorod, Ukraine

N. Boyko
RDE Center of Molecular Microbiology and Mucosal Immunology, Uzhhorod National University, Uzhhorod, Ukraine

Department of Clinical Laboratory Diagnostics and Pharmacology, Uzhhorod National University, Uzhhorod, Ukraine

Ediens LLC, Uzhhorod, Ukraine

O. Sulaieva
Medical Laboratory CSD, Kyiv, Ukraine

© The Author(s), under exclusive license to Springer Nature Switzerland AG 2023
N. Boyko, O. Golubnitschaja (eds.), *Microbiome in 3P Medicine Strategies*,
Advances in Predictive, Preventive and Personalised Medicine 16,
https://doi.org/10.1007/978-3-031-19564-8_5

disorders to understand obesity's etiology better and find modern methods of its prevention and/or treatment.

This review presents personalized medicine as a modern approach to the prevention and treatment of obesity. First, we review obesity as a disease of civilization with complex health issues (unhealthy nutrition, psychosocial, behavioral, environmental, epigenetic and genetic/genomic risk factors). Second, the relationship between obesity and gut microbiota has been presented in the context of predictive, preventive, and personalized (PPPM/3P) medicine. Finally, we discuss that pre and/ or probiotic therapy is considered the promising strategy for control of metabolic disorders by virtue of microbiota composition and/or health maintenance recovery in personalized medicine and could be additional development and standardization of innovative targeted therapies and clinical tools.

Keywords Obesity · Obesity complications · Gut microbiota · Microbiome · Probiotic and prebiotic · Predictive · Preventive · Personalized · Medicine

5.1 Introduction

Obesity and overweight are considered abnormal and/or excessive fat accumulation that may harm any human health [1]. Obesity causes different severe diseases and illnesses, e.g., cardiovascular disease [2, 3], type 2 diabetes (T2D) [4], dyslipidemia [5], hepatobiliary diseases (i.e. cholelithiasis, gallbladder dyskinesia and non-alcoholic fatty liver disease, etc.) [6] and several different types of cancers [7, 8]. Worldwide obesity has nearly doubled from 1980 to 2014. That's why the World Health Organization (WHO) has announced obesity as a global epidemic that made the WHO take it under control [9].

The fundamental cause of obesity and overweight is an energy imbalance between calories consumed and calories expended. Also, cases are an increased intake of energy-dense foods high in fat and sugars and a decrease in physical inactivity due to the increasingly sedentary nature of many forms of work, changing modes of transportation, and increasing urbanization [10]. Obesity, like metabolic syndrome, has been thoroughly studied by experts from the European Association for Predictive, Preventive and Personalized Medicine (EPMA) [11]. As a result of these studies, effective predictive, preventive and personalized medicine (PPPM) solutions have been proposed, taking into account the balance between health and disease in humans; environmental factors in epidemics and interactions between the genetic component, epigenetic regulators, and environmental factors [12–14]. To promote anti-overweight measures the EU the European Commission has created a European platform with approximately 300 initiatives. The main goal is to combat with consequences of the severe physical and mental health of obese people [11].

The recent spread of severe acute respiratory syndrome coronavirus 2 (SARS-CoV-2) infection and its associated coronavirus disease (COVID-19), in which inflammation has a crucial role, has gripped the entire international community and caused widespread public health concerns [12, 15]. The WHO has declared

SARS-Cov-2 infection a global pandemic [16, 17]. Facing the coronavirus COVID-19 pandemic, it is getting evident globally that the predictive approach, targeted prevention and personalisation of medical services is the optimal paradigm in healthcare, demonstrating the high potential to save lives and to benefit society as a whole [12, 18, 19].

Despite the high incidence and prevalence of SARS-Cov-2 infection, the leading cause of death worldwide remains the diseases associated with cardiovascular events, and at the sixth place, diabetes mellitus. It should be noted that these two conditions are closely associated and both have a strict relationship with overweight and obesity [20. 21].

Diabetes mellitus (DM) is a chronic type of disease/illness that alters the metabolism of carbohydrates, fats and proteins according to the β-cells inability to secrete insulin and/or the peripheral cells failing to give a normal response to insulin [22]. Genetic susceptibility and environmental and/or lifestyle factors may be involved in this complicated process. It will profoundly impact the demand for health care, economic costs and quality of life [23]. The gut microbiota is gaining meaningful scientific interest concerning obesity and different associated metabolic disorders to understand obesity's etiology better and finds modern methods for its prevention and/or treatment [24–26].

To study the current state of scientific research on the selected topic, namely understanding the role of the gut microbiome in the development of obesity, we analyzed our own research and searched for international scientific publications. In the scientific global medical database PubMed, when entering the keywords "Microbiome and Obesity", 7954 publications were obtained from 2000 to 2022 as of July 2022. Over the past 5 years, from 2017 to 2022, the number of publications has dynamically doubled (Fig. 5.1). Eight hundred and eighteen articles were

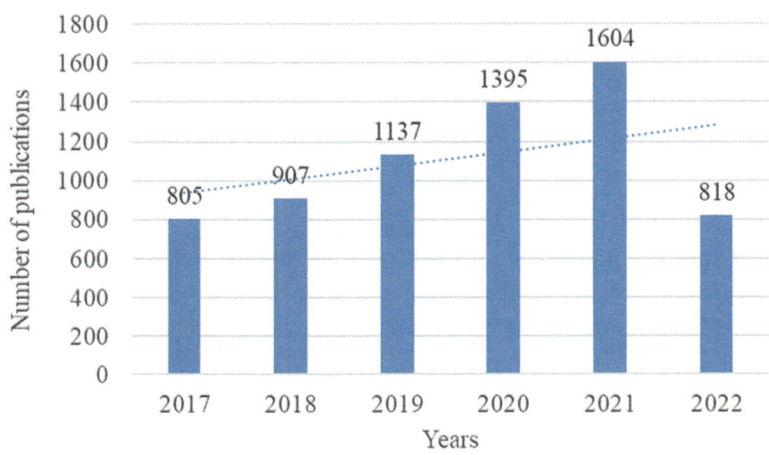

Fig. 5.1 The number of publications of the search query "Microbiome and Obesity" in the PubMed medical scientific metric database for 2017–2022

published in the last 6 months. Such a high publication activity of scientists from all over the world confirms the extreme relevance of this review. In recent years research has clearly established that gut microbiota is considered an independent organ due to its multiplexity and has a central role in alimentary, immunological and metabolism-mediated functions [27–29].

5.2   Gut Microbiota Composition in Obesity

The alteration composition of gut microbiota is one of the most fundamental aspects of obesity [27–29]. Tomasello et al. point out that the human gut microbiome is a composition of different genetic characteristics and environmental conditions [30]. Usually, adults have more than 500–1000 different bacteria species in their intestinal tract. Considering molecular phylogeny, the investigation of 16S ribosomal RNA sequences of every microorganism represented in the human intestinal tract demonstrated the presence of the following domains: archaea, bacteria and eukarya. *Bacteroidetes* and *Firmicutes* have been identified as basic phylogenetic lines when studying the microbiota of fecal bacteria [31, 32].

In most cases, the gut microbiota has *Firmicutes* (64%), including *Bacilli, Clostridia,* and *Mollicutes* (most of the microorganisms, belonging to this type are closely connected to *Clostridium* and *Streptococcus*). *Bacteroidetes* (23%) are made up of the *Proteobacteria* (8%), *Helicobacter pylori* and *Escherichia coli* are examples of Gram-negative bacteria; *Actinobacteria, Fusobacteria* and *Verrucomicrobia* (3%), which have *Bifidobacteria*. Nowadays, more than 20 genera of *Bacteroidetes* have already been studied and characterized. *Bacteroidales*, represented by *Bacteroides*, are considered to be the most analyzed. *Firmicutes* are Gram-positive bacteria with the following three groups: *Bacilli, Clostridia* and *Mollicutes* [33–35]. Compared to normal-weight mice, an augmentation in the correlation of *Firmicutes /Bacteroidetes* was supported in obese mice [22, 23]. Increasement in *Sphingomonas* and *Halomonas*, as well as declinement in *Bifidobacteria* numbers, were demonstrated in obese mice [36]. When determining the expression of intestinal microbiota by real-time PCR, a decline of *Bacteroidetes* in the obese group compared to the normal-weight group was demonstrated. On contrary, *Firmicutes* were observed to be the same for all the rest groups [37]. In a more specific study of different bacteria types in obesity was explained: in the period of obesity there was not only a decline in Bacteroides *Lactobacillus (L.)* but there were different variants of *L.*, where *L. gasseri* and *L. reuteri* notably connected with obesity, while *L. paracasei* were defined to be related to the lean position [38].

The study of cultured bacteria composition helped Zuo et al. to have demonstrated a decline in *Bacteroides*, the main of which are *Bacteroidetes* and *Clostridium perfringens* due to the obese conditions [39]. The key task in the study of microbiota was to investigate the characteristics of the impact of intestinal bacteria on host metabolism. Kobyliak et al. outlined metabolic processes mechanisms and obesity in different humans, taking into account the intestinal microbiota connection. The relationship between obesity and dysbiosis is observed. It is because of lipoprotein lipase inhibitor (also known as a diminished fasting-induced adipose factor) and key enzyme controlling cellular energy position (also known as AMP-activated protein kinase activity). It declined absorption of vitamins, disabled short-chain fatty acids (SCFAs) production, reduced inflammatory mechanisms, and reduced metabolic endotoxemia [40]. Recently, the probiotics action process on lipid metabolism and certain options for its usage as an obesity treatment are being studied from different approaches [41–43]. Taking into account different bacteria types to supplement the full composition of the gut microbiome is considered to be one of the hopeful areas for obesity treatment. The effect of different bacteria types on weight loss and other metabolic parameters has recently been investigated. The important upshot was indicated by *Bifidobacterium* and/or *Lactobacillus* strains and by the following potential bacterial candidates: *Akkermansia (A.) muciniphila, Enterobacter halii, Saccharomyces cerevisiae var. boulardii* [44].

The application of various bacteria usage significantly correlates with lipid levels and glucose; it also helps reduce body weight and chronic systemic inflammation [28]. Other papers on bacteria lyophilized probiotic strains demonstrated a significant result. The decreased visceral fat levels, leptin in adipose tissue, levels of adiponectin, and total lipid metabolism changes have been observed in obese rats that were treated with alive bacteria and/or multistrain lyophilized [45–48].

Some investigations do not show the relationships between the variety of the human major groups colonic bacteria, e.g. *Bacteroidetes*, and the body mass index (BMI). The obtained data offered that different genera or phyla had a special effect on obesity [49]. In addition, one more study on the gut microbiota ratio in infancy conducted by Kalliomäki et al. was considered a possible prerequisite for overweight. The great number of *Bifidobacteria* was demonstrated to be much higher in children with normal weight than in obese littermates. By the way, the low levels of *S. aureus* were outlined at normal weight ones [50].

5.3 Preclinical Investigations and Studies: Evidence

The results obtained from preclinical studies showed that the great effects of the probiotics that were used as a treatment for obesity and different associated diseases were received mainly while studying the strains belonging to the *Bifidobacterium (B.)* and *Lactobacillus (L.)* genera [51, 52]. *L. casei* IMV B-7280 (separately) and composition of *B. animalis* VKL/*B animalis* VKB/*L. casei* IMV B-7280 are particularly effective in reducing the weight of obese mice, lowering cholesterol levels, restoring liver morphology, and also it has been noted a positive effect on the gut microbiome in high-calorie obesity [42].

It should be noted that periodic administration of a mixture of probiotics (2:1:1 *L.* IMVB-7280, *B. animalis* VKL, *B. animalis* VKB) to rats at the dose of 5×10^9 CFU/kg (50 mg/kg) with monosodium glutamate (MSG) in newborns led to the restoration of lipid metabolism and prevention of obesity [41, 43]. The greatest effect is observed in the group of multiprobiotics, which is probably associated with the more pronounced viability of live strains and a general synergistic effect [44]. For monotherapy with lyophilized probiotics, no preventive impact on the development of obesity and non-alcoholic fatty liver disease (NAFLD) was found [18, 45–47]. Interestingly, probiotics loaded with omega-3 fatty acids are more effective in reducing liver steatosis in an animal model of obesity [15].

The study of 20 articles (2013–2014) by Cani et al. represents the following idea: about 15 different *L.* strains and 3 *B.* strains do not totally affect the metabolic parameters in animals with experimental obesity. It should be taken into account that only 10 strains out of 18 helped to reduce both the visceral adipose and tissue body weight; 12 strains led to the decline of inflammatory-necrotic alterations and 11 strains greatly influenced the liver steatosis and thyroglobulin (Tg) accumulation [53]. Different potential action characteristics possibly cause the represented differences and the outlined effects should be defined as strain-specified [54].

It is of great importance that the preclinical data showed an indication that the probiotic strains usage leads to a statistically significant reduction of body mass on various models of experimental obesity. That is why it is simultaneously connected with the following metabolic effects: glycemic control and insulin sensitivity improvement, intestinal permeability, lipopolysaccharide (LPS) translocation in the systemic cir + culation, inflammatory cytokines production, lipid profile improvement, Tg decreased accumulation in the liver parenchyma and NAFLD prevention (Table 5.1) [55].

Different probiotic strains characterizing with anti-obesity effect have both *Pediococcus pentosaceus* LP28 and lactic acid bacteria [84] and the SCFA producers, e.g. *A. muciniphila* [85, 86] and *Prevotela (P) copri* [87]. The *A. muciniphila* is considered to be mucin-degrading bacteria, which usually live and/or colonize the mucus. By the way, these bacteria are of great importance in the study of the correlation between *A. muciniphila* and different metabolic disorders [86, 88]. Scholars approved that the prebiotics (oligofructose) administration to the mice having genetically determined obesity (*ob/ob*) greatly increased the *A. muciniphila* amount by 100 times [88]. The variety of *A. muciniphila* conversely correlates with body mass in high-fat diet (HFD) obesity-induced models. Additionally, everyday administration of *A. muciniphila* to animals with diet-induced obesity (DIO) [85] in the period of about 4 weeks lead to weight reduction. It improves the proportionality of the body regardless of food consumption [85]. Moreover, *A. muciniphila* and its propionate metabolite stimulate the expression of *Gpr43, Fiaf* genes, histone deacetylases (HDACs), as well as Ppar-γ. They are essential regulators and transcription factors that help control lipolysis, cell division, satiety and insulin sensitivity [89]. Kim et al. proved that administration of *A. muciniphila* to mice with DIO leads to reducing the alanine aminotransferase (ALT) in serum and triglycerides, interleukin (IL)-6 and sterol regulatory element-binding proteins (SREBP) expression, as well as helps to normalize the composition of the gut microbiome [90]. Wu et al. investigating the same experimental obesity model showed the positive influence of *A. muciniphila* on the human body weight, glucose tolerance together with the neurodegenerative processes that are associated with existing model [91]. That

Table 5.1 Summary of animal studies on the effect of probiotic strains on obesity and associated diseases (published [55] and modified for this review)

Study	Probiotic type/obesity model	Main results
Yoo S.R. et al. [56]	*L. curvatus HY7601* *L. plantarum KY1032* **HFHCD**	↓ body mass increase ↓ hepatic lipid droplets accumulation and adipocytes size ↓ cholesterol in plasma and liver ↓ gene expression of fatty acid synthesis enzymes ↓ proinflammatory cytokines (TNF-α, IL-1β)
Park D.Y. et al. [57]	*L. curvatus HY7601* *L. plantarum KY1032* **HFD**	↓ body mass increase and fat accumulation ↓ insulin, leptin, total cholesterol and liver biomarkers ↓ proinflammatory genes (TNF-α, IL-6, IL-1β, MCP1) in adipose tissue ↓ genes associated with fatty acid oxidation (PGC1α, CPT1, CPT2 and ACOX1) in liver
Wang J. et al. [58]	*L. paracasei CNCM I-4270,* *L. rhamnosus I-3690* *Bifidobacterium/* **HFD**	↓ body mass increase ↓ infiltration of macrophages into epididymal adipose tissue ↓ liver steatosis ↑ glucose and insulin homeostasis Strain-specific attenuation of diseases associated with obesity
An H.M. et al. [59]	*B. pseudocatenulatum SPM 1204,* *B. longum SPM 1205,* *B. longum SPM 1207/***HFD**	↓ body mass increase and fat accumulation ↓ level of total cholesterol in blood serum, HDL, LDL, triglycerides, glucose, leptin ↓ biomarkers of liver toxicity (AST, ALT)
Chen J. et al. [60]	*B. adolescentis/***HFD**	↓ body mass increase and fat accumulation ↑ insulin sensitivity
Gauffin C.P. et al. [61]	*Bacteroides uniformis CECT 7771* **HFD**	↓ body mass increase, visceral fat accumulation and liver lipid content (triglycerides and cholesterol) ↓ cholesterol, triglycerides, glucose, insulin and leptin levels ↑ oral glucose tolerance ↓ absorption of fat from food (reduced number of fat micelles in enterocytes) ↑ mechanisms of immune protection
Kim S.W. et al. [62]	*L. rhamnosus GG* **HFD**	↓ weight and fat mass increase ↑ insulin sensitivity ↑ expression of genes associated with glucose metabolism (GLUT4 mRNA) ↑ production of adiponectin in adipose tissue ↑ AMPK in skeletal muscle and adipose tissue
Tabuchi M. et al. [63]	*L. rhamnosus GG* **STZ**	↓ HbA1c ↑ oral glucose tolerance

(continued)

Table 5.1 (continued)

Study	Probiotic type/obesity model	Main results
Kobyliak et al. [45, 64]	Multiprobiotic Symbiter (biomass of 14 alive strains: *Lactobacillus, Bifidobacterium, Propionibacterium, Acetobacter* genera) vs Lyophilized monostrains *L. casei* IMVB-7280, *B. animalis* VKL and VKB vs Three-strained combination **GIO**	Probiotic mixtures with preference to alive strains led to significantly lower rates of obesity, decrease in HOMA-IR, proinflammatory cytokines levels (IL-1β, IL-12Bp40) and elevation of adiponectin, TGF-β Lack of effect for monostrains
Park K.Y. et al. [65]	*L. rhamnosus GG* **ob/ob**	↑ glucose tolerance ↑ insulin-stimulated Akt phosphorylation and GLUT4 translocation ↓ endoplasmic reticulum (EP) stress of skeletal muscle ↓ M1-like activation of macrophages in white adipose tissue ↑ insulin sensitivity
Yadav H. et al. [66]	*Lactococcus lactis*/**HFrD**	↓ HbA1c ↓ fasting blood glucose, insulin, free fatty acids and triglycerides
Yadav H. et al. [67]	*L. casei/L. acidophilus* **HFrD**	↓ HbA1c, fasting insulin glucose, total cholesterol, triacylglycerol, LDL, VLDL and free blood fatty acids ↓ liver glycogen ↑ reduced glutathione in the liver and pancreas tissues
Yin Y.N. et al. [68]	*B. L66-5, B. L75-4, B. M13-4 i B. FS31-12* **HFD**	↓ liver TG and TC, (all 4 strains, most pronounced for *B. L66-5*) ↓ TG and TC in blood serum (all 4 strains, most pronounced for *B. L66-5* та *B. FS31-12*) ↓ weight gain – *B. L66-5* ↑ weight gain – *B. M13-4* No changes in body mass increase тіла *L75-4* та *FS31-12*
Reichold A. et al. [69]	*B. adolescentis* **HFD**	↓ body mass increase,liver inflammation and steatosis ↓ reactive oxygen species formation ↓ NFκB activation Does not affect the expression of LPS, TLR-4 and MyD-88 mRNA in the liver

(continued)

Table 5.1 (continued)

Study	Probiotic type/obesity model	Main results
Plaza-Diaz J. et al. [70]	*L. paracasei CNCM I-4034, B. breve CNCM I-4035* and *L. Rhamnosus CNCM* or mixture of 3 strains/ **ob/ob**	↓ liver triglyceride content (for *L. rhamnosus*, B. breve or mixtures) ↓ liver lipid content of neutral lipids (for all 4 probiotic groups) ↓ serum LPS level (for all 4 probiotic groups) ↓ the level of CNP-α in the serum (for *B. breve, L. rhamnosus* or mixture) ↓ serum IL-6 level (for L. paracasei)
Song W. et al. [71]	*L. coryniformis subsp. torquens T3, L. paracasei subsp. paracasei M5* and *L. paracasei subsp. paracasei X12/***HFD**	↓ body weight and hepatomegaly (for T3 and M5) ↓ blood TG and TC levels (for T3 and M5) ↑ antioxidant indices (MDA, SOD, glutathione peroxidase, catalase and total antioxidant capacity) – for T3 and M5
Shin J.H. et al. [72]	*"Duolak gold", containing 2 strains of lactobacilli (L. acidophilus LA1 and L. rharmnosus LR5)*, 3 bifidobacteria (*B. bifidum BF3, B. lactis BL3 and B. longum BG7*) and *Streptococcus thermophilus ST3/* **HFD**	↓ body mass ↑ *Bacteroidetes, Lactobacillus* and *Bifidobacterium* ↓ *Firmicutes* ↓ serum levels of proinflammatory cytokines and chemokines ↓ levels of fatty acids, lysophosphatidylcholine, lysophosphatidylethanolamine, phosphatidylcholine and triglycerides
Roselli M. et al. [73]	*L. rhamnosus GG, L. acidophilus LA1/K8,* mixture of *B. lactis Bi1, B. breve Bbr8 and B. breve BL10* (B. mix) mixture of *L. bulgaricus Lb2 and S. termophilus Z57/***HFD**	Only a positive effect for the mixture (B. mix) ↓ weight gain and adipose tissue accumulation ↓ adipocytes size ↓ adipose tissue infiltration by macrophages and CD4 + T lymphocytes ↓ serum lipid levels ↓ secretion of leptin and proinflammatory cytokines
Kim S. et al. [74]	*L. plantarum K10/* **HFD**	↓ weight gain, subcutaneous and visceral adipose tissue ↓ permeability of the intestinal barrier
Kim D.H. et al. [75]	*L. kefiri DH5/* **HFD**	↓ body weight and epididymal adipose tissue ↓ serum triglyceride, cholesterol and LDL levels ↓ hepatic steatosis and adipocytes diameter ↑ expression of PPAR-α, FABP4 and CPT1 in epididymal adipose tissue
Lee E. et al. [76]	*L. plantarum Ln4/* **HFD**	↓ body weight and epididymal adipose tissue ↓ total serum triglycerides level ↓ HOMA-IR and ↑ glucose tolerance

(continued)

Table 5.1 (continued)

Study	Probiotic type/obesity model	Main results
Chen Y.T. et al. [77]	*L. mali APS1/* **HFD**	↓ weight gain ↓ hepatic steatosis and the development of NAFLD by regulating the expression of SIRT-1 / PGC-1α / SREBP-1 ↑ antioxidant activity of the liver by modulating the expression of Nrf-2/HO-1 ↑ *Bacteroidetes/Firmicutes* ratio
Avolio E. et al. [78]	*Streptococcus thermophilus, L. bulgaricus,* *L. lactis subsp. lactis,* *L. acidophilus,* *L. plantarum;* *B. lactis; L. reuteri/***HFD**	↓ stress-like ↓ expression of proinflammatory factors such as IL-1β, NLRP3, caspase-1 and NF-kB in the hypothalamus ↓ levels of IL-1β, NLRP3 and caspase-1, but not the NF-kB in blood serum
Bagarolli R.A. et al. [79]	*L. rhamnosus,* *L. acidophilus,* *B. bifidumi/* **HFD**	↓ intestinal barrier permeability and LPS translocation ↓ systemic low-gradient inflammation ↑ glucose tolerance ↓ hyperphagia and hypothalamic resistance to insulin and leptin
Balakumar M. et al. [80]	*L. plantarum MTCC5690* *L. fermentum MTCC5689/* **HFD**	↑ GLP-1 level ↓ intestinal barrier permeability (occludin and ZO-1) and LPS translocation ↓ gene expression of proinflammatory cytokines (CNP-α, IL-6) in visceral fat ↑ expression of FIAF mRNA and adiponectin ↓ expression of *de novo* lipogenesis and gluconeogenesis genes in the liver
Wang G. et al. [81]	*L. casei CCFM419/* **STZ**	↓ fasting and postprandial blood glucose ↑ glucose tolerance ↑ GPP-1 level and *Bacteroidetes/Firmicutes* ratio ↓ IR and proinflammatory cytokine levels (TNF -α and IL-6) in serum
Balcells F. et al [82].	*Lactobacillus casei* CRL 431 and/ or its cell wall **BALB/c mice HFD**	↓ body weight ↑total population of enterobacteria, lactobacilli and anaerobes ↓IFN-γ, TNF-α, IL-12 ↑IL-6 Positive effect on thymus histology that reactivates the function of the thymus with the replacement of the T lymphocyte population
Mulhall H. et al. [83].	*Akkermansia muciniphila* **ob/ob**	↓TNF-α ↑IL-10 ↓periodontitis

Note: *GIO* glutamate-induced obesity, *HFD* a diet high in fat, *HFHCD* a diet high in cholesterol and fat, *HFrD* a diet high in fructose, *ob/ob* genetically determined obesity, *STZ* streptozotocin-induced diabetes

is why, the following assumption may be given: *A. muciniphila* may become the therapeutic agent for the next generation in treating both obesity and different associated diseases [86, 92].

Recent investigations and studies demonstrated that *P. copri* improves insulin sensitivity and glucose metabolism with the help of a process connected with succinate synthesis during dietary fiber fermentation [93]. Succinate is defined as the substrate. Different enterocytes use it for intestinal gluconeogenesis and, as a result, it leads to glucose homeostasis improvement [94].

De Vadder et al. proved that the diet with dietary fiber that are connected with the succinate overexpression improves insulin sensitivity and glucose tolerance [93]. Furthermore, such metabolic processes alterations can be defined in mice having DIO after colonization of *P. copri* and are offset by genetic knockout of the glucose-6-phosphatase gene (−/−). The latter is defined as a specific intestinal epithelium enzyme, which regulates gluconeogenesis rate [93, 95]. The attention should be drawn to the opposite results of the same year obtained by Pederson et al. A positive connection between serum metabolome and *P. copri* was represented in patients having insulin resistance. It is characterized by the increasement of different branched-chain amino acids. In order to experimentally confirm this relationship between glucose metabolism disorder and *P. copri*, animals with DIO were treated with the help of the placebo or a probiotic. *P. copri* strengthens glucose intolerance, helps to reduce the insulin sensitivity and to increase the branched-chain amino acids level in blood serum [96]. After gastrectomy (experimental model of spontaneously induced T2D) the administration of *P. copri* to Goto-Kakizaki mice is associated with the glucose tolerance improvement without any reference to changes in body weight through the increase level of both FXR activation and bile acids metabolism [97].

Christensenella minuta (also known as *C. minuta*) is considered to be a gram-negative bacterium of the gastrointestinal tract. It is connected with weight loss and it is represented in people having a reduced body mass index. Furthermore, the cultured bacteria *C. minuta* injection to germ-free mice is associated with the decline in weight gain [98]. Gut bacteria observed in fat mice convert the lean phenotype to associated with obesity diseases [27, 99]. Further studies showed that all skinny germ-free mice try to plump upon for receiving a fecal transplant from a human donor purporting the following idea: the bacteria is aimed at digesting, as well as metabolizing greatly [27, 99, 100]. If the fecal transplant of the human donor was added with the help *of C. minuta the recipient mice were thinner, demonstrating* an anti-obesity effect [101].

Finally, the antibacterial characteristics of probiotic yeasts were represented in the researches of Everard et al. [102]. *Saccharomyces boulardii* (Biocodex) administration to mice having genetically determined T2D (*db/db*) and DIO adipose tissue growth, reduces the body mass and greatly changes the composition of the gut microbiome with, on the one hand, *Bacteroidetes* increasement, and, on the other hand, *Firmicutes, Tenericutes* and *Proteobacteria* declinement [102].

5.4   Clinically Observed and Proved Data Material

The search for new means concerning the obesity prevention is an important problem of scientific knowledge. Gut microbiota alterations with the probiotics usage are of significant attention because it influences appetite, eating, body, and various metabolic parameters. By the way, it was demonstrated in randomized clinical trials (RCT) (Table 5.2) [55].

In the scientific works, there is research with the evaluation of probiotic influence on the children's growth and progress in the perinatal age and the decline of excess weight throughout the 10-year observation [47]. *L. rhamnosus* GG prescription for 4 weeks led to the intimidation of the excess body mass in the first years of life. The valuable changes were fixed at the age of 4 (p = 0.063) with the lack of efficiency at further stages of growth and progress [47]. The survey with 5 RCT showed the role of probiotics, most of which include the *Lactobacillus* strains, in the metabolic parameters modulation in the patients having T2D [133]. The statistically significant decline of the primary endpoints was identified in all RCT, particularly the plasma glucose on HbA1c, the empty stomach, IR, as well as insulin.

Firstly, Depommier et al. investigated the *A. muciniphila* clinical efficiency on obese patients and those who have T2D. It proved the assumption that *A. muciniphila* is considered to be the harmless strain. The intervention is tolerated in a good way and it improves the metabolic parameters [134]. It should be noted that in 3 months of consuming A. muciniphila the patients having T2D have reliable insulin resistance decline, total cholesterol, insulinemia, liver function description markers, and chronic inflammation compared with the placebo group [86, 134].

Faecalibacterium prausnitzii (also known as *F. prausnitzii*) is anaerobic. This gram-positive bacterium is among the most important and numerous in the human gut microbiota. It is also one of the key butyrate producers in the human gut with good health. Hippe et al. discovered that the *F. prausnitzii* was represented in various phylotypes in obesity conditions, T2D and patients with normal weight. The smallest number of *F. prausnitzii* was in a group of people with T2D, but the highest number was in a group of people with average weight [135]. The polyphenolic compounds epigallocatechin gallate (EGCG) and resveratrol (RES) usage has a declining effect on *F. prausnitzii* and *Bacteroidetes* level in men, while there was no effect in women. Moreover, taking into consideration EGCG + RES, the oxidation of the fatty acids demonstrated the increase among men, but not among women [136]. Da Silva et al. explained whether there is a connection between NAFLD and dysbacteriosis. She determined that *Bacteroidetes, F. prausnitzii, Firmicutes, Ruminococcus* and *Coprococcus* declined under the conditions of NAFLD compared to people with good health. The high concentrations of serum 2-hydroxybutyrate, isobutyric acid, propionate, as well as L-lactic acid were represented in the investigation of the fecal samples metabolites [137].

To understand the hypothesis whether the gut microbiota refers to age and/or obesity, Chierico et al. tried to investigate microbiota profiles in adults and adolescents with obesity; they compared it with the samples obtained from people with

Table 5.2 Summary of randomized clinical trials (RCTs) on the effects of probiotic strains and prebiotics on obesity and associated diseases (published [55] and modified for this review)

Research	Probiotic type	Research type	Main results
Kadooka Y. et al. [103]	*L. gasseri SBT2055*	Multicenter, double-blind, placebo-controlled RCT	↓ visceral (4.6%) and subcutaneous (3.3%) adipose tissue according to computer tomography (CT) ↓ body weight and BMI (by 1.5%) ↑ adiponectin in serum
Luoto R. et al. [47]	*L. rhamnosus GG*	Double-blind, promising further study	↓ weight during the first years of life – the most pronounced changes were observed at the age of 4 years (p = 0.063), with the lack of efficiency in subsequent age periods
Vrieze A. et al. [104]	Allogeneic microbiota from lean donors to recipients with metabolic syndrome (MS)	Double-blind, parallel, placebo-controlled RCT (FATLOSE study)	↑ insulin sensitivity (6 weeks after administration) ↑ intestinal microbiota that produces butyrate
Mazloom Z. et al. [105]	*L. acidophilus, L. bulgaricus, L. bifidum* and *L. casei*	Blind, parallel, placebo-controlled RCT	↓ Tg ↓ malondialdehyde (MDA) ↓ IL-6 ↓ IR All changes are not statistically significant
Malaguarnera M. et al. [106]	*B. longum* combined with fructooligosaccharides (FOS)	Open study in patients with NASH	↓ TNF-α, CRP, AST ↓ LPS in blood serum ↓ HOMA-IR ↓ steatosis and NASH activity index
Wong V.W.S. et al. [107]	Probiotic formula Lepicol	Open study in patients with histologically confirmed NASH	↓ triglycerides in the liver (IHTG) ↓ AST level Changes in BMI, waist circumference, glucose and lipid profile are not reliable
Vajro P. et al. [108]	*L. rhamnosus GG*	A double-blind, placebo-controlled pilot study of NAFLD in children	↓ ALT TNF-α and ultrasound signs of hepatic steatosis did not change significantly

(continued)

Table 5.2 (continued)

Research	Probiotic type	Research type	Main results
Mykhalchyshyn G. et al. [109]	Multiprobiotic containing concentrated biomass of 14 live probiotic bacteria	Open study in patients with NAFLD	↓ IL-6, IL-8, TNF-α, IL-1β, IFN-γ (subgroup of elevated transaminases) ↓ IL-6, IL-8, TNF-α, (subgroup of normal transaminases)
Kobyliak et al. [110]	Multiprobiotic Symbiter (biomass of 14 alive strains: *L.*, *B.*, *Propionibacterium*, *Acetobacter* genera)	Double-blind, parallel, placebo-controlled RCT in patients with NAFLD	↓ fatty liver index (FLI) ↓ AST, γ-GT ↓ IL-6, TNF-α, ↓ lipid parameters only in the intragroup analysis, in ANCOVA (intergroup) not significant
Kobyliak et al. [111]	Multiprobiotic Symbiter (biomass of 14 alive strains: *L.*, *B.*, *Propionibacterium*, *Acetobacter* genera)	Double-blind, parallel, placebo-controlled RCT in patients with obesity and T2D	↓ HOMA-IR ↓ BMI, waist circumference, weight ↓ IL-6, IL-8, TNF-α ↓ HbA1c only in responders
Kobyliak et al. [112]	Symbiter-omega (biomass of 14 alive strains: *L.*, *B*, *Propionibacterium*, *Acetobacter* genera) with 250 mg of flaxseed oil and 250 mg of wheat germ oil (concentration of omega-3 PUFA - 0.5-5)	Double-blind, parallel, placebo-controlled RCT in patients with NAFLD	↓ FLI ↓ γ-GT ↓ IL-6, IL-8, TNF-α, IL-1β and γ-INF in intragroup analysis ↓ in ANCOVA (intergroup) analysis significant only for IL-6 and TNF-α ↓ total cholesterol, triglycerides, LDL
Kobyliak et al. [32]	Symbiter-omega (biomass of 14 alive strains: *L.*, *B.*, *Propionibacterium*, *Acetobacter* genera) with 250 mg of flaxseed oil and 250 mg of wheat germ oil (concentration of omega-3 PUFA - 0.5-5)	Double-blind, parallel, placebo-controlled RCT in patients with obesity and T2D	↓ HOMA-IR ↑ insulin sensitivity (S %) ↓ HbA1c ↓ BMI, waist circumference, weight ↓ IL-6, IL-8, TNF-α, IL-1β ↓ in ANCOVA (intergroup) analysis significant only TNF-α

(continued)

Table 5.2 (continued)

Research	Probiotic type	Research type	Main results
Kobyliak et al. [113]	Symbiter-forte (biomass of 14 alive strains: *L.*, *B.*, *Propionibacterium*, *Acetobacter* genera) with 200 mg of smectite gel	Double-blind, parallel, placebo-controlled RCT in patients with NAFLD	↓ liver stiffness according to shear-wave elastography in ANCOVA (intergroup) analysis ↓ ALT and AST ↓ TNF-α and IL-1β ↓ total cholesterol
Kobyliak et al. [114]	Symbiter-Forte (biomass of 14 alive strains: *L.*, *B.*, *Propionibacterium*, *Acetobacter* genera) with 200 mg of smectite gel	Double-blind, parallel, placebo-controlled RCT in patients with obesity and T2D	↓ HOMA-IR ↑ insulin sensitivity (S %) ↓ HbA1c ↓ waist circumference ↓ IL-6, IL-8, TNF-α, IL-1β ↓ in ANCOVA (intergroup) analysis significant only TNF-α and IL-6
Shavakhi A. et al. [115]	Probiotic Protexin plus metformin 500 mg (met/pro) vs. metformin 500 mg plus placebo (met/P)	Double-blind, parallel, placebo-controlled RCT in patients with confirmed NASH histology	Met/pro improves liver aminotransferase better than metformin alone BMI, fasting blood glucose, cholesterol, and triglycerides were significantly reduced in both groups
Khalili L. et al. [116]	*L. casei*	RCT in parallel groups in patients with T2D	↓ fasting blood glucose and HbA1c ↓ insulin concentration and IR ↓ fetuin-A and ↑ SIRT1
Sabico S. et al. [117]	Multiprobiotic «Ecologic®Barrier»: *B. bifidum W23, L. lactis W19, L. acidophilus W37, L. brevis W63, L.s lactis W58, L. casei W56, B. lactis W52, L. salivarius W24*	Double-blind, placebo-controlled RCT in patients with T2D	↓ HOMA-IR, ↓ Tg, total cholesterol, LDL
Sato J. et al. [118]	*L. casei*	RCT in patients with T2D	↑ *Clostridium coccoides* ↑ *Clostridium leptum* and *L*
Mobini R. et al. [119]	*L. reuteri* DSM 17938	Double-blind, placebo-controlled RCT in patients with T2D	↑ insulin sensitivity (ISI) ↑ deoxycholic acid ↑ secondary bile acids

(continued)

Table 5.2 (continued)

Research	Probiotic type	Research type	Main results
Soleimani A. et al. [120]	*B. bifidum* and *L. casei*	Double-blind, placebo-controlled RCT in patients with T2D	↓ fasting blood glucose and HbA1c ↓ of insulinemia and HOMA-IR ↓ HOMA-B ↑ ISI ↑ total antioxidant plasma ability
Firouzi S. et al. [121]	*3 strains of the genus Lactobacillus and 3 strains of the genus Bifidobacterium*	Double-blind, placebo-controlled RCT in patients with T2D	↓ insulin on an empty stomach ↓ HbA1c
Tonucci L.B. et al. [122]	*B. animalis subsp, Lactis BB-12* and *L. acidophilus La-5*	Double-blind, placebo-controlled RCT in patients with T2D	↓ fructosamine and HbA1c ↓ TNF-α and resistin
Ejtahed H.S. et al. [123]	*B. Lactis Bb1* and *L. acidophilus La5*	Double-blind, controlled RCD in patients with T2D	↓ fasting glucose and HbA1c ↑ superoxide dismutase (SOD) of erythrocytes ↑ glutathione peroxidase ↑ antioxidant status
Pedret A. et al. [124]	*B. animalis subsp. lactis CECT 8145*	Double-blind, placebo-controlled RCT in patients with visceral obesity	↓ anthropometric data (waist, WHR, BMI) ↓ areas of visceral fat according to CT ↓ HOMA-IR ↑ population size of Akkermansia spp.
Minami J. et al. [125]	*B. breve B-3*	Double-blind, placebo-controlled RCT in overweight patients	↓ adipose tissue mass and fat percentage ↓ Tg level ↑ HDL
Kim J. et al. [126]	*L. gasseri BNR17*	Double-blind, placebo-controlled RCT in obese patients	↓ visceral adipose tissue ↓ waist
Sanchis-Chordà J. et al. [127]	*B. pseudocatenulatum CECT 7765*	Double-blind, placebo-controlled RCT in obese children and IR	↓ BMI ↓ hs-CRP and CCL2/MAP-1 ↑ HDL and omentin-1 ↑ Rikenellaceae, especially of the Alistipes genus

(continued)

Table 5.2 (continued)

Research	Probiotic type	Research type	Main results
Kaczmarczyk M. et al. [128]	Multiprobiotic: *B. bifidum* W23, *B. lactis* W51, *B. lactis* W52, *L. acidophilus*W37, *Levilactobacillus brevis* W63, *Lacticaseibacillus casei* W56, *Ligilactobacillus salivarius* W24, *Lactococcus lactis* W19 and *Lactococcus lactis* W58	Double-blind, placebo-controlled RCT in obese postmenopausal women	Probiotic intervention alters the influence of microbiota on biochemical, physiological and immunological parameters Multiprobiotic does not affect diversity and taxonomic composition
Li Y. et al. [129]	Prebiotic "Konjaku flour"	The randomized, double-blind, place-controlled trial in obese volunteers aged 25–35	↓ BMI, fat mass, P3F, Tg, HbA1c, AST, ALT ↑ the α-diversity and changed the β-diversity of intestinal microflora ↑ *Lachnospiraceae, Roseburia (R), Solobacterium, R. inulinivorans, Clostridium perfringens, Intestinimonas butyriciproducens* ↓ *Lactococcus, Bacteroides fragilis, Lactococcus garvieae, B. coprophilus, B. ovatus, B. thetaiotaomicron*
Fonvig C.E. et al. [130]	Prebiotic 2′-fucosyllactose (2′FL) and a mix of 2′FL and lacto-N-neotetraose (mix)	Double-blind, placebo-controlled RCT children with overweight (including obesity) ages 6–12 years	↑ Bifidobacteria 2′FL alone or a mix is safe and well tolerated in children
Kanazawa A. et al. [131]	Synbiotic (*Lacticaseibacillus paracasei* strain Shirota (previously *Lactobacillus casei* strain Shirota) and *Bifidobacterium breve* strain Yakult, and galactooligosaccharides)	Double-blind, placebo-controlled RCT in obese patients with T2D	↑ Total Bifidobacterium and Lactobacilli No significant changes in inflammatory markers
Rahayu ES. Et al. [132]	Probiotic *Lactobacillus plantarum* Dad-13	Double-blind, placebo-controlled RCT in overweight adults	↓ BMI No significant change in lipid profile, SCFAs (*e.g.,* butyrate, propionate, acetic acid) and pH level

(continued)

Table 5.2 (continued)

Note: *AST* aspartate aminotransferase, *BMI* body mass index, *CT* computer tomography, *CRP* C-reactive protin, *FOS* fructooligosaccharides, *HbA1c* glycated hemoglobin, *HOMA-IR* insulin resistance score, *FLI* fatty liver index, *IFN-γ* interferon-gamma, *IL* interleukin, *ISI* insulin sensitivity, *IR* insulin resistance, *MDA* malondialdehyde, *MS* metabolic syndrome, *TNF-α* tumor necrosis factor-alpha, *NASH* nonalcoholic steatohepatitis, *NAFLD* nonalcoholic fatty liver disease, *γ-GT* gamma-glutamyl transferase, *LDL* low-density lipoprotein, *PBF* percentage body fat, *PUFAs* polyunsaturated fatty acids, *Tg* triglyceride, *SIRT₁* sirtuin 1, *SOD* superoxide dismutase, *WHR* waist to hip

normal weight. It was defined that the intestinal microbiota composition differs in adults and adolescents with obesity. Where were the first group of people has high *Actinobacteria* levels, the second group of people – high *Bacteroidetes* levels. A negative correlation was found among the age, BMI, and a variety of *F. prausnitzii* [138].

As a result of a double-blind, placebo-controlled, randomized clinical trial, it was found that the administration of probiotics for 12 weeks improved markers associated with obesity in obese individuals. The intestinal microbial composition of all subjects was divided into two enterotypes according to the *Prevotella /Bacteroides* ratio. Fat, blood glucose and insulin levels were significantly increased in the *Prevotella*-rich enterotype in the placebo group. Markers that are associated with obesity, in particular waist circumference, total fat area, visceral fat, and visceral to subcutaneous fat area ratio, were significantly reduced in the probiotic group. The rate of decline in obesity-related markers was much greater in the *Prevotella*-rich enterotype than in the *Bacteroides*-rich enterotype [74].

It is of great interest that the published two analyses with the assessment of probiotic influence on the glycemic control in patients having T2D have shown the disputable results concerning HOMA-IR and HbA1c [139, 140]. Kasinska's et al. analysis on the of the data from 8 RCT (n = 438) inclusion shows that there is a statistically considerable reduction of HOMA-IR (SMD -2.10; from −3.00 to −1.20, p < 0.001; p = 0.0029 as of heterogeneity) and HbA1c (SMD -0.81; CI from −1.33 to −0.29, p = 0.0023; p = 0.0421 as of heterogeneity) after probiotic therapy course [139]. Li et al. tried to analyze data of 9 RCT (n = 368) and they did not find the possible discrepancies concerning HOMA-IR and HbA1c between control groups and intervention [140]. Conversely, Ruan Y. et al. included 17 RCT (n = 635) in analyses among the overweight groups of people, pregnancy and T2D [141]. Consumption of probiotics in comparison with the placebo are statistically and significantly lower in terms of HOMA-IR (MD = 0.48; 95% CI 0.83, 0.13; p = 0.007).

Some RCT have demonstrated hopeful effects of probiotics concerning the improvement of IR, the liver functional status and metabolism of fats in the patients having NAFLD. The first RCT on probiotics in patients having the chronic liver disease was published by the Loguercio C. group in 2005 [142]. The patients having NAFLD (n = 22), alcoholic steatohepatitis (n = 2), chronic viral hepatitis C (n = 20) and cryptogenic cirrhosis (n = 16) were involved into the research. The patients having NAFLD, taking the probiotic VLS #3 during 3 months had a significant decline of both ALT activity and lipid peroxidation markers while the proinflammatory cytokines represented in the blood serum (TNF-α, IL-6) did not alter in comparison with the patients with alcoholic hepatitis [142].

The 4 RCT involved 134 patients with NAFLD /NASH and were generalized in the analysis [143]. These results demonstrated that the probiotic therapy declines total cholesterol, the aminotransferases activity and TNF-α levels compared with IR improvement. The probiotics used did not refer to the significant alterations of BMI, LDL and fasting glycemia [143].

Various studies represented the diet with yogurt, including the following probiotic strains: B. longum, L. acidophilus, Enterococcus faecium, L. plantcrum Streptococcus thermophilus and/or B. Lactis leads to the decline of the total cholesterol level, LDL in serum and LDL: HDL ratio improvement [144, 145]. The short-term RCT analysis proves that fermented milk yogurt can be enriched with probiotic strains that reduce 4% total cholesterol and/or 5% LDL cholesterol [146].

Studies have shown that in obese patients with T2DM, the use of a multi-probiotic (biomass of 14 live probiotic strains L. + Lactococcus, B., Propionibacterium, Acetobacter) for 8 weeks is associated with a significant decrease in HOMA-IR, body weight, BMI, waist circumference and cytokine content (TNF-α, IL-1β, IL-6). Notably, in an intergroup analysis using ANCOVA (analysis of covariance), previous results were only confirmed for waist circumference and serum levels of TNF-α and IL-1β. Moreover, the parameters of glycemic control, as the main secondary endpoints of the study, were not statistically significant in intra- or intergroup analysis. On the other hand, a sub-analysis was performed in which changes in HbA1c were analyzed in each group depending on the patient's response to treatment. The effect was positive in patients with a decrease in the HOMA-IR index. A significant decrease in the HbA1c level by 0.39% (p = 0.022) was found only in respondents from the probiotic therapy group [99].

New data support the perspective of further trials to assess the potential role of probiotics in preventing viral upper respiratory tract infections (and possibly COVID-19), particularly in overweight/obese people. It was published that probiotics reduce self-reported upper respiratory tract infection symptoms in overweight and obese adults [147]. This fack could improve the severity of the course of Covid-19 in obese patients.

So we conclude that obesity is associated with a lower abundance of A. muciniphila and increasing Bacteroides in the gut microbiota than in healthy individuals. This condition leads to dysbacteriosis of the intestinal microbiota, which is accompanied by a disruption of the intestinal barrier, which, in turn, promotes the release of bacterial LPS into the bloodstream (i.e., metabolic endotoxemia). At the same time, A. muciniphila slows down the process of lipid overload, which is associated with the LDL receptor, and at the same time increases visceral adiposity and impaired insulin sensitivity in muscles and liver compared with a healthy individual. The production of SCFAs, which is stimulated by A. muciniphila, participates in host signaling by inhibiting histone deacetylase (HDAC) or by activating G protein-coupled receptors that trigger other metabolic pathways that collectively result in immune stimulation, including macrophage transmigration and changes in Treg proliferation. reduce the level of inflammatory cytokines in the blood serum of

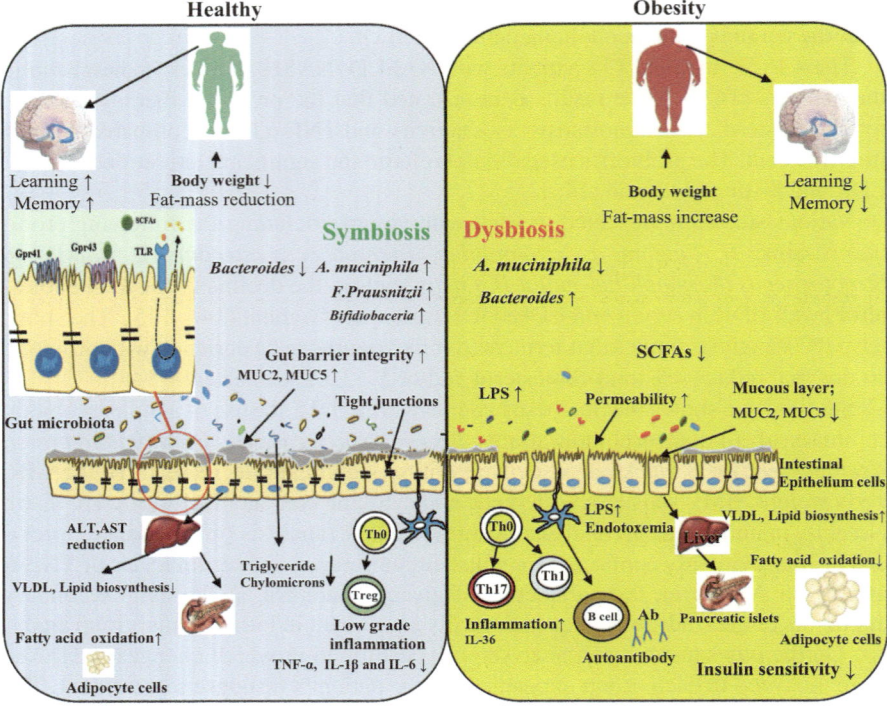

Fig. 5.2 Relationship between obesity and gut microbiota. (Published [86] and modified for this review)
TG total triglyceride, *ALT* alanine aminotransferase, *AST* aspartate aminotransferase, *LPS* lipopolysaccharide, *TLR* Toll-like receptors, *TNF* tumor necrosis factor, *IL-6* interleukin 6, *IL-1β* interleukin 1β. *Gpr43; Gpr41* G-protein-coupled receptor; glucagon-like peptide 1 (GLP-1) to increase insulin sensitivity. *SCFAs* short-chain fatty acids, *Th0* T helper cell naive, *Th2* Type 2 helper T cell, *Th1* Type 1 helper T cell, *Th17* Type 17 helper t cell, *Treg* regulatory T cell, *DC* Dendritic cells

healthy individuals. Engagement Gpr43; and gpr41 receptors by SCFAs have been shown to trigger the incretion hormone GLP-1 to increase insulin sensitivity. *A. muciniphila* coordinates the intestinal permeability and gut barrier via tight-junction proteins and improves intestinal barrier integrity (Fig. 5.2) [86].

5.5 Discussion/Conclusion/Obstacles and Limitation of Wide Implementation

The effectiveness of probiotics/prebiotics therapy has been proven in metabolic disorders (insulin resistance, hyperglycemia, type 2 diabetes, obesity, hyperlipidemia, hypertension, nonalcoholic fatty liver disease, and metabolic syndrome). On the one hand, the therapy is relatively safe, with the minimum amount of side effects, appropriate and well-tolerated for long-term usage. On the other hand, it may affect the whole body mass, fat metabolism and glucose itself, improve insulin sensitivity and decrease systematicity in chronic inflammation. This therapy is hoped to be a promising strategy for control of metabolic disorders by virtue of microbiota composition and/or health maintenance recovery with the help of influence on the above processes and mechanisms [55].

Accordingly, many scientific papers focus on the beneficial properties of probiotics in the process of regulating metabolism, while at the same time, some scientific papers question the effectiveness and safety of probiotics. In turn, postbiotics are preparations of inanimate microorganisms and/or their components, which are directly identified with the safety of their use and the health benefits of the host. Also, postbiotics, unlike probiotics, are not limited to such harsh production/storage conditions. These components are known to include peptidoglycan-derived muropeptides, exopolysaccharides, teichoic acids and surface-protruding molecules such as fimbriae, pili or flagella constituting cell wall components, secreted proteins/peptides, bacteriocins such as acidophilus, reuterin and bifidin, cell-free supernatant, organic acids such as lactic acid and acetic acid, vitamins, short chain fatty acids such as butyric acid and propionate, neurotransmitters, biosurfactants, etc. [148, 149]. Because of chemical structure, storage stability of postbiotics it is found that postbiotics have many health benefits, in particular, they have a local effect on certain tissues of the intestinal epithelium, and also influence on many other organs and tissues. It is postbiotics that create the appearance of a therapeutic effect of probiotics, which, in turn, limits the risk of introducing living microorganisms into a weakened immune defense. It should also be pointed out that postbiotics are more stable and also have a longer shelf life [150].

Recent scientific studies on the stated issues point to the benefits of postbiotics in the treatment of metabolic disorders. In this case, the so-called potential effects of postbiotics are taken into account, namely: anti-inflammatory, antibacterial, immunomodulatory, anticarcinogenic, antioxidant, antihypertensive, antiproliferative and hypocholesterolemic properties, intestinal barrier functions [150].

PPPM strategies in the field of world health care for the preservation of national health. Through the interaction of the gut microbiota and host homeostasis, gut bacteria are believed to play a significant role in various diseases, including metabolic syndrome, obesity and related disorders. Recent data indicate that the gut microbiota is involved in controlling body weight, energy homeostasis, and inflammation and, accordingly, plays an important role in the pathophysiology of obesity. The human gut microbiota has a different community structure. Significantly, probiotics alter the composition of the gut microbiota and influence food intake and appetite, body weight, composition and metabolic functions through the gastrointestinal tract and modulation of the gut bacterial community. The studies presented in this chapter have shown that the administration of probiotics improved markers associated with obesity in preclinical [30, 58–60] and clinical studies [111, 114, 115]. The most commonly used probiotics, *Lactobacillus* and *Bifidobacterium*, have different effects depending on dose, duration of treatment and route of administration. For example, it has been demonstrated that enrichment of probiotics with other representatives of functional foods, such as nutraceuticals, can enhance or summarize the individual effects of a particular composition, which may be more beneficial for the treatment or prevention of various metabolic disorders [28, 32, 114]. The development of human-related probiotic strains and intelligent low-dose physiological therapies to correct the gut microbiome is an important factor in personalized dietetics and a challenge for future medicine [14]. Considering the high biological activity and safety of postbiotics, it should be concluded that such a treatment vector will be promising in the nearest future.

5.6 Expert Recommendations in the Framework of 3PM to the Practical Medicine

PPPM/3P manages and integrates these data to apply personalized, preventive, and therapeutic approaches. This is significant because there is an emerging need to develop effective and safe means to prevent and treat obesity. Restoration of the qualitative and quantitative composition of gut microbiota could create a window opportunity for PPPM of chronic diseases of civilization, first of all, obesity:

- Prebiotics, probiotics and synbiotics could be one of the therapeutic strategies for obesity treatment these days.
- Fecal microbiota transplantation (FMT) from lean to patients with obesity should be thoroughly examined in the future to develop predictive diagnostics methods and personalized treatment in obesity.
- Postbiotics is a new way to therapy of metabolic disease probably would probably be more perspective than probiotics in the future.
- Most researchers try to focus on genomics but seldom on investigating the metabolome, proteome or transcriptome. Taking into account the genomic level, the deep shotgun sequencing is quite expensive, representing marker-based amplicon with the prevailing 16SrRNA gene. By the way, the existing sequencing and analysis technologies rarely indicate the microbes at species and/or strain levels. Whereas the functional capacity varies between strains of the same species, microbes and microbial genes identification connected with the disease is challenging.
- New approaches, e.g. Transkingdom Network Analysis [151], as well as application of the methods and techniques of Mendelian Randomization (MR) [152] have been developed. They helped to answer the following question: which microbes and microbial pathways/genes are under control concerning host physiological processes [153]. While applying MR methods, Sanna and his colleagues tried to define whether different bacteria types are grouped due to the certain functions they usually perform in the gut and whether there is such a causal effect on several metabolic characteristics [154]. All in all, it was concluded that the increase of host-genetic-driven gut production of SCFA butyrate was connected with improved insulin reaction (response) after an oral glucose tolerance test

$(P = 9.8 \times 10-5)$. On the other hand, abnormalities in the absorption or even production of one more SCFA were related to a T2D $(P = 0.004)$ increased risk. The data provide experience in the causal effect of the gut microbiome on metabolic traits. It also tries to support the MR usage as a means to explain causal relationships from different microbiome-wide association findings.

Legalislation Issues Data/Results/Biomarkers /Other approaches recommended as a Preliminary Protocols for the Prevention and Predictive Personalised Patients treatment and Relevance to the Existent Medical Protocols (Country, National, etc.)

No declare.

Founding No declare.

Other Acknowledgments No declare.

References

1. Jensen CD, Sato AF, Jelalian E (2020) Obesity. In: Encyclopedia of behavioral medicine. Springer, Cham, pp 1523–1524
2. Jiang J, Ahn J, Huang WY, Hayes RB (2013) Association of obesity with cardiovascular disease mortality in the PLCO trial. Prev Med (Baltim) 57:60–64. https://doi.org/10.1016/j.ypmed.2013.04.014
3. Fromentin S, Forslund SK, Chechi K et al (2022) Microbiome and metabolome features of the cardiometabolic disease spectrum. Nat Med 28:303–314. https://doi.org/10.1038/s41591-022-01688-4
4. Phillips CM (2013) Metabolically healthy obesity: definitions, determinants and clinical implications. Rev Endocr Metab Disord 14:219–227. https://doi.org/10.1007/s11154-013-9252-x
5. Jung DH, Kim JY, Kim JK et al (2014) Relative contribution of obesity and serum adiponectin to the development of hypertension. Diabetes Res Clin Pract 103:51–56. https://doi.org/10.1016/j.diabres.2013.09.018
6. Abenavoli L, Falalyeyeva T, Boccuto L et al (2018) Obeticholic acid: a new era in the treatment of nonalcoholic fatty liver disease. Pharmaceuticals 11:104. https://doi.org/10.3390/ph11040104
7. Eslami M, Sadrifar S, Karbalaei M et al (2020) Importance of the microbiota inhibitory mechanism on the Warburg effect in colorectal cancer cells. J Gastrointest Cancer 51:738–747. https://doi.org/10.1007/s12029-019-00329-3
8. Karimi K, Lindgren TH, Koch CA, Brodell RT (2016) Obesity as a risk factor for malignant melanoma and non-melanoma skin cancer. Rev Endocr Metab Disord 17:389–403
9. Kobyliak N, Virchenko O, Falalyeyeva T et al (2017) Cerium dioxide nanoparticles possess anti-inflammatory properties in the conditions of the obesity-associated NAFLD in rats. Biomed Pharmacother 90:608–614. https://doi.org/10.1016/j.biopha.2017.03.099
10. WHO I World Health Organization. https://www.who.int/. Accessed 11 Dec 2021
11. Golubnitschaja O, Liskova A, Koklesova L et al (2021) Caution, "normal" BMI: health risks associated with potentially masked individual underweight—EPMA position paper 2021. EPMA J 12:243–264. https://doi.org/10.1007/s13167-021-00251-4
12. Liskova A, Koklesova L, Samec M et al (2021) Targeting phytoprotection in the COVID-19-induced lung damage and associated systemic effects—the evidence-based 3PM proposition to mitigate individual risks. EPMA J 12:325–347. https://doi.org/10.1007/s13167-021-00249-y

13. Beregova TV, Neporada KS, Skrypnyk M et al (2017) Efficacy of nanoceria for periodontal tissues alteration in glutamate-induced obese rats-multidisciplinary considerations for personalized dentistry and prevention. EPMA J:8. https://doi.org/10.1007/s13167-017-0085-7
14. Bubnov RV, Spivak MY, Lazarenko LM et al (2015) Probiotics and immunity: provisional role for personalized diets and disease prevention. EPMA J 6:14. https://doi.org/10.1186/s13167-015-0036-0
15. Ather A, Patel B, Ruparel NB et al (2020) Coronavirus disease 19 (COVID-19): implications for clinical dental care. J Endod 46:584–595. https://doi.org/10.1016/j.joen.2020.03.008
16. Ludwig S, Zarbock A (2020) Coronaviruses and SARS-CoV-2: a brief overview. Anesth Analg 131:93–96. https://doi.org/10.1213/ANE.0000000000004845
17. Falalyeyeva T, Komisarenko I, Yanchyshyn A et al (2021) Vitamin D in the prevention and treatment of type-2 diabetes and associated diseases: a critical view during COVID-19 time. Minerva Biotechnol Biomol Res 33:65–75. https://doi.org/10.23736/S2724-542X.21.02766-X
18. Golubnitschaja O, Topolcan O, Kucera R et al (2020) 10th anniversary of the European Association for Predictive, Preventive and Personalised (3P) Medicine – EPMA world congress supplement 2020. EPMA J 11:1–133. https://doi.org/10.1007/s13167-020-00206-1
19. Koklesova L, Mazurakova A, Samec M et al (2022) Mitochondrial health quality control: measurements and interpretation in the framework of predictive, preventive, and personalized medicine. EPMA J 13:177–193. https://doi.org/10.1007/s13167-022-00281-6
20. Chawla R, Madhu S, Makkar B et al (2020) RSSDI-ESI clinical practice recommendations for the management of type 2 diabetes mellitus 2020. Indian J Endocrinol Metab 24:1. https://doi.org/10.4103/ijem.ijem_225_20
21. Falalyeyeva T, Pellicano R (2021) Hot topics in chronic inflammation-induced diseases at the time of COVID-19. Minerva Biotechnol Biomol Res 33:2. https://doi.org/10.23736/S2724-542X.21.02803-0
22. Al-Goblan AS, Al-Alfi MA, Khan MZ (2014) Mechanism linking diabetes mellitus and obesity. Diab Metab Syndr Obes Targets Ther 7:587–591. https://doi.org/10.2147/DMSO.S67400
23. Eslami M, Bahar A, Hemati M et al (2020) Dietary pattern, colonic microbiota and immunometabolism interaction: new frontiers for diabetes mellitus and related disorders. Diabet Med:e14415. https://doi.org/10.1111/dme.14415
24. Kobyliak N, Falalyeyeva T, Bodnar P, Beregova T (2017) Probiotics supplemented with omega-3 fatty acids are more effective for hepatic steatosis reduction in an animal model of obesity. Probiotics Antimicrob Proteins 9:123–130. https://doi.org/10.1007/s12602-016-9230-1
25. Kobyliak N, Abenavoli L, Falalyeyeva T, Beregova T (2018) Efficacy of probiotics and smectite in rats with non-alcoholic fatty liver disease. Ann Hepatol 17:153–161. https://doi.org/10.5604/01.3001.0010.7547
26. Hijová E (2022) Synbiotic supplements in the prevention of obesity and obesity-related diseases. Meta 12:313. https://doi.org/10.3390/metabo12040313
27. Castaner O, Goday A, Park Y-M et al (2018) The gut microbiome profile in obesity: a systematic review. Int J Endocrinol 2018:1–9. https://doi.org/10.1155/2018/4095789
28. Kobyliak N, Falalyeyeva T, Boyko N et al (2018) Probiotics and nutraceuticals as a new frontier in obesity prevention and management. Diabetes Res Clin Pract 141. https://doi.org/10.1016/j.diabres.2018.05.005
29. Bubnov R, Babenko L, Lazarenko L et al (2019) Can tailored nanoceria act as a prebiotic? Report on improved lipid profile and gut microbiota in obese mice. EPMA J 10:317–335. https://doi.org/10.1007/s13167-019-00190-1
30. Tomasello G, Mazzola M, Jurjus A et al (2017) The fingerprint of the human gastrointestinal tract microbiota: a hypothesis of molecular mapping. J Biol Regul Homeost Agents 31:245–249
31. Nash AK, Auchtung TA, Wong MC et al (2017) The gut mycobiome of the human microbiome project healthy cohort. Microbiome 5:153. https://doi.org/10.1186/s40168-017-0373-4

32. Kobyliak N, Falalyeyeva T, Mykhalchyshyn G et al (2020) Probiotic and omega-3 poly-unsaturated fatty acids supplementation reduces insulin resistance, improves glycemia and obesity parameters in individuals with type 2 diabetes: a randomised controlled trial. Obes Med 19:100248. https://doi.org/10.1016/j.obmed.2020.100248

33. Sommer F, Anderson JM, Bharti R et al (2017) The resilience of the intestinal microbiota influences health and disease. Nat Rev Microbiol 15:630–638. https://doi.org/10.1038/nrmicro.2017.58

34. Corfield AP (2018) The interaction of the gut microbiota with the mucus barrier in health and disease in human. Microorganisms 6:78. https://doi.org/10.3390/microorganisms6030078

35. Gagliardi A, Totino V, Cacciotti F et al (2018) Rebuilding the gut microbiota ecosystem. Int J Environ Res Public Health 15:1679. https://doi.org/10.3390/ijerph15081679

36. Waldram A, Holmes E, Wang Y et al (2009) Top-down systems biology modeling of host metabotype–microbiome associations in obese rodents. J Proteome Res 8:2361–2375. https://doi.org/10.1021/pr8009885

37. Armougom F, Henry M, Vialettes B et al (2009) Monitoring bacterial community of human gut microbiota reveals an increase in lactobacillus in obese patients and methanogens in anorexic patients. PLoS One 4:e7125. https://doi.org/10.1371/journal.pone.0007125

38. Million M, Maraninchi M, Henry M et al (2012) Obesity-associated gut microbiota is enriched in Lactobacillus reuteri and depleted in Bifidobacterium animalis and Methanobrevibacter smithii. Int J Obes 36:817–825. https://doi.org/10.1038/ijo.2011.153

39. Zuo HJ, Xie ZM, Zhang WW et al (2011) Gut bacteria alteration in obese people and its relationship with gene polymorphism. World J Gastroenterol 17:1076–1081. https://doi.org/10.3748/wjg.v17.i8.1076

40. Kyriachenko Y, Falalyeyeva T, Korotkyi O et al (2019) Crosstalk between gut microbiota and antidiabetic drug action. World J Diabetes 10:154–168. https://doi.org/10.4239/wjd.v10.i3.154

41. Lee J, Cho YK, Kang YM et al (2019) The impact of NAFLD and waist circumference changes on diabetes development in prediabetes subjects. Sci Rep 9:1–8. https://doi.org/10.1038/s41598-019-53947-z

42. Luoto R, Laitinen K, Nermes M, Isolauri E (2012) Impact of maternal probiotic-supplemented dietary counseling during pregnancy on colostrum adiponectin concentration: a prospective, randomized, placebo-controlled study. Early Hum Dev 88:339–344. https://doi.org/10.1016/j.earlhumdev.2011.09.006

43. Kobyliak N, Abenavoli L, Falalyeyeva T et al (2020) Metabolic benefits of probiotic combination with absorbent smectite in type 2 diabetes patients: a randomised controlled trial. Rev Recent Clin Trials 15:1–10. https://doi.org/10.2174/1574887115666200709141131

44. Cani PD, de Vos WM (2017) Next-generation beneficial microbes: the case of Akkermansia muciniphila. Front Microbiol 8:1765. https://doi.org/10.3389/fmicb.2017.01765

45. Kobyliak N, Falalyeyeva T, Beregova T, Spivak M (2017) Probiotics for experimental obesity prevention: focus on strain dependence and viability of composition. Endokrynol Pol 68:659–667. https://doi.org/10.5603/EP.a2017.0055

46. Cerdó T, García-Santos JG, Bermúdez M, Campoy C (2019) The role of probiotics and prebiotics in the prevention and treatment of obesity. Nutrients 11:635. https://doi.org/10.3390/nu11030635

47. Luoto R, Kalliomäki M, Laitinen K, Isolauri E (2010) The impact of perinatal probiotic intervention on the development of overweight and obesity: follow-up study from birth to 10 years. Int J Obes 34:1531–1537. https://doi.org/10.1038/ijo.2010.50

48. Kobyliak N, Falalyeyeva TM, Virchenko OV et al (2016) Probiotic use in obesity and metabolic syndrome. In: Probiotics in children. Nova Science, New York, pp 191–233

49. Duncan SH, Lobley GE, Holtrop G et al (2008) Human colonic microbiota associated with diet, obesity and weight loss. Int J Obes 32:1720–1724. https://doi.org/10.1038/ijo.2008.155

50. Kalliomäki M, Carmen Collado M, Salminen S, Isolauri E (2008) Early differences in fecal microbiota composition in children may predict overweight. Am J Clin Nutr 87:534–538. https://doi.org/10.1093/ajcn/87.3.534

51. Eslami M, Bahar A, Hemati M et al (2021) Dietary pattern, colonic microbiota and immuno-metabolism interaction: new frontiers for diabetes mellitus and related disorders. Diabet Med 38:e1441. https://doi.org/10.1111/dme.14415

52. Savcheniuk OA, Virchenko OV, Falalyeyeva TM et al (2014) The efficacy of probiotics for monosodium glutamate-induced obesity: dietology concerns and opportunities for prevention. EPMA J:5. https://doi.org/10.1186/1878-5085-5-2

53. Cani PD, Van Hul M (2015) Novel opportunities for next-generation probiotics targeting metabolic syndrome. Curr Opin Biotechnol 32:21–27

54. Kobyliak N, Conte C, Cammarota G et al (2016) Probiotics in prevention and treatment of obesity: a critical view. Nutr Metab 13:14. https://doi.org/10.1186/s12986-016-0067-0

55. Falalyeyeva T, Mamula Y, Scarpellini E et al (2021) Probiotics and obesity associated disease: an extended view beyond traditional strains. Minerva Gastroenterol 67. https://doi.org/10.23736/S2724-5985.21.02909-0

56. Yoo SR, Kim YJ, Park DY et al (2013) Probiotics L. plantarum and L. curvatus in combination alter hepatic lipid metabolism and suppress diet-induced obesity. Obesity 21:2571–2578. https://doi.org/10.1002/oby.20428

57. Park DY, Ahn YT, Park SH et al (2013) Supplementation of lactobacillus curvatus KY1032 in diet-induced obese mice is associated with gut microbial changes and reduction in obesity. PLoS One 8:e59470. https://doi.org/10.1371/journal.pone.0059470

58. Wang J, Tang H, Zhang C et al (2015) Modulation of gut microbiota during probiotic-mediated attenuation of metabolic syndrome in high fat diet-fed mice. ISME J 9:1–15. https://doi.org/10.1038/ismej.2014.99

59. An HM, Park SY, Lee DK et al (2011) Antiobesity and lipid-lowering effects of Bifidobacterium spp. in high fat diet-induced obese rats. Lipids Health Dis 10:116. https://doi.org/10.1186/1476-511X-10-116

60. Chen J, Wang R, Li XF, Wang RL (2012) Bifidobacterium adolescentis supplementation ameliorates visceral fat accumulation and insulin sensitivity in an experimental model of the metabolic syndrome. Br J Nutr 107:1429–1434. https://doi.org/10.1017/S0007114511004491

61. Gauffin Cano P, Santacruz A, Moya Á, Sanz Y (2012) Bacteroides uniformis CECT 7771 ameliorates metabolic and immunological dysfunction in mice with high-fat-diet induced obesity. PLoS One 7:e41079. https://doi.org/10.1371/journal.pone.0041079

62. Kim SW, Park KY, Kim B et al (2013) Lactobacillus rhamnosus GG improves insulin sensitivity and reduces adiposity in high-fat diet-fed mice through enhancement of adiponectin production. Biochem Biophys Res Commun 431:258–263. https://doi.org/10.1016/j.bbrc.2012.12.121

63. Tabuchi M, Ozaki M, Tamura A et al (2003) Antidiabetic effect of lactobacillus GG in streptozotocin-induced diabetic rats. Biosci Biotechnol Biochem 67:1421–1424. https://doi.org/10.1271/bbb.67.1421

64. Kobyliak N, Falalyeyeva T, Tsyryuk O et al (2020) New insights on strain-specific impacts of probiotics on insulin resistance: evidence from animal study. J Diabetes Metab Disord 19:289–296. https://doi.org/10.1007/s40200-020-00506-3

65. Park KY, Kim B, Hyun CK (2015) Lactobacillus rhamnosus GG improves glucose tolerance through alleviating ER stress and suppressing macrophage activation in db/db mice. J Clin Biochem Nutr 56:240–246. https://doi.org/10.3164/jcbn.14-116

66. Yadav H, Jain S, Sinha PR (2006) Effect of dahi containing Lactococcus lactis on the progression of diabetes induced by a high-fructose diet in rats. Biosci Biotechnol Biochem 70:1255–1258. https://doi.org/10.1271/bbb.70.1255

67. Yadav H, Jain S, Sinha PR (2007) Antidiabetic effect of probiotic dahi containing Lactobacillus acidophilus and Lactobacillus casei in high fructose fed rats. Nutrition 23:62–68. https://doi.org/10.1016/j.nut.2006.09.002

68. Yin YN, Yu QF, Fu N et al (2010) Effects of four Bifidobacteria on obesity in high-fat diet induced rats. World J Gastroenterol 16:3394–3401. https://doi.org/10.3748/wjg.v16.i27.3394

69. Reichold A, Brenner SA, Spruss A et al (2014) Bifidobacterium adolescentis protects from the development of nonalcoholic steatohepatitis in a mouse model. J Nutr Biochem 25:118–125. https://doi.org/10.1016/j.jnutbio.2013.09.011
70. Plaza-Diaz J, Gomez-Llorente C, Abadia-Molina F et al (2014) Effects of Lactobacillus paracasei CNCM I-4034, Bifidobacterium breve CNCM I-4035 and Lactobacillus rhamnosus CNCM I-4036 on hepatic steatosis in Zucker rats. PLoS One 9:e98401. https //doi.org/10.1371/journal.pone.0098401
71. Song W, Song C, Shan Y et al (2016) The antioxidative effects of three lactobacilli on high-fat diet induced obese mice. RSC Adv 6:65808–65815. https://doi.org/10.1039/c6ra06389f
72. Shin JH, Nam MH, Lee H et al (2018) Amelioration of obesity-related characteristics by a probiotic formulation in a high-fat diet-induced obese rat model. Eur J Nutr 57:2081–2090. https://doi.org/10.1007/s00394-017-1481-4
73. Roselli M, Finamore A, Brasili E et al (2018) Beneficial effects of a selected probiotic mixture administered to high fat-fed mice before and after the development of obesity. J Funct Foods 45:321–329. https://doi.org/10.1016/j.jff.2018.03.039
74. Kim S, Huarg E, Park S et al (2018) Physiological characteristics and anti-obesity effect of lactobacillus plantarum K10. Korean J Food Sci Anim Resour 38:554–569. https://doi.org/10.5851/kosfa.2018.38.3.554
75. Kim DH, Jeong D, Kang IB et al (2017) Dual function of Lactobacillus kefiri DH5 in preventing high-fat-diet-induced obesity: direct reduction of cholesterol and upregulation of PPAR-α in adipose tissue. Mol Nutr Food Res 61:1700252. https://doi.org/10.1002/mnfr.201700252
76. Lee E, Jung SR, Lee SY et al (2018) Lactobacillus plantarum strain ln4 attenuates diet-induced obesity, insulin resistance, and changes in hepatic mRNA levels associated with glucose and lipid metabolism. Nutrients 10:643. https://doi.org/10.3390/nu10050643
77. Chen YT, Lin YC, Lin JS et al (2018) Sugary kefir strain Lactobacillus mali APS1 ameliorated hepatic steatosis by regulation of SIRT-1/Nrf-2 and gut microbiota in rats. Mol Nutr Food Res 62:e1700903. https://doi.org/10.1002/mnfr.201700903
78. Avolio E, Fazzari G, Zizza M et al (2019) Probiotics modify body weight together with anxiety states via pro-inflammatory factors in HFD-treated Syrian golden hamster. Behav Brain Res 356:390–399. https://doi.org/10.1016/j.bbr.2018.09.010
79. Bagarolli RA, Tobar N, Oliveira AG et al (2017) Probiotics modulate gut microbiota and improve insulin sensitivity in DIO mice. J Nutr Biochem 50:16–25. https://doi.org/10.1016/j.jnutbio.2017.08.006
80. Balakumar M, Prabhu D, Sathishkumar C et al (2018) Improvement in glucose tolerance and insulin sensitivity by probiotic strains of Indian gut origin in high-fat diet-fed C57BL/6J mice. Eur J Nutr 57:279–295. https://doi.org/10.1007/s00394-016-1317-7
81. Wang G, Li X, Zhao J et al (2017) Lactobacillus casei CCFM419 attenuates type 2 diabetes via a gut microbiota dependent mechanism. Food Funct 8:3155–3164. https://doi.org/10.1039/c7fo00593h
82. Balcells F, Martínez Monteros MJ, Gómez AL et al (2022) Probiotic consumption boosts thymus in obesity and senescence mouse models. Nutrients 14:616. https://doi.org/10.3390/nu14030616
83. Mulhall H, DiChiara JM, Huck O, Amar S (2022) Pasteurized Akkermansia muciniphila reduces periodontal and systemic inflammation induced by Porphyromonas gingivalis in lean and obese mice. J Clin Periodontol 49:717–729. https://doi.org/10.1111/jcpe.13629
84. Zhao X, Higashikawa F, Noda M et al (2012) The obesity and fatty liver are reduced by plant-derived Pediococcus pentosaceus LP28 in high fat diet-induced obese mice. PLoS One 7:e30696. https://doi.org/10.1371/journal.pone.0030696
85. Everard A, Belzer C, Geurts L et al (2013) Cross-talk between Akkermansia muciniphila and intestinal epithelium controls diet-induced obesity. Proc Natl Acad Sci U S A 110:9066–9071. https://doi.org/10.1073/pnas.1219451110
86. Kobyliak N, Falalyeyeva T, Kyriachenko Y et al (2022) Akkermansia muciniphila as a novel powerful bacterial player in the treatment of metabolic disorders. Minerva Endocrinol. https://doi.org/10.23736/S2724-6507.22.03752-6

87. Abenavoli L, Falalyeyeva T, Pellicano R et al (2020) Next-generation of strain specific probiotics in diabetes treatment: the case of Prevotella copri. Minerva Endocrinol 45:277–279. https://doi.org/10.23736/S0391-1977.20.03376-3

88. Everard A, Lazarevic V, Derrien M et al (2011) Responses of gut microbiota and glucose and lipid metabolism to prebiotics in genetic obese and diet-induced leptin-resistant mice. Diabetes 60:2775–2786. https://doi.org/10.2337/db11-0227

89. Lukovac S, Belzer C, Pellis L et al (2014) Differential modulation by Akkermansia muciniphila and faecalibacterium prausnitzii of host peripheral lipid metabolism and histone acetylation in mouse gut organoids. MBio 5:e01438–e01414. https://doi.org/10.1128/mBio.01438-14

90. Kim S, Lee Y, Kim Y et al (2020) Akkermansia muciniphila prevents fatty liver disease, decreases serum triglycerides, and maintains gut homeostasis. Appl Environ Microbiol 86:e03004–e03019. https://doi.org/10.1128/AEM.03004-19

91. Wu F, Guo X, Zhang M et al (2020) An Akkermansia muciniphila subtype alleviates high-fat diet-induced metabolic disorders and inhibits the neurodegenerative process in mice. Anaerobe 61:102138. https://doi.org/10.1016/j.anaerobe.2019.102138

92. Xu Y, Wang N, Tan H-Y et al (2020) Function of Akkermansia muciniphila in obesity: interactions with lipid metabolism, immune response and gut systems. Front Microbiol 11. https://doi.org/10.3389/fmicb.2020.00219

93. De Vadder F, Kovatcheva-Datchary P, Zitoun C et al (2016) Microbiota-produced succinate improves glucose homeostasis via intestinal gluconeogenesis. Cell Metab 24:151–157. https://doi.org/10.1016/j.cmet.2016.06.013

94. De Vadder F, Kovatcheva-Datchary P, Goncalves D et al (2014) Microbiota-generated metabolites promote metabolic benefits via gut-brain neural circuits. Cell 156:84–96. https://doi.org/10.1016/j.cell.2013.12.016

95. Cani PD (2018) Human gut microbiome: hopes, threats and promises. Gut 67:1716–1725. https://doi.org/10.1136/gutjnl-2018-316723

96. Pedersen HK, Gudmundsdottir V, Nielsen HB et al (2016) Human gut microbes impact host serum metabolome and insulin sensitivity. Nature 535:376–381. https://doi.org/10.1038/nature18646

97. Péan N, Le Lay A, Brial F et al (2020) Dominant gut Prevotella copri in gastrectomised non-obese diabetic Goto–Kakizaki rats improves glucose homeostasis through enhanced FXR signalling. Diabetologia 63:1223–1235. https://doi.org/10.1007/s00125-020-05122-7

98. Goodrich JK, Waters JL, Poole AC et al (2014) Human genetics shape the gut microbiome. Cell 159:789–799. https://doi.org/10.1016/j.cell.2014.09.053

99. Braga RM, Dourado MN, Araújo WL (2016) Microbial interactions: ecology in a molecular perspective. Braz J Microbiol 47:86–98. https://doi.org/10.1016/j.bjm.2016.10.005

100. Festi D, Schiumerini R, Eusebi LH et al (2014) Gut microbiota and metabolic syndrome. World J Gastroenterol 20:16079. https://doi.org/10.3748/wjg.v20.i43.16079

101. Shivaji S (2017) We are not alone: a case for the human microbiome in extra intestinal diseases. Gut Pathog 9:13. https://doi.org/10.1186/s13099-017-0163-3

102. Everard A, Matamoros S, Geurts L et al (2014) Saccharomyces boulardii administration changes gut microbiota and reduces hepatic steatosis, low-grade inflammation, and fat mass in obese and type 2 diabetic db/db mice. MBio 5:e01011–e01014. https://doi.org/10.1128/mBio.01011-14

103. Kadooka Y, Sato M, Imaizumi K et al (2010) Regulation of abdominal adiposity by probiotics (Lactobacillus gasseri SBT2055) in adults with obese tendencies in a randomized controlled trial. Eur J Clin Nutr 64:636–643. https://doi.org/10.1038/ejcn.2010.19

104. Vrieze A, Van Nood E, Holleman F et al (2012) Transfer of intestinal microbiota from lean donors increases insulin sensitivity in individuals with metabolic syndrome. Gastroenterology 143:913–916. https://doi.org/10.1053/j.gastro.2012.06.031

105. Mazloom Z, Yousefinejad A, Dabbaghmanesh MH (2013) Effect of probiotics on lipid profile, glycemic control, insulin action, oxidative stress, and inflammatory markers in patients with type 2 diabetes: a clinical trial. Iran J Med Sci 38:38–43

106. Malaguarnera M, Vacante M, Antic T et al (2012) Bifidobacterium longum with fructo-oligosaccharides in patients with non alcoholic steatohepatitis. Dig Dis Sci 57:545–553. https://doi.org/10.1007/s10620-011-1887-4
107. Wong VWS, Wong GLH, Chim AML et al (2013) Treatment of nonalcoholic steatohepatitis with probiotics. A proof-of-concept study. Ann Hepatol 12:256–262. https://doi.org/10.1016/s1665-2681(19)31364-x
108. Vajro P, Mandato C, Licenziati MR et al (2011) Effects of Lactobacillus rhamnosus strain gg in pediatric obesity-related liver disease. J Pediatr Gastroenterol Nutr 52:740–743. https://doi.org/10.1097/MPG.0b013e31821f9b85
109. Mykhal'chyshyn HP, Bodnar PM, Kobyliak NM (2013) Effect of probiotics on proinflammatory cytokines level in patients with type 2 diabetes and nonalcoholic fatty liver disease. Lik Sprava 2 56–62
110. Kobyliak N, Abenavoli L, Mykhalchyshyn G et al (2018) A multi-strain probiotic reduces the fatty liver index, cytokines and aminotransferase levels in NAFLD patients: evidence from a randomized clinical trial. J Gastrointest Liver Dis 27:41–49. https://doi.org/10.15403/jgld.2014.1121.271.kby
111. Kobyliak N, Falalyeyeva T, Mykhalchyshyn G et al (2018) Effect of alive probiotic on insulin resistance in type 2 diabetes patients: randomized clinical trial. Diabetes Metab Syndr Clin Res Rev 12. https://doi.org/10.1016/j.dsx.2018.04.015
112. Kobyliak N, Abenavoli L, Falalyeyeva T et al (2018) Beneficial effects of probiotic combination with omega-3 fatty acids in NAFLD: a randomized clinical study. Minerva Med 109:418–428. https://doi.org/10.23736/S0026-4806.18.05845-7
113. Kobyliak N, Abenavoli L, Mykhalchyshyn G et al (2019) Probiotics and smectite absorbent gel formulation reduce liver stiffness, transaminase and cytokine levels in NAFLD associated with type 2 diabetes: a randomized clinical study. Clin Diabetol 8:205–214. https://doi.org/10.5603/dk.2019.0016
114. Kobyliak N, Abenavoli L, Falalyeyeva T et al (2021) Metabolic benefits of probiotic combination with absorbent smectite in type 2 diabetes patients a randomised controlled trial. Rev Recent Clin Trials 16:109–119. https://doi.org/10.2174/1574887115666200709141131
115. Shavakhi A, Minakari M, Firouzian H et al (2013) Effect of a probiotic and metformin on liver aminotransferases in non-alcoholic steatohepatitis: a double blind randomized clinical trial. Int J Prev Med 4:531–537
116. Khalili L, Alipour B, Jafar-Abadi MA et al (2019) The effects of Lactobacillus casei on glycemic response, serum Sirtuin1 and Fetuin – a levels in patients with type 2 diabetes mellitus: a randomized controlled trial. Iran Biomed J 23:68–77. https://doi.org/10.29252/.23.1.68
117. Sabico S, Al-Mashharawi A, Al-Daghri NM et al (2017) Effects of a multi-strain probiotic supplement for 12 weeks in circulating endotoxin levels and cardiometabolic profiles of medication naïve T2DM patients: a randomized clinical trial. J Transl Med 15:249. https://doi.org/10.1186/s12967-017-1354-x
118. Sato J, Kanazawa A, Azuma K et al (2017) Probiotic reduces bacterial translocation in type 2 diabetes mellitus: a randomised controlled study. Sci Rep 7:1–10. https://doi.org/10.1038/s41598-017-12535-9
119. Mobini R, Tremaroli V, Ståhlman M et al (2017) Metabolic effects of Lactobacillus reuteri DSM 17938 in people with type 2 diabetes: a randomized controlled trial. Diabetes Obes Metab 19:579–589. https://doi.org/10.1111/dom.12861
120. Soleimani A, Zarrati Mojarrad M, Bahmani F et al (2017) Probiotic supplementation in diabetic hemodialysis patients has beneficial metabolic effects. Kidney Int 91:435–442. https://doi.org/10.1016/j.kint.2016.09.040
121. Firouzi S, Majid HA, Ismail A et al (2017) Effect of multi-strain probiotics (multi-strain microbial cell preparation) on glycemic control and other diabetes-related outcomes in people with type 2 diabetes: a randomized controlled trial. Eur J Nutr 56:1535–1550. https://doi.org/10.1007/s00394-016-1199-8

122. Tonucci LB, Olbrich dos Santos KM, Licursi de Oliveira L et al (2017) Clinical application of probiotics in type 2 diabetes mellitus: a randomized, double-blind, placebo-controlled study. Clin Nutr 36:85–92. https://doi.org/10.1016/j.clnu.2015.11.011

123. Ejtahed HS, Mohtadi-Nia J, Homayouni-Rad A et al (2012) Probiotic yogurt improves antioxidant status in type 2 diabetic patients. Nutrition 28:539–543. https://doi.org/10.1016/j.nut.2011.08.013

124. Pedret A, Valls RM, Calderón-Pérez L et al (2019) Effects of daily consumption of the probiotic Bifidobacterium animalis subsp. lactis CECT 8145 on anthropometric adiposity biomarkers in abdominally obese subjects: a randomized controlled trial. Int J Obes 43:1863–1868. https://doi.org/10.1038/s41366-018-0220-0

125. Minami J, Iwabuchi N, Tanaka M et al (2018) Effects of Bifidobacterium breve B-3 on body fat reductions in pre-obese adults: a randomized, double-blind, placebo-controlled trial. Biosci Microbiota Food Health 37:67–75. https://doi.org/10.12938/bmfh.18-001

126. Kim J, Yun JM, Kim MK et al (2018) Lactobacillus gasseri BNR17 supplementation reduces the visceral fat accumulation and waist circumference in obese adults: a randomized, double-blind, placebo-controlled trial. J Med Food 21:454–461. https://doi.org/10.1089/jmf.2017.3937

127. Sanchis-Chordà J, del Pulgar EMG, Carrasco-Luna J et al (2019) Bifidobacterium pseudocatenulatum CECT 7765 supplementation improves inflammatory status in insulin-resistant obese children. Eur J Nutr 58:2789–2800. https://doi.org/10.1007/s00394-018-1828-5

128. Kaczmarczyk M, Szulińska M, Łoniewski I et al (2022) Treatment with multi-species probiotics changes the functions, not the composition of gut microbiota in postmenopausal women with obesity: a randomized, double-blind, placebo-controlled study. Front Cell Infect Microbiol 12. https://doi.org/10.3389/fcimb.2022.815798

129. Li Y, Kang Y, Du Y et al (2022) Effects of Konjaku flour on the gut microbiota of obese patients. Front Cell Infect Microbiol 12. https://doi.org/10.3389/fcimb.2022.771748

130. Fonvig CE, Amundsen ID, Vigsnæs LK et al (2021) Human milk oligosaccharides modulate fecal microbiota and are safe for use in children with overweight: a randomized controlled trial. J Pediatr Gastroenterol Nutr 73:408–414. https://doi.org/10.1097/MPG.0000000000003205

131. Kanazawa A, Aida M, Yoshida Y et al (2021) Effects of synbiotic supplementation on chronic inflammation and the gut microbiota in obese patients with type 2 diabetes mellitus: a randomized controlled study. Nutrients 13:558. https://doi.org/10.3390/nu13020558

132. Rahayu ES (2021) Correction to "Effect of probiotic Lactobacillus plantarum Dad-13 powder consumption on the gut microbiota and intestinal health of overweight adults". World J Gastroenterol 2021; 27(1):107–128 [PMID: 33505154 DOI: 10.3748/wjg.v27.i1.107]. World J Gastroenterol 27:6511–6512. https://doi.org/10.3748/wjg.v27.i38.6511

133. Razmpoosh E, Javadi M, Ejtahed HS, Mirmiran P (2016) Probiotics as beneficial agents in the management of diabetes mellitus: a systematic review. Diabetes Metab Res Rev 32:143–168. https://doi.org/10.1002/dmrr.2665

134. Depommier C, Everard A, Druart C et al (2019) Supplementation with Akkermansia muciniphila in overweight and obese human volunteers: a proof-of-concept exploratory study. Nat Med 25:1096–1103. https://doi.org/10.1038/s41591-019-0495-2

135. Hippe B, Remely M, Aumueller E et al (2016) Faecalibacterium prausnitzii phylotypes in type two diabetic, obese, and lean control subjects. Benef Microbes 7:511–517. https://doi.org/10.3920/BM2015.0075

136. Most J, Penders J, Lucchesi M et al (2017) Gut microbiota composition in relation to the metabolic response to 12-week combined polyphenol supplementation in overweight men and women. Eur J Clin Nutr 71:1040–1045. https://doi.org/10.1038/ejcn.2017.89

137. Da Silva HE, Teterina A, Comelli EM et al (2018) Nonalcoholic fatty liver disease is associated with dysbiosis independent of body mass index and insulin resistance. Sci Rep 8:1466. https://doi.org/10.1038/s41598-018-19753-9

138. Del Chierico F, Abbatini F, Russo A et al (2018) Gut microbiota markers in obese adolescent and adult patients: age-dependent differential patterns. Front Microbiol 9. https://doi. org/10.3389/fmicb.2018.01210

139. Kasińska MA, Drzewoski J (2015) Effectiveness of probiotics in type 2 diabetes: a meta-analysis. Pol Arch Med Wewn 125:803–813. https://doi.org/10.20452/pamw.3156

140. Li C, Li X, Han H et al (2016) Effect of probiotics on metabolic profiles in type 2 diabetes mellitus. Medicine (Baltimore) 95:e4088. https://doi.org/10.1097/MD.0000000000004088

141. Ruan Y, Sun J, He J et al (2015) Effect of probiotics on glycemic control: a systematic review and meta-analysis of randomized, controlled trials. PLoS One 10:e0132121. https://doi. org/10.1371/journal.pone.0132121

142. Loguercio C, Federico A, Tuccillo C et al (2005) Beneficial effects of a probiotic VSL#3 on parameters of liver dysfunction in chronic liver diseases. J Clin Gastroenterol 39:540–543. https://doi.org/10.1097/01.mcg.0000165671.25272.0f

143. Ma YY, Li L, Yu CH et al (2013) Effects of probiotics on nonalcoholic fatty liver disease: a meta-analysis. World J Gastroenterol 19:6911–6918. https://doi.org/10.3748/wjg.v19. i40.6911

144. Kießling G, Schneider J, Jahreis G (2002) Long-term consumption of fermented dairy products over 6 months increases HDL cholesterol. Eur J Clin Nutr 56:843–849. https://doi. org/10.1038/sj.ejcn.1601399

145. Xiao JZ, Kondo S, Takahashi N et al (2003) Effects of milk products fermented by Bifidobacterium longum on blood lipids in rats and healthy adult male volunteers. J Dairy Sci 86:2452–2461. https://doi.org/10.3168/jds.S0022-0302(03)73839-9

146. Agerholm-Larsen L, Bell ML, Grunwald GK, Astrup A (2000) The effect of a probiotic milk product on plasma cholesterol: a meta-analysis of short-term intervention studies. Eur J Clin Nutr 54:856–860. https://doi.org/10.1038/sj.ejcn.1601104

147. Mullish BH, Marchesi JR, McDonald JAK et al (2021) Probiotics reduce self-reported symptoms of upper respiratory tract infection in overweight and obese adults: should we be considering probiotics during viral pandemics? Gut Microbes 13. https://doi.org/10.1080/1949097 6.2021.1900997

148. Nataraj BH, Ali SA, Behare PV, Yadav H (2020) Postbiotics-parabiotics: the new horizons in microbial biotherapy and functional foods. Microb Cell Factories 19:168. https://doi. org/10.1186/s12934-020-01426-w

149. Malashree L, Angadi V, Yadav KS, Prabha R (2019) "Postbiotics" – one step ahead of probiotics. Int J Curr Microbiol Appl Sci 8:2049–2053. https://doi.org/10.20546/ijcmas.2019.801 214

150. Bourebaba Y, Marycz K, Mularczyk M, Bourebaba L (2022) Postbiotics as potential new therapeutic agents for metabolic disorders management. Biomed Pharmacother 153:113138. https://doi.org/10.1016/j.biopha.2022.113138

151. Mathur R, Amichai M, Chua KS et al (2013) Methane and hydrogen positivity on breath test is associated with greater body mass index and body fat. J Clin Endocrinol Metab 98:E698–E702. https://doi.org/10.1210/jc.2012-3144

152. García-Mantrana I, Selma-Royo M, González S et al (2020) Distinct maternal microbiota clusters are associated with diet during pregnancy: impact on neonatal microbiota and infant growth during the first 18 months of life. Gut Microbes 11:962–978. https://doi.org/10.108 0/19490976.2020.1730294

153. Gurung M, Li Z, You H et al (2020) Role of gut microbiota in type 2 diabetes pathophysiology. EBioMedicine 51:102590. https://doi.org/10.1016/j.ebiom.2019.11.051

154. Sanna S, van Zuydam NR, Mahajan A et al (2019) Causal relationships among the gut microbiome, short-chain fatty acids and metabolic diseases. Nat Genet 51:600–605. https://doi. org/10.1038/s41588-019-0350-x

Chapter 6
Pathophysiology-Based Individualized Use of Probiotics and Prebiotics for Metabolic Syndrome: Implementing Predictive, Preventive, and Personalized Medical Approach

Rostyslav Bubnov and Mykola Spivak

Abstract The modification the gut microbiota in metabolic syndrome and associated chronic diseases is among leading tasks of microbiome research and needs for clinical use of probiotics. Evidence lack for the implications for microbiome modification to improve metabolic health in particular when applied impersonalized. Probiotics have tremendous potential in personalized nutrition and medicine to develop healthy diets. **The aim** was to conduct comprehensive overview of recent updates of role of microbiota on human health and development of metabolic syndrome and efficacy of microbiota modulation considering specific properties of probiotic strain and particular aspects of metabolic syndrome and patient's phenotype to fill the gap between probiotic product and individual to facilitate development of individualized/personalized probiotic and prebiotic treatments. We discuss the relevance of using host phenotype-associated biomarkers, those based on imaging and molecular and patient's history, reliable and accessible to facilitate person-specific appication of probiotics and prebiotic substances. Microbiome phenotypes can be parameters of predictive medicine to recognize patient's predispositions and evaluate treatment responses; the number of phenotype markers can be effectively involved to monitor microbiome modulation. The studied strain-dependent properties of probiotic strains are potentially relevant for individualized treatment for gut

R. Bubnov (✉)
Zabolotny Institute of Microbiology and Virology, National Academy of Sciences of Ukraine, Kyiv, Ukraine

Clinical Hospital 'Pheophania' of State Affairs Department, Kyiv, Ukraine

M. Spivak
Zabolotny Institute of Microbiology and Virology, National Academy of Sciences of Ukraine, Kyiv, Ukraine

PJSC «SPC Diaproph-Med», Kyiv, Ukraine

© The Author(s), under exclusive license to Springer Nature Switzerland AG 2023
N. Boyko, O. Golubnitschaja (eds.), *Microbiome in 3P Medicine Strategies*,
Advances in Predictive, Preventive and Personalised Medicine 16,
https://doi.org/10.1007/978-3-031-19564-8_6

and distant sites microbiome modulation. The evidence regarding probiotic strains properties can be taken to account via pathophysiology-based approach for most effective individualized treatment via gut, oral and vaginal and other sites microbiome modulation according to phenotype of the patient providing individualized and personalized medical approaches. Preventive potential of probiotics is strong and well-documented. Recommendations for individualized clinical use of probiotics, and for probiotic studies design have been suggested.

Keywords Predictive preventive personalized medicine · Lactobacillus · Bifidobacterium · Probiotics, gut microbiota · Patient phenotype, predictive preventive personalised medicine · Metabolic syndrome

Abbreviations

MetS	metabolic syndrome
BMI	body mass index
WC	waist circumference
LAB	lactic acid bacteria
FED	fat-enriched diet
FRD	fructose-rich diet
HDL	high density lipoprotein
IL	interleukin
LPS	lipopolysaccharide
PEMs	peritoneal exudate macrophages
SCFA	short-chain fatty acid
TNF-α	tumor necrosis factor-α
DM	diabetes mellitus
T2DM	type 2 diabetes mellitus
FBG	fasting blood glucose
ACE	angiotensin converting enzyme
ROS	reactive oxygen species
COX	cyclooxygenase
ADCF	adipose-derived contracting factor
ED	endothelial dysfunction
NO	nitric oxide
EDN	endothelin
WAT	white adipose tissue
BAT	brown adipose tissue
PVAT	perivascular adipose tissue
CT	computed tomography
MRI	magnetic resonance imaging
US	ultrasound
SIBO	small intestinal bacterial overgrowth

NAFLD   non-alcoholic fatty liver disease
CKD   chronic kidney disease

6.1 Introduction

6.1.1 Microbiota and Metabolic Syndrome: Strains Stratification for Effective Personalized Probiotic Interventions

Metabolic syndrome (MetS) is a violation of metabolism including the development of obesity, liver disease, hypertension, dyslipidemia, hyperglycemia and insulin resistance and still is a large global challenge [1–3].

The **diagnosis of** "MetS" can be made if at least three of the following five criteria [2] are met:

- obesity with **abdominal fat distribution**, determined by an abdominal circumference of over 102 cm in men or over 88 cm in women;
- dyslipidemia (increasing Serum triglycerides greater than 150 mg/dL (>1.7 mmol/L);
- high density lipoprotein (HDL) cholesterol $\leq$ 40 mg/dL;
- hypertension of 130/85 mmHg or more;
- and **fasting blood sugar $\geq$ 110 mg/dL (5.6 mmol/L), or** *type 2 diabetes mellitus* (T2DM)**.**

MetS is a condition of alteration of metabolism of lipids, carbohydrates, insulin, and associated with development of inflammatory reactions. Obesity in adults and children is a global epidemic, is often associated with hyperglycemia, hypertriglyceridemia, dyslipidemia and hypertension and is considered as the main risk factor for cardiovascular diseases (CVD). WHO has predicted that CVD to remain the leading cause of death, and by 2030 [2]. The developing and continuous updating a panel of biomarkers of the MetS for diagnosis and prediction of metabolic diseases, prevention and personalized treatment is an urgent task. The development and continuous updating of MetS biomarkers is an urgent task for the diagnosis and prognosis of metabolic diseases, prevention and individual treatment. The importance of the prognostic and diagnostic value of total cholesterol and its fractions is widely demonstrated by experimental and clinical studies [4] as the main risk factor for coronary heart disease. Today, cholesterol administration requires statin therapy at a growing target level for low-density lipoprotein (LDL) -cholesterol levels of 4.9 mmol/L in patients with atherosclerotic cardiovascular disease [3].

The gut microbiota is considered an extension of the self and, together with the genetic makeup, determines the physiology of an organism, metabolism and digestion. Intestinal microbial population largely represented by Bacteroidetes and Firmicutes, has been proven to impact on human health and maintaining homeostasis [5–11]. The gut microbiota has been recognized as an important contributor to pathological conditions such as obesity and metabolic disorders.

Numerous findings on Mets and obesity support evidence for manipulation of the gut microbiota as treatment of obesity and associated health complications, both as a standalone therapy and as part of interventions such as weight loss. Modification the gut microbiota in chronic diseases and metabolic syndrome is among leading tasks of microbiome research and needs for clinical use of probiotics [4, 12–18].

The **aim** was to to conduct comprehensive overview of the recent updates of role of microbiota on human health and development of metabolic syndrome and efficacy of microbiota modulation considering specific properties of strain and particular aspects of metabolic syndrome and patient's phenotype to fill the gap between probiotic product and individual to facilitate development of individualized/personalized probiotic and prebiotic treatments.

6.1.2   Probiotics and Prebiotics

The definition of a *probiotic* as *"live microorganisms which when administered in adequate amounts confer a health benefit on the host"* defined by Food and Agriculture Organization of the United Nations (FAO) and the World Health Organization (WHO) in 2001 [19]; and was confirmed in 2014 by International Scientific Association for Probiotics and Prebiotics (ISAPP) experts [20] and later remain unchanged being agreed in the broad expert communities.

The studied strains meet such important selection criteria as antibiotic resistance according to international guidelines for probiotics like the FAO and WHO [2] and European Food Safety Authority (EFSA) [21, 22].

There is a large promising potential of using probiotics to develop healthy diets and integrated approach for immunity-related diseases treatment and prevention; are effective actors in the gut and in distant sites [9] with strong potential for applications in personalized medicine and nutrition [23–25].

Thus, the current ISAPP consensus panel now proposes the following definition of a *prebiotic*: **a substrate that is selectively utilized by host microorganisms conferring a health benefit** [26].

However, *evidence-supported knowledge* on probiotics contribution to disease pathophysiology and applicability to clinical care is *not yet sufficient*, excluding very few aspects. Thus, in cases of antibiotic- and *Clostridium difficile*-associated diarrhea, and respiratory tract infections, the effects of probiotics are considered *"evidence-based"* [27–29].

Evidence based probiotic treatment was summarized by Wilkins et al. according to the recent Cochrane and systematic reviews it was established as follows [29]:

• Probiotic use reduces the risk of antibiotic-associated diarrhea in children and adults (level of evidence A);

- Probiotic use may reduce the incidence of Clostridium difficile–associated diarrhea (level of evidence B);
- Probiotics can significantly reduce the risk of hepatic encephalopathy, however, the evidence is insufficient in respect to the effect on nonalcoholic fatty liver disease (NAFLD) and nonalcoholic steatohepatitis (level of evidence B);
- Probiotic use increases remission rates in adults with ulcerative colitis (level of evidence A);
- Probiotics can alleviate abdominal pain in children and adults with irritable bowel syndrome (level of evidence B).

Evidence supporting probiotic interventions efficacy has not been completed yet in respect to MetS, hypercholesterolemia, liver disease, hepertension treatment and the modification gut microbiota in obesity.

6.1.3   Clinical Indication Prioritization

The semi-structured interviews performed by van den Nieuwboer et al. [30] allowed the identification of nine major disease areas potentially equiring increased research attention for probiotics, as follows: *metabolic disorders,* allergies, auto-immune disorders, cancer, cardiovascular disease, gastrointestinal disorders, infections (bacterial and viral), neurological disorders and general conditions (e.g., acne).

Current review is a logical follow up on our previous in vitro [31, 32] and in vivo research on probiotic strains [11, 33] and on potential prebiotics [34–37] and disccussed in [8–10], and suggesting that cumulated evidence in regard to phenotype of the probiotic strain should be considered for most effective individualized treatment via gut, oral and vaginal and other sites microbiome modulation. This can be implenmented according to phenotype of the patient and therefore individualized and personalized medical approaches. Number of microbiome phenotype variables can be used as parameters of predictive medicine to recognize patient's predispositions and evaluate treatment responses; on the other hand, number of phenotype markers have been effectively involved during microbiome modulation.

6.2   Patophysiology: Microbiota & MetS Interplay

6.2.1   Relevance of In Vitro Research

Recently **we have studied** [31] the biological properties of LAB and Bifidobacteria probiotic strains, namely adhesive properties, resistance to antibiotics and biological fluids (gastric juice, bile, pancreatic enzymes); and *formulated potential 'secondary' effects for beneficial individualized use meeting the patient's needs.*

The studied strains of LAB and bifidobacteria have been found to be sensitive to wide range of antibiotics, however, showed different **resistance** to *gastric juice, bile and pancreatic enzymes* [31]. The most resistant to antibiotics were *L. rhamnosus* LB-3 VK6 and *L. delbrueckii* LE VK8 strains. The most susceptible to gastric juice was *L. plantarum* LM VK7, which stopped its growth at 8% of gastric juice; *L. acidophilus* IMV B-7279, *B. animalis* VKL and *B. animalis* VKB strains were resistant even in the 100% concentration. Strains *L. acidophilus* IMV B-7279, *L. casei* IMV B-7280, *B. animalis* VKL, *B. animalis* VKB, *L. rhamnosus* LB-3 VK6, *L. delbrueckii* LE VK8 and *L. delbrueckii* subsp. *bulgaricus* IMV B-7281 were resistant to pancreatic enzymes.

Adhesive properties have been detected as high in strains of *L. casei* IMV B-7280, *B. animalis* VKL and *B. animalis* VKB; were moderate in *L. delbrueckii* subsp. *bulgaricus* IMV B-7281; and were low in strains as *L. acidophilus* IMV B-7279, *L. rhamnosus* LB-3 VK6, *L. delbrueckii* LE VK8 and *L. plantarum* LM VK7.

6.2.2 Probiotic Bacterial Cell Wall Heterogeneity: A Biomarker to Predict Host–Bacteria Interaction [32]

Since the LAB are gram-positive bacteria, their cell walls is complex and include glycolipids, lipoproteins, and phosphorylated polysaccharides within a thick layer of PGN, a polymer of β linked *N*-acetylglucosamine and *N*-acetylmuramic acid, cross-linked by short peptides [38]. The Gram-positive bacteria membrane is covered by a thick cell wall consisting of multiple layers of peptidoglycan, capsular polysaccharide (CPS), lipoproteins, and teichoic acids [38]. Some of these molecules contain specific *microbe-associated molecular patterns (MAMPs)* that are recognized by specific *pattern-recognition receptors (PRRs)* expressed in host intestinal mucosa. *L. delbrueckii* subsp. *bulgaricus* IMV B-7281, that had the most elastic cell wall, caused the considerable activation of the phagocytes. According to the patterns of cytokine, some strains of lactic acid bacteria can stimulate macrophages and dendritic cells to the IL-12 synthesis, which, along with IFN-γ, play a key role in the activation of cell-mediated immunity. All the mentioned strains can significantly stimulate macrophages to the IL-12 production [32].

6.2.3 Diet and Microbiota

Nutrition is a driving factor in shaping gut microbiota composition and its functional maturation from the early stages of life, resulting alterations of the gut microbiota composition and functional properties are associated with obesity. It is strongly recommended to medical professionals to make decisions on prevention and treatment of disease by food and probiotics using *evidence-based data* [39].

Thus, as the examples, an increasing *Bifidobacterium spp.* in diet may have anti-obesity effects [40]; the recent knowledge does not support the idea that dietary *fat* or *carbohydrate* content *per se* promotes development of metabolic syndrome [41]; thus, high-fat *vs* hypercaloric-hydrocarbonate diets have not been proved as a clear causal trigger cf obesity, consuming energy via *carbohydrate* or *fat* did not differentially altered visceral adiposity and metabolic syndrome.

6.2.4   Calorie Restriction

The findings suggest that the microbiome should be largely considered as a target during antiobesity programs [42], close interplay between modulation of gut microbiota and healthy aging has been demonstrated [43]. Thus, calorie restriction can effectively increase lifespan in animal models, and has potential for and health-promoting effects in humans balancing gut microbiota via homoeostatic control of microbiota in the lower gut supporting competition between bacteria for nutrients. This so called 'oligotrophic condition' is recommended to preserve during lifespan [43].

On the other hand underestimated values of nutrition like content of *fructose and monosodium glutamate* intake were reported in resulting *hyperuricemia* [44–48].

6.2.5   Fructose Intake

Fructose is a major chemical of sweets and is one of the key, althoogh underestimated, dietary promoters of metabolic syndrome development [44–48]. Dietary fructose is converted into glucose and organic acids in small intestine, a higher doses of fructose exceed capacity of intestinal fructose absorption and clearance, resulting in reaching fructose to both the liver and colonic microbiota [44].

Diets enriched in fructose reduce bacterial colonization, lead to dysbiosis, increase numbers of mucin-degrading bacteria [44].

When fructose from dietary sources is absorbed through the fructose transporter GLUT5 within the intestinal epithelium and transported to the liver, it is rapidly phosphorylated in the liver by fructokinase, causing hepatic accumulation of fructose-1-phosphate (F-1-P) and a simultaneously increase in AMP [44].

Fructose promotes alterations in the gut microbiota profile triggering inflammation and metabolic imbalance in the gut, liver, and in visceral white adipose tissue. These obesity-related features can be experimentally reversed by treatment with antibiotics [45]. *Fructose-rich diet (FRD)* induce endocrine-metabolic alterations and dysbiosis in mice; FRD does not alter the phyla of Bacteroidetes and Firmicutes, but decreases *Lactobacillus* spp. [45]. The beneficial effects of *L. kefiri* as a probiotic was demonstrated to alleviate effects of high fructose intake [47].

Importantly, that even a single administration of fructose reduces uric acid excretion in the ileum and long-term use of fructose suppresses renal uric excretion resulting in *increased serum uric acid* levels and gout development [44].

The preventive effect of *Lactobacillus kefiri* (*L. kefiri*) administration for FRD was demonstrated in a mice model [45]. More studies of the effects of fructose intake on health and gut microbiota are needed.

6.2.6   Dietary Fibers – Fermentable Carbohydrates

The production of **short-chain fatty acids (SCFAs)** via fermentation of carbohydrates by probiotic bacteria is an example of balanced microbial ecosystem and key beneficial effects for human health [49]. A group of **acetate and butyrate**-producing bacterial strains has been identified that can be selectively promoted by increased availability of various fermentable carbohydrates in the form of dietary fibers [49].

Butyrate has been found to be a major energy source for intestial cells, and also to increase mitochondrial activity, prevent metabolic endotoxemia, improve insulin sensitivity, possess anti-inflammatory potential, increase intestinal barrier function and protect against diet-induced obesity without causing hypophagia. **Propionate** has been found to inhibit cholesterol synthesis, that is antagonizing to the cholesterol increasing activity of **acetate**, and can inhibit the expression of resistin in adipocytes [17, 50, 51].

Monosodium glutamate (MSG, C5H8NO4Na, E 621) is widely distributed and is naturally occurring in various standard foods and increase food intake. MSG can enhance the flavor of bland food, and contain purines, which are directly metabolized into **uric acid**, as guanylate (E626, E627, E628 and E629), inosinate (E630, E631, E632 and E633), and their compounds ribonucltides (E634 and E635) are metabolized to purines and lead to gdevelopment of hyperuricemia, gout [34]. Because the deleterious effects of MSG, i.e., induced overfeeding, were not seen in the animals fed the fiber-enriched diets [52].

A gluten-free diet (GFD) is the most commonly adopted special diet worldwide, positive effect of a GFD on the composition of the gut microbiome have been reported in coeliac disease patients. GFD can modify the composition of the intestinal microbiota and change the activity of microbial pathways. The most important observation in these studies is the difference in the number and variety of Lactobacilli and Bifidobacteria in treated and untreated patients [53].

The vegetarian diet that includes soy-based foods supposes increased levels of phytoestrogens beneficial for MetS and LF, however, might be associated with a higher risk of alering the male reproductive male system [54].

Genetic microbial variation trigger phenotypic diversity and influences the predisposition to metabolic syndrome altering **diet-induced metabolic phenotypes** [55]. Gut microbiome contributes to the genetic and phenotypic strains diversity and provide a link between the gut microbiome and insulin secretion. Since, microbial taxa correlate with their metabolic phenotypes, the gut microbiome is a source of broad genetic variation that determine different host-associated diet-induced *metabotypes* [55]. This impact of gut microbes on host physiology is suggested to be modulating in part by BA pool composition [55].

The promising approaches among dietary interventions to improve metabolism and microbiota seem intermitten fasting and ketonic diet (KD). Thus, KD can beneficially modify gut microbiota (increasing Akkermansia muciniphila and Lactobacillus), and improve immune and metabolic profiles and increase endothelial nitric oxide synthase (eNOS) protein expressions [56].

6.2.7 Hereditary Factors and Family Diet History

The priority effects important to human health has an origin from the ealy life according to ecological theory and circumstantial evidence [57–59]. The mechanisms, conditions and consequences of priority effects that might affect microorganisms in the gut, bacterial community remains highly conserved between corresponding body sites in human hosts, while gene transcription is much more variable [58]. Nutrition in the early life may influence the epigenome via microbial metabolites, which can contribute to the development of obesity in adults [58, 59]. *"First 1,000 days of life"* concept has been suggested describing critical windows in organism development wheres all systems and functions are largely vulnerable in particular for DNA methylation [59].

Dietary and food patterns can modulate the gut microbiota composition oand therefore its metabolites. The difference in the presence of short-chain fatty acids (for example, butyrate) and bacterial metabolites important for one-carbon metabolism (folic acid) depend on food habits and microbiota composition.

Thus, these substrates provide **epigenetic** activity, early postnatal nutrition can form the developing epigenome of target tissues, which can determins the predisposition to obesity.

As examples, following bacterial metabolites are able to modulate the epigenome:

1. *folate*, that is crucially involved in one-carbon metabolism and can influence DNA methylation to disable gene transcription;
2. *butyrate*, a SCFA and a potent inhibitor of histone deacetylases [58].

Specific synbiotics have been reported to be effective for early life protection against diet-induced obesity in early life [60].

6.2.8 Prebiotics

Prebiotics have immense ability to enhance probiotics effects and in the context of above has largely potential to modulate microbiome and metabolome by itself. A prebiotic was defined by Gibson et al. as a *"non-digestible food ingredient that beneficially affects the host by selectively stimulating the growth and/or activity of one or a limited number of bacteria in the colon, and thus improves host health"* [26]. The issue of the specificity of microbial changes has been defined as the key point to be studied.

Number food ingredients, many still underevauated, being selectively fermented, can induce specific changes in gut microbiota; prebiotics are beneficial to the host's well-being and health have a protective effect and may be useful for many conditions. The terms of prebiotic/functional food seem overly bureaucratic, since e.g., *fecal microbiota transplantation (FMT)*, although not being probiotic, could be considered a fermented food, given the microbes and nutrients present. The option of strains that are core to FMT efficacy being used as a probiotic is also being viewed as a drug, but if the strains have a safe history of use in humans, this [39].

Thus, as examples, herbal-based biopolymers as *fenugreek* have antiobesogenic properties and offer effective added value as prebiotic towards the enhancement of probiotic activity [34]. The combined use of probiotics with nanoparticle-based treatment and food supplements is promisingm in particule, nanoparticles of cerium dioxide [35–37, 61] and gold [62, 63] have been known as strong agents against oxidative damage having anti-aging activity, and can demonstrate antiviral, antibacterial, antifungal activity, cardioprotective, neurotrophic, hepato- and nephroprotective, and anti-aging effect, have potential for various biomedical applications [35–37]. Nanoceria has also therapetic and preventive perspectives in reproductive medicine, enhancing female and male fertility [37].

6.2.9 Antibiotics

The enormous use of antibiotics can alter and gut microbiota and host's phenotypes and metabolism and can increase risk of obesity and atherosclerosis [64–66]. The uncontrolled antibiotic therapy has became widespread epidemics in recent decades, this led to the formation of associations of microorganisms with increased virulence, in particular so-called "hospital strains". Gut microbiota is a potential reservoir of antimicrobial resistance (AMR) genes; microorganisms including AMR have been extensively studied within the as so called "resistome" [65–67]. The ability for the horizontal transfer to potential pathogenic bacteria within this ecosystem was demonstrated [65], this antibiotic susceptibility of probiotic strain can be a significant specific indicator, and the antibacterial resistance was studied for LAB and Bifidobacterium strains [66]. The impact of antibiotics on the establishment of the *infant gut resistome* was demonstrated [67].

6.2.10   Molecular Mechanisms of Probiotic Effects

Molecular mechanisms of health benefits of by consumption probiotics is largely unknown. Bacterial metabolites were indicated to have an **epigenetic** function. Therefore, xenobiotic metabolism of gut microbiota is essential issue fo future studies and enzymes discovery [57, 68–71].

Probiotic strains can alter host's **genes**, thus, administration of *Lactobacillus paracasei* CNCM I-4034, *Bifidobacterium breve* CNCM I-4035 and *Lactobacillus rhamnosus* CNCM I-4036 can modulate the expression of genes in the intestinal mucosa of obese Zucker rats [72].

Transcriptional networks regulate major basal mucosal processes and uncovered remarkable similarity to response profiles obtained for specific bioactive molecules and drugs [72], probiotic strains from the species *Lactobacillus acidophilus*, *L. casei*, and *L. rhamnosus* induce differential gene-regulatory networks and pathways in the human mucosa of the proximal small intestine of healthy volunteers. Thus, consumption of *L. casei* can lead to *mucosal gene-expression networks* that regulating Th1 and Th2 between and cell proliferation and balance, immune response, metabolism, and hormonal activity regulating blood pressure. The consumption of *L. rhamnosus* can lead to modify the expression of genes involved in wound repair and healing, angiogenesis, IFN response, calcium signaling, and ion homeostasis [72]. A core microbiota established in early life accompanies host's organism during human life, and decrease in abundance along with aging [73]. In this regard the breast milk containing a large amount of LAB is considered as crucial important programing factor for further human life.

6.2.11   Microbiota and Immunity – Allergy and Autoimmune Diseases

Strachan [74] described the *hygiene hypothesis* that is associated with reduced microbial contact to microbes in early life and is suggested to be one of the main mechanisms of the increasing predisposition to allergic diseases over the past decades. Today, reduced microbial exposures (and accordingly the rise in allergic conditions) have been triggered by Western diet, antibiotic use, vaccinations, smaller household size and improved hygiene [74].

Gut microbiota is involved in regulating both *Th1* and *Th2* immune response. Thus, in patients with IBD the gut microbiota has been shown to be of less diversity, an altered microbial metabolite profile with reduced number of bacteria compared to healthy individuals has been demonstrated [75]. A similar etiology is believed to exist in rheumatoid arthritis, ankylosing spondylitis, multiple sclerosis, type 1 diabetes mellitus (T1DM), and celiac disease [76].

Obesity coincides with a low-level chronic inflammation in metabolic tissues. This obesity-related 'metabolic inflammation' involving adipose tissue, liver and muscle, which are key regulators of whole-body glucose homoeostasis, drives immunological underpinnings of insulin resistance and CVD.

We hypothesized that according to the *inflammation-centred theory* the immune response and metabolic regulation are highly integrated and the proper function of each is dependent on the other [77], claiming that gut microbiota can influence immune function beyond the gut, would be crucially helpful for choosing appropriate probiotic bacteria in the personalized clinical set.

Environmental factors, *i.e.* medication (antibiotics, non-steroid anti-inflammatory drugs and hormones), dietary habits, are of living environment, and previous infections histiory have clear influence on this immune balance [75].

6.2.12   Cytokine Profiles of Toll-Like Receptors

Gram positive bacteria affect the formation of T-and B-cell immune response by altering products primarily IFN-γ and IL-12 are required for differentiation of T helper cells into Th1 subpopulation direction. But probiotic preparations are capable of activating both (Th1 and Th2) lymphocyte subpopulations, which provides a balance of cytokine production. Immunomodulatory activity of probiotic preparations most important to identify for the goods induced opposite cytokines IL-10 or IL-12 in experiments in vitro when stimulated macrophage cells [78]. The immune response against infectious diseases of probiotic drugs due to the ability to balance the body's immune status at the level of receptor-ligand interactions [78].

Induction of pro-inflammatory cytokines induced by dendritic cells (DCs) expressing pattern recognition receptors may skew naive T cells to T helper 1 polarization, which is strongly implicated in mucosal autoimmunity through a mechanism that involves IL-10 and CD4+ FoxP3+ T regulatory cells to dampen exaggerated mucosal inflammation [79]. The ability of probiotics to affect the relevant *Toll-like receptors* (TLRs) can promote effective immune response and the initiation of an effective immune defense.

Interleukin (IL)-10 is an anti-inflammatory cytokine. cytokine profile of IL-10 is associated with the gut-associated lymphoid tissue (GALT) most pronounced changes in the Peyer's patch. Probiotic-mediated immune modulation in IL-10 knock-out mice demonstrated a probiotic mechanism of treatment of gastrointestinal inflammation independent of IL-10 [80].

Interleukin-22 (IL-22) has a crucial role in the early phase of host defense against *C. rodentium*. Innate immune function for IL-22 in regulating early defense mechanisms against A/E bacterial pathogens [81].

6.2.13 Defining Causality vs Correlation – Is an Inflammation in Focus?

The identifying the causative associations of obesity and the human microbiota is still a challenge [73, 77, 82]. The communication between the microbiota and immunity alter the metabolic responses during obesity and MetS. The beneficial bacteria can induce pro-inflammatory or regulatory immune responses, depending on the individual phenotype of gut microbiome, and dietary habits [82]. The associations between *immune modulatory and hyposholesterolemic properties* of *L. reuteri* ATCC PTA 4659 probiotic strain demonstrated [82].

Lipopolysaccharide, the cell wall component of gram-negative bacteria in the gut, are supposed as an important trigger of chronic inflammation associated with obesity [83]. Gram-positive bacteria are potent inducers of monocytic proinflammatory interleukin-12 (IL-12) with immunoregulatory functions, while gram-negative bacteria preferentially stimulate anti-inflammatory IL-10 production [84].

6.2.14 Infections

Many routinely-used antibiotics are already ineffective in the clinic; some even speculate that the twenty-first century will come to be known as the 'post-antibiotic' era [84]. However, the use of probiotics might have several potential disadvantages; namely, the introduction of foreign microorganisms induces antagonistic activity against pathogenic and indigenous microorganisms and rapid elimination of probiotic strains. Therefore, to achieve a personalized approach, products developed and applied from the own strains of the body, appear promising. For this reason, some individual microorganisms can be grown on artificial nutrients, studying their ecological compatibility, establishing the antagonistic effects of the spectrum on the body. A potential alternative to probiotics may be proposed by lysates of probiotic strains, which can also maintain immunomodulatory activity [33].

The broad associations have beenillustrated among **virus action** during metabolic sydrome and T2DM development, including HPV infection, cellular oxidative stress, gene damage, multiple microbiota-related immune pathways and proteomic changes leading cancer and chronic disorders genesis [85, 86].

6.2.15 Intestinal Permeability

The interrelated parameters of the metabolic disease, such as fatty liver disease, high values of homeostatic model assessment (HOMA), high waist circumference, and subclinical inflammation, ha been known associated with *intestinal permeability*. Recent data show that by successfully treated overweight, increased intestinal permeability may be altered to normal levels [87]. A similar effect has been found in obese people who have undergone a dietary intervention based on traditional Chinese medicines and prebiotics [88].

6.2.16 Oxidative Stress: Emerging Role of Nanomedicine

It is known that oxidative stress has been postulated as one of the principle physio-pathological mechanisms of number of chronic diseases, includiong the pathogenesis of obesity-related diseases [89, 90]. The cellular imbalance between endogenous antioxidant defenses and reactive oxygen species (ROS) is one of its primary characteristics [89, 90]. Several mechanisms have been suggested to explain the enhanced oxidative stress observed in obese subjects, including altered lipid and glucose metabolism, chronic inflammation, tissue dysfunction, hyperleptinemia, and abnormal post-prandial ROS generation [89]. Thus, the nanoparticles of gold [62, 63] and cerium dioxide [37, 61] were reported to be effective agents against oxidative damage having anti-aging activity, and potential for prebiotic activity via modifiing intracellular ROS generation in bacteria.

However, only a few studies have been conducted on the oxygen tolerance of probiotic bacteria. Most of these studies have focused on *Bifidobacterium* spp. Little is known about the effect of oxygen on the physiology of *L. acidophilus*. *L. rhamnosus* GG can potentiate intestinal hypoxia-inducible factor [91].

6.2.17 Microbiota Profile & Microorganism-Based Biomarkers

The search for reliable phenotypic microbial markers is essential for longitudinal observation and reproduced in large populations, is the most important task for the study of microbial and probiotics *in clinico*. Prebiotic and probiotic therapy is aimed at the formation of microbiota for the improvement of health. However, the gut contains a large number of different microorganisms that are difficult to calculate. Out of these, **three phyla, *Bacteroidetes* (Gram negative), *Firmicutes* (Gram positive) and *Actinobacteria* (Gram positive),** are most common and they determine the dominant role in the pathophysiology of metabolic disorders, in particular in obesity. Other fillets also contribute, but to a lesser extent [17, 92].

Arumugam et al. [93] even identified some typical clusters of of fecal microbial compositions called "*enterotypes*"composition that are recurring in the healthy population and partly depend on dietary habits. **Enterotypes** were allocated primarily by levels of **Bacteroides (B)** and **Prevotella (P) that** were associated with long-term diets, particularly protein and animal fat (Bacteroides) versus simple carbohydrates (Prevotella). It was suggested that the ratio of Bacteroides/Prevotella (P/B) may be a tool for stratification of subjects when studying the effect of interfering with intestinal microbiote [94]. *Stratification* of humans based simply on their *P/B* ratio could allow better assessment of possible effects of interventions on the gut microbiota and physiological biomarkers [94].

For example, the *Prevotella* enterotype with a high representation of Prevotella spp., has been associated with **high-carbohydrate, high-fiber diets.**

6.2.18   Plant- vs Animal-Based Diets

High-fat diets have been associated with harmful effects on the gut microbiota. These diets generally promote decreasing in Bacteroidetes representation and overgrowth of Firmicutes, including a wide range of opportunistic pathogens (such as LABs).

Adherence to the **Mediterranean diet** is associated with beneficial microbiological effects in the intestine, including higher biological diversity, excessive Prevotella's presence, and lack of opportunistic pathogens.

The **animal-based** diet increased a large number of bile tolerant microorganisms (Alistipes, Bilophila, Bacteroides) and decreased levels of Firmicutes that metabolize polysaccharides of dietary plants (Roseburia, Eubacterium rectale, and Ruminococcus bromii). Microbial activity is a mirror of difference between herbivorous and carnivorous mammals, reflecting compromises between carbohydrates and protein fermentation [95, 96].

The remarkable differences were observed in transcriptional responses and in gene abundance between the intestinal microbiomes elicited by plant- and animal-based diets [95]; catabolism of amino acids against biosynthesis, as well as the relationships of phosphoenolipyruvate (PEP) and oxaloacetate in herbivorous and carnivorous mammals respectively [95].

Microbial communities that could quickly and properly self-modify their functional repertoire in response to a diet change will eventually improved human flexibility in diet [95, 96].

The degradation of polysaccharides by the intestinal microbiota and its influence on human health [51, 97].

The microbial community of the gut is one of the sources of human genetic and metabolic diversity, which are different among human populations, and, depend on age, geography and cultural traditions and is unique to different locations and lifestyles, in particular differ for modern western diet and a rural diet, and correlates with westernization [98, 99].

Recently, it has been observed that the composition of gut microbiota of healthy persons is different from that of obese diabetes, T2DM patients. Such observations suggested a possible relationship between the compositional pattern of gut microbiota and pathology of metabolic disorders.

Since human colon harbours a vast number of microorganisms which are extremely diverse [17, 92], the *metagenomics* analysis of microbiome divided human **into three groups**, namely: Enterotype 1 (Bacteriodes), enterotype 2

(Prevotella), and enterotype 3 (Ruminococcus) according to bacteria population found to be dominant [100].

The *Firmicutes*-to-*Bacteroidetes* (F/B) ratio was linked to body-weight and BMI [101] and was reported to be higher in obese subjects with metabolic syndrome. Louis et al. calculated the F/B ratio for each sample and found a high variability between individuals and time-points without correlation with BMI or other clinical parameters [102].

Successful weight reduction in the obese is accompanied with increased Akkermansia levels in feces. Metabolic co-morbidities are associated with a higher Firmicutes/Bacteroidetes ratio, *microbiota differences might allow discrimination between successful and unsuccessful weight loss prior to intervention* [102].

Probiotics have a significant capacity to remodel the microbiome of an individual recovering from antibiotic therapy during the recovery phase the probiotic cause a suppression of Enterobacteriaceae downgrowth (Shigella and Escherichia) and can promote a growth of Firmicutes, particularly from the Anaerotruncus genus [103]. *L. reuteri* significantly decrease the intestinal inflammation and reduce in proteobacterial populations [104].

Microbial diversity is an important parameter of intestinal health [105–108]. Thus, lower richness of gut microbiota compositions, was found in Western diet consumers shapes the microbial ecosystem [98, 99] and in the populations under the burden of obesity and metabolic disease [81, 82]. Individuals with higher diversity were reported to have a healthier dietary pattern [107, 108].

The lower diversity was associated with greater abdominal adiposity. Meta-analyses across the replication in independent samples from three population-based cohorts including American Gut, Flemish Gut Flora Project and the extended TwinsUK cohort using BMI as a surrogate phenotype, demonstrtaed significant associations of adiposity-OTU abundances with host genetic variants in the *FHIT*, *TDRG1* and *ELAVL4* genes, suggesting a potential role for host genes to mediate the link between the fecal microbiome and obesity [108]. Variety of metabolites are modulated by the action of gut microbiota richness, number of recently discovered crosslinks between gut microbes and different circulating metabolites with high predictive and diagnostic potential have been recently identified.

Individuals who have a low bacterial richness (23% of the population) characterized by more expressed overall obesity, insulin resistance and dyslipidaemia and a more pronounced inflammatory phenotype compared with individuals of high bacterial richness [109].

Metabolically active and safe *Lactobacillus* species and specific strains with particular functional properties increase the biodiversity of the whole intestinal microbiota [110].

Focused primarily on bacteria, but priority effects are also possible across domains of life (that is, between bacteria and archaea and/or eukaryotic microorganisms) [111].

The parameter as *Alpha and Beta diversity* are useful tools to evaluate microbiota. Thus, **Alpha diversity** *indicates* microbial species richness – number of taxa within a single microbial ecosystem.

Beta diversity – is a parameter of diversity in microbial community between different environments (difference in taxonomic abundance profiles from different samples).

Recently **mycobiome** has been suggested as a factor of the protective benefits via intestinal colorization by commensal fungi [112, 113] that functionally replace intestinal bacteria and alleviate tissue injury by positive activation of protective CD8 T cells. Thus, commensal gut fungi protect local and systemic immunity reactivity by providing tonic microbial stimulation that can functionally replace intestinal bacteria.

Fungi are transmitted from mother to infant in early life, their dispersal history can be highly variable among infants, and once immigrated, they can interact strongly with bacteria [114].

In particular, diverse fungal communities are present in infants [114–116].

6.2.19 Vaginal, Oral and Dermal Microbial Profiles in Distant Sites [9]

Vaginal microbiota has been known to have extensive links with the gut microbiome and metabolic syndrome development [117–120]. *Lactobacillus* species dominate in vaginal microbiota in the most of of pre and post-menopausal women being an indicator of vaginal health.

The recent study reports using interactomic approach required for vaginal probiotic administration in post-menopausal women to detect the subtle molecular changes induced by probiotic instillation [119]. Marked diversity in microbial composition was detected between women with bacterial vaginosis (BV) and those with normal flora in pregnancy [120].

Vaginal dryness and atrophy have been reported to be associated with down-regulation of human genes in epithelial structure involving changes in barrier function, up-regulated inflammation due to reducing lactobacilli in menopause [118].

Current knowledge of the **male genitalia microbiome** is very limited. *Gardnerella vaginalis* is predominant in half of the women whose partners had significant leukocytospermia [121]. Vaginal microbiome was reported to drive the chronic inflammation-malignant development of prostatic adenocarcinoma in couples [122].

Studies of structure of vaginal microbiota in regards to inflammatory conditions via analysis of samples collected in the various stages of disease and in different at-risk populations, in regards to the role of host genotype, involvement hormonal receptors might suggest promising approach for understanding pathogenesis of chronic gender-related inflammatory diseases, development personalized treatments, diet and lifestyle corrections.

The ability of LAB and bifidobacteria strains to adhere to epithelial cells in vitro is one of the most important criteria for the selection of potentially probiotic strains for intravaginal use, since it indicates their ability to attach and colonize the vaginal surfaces [31].

Vaginal and *male genital tract ecosystems* as the functional interaction between the genital microbiota and the host, and the association of semen and vaginal microbiomes are still poorly studied [121].

Combined oral and topical treatment of male partners of women with BV is acceptable and well tolerated. The combined acceptability and microbiological data presented in this paper supports the need for larger studies with longer follow up to characterize the sustained effect of dual partner treatment on the genital microbiota of couples and assess the impact on BV recurrence [123].

Thus, *neither clinical criteria, nor microbial composition can fully explain symptomatic bacterial vaginosis.*

Recen0tly the term bacterial vaginosis was suggested be dropped, as it currently offers no adequate description of a single condition [124]. The new definition will require precise definitions, diagnosis, and management options. In some case, the use of probiotics and/or prebiotics may help to restore and maintain a vaginal and male genitalia microbiome health.

6.2.20 Microbiome of Oral Cavity

The various analysis methods reveal Firmicutes, Actinobacteria, Proteobacteria, Fusobacteria, Bacteroidetes and Spirochaetes as the the dominant genus a healthy microbiome of oral cavity constituting 96% of total oral bacteria [35, 125, 126].

Recently *metatranscriptome* sequencing indicated overexpression of a number of virulence-related transcripts in oral bacterial composition during the early stages of transition to gingivitis, and the upregulation genes including those involved in proteolytic and nucleolytic processes [126].

Core oral microbiome may иу significantly different under carbohydrate and protein-rich diet consumpotion [127].

Future research dedicated to the oral bacteria involved in the pathology and leading to obesity is needed addressing the question – *how the salivary microbiology affects gastrointestinal microbiology.* The great interest is about how orally administered probiotic therapy influence on bot oral and gut microbiota.

Oral bacteria are known to contribute to the weight increase and development of obesity by at least three mechanisms [127]: (1) the oral bacteria may contribute to increased metabolic efficiency, (2) by increasing appetite, and (3) energy metabolism by facilitating insulin resistance through TNFα increasing levels or reducing levels of adiponectin.

MSG-induced obesity triggers periodontal tissue alterations in the rat model. Nanoceria contributes to the corrections of pathological changes in periodontal tissues in glutamate-induced obese rats via balancing protein-inhibitory capacity and reducing the depolymerization of fucosylated proteins and proteoglycans and antioxidative activity [35].

Lactobacillus crispatus KT-11 strain intake can prevent periodontal disease through the improvement of oral conditions, decreased plaque scores, reddish tinge, and gingival swelling scores in female participants and increased oral mucosa fluid scores in male participants [128].

6.2.21   Skin Microbiome

Interactions of skin microbial communities with host immunity and imbalance of microorganisms, termed skin dysbiosis plays crucial role in diseases of the skin [129–131]. Skin *mycobiome* plays importnt role in shaping innate and adaptive immunity in health and disease [114, 132]. Recent studies in the unique setting of the Antarctic have shown an increase in fungi on the skin in expedition participants,believed to be due to interferences with local immunity and dysbiosis of the normal skin microbiome due to stress, recycled air and antiseptic agents [133]. Akkermansia muciniphila is believed to have an important function in the pathogenesis of IBD and obesity; therefore, Akkermansia muciniphila, which is an indicator of health status, may be a key node for psoriasis as well as IBD and obesity [134].

6.2.22   Wound Healing

Wound healing is involved in metabolic disease and is remarkable a marker of health, strongly depending on the phenotype including such opponent condition as MetS and obesity and *Flammer syndrome* [135]. *Lean body mass* (LBM) is the parameter important for prediction and prognosis of the physiological wound healing. *Matrix metalloproteases (MMPs)* and inhibitors are secreted as inactive prcenzymes (zymogens) neutrophils, macrophages, fibroblasts and keratinocyt and get activated as the extracellular component [135].

Probiotics have been associated with improved healing of intestinal ulcers, and healing of infected cutaneous wounds. LAB and bifidobacteria utilize their association with gut to directly inhibit pathogens' growth and ability to induce host mucosal defense systems and tissue repair mechanisms [136]. data demonstrate that L. rhamnosus GG lysate accelerates reepithelialization of keratinocyte scratch assays, potentially via chemokine receptor pairs that induce keratinocyte migration [137]. *Lactobacillus reuteri* enhances wound-healing properties through upregulation of the neuropeptide hormone *oxytocin*, a factor integral in social bonding and reproduction, by a vagus nerve-mediated pathway. Bacteria-triggered oxytocin serves to activate host CD4 + Foxp3 + CD25+ immune T regulatory cells conveying transplantable wound healing capacity to naive Rag2-deficient animals [138].

6.2.23   The Gut Microbiota in Aging and Longevity

A core microbiota accompanies human life, decreasing in abundance along with aging [73]. Aging is thus associated with specific changes in gut microbiota. After the age of 65, overall gut microbiota composition resilience is generally reduced, so that its is more vulnerable to lifestyle changes, antibiotics treatments, and diseases [58, 59]. As a result, species biodiversity/richness (i.e., the number of taxa that metagenomic analyses are able to identify in fecal samples) is reduced, and interindividual variability is enhanced [139, 140].

In an Irish population-based study, Claesson et al. [140] showed that gut microbiota biodiversity is inversely correlated with physical function and the institutionalization of older individuals [58]. The same authors also showed a dramatic interindividual variability in the fecal microbiota of elderly subjects.

In cases of **longevity**, the age-related enrichment of subdominant taxa is boosted. The microbiota of longevous hosts accommodates allochthonous bacteria. In longevity, the age-related content of sub-dominant species increases, including pro-inflammatory species, as well as health-related taxa that can support extreme aging [73]. "Adaptation to longevity" seems to enrich the health-related bacteria [73].

6.3   Disease- and Person-Specific Application of Probiotics

6.3.1   Obesity

A broad evidence demonstrate associations between the human and microbiota and immunity altering the metabolic responses during obesity and MetS [4, 12–17, 73, 77, 82]. The beneficial bacteria can induce pro-inflammatory or regulatory immune responses, depending on the individual phenotype of gut microbiome, and dietary habits [82]. Bacterial strains of the same species showed different effects on adiposity and insulin sensitivity, illustrating the complexity of hostbacterial cross-talk and the importance of investigating specific bacterial strains. Thus, the study by Fåk et al. demonstrated associations between *immune modulatory and hyposholesterolemic properties* of *L. reuteri* ATCC PTA 4659 probiotic strain which partly prevented diet-induced obesity in Apoe−/− mice, yet, induced *no effects on blood cholesterol* or atherosclerosis and likewise *no effect on inflammatory markers* (on macrophages or T-cell numbers in plaques) [82]. L. reuteri was associated with increased liver β-oxidation, reduction of the adipose and liver weights [82].

In animal model, the weight of obese mice that received *L. casei* IMV B-7280, *L. delbrueckii* subsp. *bulgaricus* IMV B-7281, *B. animalis* VKB, *B. animalis* VKL (separately) or *B. animalis* VKL/*B. animalis* VKB/*L. casei* IMV B-7280 and *L. casei* IMV B-7280/*L. delbrueckii* subsp. *bulgaricus* IMV B-7281 probiotic compositions was decreased [11].

The changes in the host immune system composition into a more anti-inflammatory profile, which may explain the decrease in body fat [141]. Randomized controlled trial demonstrated some evidence that a 3-month synbiotic supplementation (L. reuteri with partially hydrolyzed guar gum and inulin) in addition to lifestyle modification is superior to lifestyle modification alone for the reduction of body weight, BMI and *waist circumference* and treatment of NASH.

Synbiotic did not improve intestinal **permeability** or *small intestinal bacterial overgrowth (SIBO)* and lipopolysaccharide (LPS) serum levels [142].

Synbiotics use can resultin reduction in steatosis, lost weight, diminished BMI and waist circumference (WC) measurement.

The double-blind randomized controlled clinical trial showed that probiotic and prebiotic supplementation along with lifestyle intervention creates favorable changes in glycemic parameters and leptin levels compared with the lifestyle intervention alone [143]; *oligofructose* dietary fiber intake hase been demonstrated to be as effective as probiotic supplementation for insulinemia and adipokines [143].

6.3.2 CVD, Hypertension & Hypercholesterolemia

Obesity-induced endotoxemia and liver dysfunction might be modulated by beneficial microbes via immune response, e.g., by TLR to inhibit cholesterol synthesis signaling pathway in the liver. *However, the associations between immune modulatory vs **hypocholesterolemic** activity has not been finally not elucidated yet.* Based on our preliminary data we hypothesized that *the ability of the strain with its immune-modulatory properties to decrease cholesterol may be for treatment CVD.*

Hypertension is a part of MetS [2] and is as a major risk factor for number of complication and *heart failure.* CVD affects one billion adults globally and leads to nine million deaths every year according to estimates by the World Health Organization (WHO, 2013) [144].

Daily ingestion of *L. plantarum* DSM 15313 or blueberries fermented by this strain for 3 months did not, in the current study set up, reduce the blood pressure of hypertensive subjects and did not affect either the diversity or the composition of the oral and the faecal microbiota during the intervention period [145]. Authors observed that both the oral and the faecal microbiota were highly stable within the individuals, compared to the faecal microbiota, the oral one fluctuated more and varied more between individuals. It was demonstrated that *Lactobacillus helveticus* are capable of releasing antihypertensive peptides [146].

To enhance the research power in order to predict outcomes for probiotic studies in clinical set for CVD and smart utilizing *in vivo* data to develop microbiota-related biomarkers and associated individualized treatment is an important task.

The probiotic composition VSL#3 can decrease TNF-alpha levels, MMP-2 and MMP-9 activities, and expression of iNOS and COX-2 in rats, fed the HFD diet [147]. Nanogold demonstrated prebiotic properties and is effective heart failure treatment [62, 63] that is largely associated with metabolic syndrome.

6.3.3 Diabetes Mellitus

Recently, it has been observed that the composition of gut microbiota of healthy persons is different from that of obese T2DM patients. Such observations suggested a possible relationship between the compositional pattern of gut microbiota and pathology of metabolic disorders [17].

Data from the meta-analysis conducted by Zhang et al. [148] show that probiotic consumers can modestly improve glucose metabolism with a potentially greater effect if the duration of the intervention is ≥8 weeks, or several types of probiotics are consumed.

Gu et al. [149] suggested that gut microbiota and plasma bile acids allow stratification of patients for antidiabetic treatment via for the treatment of antidiabetic drugs by means of the so-called *acarbose-gut microbiota-BA axis* and distinguished two microbiome clusters (Bacteroides and Prevotella clusters) interacting with BA metabolism. Highly relevant biomarkers of T2DM, like *bile acid metabolism* [149] and signs of diabetic neuropathy [150] will help to effectively stratify patients with MetS- T2DM for appropriate management also using individualized probiotic therapy.

Recently we have demonstrated [151], that probiotic strain L. casei IMV B-7280 (separately) and composition L. casei IMV B-7280/B. animalis VKB/B. animalis VKL can re-equilibrate metabolic and inflammation indices in mouse obesity model, induced by fat-enriched diet (FED). Probiotics were effective in reducing mice weight and visceral fat, normalization of tumor necrosis factor- alpha (TNF-alpha) and functional activity of PEMs. *L. casei* IMV B-7280 alone was more efficient in decreasing glucose levels than composition of strains [151].

6.3.4 Liver Disease and MetS

Nonalcoholic fatty liver disease (NAFLD) is a worldwide health problem characterized by ectopic accumulation of triacylglycerols in the liver, represents a hepatic metabolic syndrome and includes fatty liver (simple steatosis), steatohepatitis (NASH), liver fibrosis (LF), and cirrhosis [1–3]. The disease was more common in women, obese, with diabetes mellitus, cholestasis, gallstones and thyroid disease and largely associated with microbiota [46, 152, 153]. Beneficial microbes-based treatment have huge potential for correction MetS and NAFLD, the knowledge has been cumulated supporting probiotic therapy as a safe, inexpensive, and a noninvasive strategy that can reduce pathophysiological symptoms and improve different types of liver diseases without side effects [154–157].

Furthermore, serum *ghrelin* levels positivey correlated with *Bacteroides* and *Prevotella*, serum leptin concentrations positivey correlated with the quantity of Bifidobacterium and Lactobacillus, and negatively correlated with *Clostridium, Bacteroides* and *Prevotella* [154].

L. rhamnosus CCFM1107 decreased the level of cholesterol in the liver and serum of mice with alcoholic affection of liver [158]. After administration *L. acidophilus* to obese mice with damaged liver after cholesterol-enriched diet the reduction of cholesterol level both in serum and liver was observed [159]; and *L. plantarum* CAI6 and *L. plantarum* SC4 had a protective effect in models of CVD in hyperlipidemic mice by reducing the level of total and low-density lipoprotein cholesterol [160].

In the recent study [11] we revealed that *L. casei* IMV B-7280, *B. animalis* VKL or *B. animalis* VKL – *B. animalis* VKB – *L. casei* IMV B-7280 composition recovered the liver structure of obese mice [11]. After administration of this probiotic composition in obese mice, degenerative changes in the liver were not detected, fatty degeneration and hepatocyte necrosis are reduced after treatment. with these probiotic bacteria or probiotic compositions. Yet, hemorrhages in the liver were not found in obese mice treated with *L. casei* IMV B-7280 or *B. animalis* VKL/*B. animalis* VKB/*L. casei* IMV B-7280 composition. However, after injection of *B. animalis* VKB, *L. delbrueckii* subsp. *bulgaricus* IMV B-7281or *B. animalis* VKL/*B. animalis* VKB composition to obese mice, we found necrosis and fatty degeneration of hepatocytes. The treatment with *B. animalis* VKL/*B. animalis* VKB/*L. casei* IMV B-7280 composition effectively recovered the liver morphological structure in obese mice. *L. casei* IMV B-7280 and *B. animalis* VKL (separately) restored the liver morphological structure of obese mice to a lesser degree. *B. animalis* VKB or *L. delbrueckii* subsp. *bulgaricus* IMV B-7281 (separately) and *B. animalis* VKL/*B. animalis* VKB composition were ineffective.

Probiotic use significantly reduces the risk of hepatic encephalopathy, but there is insufficient evidence regarding the effect on nonalcoholic fatty liver disease and nonalcoholic steatohepatitis (level of evidence B) [29].

However, the therapeutic use of probiotics and prebiotics treatment and prevention of patients with obesity-related NAFLD is not supported by high-quality clinical studies [161].

The complexity and **gender aspects** of liver fibrosis development and liver potential to regenerate associations with reproductive system was demonstrated [162].

The **non-invasive markers** like FIB-4, aspartate aminotransferase (AST) to alanine aminotransferase (ALT) ratio (AAR), AST to platelet count ratio (APRI), and platelet count to spleen diameter (PC/SD) ratio), etc. are definitely underestimated in the clinical set [163] and can be effectively used to evalutae metyabolic syndrome case for prescribtion probiotic treatment.

Substances with prebiotic properties have large potential to be used with probiotic strain for liver disase. Nanoceria demonstrate liver-protective properties [163]; citrulline a non-essential amino acid that helps to maintain healthy protein balance, facilitates protein synthesis for muscle tissue retention, can improve Western diet-induced liver injuries via decreased lipid deposition, increased insulin sensitivity, lower inflammatory process and preserved antioxidant status [164].

Cholestasis is an important and underestimated in clinical set diet-related issue for non-alcoholic fatty liver disease (NAFLD) development. Relationship between adipose tissue and fatty liver and its possible evolution in fibrosis, multifactorial pathogenesis of NAFLD, and treatments for various contributory risk factors are well supported by clinical and research experience [165–168]. The upper limit of normality measured diameters of common bile ductwas reported to be 7.9 mm (from 3.9 mm among those aged 18–25 years to 4.7 mm in aged more than 55 years) [169].

Well-designed unbiased multicenter studies on evaluation of the gut-microbiota-liver metabolic network and the intervention of these relationships using probiotics, synbiotics, and prebiotics, and personazlied nutrition are strongly requires in the field.

Thus, bacterial strains have different probiotic effects on metabolic disease and obesity.

Probiotics affect on physiological functions and metabolic processes directly or through the normalization of microbiocenosis of mucous membranes of various organs and body systems, however, the range of their biological activity is a *strain-dependent characteristic* [31].

For example, in clinical and experimental studies probiotic bacteria *L. plantarum* and *L. gasseri* reduced the body weight [170] and cholesterol level [171], but, on the contrary, *L. acidophilus*, *L. fermentum* or *L. ingluviei* affect increase the body weight [172], and *L. acidophilus* NCDC 13 had no impact on obesity [173].

In the recent study [11] we defined that the probiotic bacteria *L. casei* IMV B-7280, *L. delbrueckii* subsp. *bulgaricus* IMV B-7281, *B. animalis* VKB and *B. animalis* VKL (separately) or *B. animalis* VKL/*B. animalis* VKB/*L. casei* IMV B-7280 and *B. animalis* VKB/*B. animalis* VKL compositions were capable to decrease the weight of obese BALB/c mice and cholesterol level in serum and partially normalized intestinal dysbiosis, that was manifested in the increased number *Lactobacillus* spp., *Bifidobacterium* spp. and coliform bacteria. A decreasing of the liver size and a mesenteric fat thickness measured in obese mice by ultrasound was also observed under the effect of mentioned probiotics [11].

Recently Vinderola et al. [174] noted that the value of in vitro tests as predictors of probiotic therapeutic capacity is still uncertain, and a lack of standardized in vitro protocols for strain selection. Nevertheless, studied criteria can allow narrowing the list of potential strain candidates.

Collected evidence during last two to three decades naming Lactobacillus and Bifidobacterium as best genera with probiotic properties, has been revisited and assessed through meta-analyses, several demonstrated that Lactobacilli and Bifidobacteria are effective, always in a **strain-dependent** manner, against different microbiota-associated diseases [175].

The claimed beneficial chracteristics are **strain-dependant** however, can be found within a genus and considered where appropriate as **genus-specific** if evidence on strain-specific action lacking. *Species-* and *genus-specific* health claims were documented [175].

6.3.5 Kidney and MetS

Obesity co-morbidities include insulin resistance, diabetes mellitus type 2, dislipidemia, which are the most frequent contributing factors for the inception of metabolic syndrome (MetS), and non-alcoholic fatty liver disease (NAFLD) that includes steatosis and steatohepatitis and liver fibrosis and increase the risk of developing chronic kidney disease (CKD) [176].

Endogenous intoxication syndrome (EIS) has several non-specific displays in organism in pathological conditions with inflammatory effects and metabolome changes. Biological fluids of organism in pathological processes have high contents of lipids and carbohydrates metabolites and when altered demonstrate toxic effects on the liver, kidneys and brain cells [177, 178]. Most of these toxins belong to the *middle mass molecules* (middle molecules, MM).

The *Middle Molecule Hypothesis* was suggested decades ago by Babb et al. [177] and has been rediscovered recently in personalized medicine via developing unbiased techniques in the proteomic, genomic and metabonomic [177, 178].

specific charges in the gut microbiota in CKD an increase in bacterial species prone to proteolytic fermentation, such as *Clostridium* and *Bacteroides* and/or a decrease in bacteria that may be protective or release potentially nephroprotective molecules (e.g., short chain fatty acids), such as *Lactobacillus* [179].

Fenugreek can be considered a potentially effective prebiotic for a number of beneficial applications and advances in development of treatments of immune-related disorders and decrease MM content to the normal level levels of uric acid and urea in blood in high-calorie diet induced obesity rat model [34].

6.3.5.1 Renal Doppler

Ultrasound (US) is a well-acknnowledged source to provide number of relevant biomarkers of disease and phenotype. Thus, type 2 diabetic patients have higher values of resistive index (RI) on Doppler ultrasound as compared to non-diabetics and this increment is proportional to the duration of diabetes. An intrarenal RI value of >0.7 identifies diabetic patients at risk of progressive renal disease. Higher RI correlates to higher protein in urine and duration of diabetes in diabetic patients [180].

Renadyl probioic composition (*S. thermophilus* KB 19, *L. acidophilus* KB 27, and *B. longum* KB 31) was reported to be safe to administer to end-stage renal disease patients on hemodialysis. Stability in QOL assessment is an encouraging result for a patient cohort in such advanced stage of kidney disease [181].

6.3.6 Hyperuricemia and Gout

Ultrasound can be an effective method for early detection of liver and kidneys involvement in gout patients for facilitate performing personalized treatment. The sensitivity, specificity, positive and negative predictive value and accuracy the gout involvement of liver and kidneys using complex ultrasonography diagnostic criteria hasve beenm known as high as 92.6%, 84.4%, 80%, 95%, and 91.9% respectively. Nephropathy appearance correlates with diffuse liver involvement. Integrated index is reliable for disease staging and control treatment follow up [182].

Probiotic therapy alleviates hyperuricemia in C57BL/6 mouse model [183]. Probiotics supplementation administration including compositions of *L acidophilus* KB27 + *L rhamnosus* KB79 or *L acidophilus* KB27+ *L rhamnosus* KB79 compositions prevented renal alterations, oxidative stress induced by hyperuricemia [184]. The probiotic strain *Bifidobacterium longum* 5(1A) ameliorate monosodium urate crystal (MSU)-induced inflammation in a murine model of gout, evoke inhibition of the production of CXCL1 and interleukin(IL)-1β in joints as seen by reduced hypernociception, reduced neutrophil accumulation in the joint and myeloperoxidase activity in periarticular tissue; and increase levels of the anti-inflammatory cytokine IL-10 [185].

6.3.7 Asthma

Recently we performed focused study to evaluate health metabolic parameters associated with asthma and potential external triggers affecting life quality and observed significantly higher incidence in patients with asthma [186]: younger age (20–40 years); female gender; the predominant months of birth in patients with were *January, April and July;* appendectomy and/or tonsillectomy in anamnesis strongly correlated with asthma incidence. Among asthma-associated diseases an allergy occurred in 69% patients with asthma; obesity – in 32%; gout – in 18%; T2DM – in 28%;intestinal disorders (reflux, IBD) – in 58%; microsplenia – in 54%; fungal sensitization – in 15% patients respectively. Physical and intellectual exertion, alcohol consumption, sauna, long stay in cold and damp room were most relevant parameters affecting life quality and provoking exacerbations. Are, significantly associated with asthma, risk factors, affecting exacerbations [186].

Recent data show that *C. butyricum* (CB) administration significantly increased the therapeutic effect of allergy immunotherapy (AIT) on asthma, in which the allergen-specific B10 cells were generated via inducing the chromatin remolding at the IL-10 gene locus in the B cells [187].

Lactobacillus strains were reporeted to improve outcomes of respiratory infections. Mucosal adhesion is incorrectly taught as essential for both non-immune and mucosal immune defense mechanisms. For example, noncolonizing probiotics, such as *Lactobacillus casei*, may exert their functions in a transient manner or by influencing the existing microbial communities [188].

6.3.8 Role of Spleen-Associated Biomarkers in Patient Stratification for Microbiota Modulating

Our preliminary results demonstrated changes in the spleen size in all participants after 1-year Antarctic expeditions with a tendency to deacrese after returning (this was also observed in the liver and thyroid gland size) [133]. Inordinate splenic erythropoiesis can be initiated e.g. during the development of chronic mountain sickness in chronic hypoxia [189].

The spleen and intestine are two major immune organs involved in the innate immune response to infection [190]. Spleen structure and size might be supposed as promising imaging biomarker for immunity- and stress-related conditions. Spleen structure and function are underestimated in medical profiling, since the bone marrow remains the most important erythropoietic organ under both resting and stimulated states.

LAB strains properly selected according to their antagonistic activity against pathogenic bacteria, resistance to low pH and milieu of bile salts can affect cytokine Th1/Th2 balance toward nonallergic Th1 response [191].

6.3.9 Probiotics for Neuroendocrine Applications, APUD Cells, Serotonin, Glutamate Signaling

Neuroendocrine, amine precursor uptake decarboxylase (*APUD*) cells signaling, *serotonin* are important and not sufficiently studied mechanisms for a number of pathologies of different localization and link among series of pathological processes as obesity, gut motility, cancer, etc. Serotonin is a primal signaling molecule conserved across phyla that is implicated in the control of energy balance [192–194].

As obesity increases peripheral serotonin, the inhibition of serotonin signaling or its synthesis in adipose tissue may be an effective treatment for obesity and its comorbidities [193].

Crane et al. [193] have found that genetic or chemical inhibition of Tph1 protects or reverses the development of FED-induced obesity and dysglycemia via activation of UCP1-mediated thermogenesis. Thus, inhibiting Tph1-derived serotonin may be effective in reversing obesity and related clinical disorders such as NAFLD and type 2 diabetes [194].

APUD-system play important role in apoptosis signalling and interreaction among health normal and pathological conditions cycle changes in the endometrium [195].

MSG induce development insulin resistance to peripheral glucose uptake, induces hyperinsulinemia and the obesity disrupt the regulation of the hypothalamic-pituitary-adrenal axis resulting in the hyperfunctional state of adrenals [33]. MSG evoke metabolic alteration characterized by an enhanced adipocyte capacity to transport glucose and to synthetize lipids resulting in increased insulin sensitivity. It was supposed that the central lesions produced by MSG treatment. Probiotics mixture (2:1:1 *Lactobacillus casei* IMVB-7280, *Bifidobacterium animalis* VKL, *B. animalis* VKB) was effective for MSG-induced obesity [33].

6.3.10 Collateral Pathologies Associated with the Obesity in Women

Metabolic disturbances in obesity causes a number of diseases, namely CVD, and a number of tumor sites of lung cancer, breast cancer, uterine cancer, and ovarian cancer; in women, there is a violation of ovarian menstrual cycle called dyslipidemia [196].

6.3.10.1 Progesteron

Evidence indicates that obesity is associated with hormonal (estrogen/progesterone) imbalance and also with inflammation not only in adipose tissue, but with systemic inflammation. Primary studies of experimental obesity have unfolded that progesterone promotes the growth of adipose mass of female rats [197, 198]. Progesterone replacement therapy has demonstrated the increased uptake of glucose and elevated protein level in the tissues of aging animals, increase of natural killer's activity and the with restoration of lipid and hormone levels as well [198].

The neuropeptide hormone **oxytocin** plays role in up-regulation of wound-healing enhancement using *Lactobacillus reuteri* probiotic [138].

6.3.10.2 Thyroid Hormones

Probiotics are recommended for autoimmune diseases [8, 76], both thyroiditis and Graves' disease are autoimmune thyroid conditions.

Decreased metabolism can be a result of thyroid hormone deficiency – **hypothyroidism, in majority induced by** autoimmunity and manifesting by fatigue, cold intolerance, constipation, dryness of skin and mucosda and weight gain. Probiotics-have not been known to directly affect thyroid hormones parameters in hypothyroid patients, however influence on thyroid hormones homeostasis is suggested since probiotics supplementation could be able to prevent serum hormonal fluctuations [199]. Hypothyroidism is associated with altered gut motility and *small intestinal* *bacterial overgrowth* (SIBO) [200]. *Bacillus clausii* was reported to beeffective for SIBO [201].

6.3.11 Gut Microbiota and Gut Motility

The disrupted microbiome in patients with constipation could be a potential therapeutic target. Many studies support the effects of different probiotics intervention with as a feasible way to ameliorate constipation, clinical trials show promising results in the application of probiotics (Kim et al., 2015; Wojtyniak et al., 2017) [202, 203].

The genus *Bacteroides* and proteins involved in **iron acquisition** and metabolism, cell wall, capsule, virulence and mucin degradation were enriched at the end of HBR suggest that both constipation and EC decreased intestinal metal availability leading to modified expression of co-regulated genes in *Bacteroides* genomes [204, 205]. Exercise prevent the crosstalk between the microbial physiology, mucin degradation and proinflammatory immune activities in the host [204].

We recently reported effects of CeO_2 nanoparticles affecting gastrointestinal motility on rat model and reviewed data supporting their perspectives to be applied as effective laxatives [36].

6.3.12 Probiotics for Musculoskeletal Diseases and Pain: Gut–Muscle Axis

The regulatory role of the gut microbiota in immune and inflammatory activity and the metabolic potential that it harbors provide a novel avenue of research for musculoskeletal diseases with potentially novel treatment options. The number of studies support the idea of significant associations among gut microbiota, physical activity and health [206–209].

Regular physical exercise performed at the moderate doses are recommended by the World Health Organization (WHO) [210], such physical activity as walking, cycling, or participating in sports can reduce the risk of CVD, diabetes, colon and breast cancer, and depression.

The human link with bacteria lasts over billion years and is explained by *endosymbiosis theory*. The similarity of mitochondria with Proteobacteria (gram negative bacteria) is a clear evidence for such link [211]. *Mitochondrial (MT) dysfunction* has been implicated in the aetiology of many complex diseases, as well as the ageing process. Much of the research on mitochondrial dysfunction has focused on how mitochondrial damage may potentiate pathological phenotypes [212, 213] also during physical activity. The potential for precise therapeutic microbiome interventions can target microbial-mitochondrial metabolic communication [207]. Thus, the microbiome can be an essential supplier of metabolites that act at the level of resident mitochondria of host in skeletal muscle to stabilize host metabolism [207].

6.3.12.1 Muscle Aging and Gut Microbiota

Frailty is the age-related loss of reserve capacity in multiple systems simultaneously, which results in reduced resistance to stressors at increasing age, sarcopenia is a condition of muscle loss and decreased performance and also with bone and joint disease in elderly. **Frailty** has been associated with alterations in the microbiome, in particular with butyrate producing microorganisms.

The use of novel therapeutic approaches influencing the gut *microbiota-muscle-brain axis* was considered for treatment of the frailty syndrome [214–216].

Lactobacillus strains appear to be effective for sarcopenia on a mouse model [214]. *L. reuteri* 6475 could impact the suppression of bone in a menopausal ovariectomized (Ovx) mouse model by possibly alteration of the immune response by changing intestinal microbial communities found in Ovx animals [216].

A small number of human studies have examined the impact of **exercise** on gut microbiota [217, 218]. Professional athletes had lower levels of inflammatory cytokines than the controls. In addition, they had increased microbial diversity (a positive indicator of gut health) [219, 220].

Accumulation of metabolites in m scles and in organism as a whole (like Pyruvate and Lactate) during exercise in normoxic and severe acute hypoxic conditions can be a target for microbiota associsted interventions [205]. Gut microbiota effects via by regulating gut mucosal pro-inflammatory and anti-inflammatory actions through the activity of reactive oxygen species (ROS) required for normal cellular homeostasis and physiological function including muscles [205, 206].

Multiple studies suggest a relationship between gut microbiota and inflammatory conditions such as **rheumatoid arthritis** (RA)**, spondyloarthropathies and gout** [214]. Alterations in the gut microbiome, in particular in Prevotella spp., associate with RA, but disease stage and genotype appear to moderate associations seen [214].

RA has long been associated with periodontal disease and oral microbiome [221].

A crucial molecular mechanism underlying **autoimmune and inflammatory diseases** like psoriasis, rheumatoid arthritis, and multiple sclerosis were doscovered recently. Bloch et al. [221] observed that the activity of the proinflammatory cytokine **IL-23** relies on the structural activation of its receptor **IL-23R**. The researchers involved hope that this information will support the development of **new therapies** targeting IL-23 [221].

L. casei appeared to have synergistic action with alone or alongside type II collagen (CII) and glucosamine (GS) (a candidate prebiotic) for effective reducing pain, cartilage destruction, and lymphocyte infiltration in an animal model of osteoarthritis [222]. Oral administration of L. casei together with CII and Gln more effectively reduced pain, cartilage destruction, and lymphocyte infiltration than the treatment of Gln or L. casei alone. This co-administration also decreased expression of various pro-inflammatory cytokines (interleukin-1β (IL-1β), IL-2, IL-6, IL-12, IL17, tumor necrosis factor-α (TNF-α), and interferon-γ (IFN-γ)) and matrix metalloproteinases (MMP1, MMP3, and MMP13), while up-regulating anti-inflammatory cytokines (IL-4 and IL-10). These results are concomitant with reduced translocation of NF-κB into the nucleus and increased expression of the tissue inhibitor of MMP1 (TIMP1) and CII in chondrocytes [222].

Obesity-associated inflammation can affect **osteoarthritis** progression independent of mechanical stress due to excess weight.

MetS has a cumulative and negative effect on hand osteoarthritis occurrence, independent of weight. Controlling metabolic comorbidities may have a beneficial effect on osteoarthritis, especially in obese patients [223–225].

Substantially, exercise can increase levels of Bacteroidetes and reduced *Firmicutes*. Appetite-regulating hormones (therefore the nutritional status) and exercise importantly affected the gut microbiota composition [154, 226].

The profound analysis of the regulatory pathways and mutual links between immune mechanics in tendon and muscles and skeletal muscle and their spasticity evoking myofascial pain. Chronic tension are associated with inflammation in tendons [227] and in muscles involving both immune and non-immune pathways contributing to muscle damage and weakness in myosistis [228].

The concept of of *repetition strain injury* (RSI) syndrome [229], and the evaluation of trigger points phenomena [230], and nervous phenomena evoking visceral pain can justify integrated multiparameter approach [231] in the field. This might give important pathogenesis clues to understanding this gut-brain-circulation-pain interaction as a whole for prevention of wide spectrum of MetS-associated collateral diseases and suggesting new health care policy, smart decision-making, and advances in education for economic benefits for aging society and working population.

It is essential to make efforts in increasing the level of evidence of individualized/personalized procedures of biological therapy Interventions like platelet rich plasma(PRP) [232] and/or stem cells [233], develop reliable self-assessment, development of relevant questionnaires for participating medicine, and set the studies of mutual impact of pain, lifestyle, metabolism, nutrition, gut-brain axis (GBA).

The correlation between MetS parameters like insulin resistance and blood pressure with anthropometric measures in adolescents (like *WC*, and others) were demonstrated [234, 235]. Thus, development and validation of neuromuscular, anatomy-based, movement assessemtn-based and pain biomarkers for predictive approach and for measuring outcomes can help their effective use. Extensive multilevel evaluation of motion posture is feasible and informative protocol using CAREN, static & dynamic balance tests, pressure analysis, US patterns of movement analysis to detect fittnes muscle, tendonds of fasciarelevant to metabolic disorders.

6.3.13 Vascular Regulation in Obesity, Congestion, Hypoxia and Ischemic Conditions

Recent studies have shown that adipose tissue is an active endocrine and paracrine organ secreting several mediators called *adipokines* [236]. Adipokines include hormones, inflammatory cytokines and other proteins [236]; namely: circulatory hormones (leptin, adiponectin, omentin, visfatin, angiotensin II, resistin, tumor necrosis factor-α, interleukin-6, apelin) and/or via local paracrine factors (perivascular adipocyte-derived relaxing and contractile factors). In obesity, adipose tissue becomes dysfunctional, resulting in an overproduction of proinflammatory adipokines and a lower production of anti-inflammatory adipokines. The pathological accumulation of dysfunctional adipose tissue that characterizes obesity is a major risk factor for many other diseases, including type 2 diabetes, CVD and hypertension.

Dysregulated synthesis of the vasoactive and proinflammatory adipokines may underlie the compromised vascular reactivity in obesity and obesity-related disorders.

Arterial tone can be controlled through the release of ROS, leptin, adiponectin, TNFα, IL-6, Ang II, omentin, resistin, visfatin, apelin and ADRF. The regulation of arterial tone might be compromised in obesity and obesity-related disorders (for example, T2DM, CVD and hypertension) because of alterations in the secretion of vasoactive adipokines by dysfunctional adipose tissue. Circulating levels of adiponectin and are decreased, while levels of leptin, resistin, apelin and proinflammatory cytokines are increased [236, 237].

Different depots of adipose tissue include white adipose tissue (WAT), brown adipose tissue (BAT) and thoracic and abdominal perivascular adipose tissue (PVAT). The phenotype of thoracic PVAT resembles BAT, whereas abdominal PVAT is more like WAT [238].

Perivascular adipose tissue (PVAT) was suggested to determine the inflammatory phenotype depending on species, anatomic location, and environmental factors, and that these differences are fundamentally important in determining a pathogenic versus protective role of PVAT in a vascular disease [239]. Dysfunction of perivascular adipose tissue induced by fat feeding suggests that this unique adipose depot is capable of linking metabolic signals to inflammation in the blood vessel wall [240].

Meyer et al. [241] noted that perivascular fat cells in the aorta of obese mice potentiate vascular contractility to serotonin and phenylephrine, indicating activity of a factor formed by a perivascular fat cell, which was designated as 'adipose-derived contracting factor' (ADCF) [241] Inhibition cyclooxygenase (COX) completely prevented ADCF-mediated reductions, whereas selective inhibition of COX-1 or COX-2 was only partially effective. In contrast, the inhibition of superoxide anions, NO-synthase or endothelin receptors did not affect the activity of ADCF [241].

Endothelial dysfunction (ED) is a major risk factor that affects blood flow control in various organs. Obesity impairs the microvascular function in several ways. ED is the result of an imbalance between nitric oxide (NO) and endothelin (EDN), a vascular function regulators. ED is associated with a decrease in NO production due to impaired activity and expression of endothelial NO synthase and increased production of superoxide anion and an endogenous NOS, ADMA inhibitor, along with increased vasoconstrictor factors, such as activation of endothelin-1 and sympathetic nerve [242].

In obesity, a mixed-food drink reduces skin perfusion mainly and causes acetylcholine-associated vasodilatation but does not affect the density of the capillary [243]. The acetylcholine-mediated vasodilation after eating can be impaired in obesity, the latter findings detected with a deterioration of the postprandial microvascular function in obesity [243]. Genetic variants in NO synthase and isoforms EDN and its receptors (EDNRA and EDNRB) appear to take into account important components of dispersion in ED, especially if there are simultaneous risk factors such as obesity. The analysis of genotype-phenotype interactions is critical for formulating a potentially variable susceptibility to CVD [244]. NO synthase and endothelin genes are associated with many diseases, such as asthma [245], which makes them a potential biomarker for numerical pathologies of obesity.

Insulin-resistance participates in the development of endothelial dysfunction and interferes with vascular homeostasis in patients with metabolic syndrome [246].

MetS involve large conductance vessels, promoting atherosclerosis, but also occurs at a microcirculation level, suggesting an important role for insulin in controlling vascular resistance and, finally, organ perfusion.

Early vascular changes the liver microcirculation are induced by insulin-resistance in non-alcoholic fatty liver disease and in chronic hepatitis with insulin-resistance [246].

intestinal inflammation associated with changes in the underlying mesenteric fat depots as venular dilatation and **congestion**, and perivascular accumulation of neutrophils [247].

Congestive mesenteric and/or *pelvic syndromes* are the condition characterized by the presence of venous congestion and varicose veins in the mesenteric and pelvic region, and play important role for dysregulation of intestinal and systemic microcirculation mechanisms leading to ED and have potential risk for the development of many vascular and hormonal disorders [36].

Systemic *congestive* phenomena due to heart failure associated with distinct gut microbiota dysbiosis [248].

Doppler techniques for assessment of vascular responses following cuff-induced arterial occlusion allow determinations of the kinetics of post-ischemic reperfusion and provides an accurate reporter of NO-mediated physiological recruitment [249]. At present, the reference diagnostic modality for intestinal ischaemia is contrast-enhanced *computed tomography (CT)* [250]. However, there are some disadvantages associated with these techniques, such as radiation exposure, potential nephrotoxicity and the risk of an allergic reaction to the contrast agents. Thus, not all patients with suspected bowel ischaemia can be subjected to these examinations. Despite its limitations, US could constitutes a good imaging method as first examination in acute settings of suspected mesenteric ischemia [250].

6.3.14   Hypoxia in the Gut

The epithelium overlying all mucosal tissues is supported by a rich vasculature. In these settings, even small perturbations in blood flow can result in relatively large decreases in O2 delivery (hypoxia) to the supporting epithelium [204, 205].

Hypoxia, and specifically HIF-target pathways that are strongly associated with tissue barrier function and metabolism that contribute fundamentally to inflammatory resolution [251].

Tissue (NBR) and combination of tissue and systemic hypoxia (HBR) increased inflammatory responses in inactive variants were recently linked to central inflammatory mediators nuclear factor kappa B (NF-kB) and transcription factor hypoxia inducible factor 1 (HIF-1) as a regulator of the cellular response to low oxygen levels to shape nutritional-immunity status of the gut and induce the release of reactive oxygen and nitrogen species [252].

However, it was reported [205] that a short-term modifications in host exercise levels and constipation or systemic hypoxia do not change significantly gut permeability, concentration of crucial intestinal metabolites, structure and abundance of butyrate producing microbial community; but progressive constipation (decreased intestinal motility) and increased local inflammation markers suggest that changes in microbial colonization and metabolism were taking place at the location of small intestine [205].

According to our recent observations a long stay in extreme conditions of Antarctica evoke adaptive reactions associated with hypoxia and mitochondrial dysfunction, determined by a set of molecular-genetic mechanisms that trigger the expression of the corresponding genes and alter the mitochondria ultrastructure, leading to the death of organelles, and subsequently the cells, and are associated with pronounced oxidative stress [253].

The mesenteric blood flow redistribution can impact on the gut microbiota and potential probtioc effect [254]. The higher release of short-chain fatty acids (SCFAs) was reported by the distal intestines relative to the proximal intestines. SCFAs concentrations were measured highest in the inferior mesenteric vein and the portal vein and lowest in the radial artery. The mucosa of the proximal intestines may metabolise a relatively larger fraction of SCFA and the differences in local SCFA production may play a role [254]. Since arterial acetate concentrations correlate with those in the mesenteric vein, the last value can serve as biomarker for evaluating efficacy of probiotic strain.

The development of adipose tissue involves remodelling of the extracellular matrix (ECM), which requires *matrix metalloproteinase* (MMP) activity, the potential of MMP inhibitor (*tolylsam*) to inhibit adipose tissue-derived MMP-2 and MMP-9 was confirmed. Paradoxically, gelatinase A (MMP-2) and gelatinase B (MMP-9) mRNA expression in adipose tissues was enhanced following inhibitor treatment [255].

Strains VSL#3 impact on MMP activity, MMP-2 and MMP-9 activities, and expression of iNOS and COX-2 in the rats receiving FED [147].

Peripheral microcirculation assessment might be considered to support a supplementary information for obese patients, including imaging laboratory biomarkers and capillaroscopy [particularly for vasospasm assessment and also for *Flammer syndrome* [256–258].

Probiotic VSL#3 ingestion prevents endothelial dysfunction in the mesenteric artery of CBDL rats, and this effect is associated with an improved vascular oxidative stress most likely by reducing bacterial translocation and the local angiotensin system [259].

The oxygen tolerance of probiotic bacteria can provide promising insights in the matter. Little is known about the effect of oxygen and hypoxia on the physiology of probiotic bacteria and microbe-host interactions. *Bifidobacterium* spp., *L. acidophilus. L. rhamnosus* GG can potentiate intestinal hypoxia-inducible factor (HIF) [91].

The relevance of vascular componen during microbiota modulating MetS is underestimated but has to be considered in following context:

- Adipose tissue produce and secrete several adipokines. Some of these adipokines possess vasoactive properties;
- Arterial tone and congestive phenomena provide different vascular patterns;
- Vascular phenomena impact on permeability and absorption of metabolites digestion in different part of intestine;
- Hypoxia can impact on microbes and specific straind have different properties;
- The role of microbiota in vascular dysregulation development via genetic predisposition and mutual affecting is still unclear;
- Probiotics can boost antiinflammatory PVAT, affect endothelin, HIF [91] signaling, etc.

6.3.15 Cancer, Gut Microbiota and MetS

Various prognostic and etiological factors, biomarkers, and molecular pathways of development and progression of the disease, common to MetS, atherosclerosis and cancer, suggest that the two most common diseases globally are significantly more aligned than previously thought. Both diseases have common etiological factors: genetic predisposition, age, sex hormones, smoking cigarettes, high intake of dietary fat, toxins and mutagens. The consequences of the aforementioned etiologic factor actions are deregulation of the cell cycle, oxidative stress, chronic inflammation, endothelial dysfunction, dysregulation of apoptosis and angiogenesis, instability of DNA and damage to DNA repair [260].

The TGF-ß signaling pathway, other growth factors, cell adhesion molecules, the Wnt-ß-catenin signaling pathway, excess matrix digestion associated with matrix metalloproteases, and NF-kB signaling pathway represent other common molecular progression pathways shared by both diseases [260].

In addition, the associations between microbiota and metastatic cancer, hypoxia in particular for Flammer syndromne phenotype individuals is a challenging task

[258]. A novel hypoxia-based mechanism of regulation of homeostasis and metastasis, leading to the formation of focal pre-metastatic lesions, and these lesions subsequently provide a platform for circulating tumour cells to colonise and form metastases [258, 261].

Many of the bacterial species of the phylum Firmicutes (LAB) produce butyrate, and a decreased abundance of these bacteria was observed in patients with colorectal cancer [262]. It has become evident that microbiota, and particularly the gut microbiota, modulates the response to cancer therapy and susceptibility to toxic side effects [263, 264]. Finally, many probiotic properties should be implemented to cancer case management as supoprtive therapy and to faciliate symptoms, associated with treatment [264].

Lactobacillus rhamnosus GG probiotic strain have been shown in mice to protect the intestinal mucosa against chemotherapy- or radiotherapy-induced toxicity by relocating cyclooxygenase 2 (COX2)-expressing cells from the villi to the base of the intestinal crypts [265]. *Bifidobacterium spp.* in the gut microbiota promotes antitumour immunity in mice that is received anti-PDL1 therapy [266].

However, the translation from mouse models as a main source of evidence to humans is a challenge. Thus, it is difficult yet to conclude that activation of TLR9 in humans by Bifidobacterium spp. has the same immunostimulating activity as observed in the mouse, and detailed clinical data are required to determine whether Bifidobacterium-containing probiotics would stimulate antitumour activity also in patients [263].

6.3.16 Gender-Specific Approach for Microbiota Modulation

Age and gender aspects are important issues for the selection of probiotic species for individual use.

Gender-specific integrated *Women* and *Men health* concepts have been widely appreciated as part of a large range of factors that affect fertility and general health that are associated with lifestyles, nutrition, obesity, and gender, with pathology [196, 267].

There is no consensus on what would be characteristic and consistent discrepancies between the microbiote of women and men still exists.

Differential metabolic responses to weight loss diets, with lower abdominal fat loss for women, better response to high levels of protein compared to high carbohydrate diets, higher seizure-risk behaviors compared to the benefits of physical exercise, as well as the tendency to slow down central manifestations obesity, MetS, T2DM, cardiovascular disease and some types of cancers before menopause, but then accelerates-do not foresee the need for different metabolic and chronological perspectives for the prevention running/interference [267].

A large number of bacterial genes was smaller in men than in women. In fact, a large number of this type has decreased in men with an increase in BMI [268].

Thus, the use of antibiotics like vancomycin can seriously affect the host microbiota and metabolism, especially in the risk groups of obesity prediabetes, in men, can reduce the bacterial diversity and reduce Firmicutes, which are involved in the metabolism of the short chain fatty acids and bile acids, and also activatethe expression of genes in adipose tissue of the oxidative pathway and associated with the immune pathway [269].

Among the factors that most likely mediate gender-dependent interactions are **sex hormones** [270–272].

Org et al. [270] showed gender-specific differences in gut microbiota composition and bile acids Interestingly, the hormonal status of male mice clearly affected the composition of microbiota on chow and high fat diets, whereas *in females this effect was more prevalent in response to the high-fat diet.* Testosterone treatment after gonadectomy prevented the significant changes that were seen in untreated males. Hormonal changes can also strongly affect bile acid profiles and that significant gender-specific differences in bile acid profiles become more prominent in response to a high-fat high-sugar diet [270].

Sex-specific changes in glucose–insulin homeostasis, can be ameliorated in males treated with estrogen [271, 272].

Compared with males, female mice demonstrate increased capacity for adipocyte enlargement in response to a long-term high-fat feeding, which is associated with reduced adipose tissue macrophage infiltration and lower fat deposition in the liver, and with better insulin sensitivity [273]. The extensibility of adipose tissue linked to adiponectin secretion might determine the sex differences in obesity-associated metabolic disorders [273].

The asscociations between liver function and reproductive system as well as sex-dependent aspects of liver fibrosis were demonstrated [162].

A high incidence of hyperandrogenism, polycystic ovarian morphology (PCOM) and polycystic ovary syndrome (PCOS) has been reported in T1DM, which is thought to be due to intensive insulin therapy [274]. Patients with PCOS have less diversity and altered phylogenetic profile in the microbioma of the stool, due to clinical parameters. Intestinal barrier and endotoxin dysfunction are not the driving factors in this cohort of patients, but may contribute to the clinical phenotype in some patients with PCOS [275]. Women with PCOM have changed α diversity, which was an intermediate between the two other groups. Below, α-diversity is observed in women with PCI compared with healthy women. The results show that hyperandrogenicity can play a decisive role in the change of intestinal microbial in women with PCOS [276]. The probiotic supplementation for women with PCOS for 12 weeks favorably affects the total testosterone, TAC and MDA, SHBG, mFG scores, hs-CRP, but did not affect other metabolic profiles [277].

6.3.17 Age

This integrated vision of theory of aging, and longevity under "optimistic conception of prolongation of human life" under using probiotics, developed and foreseen by Ukrainian scientist Elie Metchnikoff, the founder of concepts of probiotics, phagocytosis, and gerontology [278], and more, the Nobel prize winner in 1908, who, created and developed the concept for diet-driven microbiota modulation and probiotic treatments, beneficial for health decline upon ageing that becomes to a reality today over 100 years after [279].

Three problems common in the elderly, namely, undernutrition, constipation, and the decline in efficiency of the immune system may all be beneficially affected by appropriate probiotic organisms [280]. Collectively, the data support a relationship between diet, microbiota and health status, and indicate a role [280, 281].

The loss of community-related microbiota correlates with increased muscle **frailty**. In general, the data support the relationship between diet, microbiote and health, and points to the role of nutritional changes in microbiota of varying degrees in the reduction of aging [281]. During aging, the microbiological compartment significantly correlates with indicators of weakness, concomitant illness, nutrition, inflammatory markers and metabolites in fecal water. The individual microbiota of long-term care was much less diverse than that of the community. The loss of microbiota associated with the community correlates with increased deficits [281].

Women in menopause is specific case of aging strongly associated with gut microbiota changes. However, the scientific evidence up to date still *do not definitively demonstrate* how non-vaginal microbiota interplay with the health of menopausal women [282].

Reproductive aging negatively affects diabetes [283]. Women with T1DM have have shorter than average reproductive life through later menarche and earlier menstruation. Reproductive aging among women with T2DM is more diverse; early menopause may occur more often [283].

Lignans, which are the major phytoestrogens occurring in Western diets are recommended for people in age [284]. Consumption of *Bifidobacterium lactis, Lactobacillus rhamnosus and Lactobacillus acidophilus* demonstrated increasing the ability to fight infections in elderly patients [284].

6.3.18 Ethnicity

The evidence about ethnicity or population-specific microbiome compositional variations rise questions on the universality of microbime modulations and suppose to recommend geographically adapted approaches for therapeutic strategies. General microbiological manipulations, developed on the basis of research in Western societies, might have unexpected and even adverse effects for non-western groups [74, 98, 99, 285].

6.3.19 Environment

The gut microbiome is not significantly associated with **genetic** ancestry, and host genetics have a minor role in determining microbiome composition, rather **environment** is supposed to be a main trigger modulating human microbiome [111, 286]. On the other hand reciprocallymicrobiome data can significantly improve the prediction accuracy for many human diseases like MetS, compared to models that use only host genetic and environmental data [286].

A notable beneficial mutualistic relationship of the host with gut microbiota, effort should be given to the identification of the conditions that change the expression and maintenance of the probiotic effector compounds mediating host–microbe interactions in the gut [287].

Recently the role of structure of the surrounding microbial ecology, its biodiversity, has been emphasized for implications on human microbiome and public health [288], in particlule the indoor microbiome as a complex microbial ecosystem is largely dependent on the human-associated habitats, and environmental factors like geography and building type [288], these interrelationship maintaining critical.

Differential gut microbial community assembly scenarios in **rural** and **urban** settings were demonstrated [289]. Thus, Western diets, antibiotics and food additives (high variable selection) lead to low α-diversity (species richness within a single microbial ecosystem) and high β-diversity (diversity in microbial community between different) [289]. High α-diversity and low β-diversity is observed under low dispersal limitation (poor hygiene and sanitation) [289].

Therefore promoting low homogeneous selection (visiting rural area, in particular milk farms, more farming and labour work, and contact to domestic animals,etc), low variable selection (diet rich in fibre and natural foods) might be hypothesized to increase environmental and human diversity and improve multisite microbial health.

6.4 Endnotes and Recommendations

6.4.1 Recommendation for Individualized Clinical Use of Probiotics

Interindividual differences in the risk of developing MetS, disease manifestations, and responses to diet and medical treatment are often ascribed to human genetics and lifestyle [24]

6.4.1.1 Recommendations

- High product quality;
- Effectiveness should be proven on the basis of evidence-based medicine for routine use in the clinical setting;
- Personalized (or individualized) approach needed in prescribing probiotic according to the disease, clinical case and phenotype of the patient;
- The probiotic properties should be considered in strain-dependent approach [31] and /or genus-specific [175] in case if the evidence on strain-specific effects is lacking;
- Using live microorganisms is essential for therapeutic efect (however, dead microorganisms also might demonstrate fair therapethic effect);
- Selection the 'best' strain for particular case (for example, the *L casei* strain has strongest properties in most characteristics);
- The higher effectiveness of *multiprobiotics* has not been finally proved – best *"single* strain" concept is preferrable for the personalized use;
- The dose should be at least 10^9 microbial bodies;
- Use of right prebiotics and right combination with probiotic;
- The appropriate route of delivering a probiotic drug (capsule, gel, novel encapsulation technologies);
- Crucially important is combination with a appropriate diet.

6.4.2 Dose & Periodicity of Probiotics Treatments

The recent review of dose-responses of probiotics during studies and antiobesity programs suggests that

- The studying higher doses for this end-point would be most worthwhile;
- the lack of a clear dose-response on lower doses (less than 10^8 CFU/day) [290];
- are lacking and may explain why a non-effective dose is not commonly identified
- evidence-based recommendations for treatment indications for probiotics suggested the dose 10^9 or higher [19], in some cases dose can be increased;
- in a volunteer study by Larsen et al. [106] the recovery was demonstrated in group receiving 10^{11} CFU/day of probiotic strain. **High doses of probiotics** in humans *are well tolerated* [291].

The recent findings suggested that the microbiome should be targeted during antiobesity programs, close interplay between nutritional modulation of gut microbiota for healthy aging. E.g., calorie restriction can effectively lengthen lifespan has health-promoting potential. However, these option should be treatly person-related. A correct selection of an optimal time-frame for intervention during antiobesity program is critical point effecting clinical success. In our study the metabolic disorders (e.g., increased glucose, cholesterol levels) remained long after receiving FED, even on the standard diet.

Recommendations on a probiotic treatment **duration**, breaks between sessions and dietary regime during and after treatment [292] have not been finalized.

The beneficial changes of both gut microbiome diversity and metabolism in obese humans under weight loss intervention were not sustained during weight maintenance (Heinsen et al. [42]).

6.4.3   Recommendation for Probiotic Studies Design

The major of strains demonstrating beneficial properties for health *in vivo* have to be supposed to be clinically effective and chosen for further studies to be tested more precisely. This approach to choose appropriate strain would be helpful considering strong bias in the clinical trials.

Correlation between in vitro and in vivo assays in selection strains has been debated [174]. Some common in vitro tests in the selection of potential probiotic strains used globally include evaluation of resistance to gastrointestinal digestion, adhesion to cell lines and prokaryotic-eukaryotic co-culture for immunomodulation. Some common in vitro tests in the selection of potential probiotic strains used globally include evaluation of resistance to gastrointestinal digestion, adhesion to cell lines and prokaryotic-eukaryotic co-culture for immunomodulation.

The associations between in vitro properties and potential probiotic application were hypothesized and illustrated [31]. The study by Larsen [293] indicates that pectins have a potential to protect probiotic bacteria of Lactobacillus species through the gastro-intestinal transit. Thus, pectins have a potential to improve survival of probiotic Lactobacillus species exposed to the gastro-intestinal stresses, and identifies the features linked to their functionality [293].

Recent results indicated that *Lactobacillus plantarum* strains preferred to metabolize malic acid and reducing sugar in non-pH-adjusted juice (NJ, pH 2.65) [294].

Animal studies need to be closer to real digestion, focus on evironmental models over genetic as more realistic.

Microbiome data significantly improve the prediction accuracy for many human traits, such as glucose and obesity measures, compared to models that use only host genetic and environmental data [286].

Microbiome interventions improving clinical outcomes may be carried out across diverse genetic backgrounds [286]. Using algorithm integrating information of omics-based matrices [295] including study epigenetics transciptome, etc. [56, 57]. more predictable for human intervention studies [174].

Novel protocols are needed to render the selection of potential future probiotics more rational and the fact that changes in gastric pH and gastric emptying along digestion [174], using parameters of the microbiota diversity, like Alpha and Beta diversity, (P/B) ratio, Firmicutes-to-Bacteroidetes (F/B) mycobiome, etc. [92–94].

Some concerns about these tests include the fact that changes in gastric pH and gastric emptying along digestion are difficult to mimic in simple in vitro tests, unless more sophisticated approaches (for example, SHIME) should be used [174].

Recently we discussed the role of Simulator of Human Intestinal Microbial Ecosystem (SHIME) to study diet and microbiota and suggested as follows [296]: (1) Direct coupling of the SHIME technology with cell culture models required for evaluation of the gut barrier and endothelial function; (2) Clinical intervention study with SGM sequencing data before and after defined diets implementation for chronic diseases treatment is necessary; (3) Comparison of SHIME integrated technology with results obtained on cells/animal experiments and in silico model data for evaluation of adequacy of pre-clinical and clinical tools for the following implementation of patient stratification strategy in health care system [296].

Using of **preclinical imaging** (in analogue with the setting *in clinico*) can strongly extend results of experiment.

The bacterial wall elasticity evaluation as a fast and accurate method to assess parameters of probiotic strains to predict their immune-modulatory properties.

According to our observations, strains with most pronounced immune-modulatory properties demonstrate also a high efficacy in decreasing cholesterol levels, the correlation between *in vitro* – *in vivo* studies in decreasing cholesterol levels has been showed e.g., for *L. casei* IMV B-7280. There are examples of successful clinical implementation [23].

Human Studies – Personalized Approach for Microbiome-Modulating Interventions Needed for Searching Evidence

Many novel treatment although usual every day practice treatments found to be effective are still not supported by *level-I evidence*.

A case of probiotic research and the translation is a cornerstone to solve, possible only via changing health care and extensive public–private partnerships and regulatory bodies [297].

Importantly to consider appropriate *designs* for conducting, publishing, and communicating results of clinical studies involving probiotic applications in human participants [298, 299].

The recommendations of International Scientific Association for Probiotics and Prebiotics (ISAPP) [299] suggest to follow four recommendations to conduct clinical studies of probiotic and/or prebiotic use: to define the end goal to reach a highest clinical effect and impact; design the study to maximize the chance of a positive response; choose which strain(s) and/or product should be used and why; and carefully select the study cohort.

- *Nevertheless, it is realistic, that proper design of probiotic clinical trials is rather unfeasible or largely limited in large cohorts, especially done unpersonalized.*
- *Selection of most effective strain needs an effective research agenda for translation require high validity for prediction results in clinical set based on studies in vivo.*
- *Evidence might lack, when personalized approach (or at least individualized or person-centred) should be initially supposed, but not applied.*

The recent advances in *predictive, preventive, and personalized medicine* (PPPM) open new era in utilization of the microbiome in human health for patient-tailored preventative or early treatment measures. Personalized modulation of the microbiome via nutritional and *pre-, pro-, and post-biotic* intervention, suppose dramatic increasing of their efficacy and level of evidence [8, 24, 39, 300].

We believe that a comprehensive approach for evaluating efficacy of probiotic strains on obesity model allows to select the strains for creation effective probiotic preparations for prevention and treatment of metabolic diseases, which could be recommended for further preclinical and clinical studies.

The microbiome-wide association studies, which are analogous to genome-wide association studies *are the best option to follow up* current research with multiparameter stratification patients with MetS, including data of lipid, carbohydrates metabolism, antioxidant system, inflammatory response, etc. on the largest cohorts possible [301].

In order to achieve this ambitious goal a **diagnostic and predictive panel** with reliable model for stratification MetS is needed to be created via host profiling using dynamic monitoring of a set of translational biomarkers. A basic panel should include data of host's sex, age, phenotype, and *metabolic profile* with estimation of levels of cholesterol, lipids, glucose, insulin resistance, uric acid, leptin, adiponectin, plasminogen activator inhibitor-1, interleukin-6, −10, −12, −22, tumor necrosis factor-α, oxidized LDL, paraoxonase-1; imaging data on liver, kidney structure/function, organs vascularity patterns, etc.

Microbiome biomarkers, those related to the etiological role of gut microbiota, like lipopolysaccharide binding protein (LBP), C-reactive protein (CRP), fasting insulin, and homeostasis model assessment of insulin resistance (HOMA-IR), and other host-associated factors influencing the gut microbiota.

Flammer syndrome biomarkers (including NO, endothelin-1, questionnaire data), physical activity patterns and a broad data on dietary experience [256–258] should be considered.

Gender aspects for the use of probiotics are unclear, immune response was reported to have differences in both sexes, as well as gut microbiota differ in men and women and its impact on insulin sensitivity, therefore women are considered to be less sensitive to gut microbiota-associated metabolic diseases than men, yet is efficacious in premenopausal women [302].

Imaging biomarkers using non-invasive imaging techniques, such as computed tomography (*CT*), magnetic resonance imaging (*MRI*), and *US are largely underevaluated during micriobiome modulating.* The information regarding colonic microbiota and the colonic mucosa; muscles and nerves, vasularisation continue microbiota-related inflammatory morphologic changes of tissues particular in the colon can be obtained.

Study of microbiome under stress, physical and psychical exercises should provide a source of potential biomarkers.

This early detection, stratification patients with MetS will support treatment and prevention via nutritional and lifestyle modulation.

6.4.4   Diet, Food and Prebiotics

Study of probiotics of consumption should be studied and implemented with strong agreement on beneficial and functional foods patterns implemented by personalized approach, provided by properly applied and interpreted *dietary biomarkers*, evidence on probiotic-nutrients interactions and assessed with proper data collection tools.

The study should keep the focus on the potential increase in the efficiency and level of evidence through the use of potential effects of probiotic compositions (mixtures) detection the best strains and additional use of prebiotics.

The new definition of a **prebiotic** as '*a substrate that is selectively utilized by host microorganisms conferring a health benefit*' opens an opportunity to test substances that were not previously considered as prebiotics and can be suggested for use with probiotic strains with synergized activity.

Using mathematical modeling, e.g.,Bayesian network analysis was used to derive the first hierarchical model of initial inactivity mediated deconditioning steps over time [205]; considering use of alpha, beta, and gamma diversities (alpha x beta = gamma) among the fundamental descriptive variables of ecology [303]. Shannon measures were shown to be the only standard diversity measures which can be decomposed into meaningful independent alpha and beta components when community weights are unequal [303].

6.4.5   Legislative Issues of Microbiome

A successful translation of microbiome research is needed for recognition of the microbial effects of food products and their ingredients on health; relevant regulations; and reliable products with clear consumer health [93].

The use of probiotics is governed by the guidelines of a number of organizations including WHO and Food and Agriculture organisation (FAO) [19], World Gastroenterology Organisation (WGO) [304], ISAPP [20, 26], European Food Safety Authority (EFSA) [21, 22], United European Gastroenterology organization (UEG) and EPMA [8, 23] and others. *The legislative process* is complex and has been recently criticized in particular for EU to be 'adjudicate claims for probiotics is severely flawed, as has been stated by many outstanding scientists, companies and organisations' [305]. Taking into account the expected rapid progress in conducting research on microbiomes and probiotics within the framework of predictive preventive personalized medicine, it is necessary to combine interdisciplinary approaches.

6.4.6 Ethical Issues of Microbiome

All interventions should adhere World Medical Association's Helsinki Agreement [306]. However, novel reality of microbiome study challenges new demands also in ethics [307], considering e.g., psychological aspects of personal identity the concepts of "confidentiality" and "privacy". In medical practice, including microbiota study patients need preserve their medical history, diagnosis, and prognosis only to be shared among the health professionals who need it for providing care [308].

This is of great importance for the development of **biobanks** in the context of the study of probiotics and fecal transplantation [309, 310].

The task of translating human microbime research results into practical applications requires further understanding of the number of scientific, clinical, political and public interests and concerns [309].

Returning individual results in human microbiome research can provide a valuable clinical tool for patient care management, but highlight the need to address how to manage the processes ethically and consider contextual factors that may be unique to human microbiome research [309].

The issues highly relevant to microbiome biobanking were suggested [310] and should be addressed early on in microbiome research projects and also call for adjusting or developing new governance mechanism to better accommodate these changes: the nature of human microbiome samples and how different understandings have an impact on benefit/risk evaluation, privacy, informed consent, and returning the result to participants [310].

6.4.7 Business Model Aspect of Probiotic Use: Guarantees & Warranties of Quality of Probiotic Products

It has recently been reported that the content of many bifidobacterial probiotic products in the United States is different from the list of ingredients, sometimes at subspecies level. Only one out of 16 probiotics perfectly matches its labels in all samples tested [311].

Given the development of sophisticated business models in personalized medicine [312], probiotic treatment is strongly needed.

Author Contributions RVB suggested the idea, did did the literature analysis, prepared discussion, formulated future outlooks, prepared the first draft and performed the second and final article drafting.

MYS did the revision manuscript and data interpretation, did the contribution to the overall development of the studied topic. Both authors read and approved the final manuscript.

Conflicts of Interest Declare conflicts of interest or state "The authors declare no conflict of interest."

Ethics No human subjects or animals were included to the study. This study has been approved by the ethics committee of institutional review board and Special Academic Council on Doctoral

Thesis of D.K. Zabolotny Institute of Microbiology and Virology of the National Academy of Sciences of Ukraine (protocol N 7 issued 03.07.2018).

Conflicts of Interest The authors declare no conflict of interest.

References

1. WHO. Obesity and overweight: Fact sheet N. 311. http://www.who.int/mediacentre/factsheets/fs311/en/. Accessed 19 Mar 2018
2. WHO (2009) Cardiovascular disease, Fact sheet no. 317. WHO, Geneva. http://www.who.int/mediacentre/factsheets/fs317/en/print.html
3. Harsch IA, Konturek PC (2018) The role of gut microbiota in obesity and type 2 and type 1 diabetes mellitus: new insights into "old" diseases. Med Sci (Basel) 6(2):pii: E32. https://doi.org/10.3390/medsci6020032
4. Park S, Bae JH (2015) Probiotics for weight loss: a systematic review and meta-analysis. Nutr Res 35:566–575
5. Human Microbiome Project Consortium (2012) Structure, function and diversity of the healthy human microbiome. Nature 486:207–214
6. Parekh PJ, Balart LA, Johnson DA (2015 Jun) The influence of the gut microbiome on obesity, metabolic syndrome and gastrointestinal disease. Clin Transl Gastroenterol 18(6):e91. https://doi.org/10.1038/ctg.2015.16
7. Aron-Wisnewsky J, Clément K (2016) The gut microbiome, diet, and links to cardiometabolic and chronic disorders. Nat Rev Nephrol 12(3):169–181. https://doi.org/10.1038/nrneph.2015.191
8. Bubnov RV, Spivak MY, Lazarenko LM, Bomba A, Boyko NV (2015) Probiotics and immunity: provisional role for personalized diets and disease prevention. EPMA J 6:14
9. Reid G, Abrahamsson T, Bailey M, Bindels LB, Bubnov R, Ganguli K, Martoni C, O'Neill C, Savignac HM, Stanton C, Ship N, Surette M, Tuohy K, van Hemert S (2017 Jul) How do probiotics and prebiotics function at distant sites? Benef Microb 20:1–14. https://doi.org/10.3920/BM2016.0222
10. Lazarenko LM, Babenko LP, Bubnov RV, Demchenko OM, Zotsenko VM, Boyko NV et al (2017) Imunobiotics are the novel biotech drugs with antibacterial and immunomodulatory properties. Mikrobiol Z 79(1):66–75
11. Bubnov RV, Babenko LP, Lazarenko LM, Mokrozub VV, Demchenko OA, Nechypurenko OV, Spivak MY (2017) Comparative study of probiotic effects of lactobacillus and Bifidobacteria strains on holesterol levels, liver morphology and the gut microbiota ino bese mice. EPMA J 8(4):357–376. https://doi.org/10.1007/s13167-017-0117-3
12. Dao MC, Clément K (2018 Feb) Gut microbiota and obesity: concepts relevant to clinical care. Eur J Intern Med 48:18–24. https://doi.org/10.1016/j.ejim.2017.10.005
13. Khan MJ, Gerasimidis K, Edwards CA, Shaikh MG (2016) Role of gut microbiota in the aetiology of obesity: proposed mechanisms and review of the literature. J Obes 15(2016):7353642
14. Shen J, Obin MS, Zhao L (2013) The gut microbiota, obesity and insulin resistance. Mol Asp Med 34(1):39–58. Epub 2012 Nov 16
15. Arora T, Singh S, Sharma RK (2013) Probiotics: interaction with gut microbiome and anti-obesity potential. Nutrition 29(4):591–596. https://doi.org/10.1016/j.nut.2012.07.017
16. Lumeng CN (2013) Innate immune activation in obesity. Mol Asp Med 34(1):12–29. https://doi.org/10.1016/j.mam.2012.10.002
17. Chakraborti CK (2015) New-found link between microbiota and obesity. World J Gastrointest Pathophysiol 6(4):110–119. https://doi.org/10.4291/wjgp.v6.i4.110

18. Luca F, Kupfer SS, Knights D, Khoruts A, Blekhman R (2018) Functional genomics of host-microbiome interactions in humans. Trends Genet 34(1):30–40. https://doi.org/10.1016/j.tig.2017.10.001
19. WHO/FAO scientific document. http://who.int/foodsafety/fs_management/en/probiotic_guidelines.pdf. Accessed 11 Feb 2018
20. Hill C, Guarner F, Reid G, Gibson GR, Merenstein DJ, Pot B, et al. Expert consensus document. The international scientific Association for Probiotics and Prebiotics consensus statement on the scope and appropriate use of the term probiotic. Nat Rev Gastroenterol Hepatol 2014(8):506–514. https://doi.org/https://doi.org/10.1038/nrgastro.2014.66
21. EFSA NDA Panel (EFSA Panel on Dietetic Products, Nutrition and Allergies) (2016) General scientific guidance for stakeholders on health claim applications. EFSA J 14(1):4367. [38 pp.]. https://doi.org/10.2903/j.efsa.2016.4367
22. European Food Safety Authority (EFSA) (2008) Technical guidance – update of the criteria used in the assessment of bacterial resistance to antibiotics of human or veterinary importance. EFSA J 732:1–15. https://doi.org/10.2903/j.efsa.2008.732
23. Golubnitschaja O, Baban B, Boniolo G, Wang W, Bubnov R, Kapalla M et al (2016) Medicine in the early twenty-first century: paradigm and anticipation – EPMA position paper 2016. EPMA J 7:23
24. Jobin C (2018) Precision medicine using microbiota. Science 359(6371):32–34. https://doi.org/10.1126/science.aar2946
25. Sanders ME, Merenstein DJ, Ouwehand AC, Reid G, Salminen S, Cabana MD, Paraskevakos G, Leyer G (2016) Probiotic use in at-risk populations. J Am Pharm Assoc 56:680–686
26. Gibson GR, Hutkins R, Sanders ME, Prescott SL, Reimer RA, Salminen SJ et al (2017 Aug) Expert consensus document: the International Scientific Association for Probiotics and Prebiotics (ISAPP) consensus statement on the definition and scope of prebiotics. Nat Rev Gastroenterol Hepatol 14(8):491–502. https://doi.org/10.1038/nrgastro.2017.75
27. Rondanelli M, Faliva MA, Perna S, Giacosa A, Peroni G, Castellazzi AM (2017) Using probiotics in clinical practice: where are we now? A review of existing meta-analyses. Gut Microbes 8(6):521–543. https://doi.org/10.1080/19490976.2017.1345414
28. Parker EA, Roy T, D'Adamo CR, Wieland LS (2018 Jan) Probiotics and gastrointestinal conditions: an overview of evidence from the Cochrane Collaboration. Nutrition 45:125. e11–134.e11. https://doi.org/10.1016/j.nut.2017.06.024
29. Wilkins T, Sequoia J (2017) Probiotics for gastrointestinal conditions: A summary of the evidence. Am Fam Physician 96(3):170–178
30. van den Nieuwboer M, Browne PD, Claassen E (2016) Patient needs and research priorities in probiotics: A quantitative KOL prioritization analysis with emphasis on infants and children. Pharm Nutr 4(1):19–28
31. Bubnov RV, Babenko LP, Lazarenko LM, Mokrozub VV, Spivak MY (2018) Specific properties of probiotic strains: relevance and benefits for the host. EPMA J 9(2):205–223. https://doi.org/10.1007/s13167-018-0132-z
32. Mokrozub VV, Lazarenko LM, Sichel LM, Bubnov RV, Spivak MY (2015) The role of beneficial bacteria wall elasticity in regulating innate immune response. EPMA J 6:13
33. Savcheniuk OA, Virchenko OV, Falalyeyeva TM, Beregova Tetyana V, Babenko LP, Lazarenko LM, Demchenko OM, Bubnov RV, Spivak MY (2014) The efficacy of probiotics for monosodium glutamate-induced obesity: dietology concerns and opportunities for prevention. EPMA J 5:2
34. Konopelniuk VV, Goloborodko II, Ishchuk TV, Synelnyk TB, Ostapchenko LI, Spivak MYa, Bubnov RV. (2017) Efficacy of fenugreek-based bionanocomposite on renal dysfunction and endogenous intoxication in high-calorie diet-induced obesity rat model—comparative study. EPMA J. https://doi.org/10.1007/s13167-017-0098-2
35. Beregova TV, Neporada KS, Skrypnyk M, Falalyeyeva TM, Zholobak NM, Shcherbakov OB, Spivak MY, Bubnov RV (2017) Efficacy of nanoceria for periodontal tissues alteration in glutamate-induced obese rats-multidisciplinary considerations for personalized dentistry and prevention. EPMA J 8(1):43–49. https://doi.org/10.1007/s13167-017-0085-7

36. Yefimenko OY, Savchenko YO, Falalyeyeva TM, Beregova TV, Zholobak NM, Spivak MY, Shcherbakov OB, Bubnov RV (2015) Nanocrystalline cerium dioxide efficacy for gastrointestinal motility: potential for prokinetic treatment and prevention in elderly. EPMA J 6:6

37. Kobyliak NM, Falalyeyeva TM, Kuryk OG, Beregova TV, Bodnar PM, Zholobak NM, Shcherbakov OB, Bubnov RV et al (2015) Antioxidative effects of cerium dioxide nanoparticles ameliorate age-related male infertility: optimistic results in rats and the review of clinical clues for integrative concept of men health and fertility. EPMA J 6:1

38. Polak-Berecka M, Waśko A, Paduch R, Skrzypek T, Sroka-Bartnicka A (2014) The effect of cell surface components on adhesion ability of lactobacillus rhamnosus. Antonie Van Leeuwenhoek 106(4):751–762. https://doi.org/10.1007/s10482-014-0245-x

39. Reid G (2018) Microbes in food to treat and prevent disease. Expert Rev Precis Med Drug Dev. https://doi.org/10.1080/23808993.2018.1429217

40. Kovatcheva-Datchary P, Arora T (2013 Feb) Nutrition, the gut microbiome and the metabolic syndrome. Best Pract Res Clin Gastroenterol 27(1):59–72. https://doi.org/10.1016/j.bpg.2013.03.017

41. Veum VL, Laupsa-Borge J, Eng Ø, Rostrup E, Larsen TH, Nordrehaug JE, Nygård OK, Sagen JV, Gudbrandsen OA, Dankel SN, Mellgren G (2017 Jan) Visceral adiposity and metabolic syndrome after very high-fat and low-fat isocaloric diets: a randomized controlled trial. Am J Clin Nutr 105(1):85–99. https://doi.org/10.3945/ajcn.115.123463

42. Heinsen FA, Fangmann D, Müller N, Schulte DM, Rühlemann MC, Türk K, Settgast U, Lieb W, Baines JF, Schreiber S, Franke A, Laudes M (2016) Beneficial effects of a dietary weight loss intervention on human gut microbiome diversity and metabolism are not sustained during weight maintenance. Obes Facts 9(6):379–391. https://doi.org/10.1159/000449506

43. Zhang C, Li S, Yang L, Huang P, Li W, Wang S, Zhao G, Zhang M, Pang X, Yan Z, Liu Y, Zhao L (2013) Structural modulation of gut microbiota in life-long calorie-restricted mice. Nat Commun 4:2163. https://doi.org/10.1038/ncomms3163

44. Jang C, Hui S, Lu W, Cowan AJ, Morscher RJ, Lee G, Liu W, Tesz GJ, Birnbaum MJ, Rabinowitz JD (2018) The small intestine converts dietary fructose into glucose and organic acids. Cell Metab 27(2):351–361.e3. https://doi.org/10.1016/j.cmet.2017.12.016

45. Wei X, Song M, Yin X, Schuschke DA, Koo I, McClain CJ et al (2015) Effects of dietary different doses of copper and high fructose feeding on rat fecal metabolome. J Proteome Res 14:4050–4058. https://doi.org/10.1021/acs.jproteome.5b00596

46. Lambertz J, Weiskirchen S, Landert S, Weiskirchen R (2017 Sep) Fructose: A dietary sugar in crosstalk with microbiota contributing to the development and progression of non-alcoholic liver disease. Front Immunol 19(8):1159. https://doi.org/10.3389/fimmu.2017.01159

47. Zubiría MG, Gambaro SE, Rey MA, Carasi P, Serradell MLÁ, Giovambattista A (2017) Deleterious metabolic effects of high fructose intake: the preventive effect of lactobacillus kefiri administration. Nutrients 9(5) pii: E470. https://doi.org/10.3390/nu9050470

48. Crescenzo R, Mazzoli A, Di Luccia B, Bianco F, Cancelliere R, Cigliano L et al (2017) Dietary fructose causes defective insulin signalling and ceramide accumulation in the liver that can be reversed by gut microbiota modulation. Food Nutr Res 61(1):1331657. https://doi.org/10.1080/16546628.2017.1331657

49. Zhao L, Zhang F, Ding X, Wu G, Lam YY, Wang X, Fu H, Xue X, Lu C et al (2018) Gut bacteria selectively promoted by dietary fibers alleviate type 2 diabetes. Science 359:1151–1156. https://doi.org/10.1126/science.aao5774

50. Flint HJ, Scott KP, Duncan SH, Louis P, Forano E (2012) Microbial degradation of complex carbohydrates in the gut. Gut Microbes 3(4):289–306. Epub 2012 May 10. Review

51. Morrison DJ, Preston T (2016) Formation of short chain fatty acids by the gut microbiota and their impact on human metabolism. Gut Microbes 7:189–200

52. Diniz YS, Faine LA, Galhardi CM, Rodrigues HG, Ebaid GX, Burneiko RC, Cicogna AC, Novelli EL (2005) Monosodium glutamate in standard and high-fiber diets: metabolic syndrome and oxidative stress in rats. Nutrition 21(6):749–755

53. Bonder MJ, Tigchelaar EF, Cai X, Trynka G, Cenit MC, Hrdlickova B, Zhong H, Vatanen T, Gevers D, Wijmenga C, Wang Y, Zhernakova A (2016) The influence of a short-term gluten-free diet on the human gut microbiome. Genome Med 8(1):45. https://doi.org/10.1186/s13073-016-0295-y

54. Cederroth CR, Zimmermann C, Nef S (2012) Soy, phytoestrogens and their impact on reproductive health. Mol Cell Endocrinol 355(2):192–200. https://doi.org/10.1016/j.mce.2011.05.049

55. Kreznar JH, Keller MP, Traeger LL, Rabaglia ME, Schueler KL, Stapleton DS, Zhao W, Vivas EI, Yandell BS, Broman AT et al (2017) Host genotype and gut microbiome modulate insulin secretion and diet-induced metabolic phenotypes. Cell Rep 18(7):1739–1750

56. Ma D, Wang AC, Parikh I, Green SJ, Hoffman JD, Chlipala G, Murphy MP, Sokola BS, Bauer B, Hartz AMS, Lin AL (2018) Ketogenic diet enhances neurovascular function with altered gut microbiome in younghealthy mice. Sci Rep 8(1):6670. https://doi.org/10.1038/s41598-018-25190-5

57. Mischke M, Plösch T (2016) The gut microbiota and their metabolites: potential implications for the host Epigenome. Adv Exp Med Biol 902:33–44. https://doi.org/10.1007/978-3-319-31248-4_3

58. Sprockett D, Fukami T, Relman DA (2018) Role of priority effects in the early-life assembly of the gut microbiota. Nat Rev Gastroenterol Hepatol. https://doi.org/10.1038/nrgastro.2017.173

59. Steegenga WT, Mischke M, Lute C, Boekschoten MV, Lendvai A, Pruis MG, Verkade HJ, van de Heijning BJ, Boekhorst J, Timmerman HM, Plösch T, Müller M, Hooiveld GJ (2017) Maternal exposure to a Western-style diet causes differences in intestinal microbiota composition and gene expression of suckling mouse pups. Mol Nutr Food Res 61(1). https://doi.org/10.1002/mnfr.201600141. Epub 2016 Jul 12

60. Mischke M, Arora T, Tims S, Engels E, Sommer N, van Limpt K, Baars A, Oozeer R, Oosting A, Bäckhed F, Knol J (2018) Specific synbiotics in early life protect against diet-induced obesity in adult mice. Diabetes Obes Metab. https://doi.org/10.1111/dom.13240

61. Zholobak NM, Sherbakov AB, Babenko LS, Bogorad-Kobelska OS, Bubnov RV, Ivanov VK, Spivak MY (2014) The perspectives of biomedical application of the nanoceria. EPMA J 5(Suppl 1):A136. https://doi.org/10.1186/1878-5085-5-S1-A136

62. Spivak MY, Bubnov RV, Yemets IM, Lazarenko LM, Timoshok NO, Ulberg ZR: Gold nanoparticles – the theranostic challenge for PPPM: nanocardiology application. EPMA J 2013, 4 (1): 18–10.1186/1878-5085-4-18

63. Spivak MY, Bubnov RV, Yemets IM, Lazarenko LM, Tymoshok NO, Ulberg ZR (2013) Development and testing of gold nanoparticles for drug delivery and treatment of heart failure: a theranostic potential for PPP cardiology. EPMA J 4(1):20. https://doi.org/10.1186/1878-5085-4-20

64. Rodrigues RR, Greer RL, Dong X, DSouza KN, Gurung M, Wu JY, Morgun A, Shulzhenko N (2017 Nov) Antibiotic-induced alterations in gut microbiota are associated with changes in glucose metabolism in healthy mice. Front Microbiol 22(8):2306. https://doi.org/10.3389/fmicb.2017.02306

65. Penders J, Stobberingh EE, Savelkoul PHM, Wolffs PFG (2013) The human microbiome as a reservoir of antimicrobial resistance. Front Microbiol 4:87. https://doi.org/10.3389/fmicb.2013.00087

66. D'Aimmo MR Modesto M, Biavati B (2007) Antibiotic resistance of lactic acid bacteria and Bifidobacterium spp. isolated from dairy and pharmaceutical products. Int J Food Microbiol 115(1):35–42

67. Nogacka AM, Salazar N, Arboleya S, Suárez M, Fernández N, Solís G, de Los Reyes-Gavilán CG, Gueimonde M (2018 Jan) Early microbiota, antibiotics and health. Cell Mol Life Sci 75(1):83–91. https://doi.org/10.1007/s00018-017-2670-2

68. Koppel N, Maini Rekdal V, Balskus EP (2017) Chemical transformation of xenobiotics by the human gut microbiota. Science 356(6344)

69. Flint HJ (2016) Gut microbial metabolites in health and disease. Gut Microbes 7(3):187–188. https://doi.org/10.1080/19490976.2016.1182295
70. Lebeer S, Bron PA, Marco ML, VanPijkeren JP, O'Connell Motherway M, Hill C, Pot B, Roos S, Klaenhammer T (2017 Nov) Identification of probiotic effector molecules: present state and future perspectives. Curr Opin Biotechnol 15(49):217–223. https://doi.org/10.1016/j.copbio.2017.10.007
71. Plaza-Díaz J, Robles-Sánchez C, Abadía-Molina F, Sáez-Lara MJ, Vilchez-Padial LM, Gil Á, Gómez-Llorente C, Fontana L (2017) Gene expression profiling in the intestinal mucosa of obese rats administered probiotic bacteria. Sci Data 4:170186. https://doi.org/10.1038/sdata.2017.186
72. van Baarlen P, Troost F, van der Meer C, Hooiveld G, Boekschoten M, Brummer RJ, Kleerebezem M (2011) Human mucosal in vivo transcriptome responses to three lactobacilli indicate how probiotics may modulate human cellular pathways. Proc Natl Acad Sci U S A 108(Suppl 1):4562–4569. https://doi.org/10.1073/pnas.1000079107
73. Biagi E, Franceschi C, Rampelli S, Severgnini M, Ostan R, Turroni S, Consolandi C, Quercia S, Scurti M, Monti D, Capri M, Brigidi P, Candela M (2016) Gut microbiota and extreme longevity. Curr Biol 26:1480–1485
74. Strachan DP (1989) Hay fever, hygiene, and household size. BMJ 299:1259–1260
75. Halling ML, Kjeldsen J, Knudsen T, Nielsen J, Hansen LK (2017) Patients with inflammatory bowel disease have increased risk of autoimmune and inflammatory diseases. World J Gastroenterol 23(33):6137–6146. https://doi.org/10.3748/wjg.v23.i33.6137
76. Vieira SM, Pagovich OE, Kriegel MA (2014) Diet, microbiota and autoimmune diseases. Lupus 23:518–526
77. Hotamisligil GS (2006) Inflammation and metabolic disorders. Nature 444(7121):860–867
78. Tymoshok NO, Lazarenko LM, Bubnov RV, Shynkarenko LN, Babenko LP, Mokrozub VV et al (2014) New aspects the regulation of immune response through balance Th1/Th2 cytokines. EPMA J 5(Suppl 1):A134
79. Mohamadzadeh M, Pfeiler EA, Brown JB, Zadeh M, Gramarossa M, Managlia E, Bere P, Sarraj B, Khan MW, Pakanati KC, Ansari MJ, O'Flaherty S, Barrett T, Klaenhammer TR (2011) Regulation of induced colonic inflammation by lactobacillus acidophilus deficient in lipoteichoic acid. Proc Natl Acad Sci U S A 108(Suppl 1):4623–4630. https://doi.org/10.1073/pnas.1005066107
80. Sheil B, MacSharry J, O'Callaghan L, O'Riordan A, Waters A, Morgan J, Collins JK, O'Mahony L, Shanahan F (2006) Role of interleukin (IL-10) in probiotic-mediated immune modulation: an assessment in wild-type and IL-10 knock-out mice. Clin Exp Immunol 144(2):273–280
81. Zheng Y, Valdez PA, Danilenko DM, Hu Y, Sa SM, Gong Q, Abbas AR, Modrusan Z, Ghilardi N, de Sauvage FJ, Ouyang W (2008) Interleukin-22 mediates early host defense against attaching and effacing bacterial pathogens. Nat Med 14(3):282–289. https://doi.org/10.1038/nm1720
82. Fåk F, Bäckhed F (2012) Lactobacillus reuteri prevents diet-induced obesity, but not atherosclerosis, in a strain dependent fashion in Apoe−/− mice. PLoS One 7(10):e46837. https://doi.org/10.1371/journal.pone.0046837
83. Cani PD, Delzenne NM (2009 Dec) Interplay between obesity and associated metabolic disorders: new insights into the gut microbiota. Curr Opin Pharmacol 9(6):737–743
84. Hessle C, Andersson B, Wold AE (2000) Gram-positive bacteria are potent inducers of monocytic interleukin-12 (IL-12) while gram-negative bacteria preferentially stimulate IL-10 production. Infect Immun 68(6):3581–3586
85. Cebioglu M, Schild HH, Golubnitschaja O (2010) Cancer predisposition in diabetics: risk factors considered for predictive diagnostics and targeted preventive measures. EPMA J 1(1):130–137. https://doi.org/10.1007/s13167-010-0015-4
86. Lazarenko LM, Nikitina OE, Nikitin EV, Demchenko OM, Kovtonyuk GV, Ganova LO, Bubnov RV, Shevchuk VO, Nastradina NM, Bila VV, Spivak MY (2014) Development of

biomarker panel to predict, prevent and create treatments tailored to the persons with human papillomavirus-induced cervical precancerous lesions. EPMA J 5(1):1. https://doi.org/10.118 6/1878-5085-5-1

87. Damms-Machado A, Louis S, Schnitzer A, Volynets V, Rings A, Basrai M, Bischoff SC (2017) Gut permeability is related to body weight, fatty liver disease, and insulin resistance in obese individuals undergoing weight reduction. Am J Clin Nutr 105(1):127–135. https://doi.org/10.3945/ajcn.116.131110

88. Xiao S, Fei N, Pang X, Shen J, Wang L, Zhang B, Zhang M, Zhang X, Zhang C, Li M et al (2014) A gut microbiota-targeted dietary intervention for amelioration of chronic inflammation underlying metabolic syndrome. FEMS Microbiol Ecol 87:357–367

89. Furukawa S, Fujita T, Shimabukuro M et al (2004) Increased oxidative stress in obesity and its impact on metabolic syndrome. J Clin Invest 114(12):1752–1761. https://doi.org/10.1172/JCI200421625

90. Huang C-J, McAllister MJ, Slusher AL, Webb HE, Mock JT, Acevedo EO (2015) Obesity-related oxidative stress: the impact of physical activity and diet manipulation. Sports Med Open 1:32. https://doi.org/10.1186/s40798-015-0031-y

91. Wang Y, Kirpich I, Liu Y, Ma Z, Barve S, McClain CJ, Feng W (2011) Lactobacillus rhamnosus GG treatment potentiates intestinal hypoxia-inducible factor, promotes intestinal integrity and ameliorates alcohol-induced liver injury. Am J Pathol 179(6):2866–2875. https://doi.org/10.1016/j.ajpath.2011.08.039

92. Abdallah Ismail N, Ragab SH, Abd Elbaky A, Shoeib AR, Alhosary Y, Fekry D (2011) Frequency of Firmicutes and Bacteroidetes in gut microbiota in obese and normal weight Egyptian children and adults. Arch Med Sci 7(3):501–507

93. Arumugam M, Raes J, Pelletier E, Le Paslier D, Yamada T, Mende DR, Fernandes GR et al (2011) Enterotypes of the human gut microbiome. Nature 473(7346):174–180. https://doi.org/10.1038/nature09944. Epub 2011 Apr 20. Erratum in: Nature. 2011 Jun 30;474(7353):666. Nature. 2014 Feb 27;506(7489):516

94. Roager HM, Licht TR, Poulsen SK, Larsen TM, Bahl MI (2014 Feb) Microbial Enterotypes, inferred by the Prevotella-to-Bacteroides ratio, remained stable during a 6-month randomized controlled diet intervention with the new Nordic diet. Appl Environ Microbiol 80(3):1142–1149. https://doi.org/10.1128/AEM.03549-13

95. David LA, Maurice CF, Carmody RN et al (2014) Diet rapidly and reproducibly alters the human gut microbiome. Nature 505(7484):559–563. https://doi.org/10.1038/nature12820

96. De Filippo C, Cavalieri D, Di Paola M, Ramazzotti M, Poullet JB, Massart S, et al. Impact of diet in shaping gut microbiota revealed by a comparative study in children from Europe and rural Africa

97. Cockburn DW, Koropatkin NM (2016) Polysaccharide degradation by the intestinal microbiota and its influence on human health and disease. J Mol Biol. https://doi.org/10.1016/j.jmb.2016.06.02

98. Yatsunenko T, Rey FE, Manary MJ, Trehan I, Dominguez-Bello MG, Contreras M et al (2012) Human gut microbiome viewed across age and geography. Nature 486:222–227. https://doi.org/10.1038/nature11053

99. De Filippo C, Cavalieri D, Di Paola M et al (2010) Impact of diet in shaping gut microbiota revealed by a comparative study in children from Europe and rural Africa. Proc Natl Acad Sci U S A 107:14691–14696. https://doi.org/10.1073/pnas.1005963107

100. Jazayeri O, Daghighi SM, Rezaee F (2017) Lifestyle alters GUT-bacteria function: linking immune response and host. Best Pract Res Clin Gastroenterol. https://doi.org/10.1016/j.bpg.2017.09.009

101. Ley RE, Bäckhed F, Turnbaugh P, Lozupone CA, Knight RD, Gordon JI (2005) Obesity alters gut microbial ecology. Proc Natl Acad Sci U S A 102(31):11070–11075

102. Louis S, Tappu RM, Damms-Machado A, Huson DH, Bischoff SC (2016) Characterization of the gut microbial Community of Obese Patients Following a weight-loss intervention using

whole metagenome shotgun sequencing. PLoS One 11(2):e0149564. https://doi.org/10.1371/journal.pone.0149564
103. Grazul H, Kanda LL, Gondek D (2016) Impact of probiotic supplements on microbiome diversity following antibiotic treatment of mice. Gut Microbes 7(2):101–114. https://doi.org/10.1080/19490976.2016.1138197
104. del Campo R, Garriga M, Pérez-Aragón A, Guallarte P, Lamas A, Máiz L, Bayón C, Roy G, Cantón R, Zamora J, Baquero F, Suárez L (2014 Dec) Improvement of digestive health and reduction in proteobacterial populations in the gut microbiota of cystic fibrosis patients using a lactobacillus reuteri probiotic preparation: a double blind prospective study. J Cyst Fibros 13(6):716–722. https://doi.org/10.1016/j.jcf.2014.007
105. Morgan XC, Segata N, Huttenhower C (2013) Biodiversity and functional genomics in the human microbiome. Trends Genet 29(1):51–58. Epub 2012 Nov 7
106. Kong LC, Holmes BA, Cotillard A, Habi-Rachedi F, Brazeilles R, Gougis S et al (2014) Dietary patterns differently associate with inflammation and gut microbiota in overweight and obese subjects. PLoS One 9:e109434. https://doi.org/10.1371/journal.pone.0109434
107. Cotillard A, Kennedy SP, Kong LC, Prifti E, Pons N, Le Chatelier E et al (2013) Dietary intervention impact on gut microbial gene richness. Nature 500:585–588. https://doi.org/10.1038/nature12480
108. Beaumont M, Goodrich JK, Jackson MA, Yet I, Davenport ER, Vieira-Silva S et al (2016) Heritable components of the human fecal microbiome are associated with visceral fat. Genome Biol 17:189. https://doi.org/10.1186/s13059-016-1052-7
109. Le Chatelier E, Nielsen T, Qin J, Prifti E, Hildebrand F, Falony G, Almeida M, Arumugam M, Batto JM, Kennedy S, Leonard P et al (2013) Richness of human gut microbiome correlates with metabolic markers. Nature 500:541–546. https://doi.org/10.1038/nature12506
110. Mikelsaar M, Sepp E, Štšepetova J, Songisepp E, Mändar R (2016) Biodiversity of intestinal lactic acid bacteria in the healthy population. Adv Exp Med Biol 932:1–64. Review
111. Tucker CM, Fukami T (2014) Environmental variability counteracts priority effects to facilitate species coexistence: evidence from nectar microbes. Proc Biol Sci 281(1778):20132637. https://doi.org/10.1098/rspb.2013.2637
112. Jiang TT, Shao TY, Ang WXG, Kinder JM, Turner LH, Pham G, Whitt J, Alenghat T, Way SS (2017) Commensal fungi recapitulate the protective benefits of intestinal bacteria. Cell Host Microb 22(6):809–816.e4. https://doi.org/10.1016/j.chom.2017.10.013
113. Ilavenil S, Park HS, Vijayakumar M, Arasu MV, Kim DH, Ravikumar S, Choi KC (2015) Probiotic potential of lactobacillus strains with antifungal activity isolated from animal manure. ScientificWorldJournal 2015:802570. https://doi.org/10.1155/2015/802570
114. Iliev ID, Leonardi I (2017) Fungal dysbiosis: immunity and interactions at mucosal barriers. Nat Rev Immunol 17(10):635–646. https://doi.org/10.1038/nri.2017.55
115. Ward TL, Knights D, Gale CA (2017) Infant fungal communities: current knowledge and research opportunities. BMC Med 15:30
116. Rizzetto L, De Filippo C, Cavalieri D (2014) Richness and diversity of mammalian fungal communities shape innate and adaptive immunity in health and disease. Eur J Immunol 44(11):3166–3181. Epub 2014 Oct 30
117. Babenko LP, Lazarenko LM, Shynkarenko LM, Mokrozub VV, Pidgorskyi VS, Spivak MY (2012) The effect of lacto- and bifidobacteria compositions on the vaginal microflora in cases of intravaginal staphylococcosis. Mikrobiol Z 74(6):80–89
118. Hummelen R, Macklaim JM, Bisanz JE, Hammond J-A, McMillan A et al (2011) Vaginal microbiome and epithelial gene Array in post-menopausal women with moderate to severe dryness. PLoS One 6(11):e26602. https://doi.org/10.1371/journal.pone.0026602
119. Bisanz JE, Seney S, McMillan A, Vongsa R, Koenig D et al (2014) A systems biology approach investigating the effect of probiotics on the vaginal microbiome and host responses in a double blind, placebo-controlled clinical trial of post-menopausal women. PLoS One 9(8):e104511. https://doi.org/10.1371/journal.pone.0104511

120. Subramaniam A, Kumar R, Cliver SP et al (2016) Vaginal microbiota in pregnancy: evaluation based on vaginal Flora, birth outcome, and race. Am J Perinatol 33(4):401–408. https://doi.org/10.1055/s-0035-1565919

121. Mändar R, Punab M, Borovkova N, Lapp E, Kiiker R, Korrovits P et al (2015) Complementary seminovaginal microbiome in couples. Res Microbiol 166(5):440–447

122. Reece AS (2017) Dying for love: Perimenopausal degeneration of vaginal microbiome drives the chronic inflammation-malignant transformation of benign prostatic hyperplasia to prostatic adenocarcinoma. Med Hypotheses 101:44–47. https://doi.org/10.1016/j.mehy.2017.02.006

123. Plummer EL, Vodstrcil LA, Danielewski JA, Murray GL, Fairley CK, Garland SM, Hocking JS, Tabrizi SN, Bradshaw CS (2018) Combined oral and topical antimicrobial therapy for male partners of women with bacterial vaginosis: acceptability, tolerability and impact on the genital microbiota of couples – a pilot study. PLoS One 13(1):e0190199. https://doi.org/10.1371/journal.pone.0190199

124. Reid G (2018 Jan) Is bacterial vaginosis a disease? Appl Microbiol Biotechnol 102(2):553–558. https://doi.org/10.1007/s00253-017-8659-9

125. Verma D, Garg PK, Dubey AK (2018) Insights into the human oral microbiome. Arch Microbiol. https://doi.org/10.1007/s00203-018-1505-3

126. Nowicki EM, Shroff R, Singleton JA, Renaud DE, Wallace D, Drury J, Zirnheld J, Colleti B, Ellington AD, Lamont RJ, Scott DA, Whiteley M (2018) Microbiota and metatranscriptome changes accompanying the onset of gingivitis. mBio 9:e00575–18. https://doi.org/10.1128/mBio.00575-18

127. Goodson JM, Groppo D, Halem S, Carpino E (2009) Is obesity an oral bacterial disease? J Dent Res 88(6):519–523. https://doi.org/10.1177/0022034509338353

128. Tobita K, Watanabe I, Tomokiyo M, Saito M (2018) Effects of heat-treated lactobacillus crispatus KT-11 strain consumption on improvement of oral cavity environment: a randomised double-blind clinical trial. Benef Microb 10:1–8. https://doi.org/10.3920/BM2017.0137

129. Grice EA, Segre JA (2011) The skin microbiome. Nat Rev Microbiol 9(4):244–253. https://doi.org/10.1038/nrmicro2537

130. Sanford JA, Gallo RL (2013) Functions of the skin microbiota in health and disease. Semin Immunol 25(5):370–377. https://doi.org/10.1016/j.smim.2013.09.005

131. Opazo MC, Ortega-Rocha EM, Coronado-Arrázola I, Bonifaz LC, Boudin H, Neunlist M, Bueno SM, Kalergis AM, Riedel CA (2018) Intestinal microbiota influences non-intestinal related autoimmune diseases. Front Microbiol 9:432

132. Rizzetto L, De Filippo C, Cavalieri D (2014) Richness and diversity of mammalian fungal communities shape innate and adaptive immunity in health and disease. Eur J Immunol 44(11):3166–3181

133. Moiseyenko YV, Sukhorukov VI, Pyshnov GY, Mankovska IM, Rozova KV, Miroshnychenko OA, Kovalevska OE, Madjar SA, Bubnov RV, Gorbach AO, Danylenko KM, Moiseyenko OI (2016) Antarctica challenges the new horizons in predictive, preventive, personalized medicine: preliminary results and attractive hypotheses for multidisciplinary prospective studies in the Ukrainian "Akademik Vernadsky" station. EPMA J 31(7):11. https://doi.org/10.1186/s13167-016-0060-8

134. Tan L, Zhao S, Zhu W, Wu L, Li J, Shen M, Lei L, Chen X, Peng C (2018 Feb) The Akkermansia muciniphila is a gut microbiota signature in psoriasis. Exp Dermatol 27(2):144–149. https://doi.org/10.1111/exd.13463

135. Stolzenburg-Veeser L, Golubnitschaja O (2017) Mini-encyclopaedia of the wound healing – opportunities for integrating multi-omic approaches into medical practice. J Proteome S1874-3919(17):30261-0. https://doi.org/10.1016/j.jprot.2017.07.017

136. Lukic J, Chen V, Strahinic I, Begovic J, Lev-Tov H, Davis SC, Tomic-Canic M, Pastar I (2017) Probiotics or pro-healers: the role of beneficial bacteria in tissue repair. Wound Repair Regen 25(6):912–922. https://doi.org/10.1111/wrr.12607

137. Mohammedsaeed W, Cruickshank S, McBain AJ, O'Neill CA (2015) Lactobacillus rhamnosus GG lysate increases re-epithelialization of keratinocyte scratch assays by promoting migration. Sci Rep 5:16147
138. Poutahidis T, Kearney SM, Levkovich T, Qi P, Varian BJ, Lakritz JR et al (2013) Microbial symbionts accelerate wound healing via the neuropeptide hormone oxytocin. PLoS One 8(10):e78898. https://doi.org/10.1371/journal.pone.0078898
139. Cusack S, O'Sullivan O, Greene-Diniz R, de Weerd H, Flannery E, Marchesi JR, Falush D, Dinan T, Fitzgerald G et al (2011) Composition, variability, and temporal stability of the intestinal microbiota of the elderly. Proc Natl Acad Sci U S A 108(Suppl. S1):4586–4591
140. Claesson MJ, Jeffery IB, Conde S, Power SE, O'Connor EM, Cusack S, Harris HM, Coakley M, Lakshminarayanan B, O'Sullivan O et al (2012) Gut microbiota composition correlates with diet and health in the elderly. Nature 488:178–184
141. Poutahidis T, Kleinewietfeld M, Smillie C, Levkovich T, Perrotta A, Bhela S, Varian BJ, Ibrahim YM, Lakritz JR, Kearney SM, Chatzigiagkos A, Hafler DA, Alm EJ, Erdman SE (2013) Microbial reprogramming inhibits Western diet-associated obesity. PLoS One 8:e68596
142. Ferolla SM, Couto CA, Costa-Silva L, Armiliato GN, Pereira CA, Martins FS, Ferrari Mde L, Vilela EG, Torres HO, Cunha AS, Ferrari TC (2016) Beneficial effect of synbiotic supplementation on hepatic steatosis and anthropometric parameters, but not on gut permeability in a population with nonalcoholic steatohepatitis. Nutrients 8(7):pii: E397. https://doi.org/10.3390/nu8070397
143. Behrouz V, Jazayeri S, Aryaeian N, Zahedi MJ, Hosseini F (2017) Effects of probiotic and prebiotic supplementation on leptin, adiponectin, and Glycemic parameters in non-alcoholic fatty liver disease: a randomized clinical trial. Middle East J Dig Dis 9(3):150–157. https://doi.org/10.15171/mejdd.2017.66
144. WHO, Global brief on hypertension. 2013. http://www.who.int/cardiovascular_diseases/publications/global_brief_hypertension/en/. Accessed 20 Aug 2018
145. Xu J, Ahrén IL, Olsson C, Jeppsson B, Ahrné S, Molin G (2015) Oral and faecal microbiota in volunteers with hypertension in a double blind, randomised placebo controlled trial with probiotics and fermented bilberries. J Funct Foods 18:275–288
146. Tuomilehto J, Lindström J, Hyyrynen J, Korpela R, Karhunen ML, Mikkola L, Jauhiainen T, Seppo L, Nissinen A (2004) Effect of ingesting sour milk fermented by lactobacillus helveticus bacteria on blood pressure in subjects with mild hypertension. J Human Hyper 18:795–802
147. Esposito E, Iacono A, Bianco G, Autore G, Cuzzocrea S, Vajro P, Canani RB, Calignano A, Raso GM, Meli R (2009) Probiotics reduce the inflammatory response induced by a high-fat diet in the liver of young rats. J Nutr 139(5):905–911. https://doi.org/10.3945/jn.108.101808
148. Zhang Q, Wu Y, Fei X (2016) Effect of probiotics on glucose metabolism in patients with type 2 diabetes mellitus: A meta-analysis of randomized controlled trials. Medicina (Kaunas) 52(1):28–34. https://doi.org/10.1016/j.medici.2015.11.008
149. Gu Y, Wang X, Li J, Zhang Y, Zhong H, Liu R et al (2017) Analyses of 1149 gut microbiota and plasma bile acids enable stratification of patients for antidiabetic treatment. Nat Commun 8(1):1785. https://doi.org/10.1038/s41467-017-01682-2
150. Bubnov RV, Ostapenko TV Ultrasound diagnosis for diabetic neuropathy – comparative study. EPMA J 7(Suppl 1):A12
151. Lazarenko L, Melnikova O, Babenko L, Bubnov R, Beregova T, Falalyeyeva T, Spivak M (2018) Lactobacillus and Bifidobacteria probiotic strains improve Glycemic and inflammation profiles in obesity model in mice. Preprints:2018080169. https://doi.org/10.20944/preprints201808.0169.v1
152. Tilg H, Cani PD, Mayer EA (2016) Gut microbiome and liver diseases. Gut 65:2035–2044
153. Qin N, Yang F, Li A, Prifti E, Chen Y, Shao L, Guo J, Le Chatelier E, Yao J, Wu L, Zhou J, Ni S, Liu L, Pons N, Batto JM, Kennedy SP, Leonard P, Yuan C, Ding W, Chen Y, Hu X, Zheng

B, Qian G, Xu W, Ehrlich SD, Zheng S, Li L (2014) Alterations of the human gut microbiome in liver cirrhosis. Nature 513(7516):59–64. https://doi.org/10.1038/nature13568

154. Codella R, Luzi L, Terruzzi I (2018 Apr) Exercise has the guts: how physical activity may positively modulate gut microbiota in chronic and immune-based diseases. Dig Liver Dis 50(4):331–341. https://doi.org/10.1016/j.dld.2017.11.016

155. Rincón D, Vaquero J, Hernando A, Galindo E, Ripoll C, Puerto M, Salcedo M, Francés R, Matilla A, Catalina MV et al (2014) Oral probiotic VSL#3 attenuates the circulatory disturbances of patients with cirrhosis and ascites. Liver Int 34(10):1504–1512. Epub 2014 Apr 4

156. Marlicz W, Wunsch E, Mydlowska M, Milkiewicz M, Serwin K, Mularczyk M, Milkiewicz P, Raszeja-Wyszomirska J (2016 Dec) The effect of short term treatment with probiotic VSL#3 on various clinical and biochemical parameters in patients with liver cirrhosis. J Physiol Pharmacol 67(6):867–877

157. Kondo S, Kamei A, Xiao JZ, Iwatsuki K, Abe K (2013) Bifidobacterium breve B-3 exerts metabolic syndrome-suppressing effects in the liver of diet-induced obese mice: a DNA microarray analysis. Benef Microb 3:247–251. https://doi.org/10.3920/BM2012.0019

158. Tian F, Chi F, Wang G, Liu X, Zhang Q, Chen Y et al (2015) Lactobacillus rhamnosus CCFM1107 treatment ameliorates alcohol-induced liver injury in amouse model of chronic alcohol feeding. J Microbiol 53(12):856–863. https://doi.org/10.1007/s12275-015-5239-5

159. Nido SA, Shituleni SA, Mengistu BM, Liu Y, Khan AZ, Gan F et al (2016) Effects of selenium-enriched probiotics on lipid metabolism, Antioxidative status, histopathological lesions, and related gene expression in mice fed a high-fat diet. Biol Trace Elem Res 171(2):399–409. https://doi.org/10.1007/s12011-015-0552-8

160. Wang LX, Liu K, Gao DW, Hao JK (2013) Protective effects of two lactobacillus plantarum strains in hyperlipidemic mice. World J Gastroenterol 19(20):3150–3156. https://doi.org/10.3748/wjg.v19.i20.3150

161. Tarantino G, Finelli C (2015) Systematic review on intervention with prebiotics/probiotics in patients with obesity-related nonalcoholic fatty liver disease. Future Microbiol 10(5):889–902. https://doi.org/10.2217/fmb.15.13

162. Bubnov RV, Drahulian MV, Buchek PV, Gulko TP (2017) High regenerative capacity of the liver and irreversible injury of male reproductive systemin carbon tetrachloride-induced liver fibrosis rat model. EPMA J 9(1):59–75. https://doi.org/10.1007/s13167-017-0115-5

163. Kobyliak N, Virchenko O, Falalyeyeva T, Kondro M, Beregova T, Bodnar P, Shcherbakov O, Bubnov R, Caprnda M, Delev D, Sabo J, Kruzliak P, Rodrigo L, Opatrilova R, Spivak M (2017) Cerium dioxide nanoparticles possess anti-inflammatory properties in the conditions of the obesity-associated NAFLD in rats. Biomed Pharmacother 90:608–614. https://doi.org/10.1016/j.biopha.2017.03.099

164. Jegatheesan P, Beutheu S, Freese K, Waligora-Dupriet AJ, Nubret E, Butel MJ et al (2016) Preventive effects of citrulline on Western diet-induced non-alcoholic fatty liver disease in rats. Br J Nutr 116:191–203. https://doi.org/10.1017/S0007114516001793

165. Thomas H (2017) NAFLD: A gut microbiome signature for advanced fibrosis diagnosis in NAFLD. Nat Rev Gastroenterol Hepatol. https://doi.org/10.1038/nrgastro.2017.67

166. Nie YF, Hu J, Yan XH (2015) Cross-talk between bile acids and intestinal microbiota in host metabolism and health. J Zhejiang Univ Sci B 16(6):436–446. https://doi.org/10.1631/jzus.B1400327

167. Park MY, Kim SJ, Ko EK, Ahn SH, Seo H, Sung MK (2016) Gut microbiota-associated bile acid deconjugation accelerates hepatic steatosis in ob/ob mice. J Appl Microbiol 121(3):800–810

168. Gu Y, Wang X, Li J, Zhang Y, Zhong H, Liu R, Zhang D, Feng Q, Xie X, Hong J, Ren H, Liu W, Ma J, Su Q, Zhang H, Yang J, Wang X, Zhao X, Gu W, Bi Y, Peng Y, Xu X, Xia H, Li F, Xu X, Yang H, Xu G, Madsen L, Kristiansen K, Ning G, Wang W (2017) Analyses of gut microbiota and plasma bile acids enable stratification of patients for antidiabetic treatment. Nat Commun 8(1):1785. https://doi.org/10.1038/s41467-017-01682-2

169. Lal N, Mehra S, Lal V (2014) Ultrasonographic measurement of normal common bile duct diameter and its correlation with age, sex and anthropometry. J Clin Diagn Res 8(12):AC01-4. https://doi.org/10.7860/JCDR/2014/8738.5232
170. Wu CC, Weng WL, Lai WL, Tsai HP, Liu WH, Lee MH et al (2015) Effect of lactobacillus plantarum strain K21 on high-fat diet-fed obese mice. Evid Based Complement Alternat Med 2015:391767. https://doi.org/10.1155/2015/391767
171. Million M, Angelakis E, Paul M, Armougom F, Leibovici L, Raoult D (2012) Comparative meta-analysis of the effect of lactobacillus species on weight gain in humans and animals. Microb Pathog 53(2):100–108. https://doi.org/10.1016/j.micpath.2012.05.007
172. Michael DR, Davies TS, Moss JWE, Calvente DL, Ramji DP, Marchesi JR, Pechlivanis A, Plummer SF, Hughes TR (2017) The anti-cholesterolaemic effect of a consortium of probiotics: an acute study in C57BL/6J mice. Sci Rep 7(1):2883. https://doi.org/10.1038/s41598-017-02889-5
173. Arora T, Anastasovska J, Gibson G, Tuohy K, Sharma RK, Bell J et al (2012) Effect of lacto-bacillus acidophilus NCDC 13 supplementation on the progression of obesity in diet-induced obese mice. Br J Nutr 108(8):1382–1389. https://doi.org/10.1017/S0007114511006957
174. Vinderola G, Gueimonde M, Gomez-Gallego C, Defederico L, Salminen S (2017) Correlation between in vitro and in vivo assays in selection of probiotics from traditional species of bacteria. Trends Food Sci Technol 68:83–90. https://doi.org/10.1016/j.tifs.2017.08.005
175. Fijan S (2014) Microorganisms with claimed probiotic properties: an overview of recent literature. Int J Environ Res Public Health 11(5):4745–4767. https://doi.org/10.3390/ijerph110504745
176. Câmara NO, Iseki K, Kramer H, Liu ZH, Sharma K (2017) Kidney disease and obesity: epidemiology, mechanisms and treatment. Nat Rev Nephrol 13(3):181–190. https://doi.org/10.1038/nrneph.2016.191. Review
177. Babb AL, Ahmad S, Bergström J, Scribner BH (1981) The middle molecule hypothesis in perspective. Am J Kidney Dis 1(1):46–50
178. Vanholder R, Van Laecke S, Glorieux G (2008) The middle-molecule hypothesis 30 years after: lost and rediscovered in the universe of uremic toxicity? J Nephrol 21(2):146–160. Review
179. Castillo-Rodriguez E, Fernandez-Prado R, Esteras R, Perez-Gomez MV, Gracia-Iguacel C, Fernandez-Fernandez B, Kanbay M, Tejedor A, Lazaro A, Ruiz-Ortega M, Gonzalez-Parra E, Sanz AB, Ortiz A, Sanchez-Niño MD (2018) Impact of altered intestinal microbiota on chronic kidney disease progression. Toxins (Basel) 10(7):E300. https://doi.org/10.3390/toxins10070300
180. Youssef DM, Fawzy FM (2012) Value of renal resistive index as an early marker of diabetic nephropathy in children with type-1 diabetes mellitus. Saudi J Kidney Dis Transpl 23(5):985–992. https://doi.org/10.4103/1319-2442.100880
181. Natarajan R, Pechenyak B, Vyas U, Ranganathan P, Weinberg A, Liang P, Mallappallil MC, Norin AJ, Friedman EA, Saggi SJ (2014) Randomized controlled trial of strain-specific probiotic formulation (Renadyl) in dialysis patients. Biomed Res Int 2014:568571. https://doi.org/10.1155/2014/568571
182. Bubnov RV, Melnyk IM Evaluation of biomarkers for diagnosnostic decision making in patients with gout using novel mathematical model. Complex PPPM approach. EPMA J 5(Suppl 1):A58
183. Cao T, Li X, Mao T, Liu H, Zhao Q, Ding X, Li C, Zhang L, Tian Z (2017) Probiotic therapy alleviates hyperuricemia in C57BL/6 mouse model. Biomed Res 28(5):2244–2249
184. García-Arroyo FE, Gonzaga G, Muñoz-Jiménez I, Blas-Marron MG, Silverio O, Tapia E, Soto V, Ranganathan N, Ranganathan P, Vyas U, Irvin A, Ir D, Robertson CE, Frank DN, Johnson RJ, Sánchez-Lozada LG (2018) Probiotic supplements prevented oxonic acid-induced hyper-uricemia and renal damage. PLoS One 13(8):e0202901. https://doi.org/10.1371/journal.pone.0202901

185. Vieira AT, Galvão I, Amaral FA, Teixeira MM, Nicoli JR, Martins FS (2015) Oral treatment with Bifidobacterium longum 51A reduced inflammation in a murine experimental model of gout. Benef Microb 6(6):799–806. https://doi.org/10.3920/BM2015.0015

186. Bubnov R, Petrenko L (2016) Asthma-associated factors – potential predictive markers for patients stratification, personalized treatments and prevention. Eur Respir J 48(Suppl. 60):3366. https://doi.org/10.1183/13993003.congress-2016.PA3366

187. Liu J, Chen FH, Qiu SQ, Yang LT, Zhang HP, Liu JQ, Geng XR, Yang G, Liu ZQ, Li J, Liu ZG, Li HB, Yang PC (2016) Probiotics enhance the effect of allergy immunotherapy on regulating antigen specific B cell activity in asthma patients. Am J Transl Res 8(12):5256–5270

188. Tapiovaara L, Pitkaranta A, Korpela R (2016) Probiotics and the upper respiratory tract – a review. Pediatric Infect Dis 1:19. https://doi.org/10.21767/2573-0282.100019

189. Kam HY, Ou LC, Thron CD, Smith RP, Leiter JC (1985) Role of the spleen in the exaggerated polycythemic response to hypoxia in chronic mountain sickness in rats. J Appl Physiol 87(5):1901–1908

190. Khailova L, Baird CH, Rush AA, Barnes C, Wischmeyer PE (2016) Lactobacillus rhamnosus GG treatment improves intestinal permeability and modulates inflammatory response and homeostasis of spleen and colon in experimental model of Pseudomonas aeruginosa pneumonia. Clin Nutr S0261-5614(16):31265–31261. https://doi.org/10.1016/j.clnu.2016.09.025

191. Cukrowska B, Motyl I, Kozáková H, Schwarzer M, Górecki RK, Klewicka E, Slizewska K, Libudzisz Z (2009) Probiotic lactobacillus strains: in vitro and in vivo studies. Folia Microbiol (Praha) 54(6):533–537. https://doi.org/10.1007/s12223-009-0077-7

192. Li Z, Chalazonitis A, Huang YY, Mann JJ, Margolis KG, Yang QM et al (2011) Essential roles of enteric neuronal serotonin in gastrointestinal motility and the development/survival of enteric dopaminergic neurons. J Neurosci 31:8998–9009

193. Crane JD, Palanivel R, Mottillo EP, Bujak AL, Wang H, Ford RJ et al (2015) Inhibiting peripheral serotonin synthesis reduces obesity and metabolic dysfunction by promoting brown adipose tissue thermogenesis. Nat Med 21(2):166–172

194. Khan WI, Ghia JE (2010) Gut hormones: emerging role in immune activation and inflammation. Clin Exp Immunol 161:19–27

195. Benyuk VO, Kalenskaya OV, Goncharenko VM, Strokan AM, Bubnov RV (2016) Immunohistological chemichal research of the apoptosis and endometrium APUD-system state interreaction in normal and pathological conditions. Women Health 1:63–66. Accessed 20.04.2018 http://nbuv.gov.ua/UJRN/Zdzh_2016_1_12

196. Goncharenko VM, Beniuk VA, Demchenko OM, Spivak MY, Bubnov RV (2013) Predictive diagnosis of endometrial hyperplasia and personalized therapeutic strategy in fertile age women. EPMA J 4:24. https://doi.org/10.1186/1878-5085-4-24

197. Carlson MJ, Thiel K, w., Yang S, Leslie KK. (2012) Catch it before it kills: progesterone, obesity, and the prevention of endometrial cancer. Discov Med 14(76):215–222

198. Moorthy K, Yadav UC, Mantha AK, Cowsik SM, Sharma D, Basir SF, Baquer NZ (2004) Estradiol and progesterone treatments change the lipid profile in naturally menopausal rats from different age groups. Biogerontology 5(6):411–419

199. Spaggiari G, Brigante G, De Vincentis S, Cattini U, Roli L, De Santis MC, Baraldi E, Tagliavini S, Varani M, Trenti T, Rochira V, Simoni M, Santi D (2017) Probiotics ingestion does not directly affect thyroid hormonal parameters in hypothyroid patients on levothyroxine treatment. Front Endocrinol (Lausanne) 14(8):316. https://doi.org/10.3389/fendo.2017.00316

200. Patil AD (2014) Link between hypothyroidism and small intestinal bacterial overgrowth. Indian J Endocrinol Metab 18(3):307–309. https://doi.org/10.4103/2230-8210.131155

201. Gabrielli M, Lauritano EC, Scarpellini E, Lupascu A, Ojetti V, Gasbarrini G, Silveri NG, Gasbarrini A (2009 May) Bacillus clausii as a treatment of small intestinal bacterial overgrowth. Am J Gastroenterol 104(5):1327–1328

202. Wojtyniak K, Horvath A, Dziechciarz P, Szajewska H (2017) Lactobacillus casei rhamnosus Lcr35 in the management of functional constipation in children: a randomized trial. J Pediatr 184:101–105

203. Kim SE, Choi SC, Park KS, Park MI, Shin JE, Lee TH, Jung KW, Koo HS, Myung SJ (2015) Change of fecal flora and effectiveness of the short-term VSL#3 probiotic treatment in patients with functional constipation. J Neurogastroenterol Motil 21:111–120
204. Šket R, Debevec T, Kublik S, Schloter M, Schoeller A, Murovec B, Vogel Mikuš K, Makuc D, Pečnik K, Plavec J, Mekjavić IB, Eiken O, Prevoršek Z, Stres B (2018) Intestinal metage-nomes and metabolomes in healthy young males: inactivity and hypoxia generated negative physiological symptoms precede microbial Dysbiosis. Front Physiol 9:198. https://doi.org/10.3389/fphys.2018.00198
205. Šket R, Treichel N, Debevec T, Eiken O, Mekjavic I, Schloter M, Vital M, Chandler J, Tiedje JM, Murovec B, Prevoršek Z (2017) Stres B hypoxia and inactivity related physiological changes (constipation, inflammation) are not reflected at the level of gut metabolites and butyrate producing microbial community: the PlanHab study. Front Physiol 8:250
206. Vitetta L, Coulson S, Linnane AW, Butt H (2013) The gastrointestinal microbiome and mus-culoskeletal diseases: a beneficial role for probiotics and prebiotics. Pathogens 2(4):606–626. https://doi.org/10.3390/pathogens2040606
207. Falvey E, Shanahan F, Cotter PD (2014) Exercise and associated dietary extremes impact on gut microbial diversity. Gut 63(12):1913–1920. https://doi.org/10.1136/gutjnl-2013-306541
208. Moloney RD, Desbonnet L, Clarke G, Dinan TG, Cryan JF (2014) The microbiome: stress, health and disease. Mamm Genome 25(1–2):49–74. https://doi.org/10.1007/s00335-013-9488-5. Epub 2013 Nov 27. Review
209. Cerdá B, Pérez M, Pérez-Santiago JD, Tornero-Aguilera JF, González-Soltero R, Larrosa M (2016) Gut microbiota modification: another piece in the puzzle of the benefits of physical exercise in health? Front Physiol 7:51. https://doi.org/10.3389/fphys.2016.00051
210. WHO: Physical activity fact sheet, Updated February 2018. http://www.who.int/mediacentre/factsheets/fs385/en/
211. Martin W, Roettger M, Kloesges T, Thiergart T, Woehle C, Gould S et al (2012) Modern endosymbiotic theory: getting lateral gene transfer into the equation. Endocytobiosis & Cell Research:23
212. Hu F, Liu F (2011) Mitochondrial stress: A bridge between mitochondrial dysfunction and metabolic diseases? Cell Signal 23:1528–1533
213. Franco-Obregón A, Gilbert JA (2017) The microbiome-mitochondrion connection: common ancestries, common mechanisms, common goals. mSystems 2(3):pii: e00018-17. https://doi.org/10.1128/mSystems.00018-17
214. Steves CJ, Bird S, Williams FM, Spector TD (2016) The microbiome and musculoskeletal conditions of aging: A review of evidence for impact and potential therapeutics. J Bone Miner Res 31(2):261–269. https://doi.org/10.1002/jbmr.2765
215. Buigues C, Fernandez-Garrido J, Pruimboom L, Hoogland AJ, Navarro-Martinez R, Martinez-Martinez M, Verdejo Y, Mascaros MC, Peris C, Cauli O (2016) Effect of a pre-biotic formulation on frailty syndrome: A randomized, double-blind clinical trial. Int J Mol Sci 17:932
216. Britton RA, Irwin R, Quach D et al (2014) Probiotic L. reuteri treatment prevents bone loss in a menopausal ovariectomized mouse model. J Cell Physiol 229(11):1822–1830
217. Rankin A, O'Donavon C, Madigan SM, et al (2017) 'Microbes in sport' – the potential role of the gut microbiota in athlete health and performance Br J sports med published online first: 25 January 2017. https://doi.org/10.1136/bjsports-2016-097227
218. O'Sullivan O, Cronin O, Clarke SF et al (2015) Exercise and the microbiota. Gut Microb 6:131–136
219. Clarke SF, Murphy EF, O'Sullivan O, Lucey AJ, Humphreys M, Hogan A, Hayes P, O'Reilly M, Jeffery IB, Wood-Martin R, Kerins DM, Quigley E, Ross RP, O'Toole PW, Molloy MG, Falvey E, Shanahan F, Cotter PD (2014) Exercise and associated dietary extremes impact on gut microbial diversity. Gut 63:1913–1920. https://doi.org/10.1136/gutjnl-2013-306541
220. Morales-Alamo D, Guerra B, Santana A, Martin-Rincon M, Gelabert-Rebato M, Dorado C, Calbet JAL (2018) Skeletal muscle pyruvate dehydrogenase phosphorylation and lactate

accumulation during Sprint exercise in Normoxia and severe acute hypoxia: effects of anti-oxidants. Front Physiol 9:188. https://doi.org/10.3389/fphys.2018.00188

221. Bloch Y, Bouchareychas L, Merceron R, Składanowska K, Van den Bossche L, Detry S, Govindaraja S, Elewaut D, Haerynck F, Dullaers M, Adamopoulos IE, Savvides SN (2018) Structural activation of pro-inflammatory human cytokine IL-23 by cognate IL-23 receptor enables recruitment of the shared receptor IL-12Rβ1. Immunity 48(1):45–58.e6. https://doi.org/10.1016/j.immuni.2017.12.008

222. So JS, Song MK, Kwon HK et al (2011) Lactobacillus casei enhances type II collagen/glucosamine-mediated suppression of inflammatory responses in experimental osteoarthritis. Life Sci 88(7–8):358–366

223. Courties A, Sellam J, Berenbaum F (2017) Metabolic syndrome-associated osteoarthritis. Curr Opin Rheumatol 29(2):214–222. https://doi.org/10.1097/BOR.0000000000000373

224. Sun AR, Parchal SK, Friis T et al (2017) Obesity-associated metabolic syndrome spontaneously induces infiltration of pro-inflammatory macrophage in synovium and promotes osteoarthritis. Gualillo O, ed. PLoS One 12(8):e0183693. https://doi.org/10.1371/journal.pone.0183693

225. Collins KH, Paul HA, Reimer RA, Seerattan RA, Hart DA, Herzog W (2015) Relationship between inflammation, the gut microbiota, and metabolic osteoarthritis development: studies in a rat model. Osteoarthr Cartil 23(11):1989–1998. https://doi.org/10.1016/j.joca.2015.03 014

226. Queipo-Ortuño MI, Seoane LM, Murri M, Pardo M, Gomez-Zumaquero JM, Cardona F et al (2013) Gut microbiota composition in male rat models under different nutritional status and physical activity and its association with serum leptin and ghrelin levels. PLoS One 8:e65465. https://doi.org/10.1371/journal.pone.0065465

227. Fouda MB, Thankam FG, Dilisio MF, Agrawal DK (2017) Alterations in tendon microenvironment in response to mechanical load: potential molecular targets for treatment strategies. Am J Transl Res 9(10):4341–4360

228. Miller FW, Lamb JA, Schmidt J, Nagaraju K (2018) Risk factors and disease mechanisms in myositis. Nature reviews. Rheumatology 14:255–268. https://doi.org/10.1038/nrrheum.2018.48

229. Quintner J (1991) The RSI syndrome in historical perspective. Int Disabil Stud 13(3):99–104

230. Bubnov RV (2010) The use of trigger point 'dry' needling under ultrasound guidance for the treatment of myofascial pain (technological innovation and literature review). Lik Sprava 5(6):56–64

231. Bubnov RV (2012) Evidence-based pain management: is the concept of integrative medicine applicable? EPMA J 3(1):13. https://doi.org/10.1186/1878-5085-3-13

232. Bubnov R, Yevseenko V, Semeniv I (2013) Ultrasound guided injections of platelets rich in plasma for muscle injury in professional athletes: comparative study. Med Ultrasound 15(2):101–105

233. Centeno CJ, Al-Sayegh H, Freeman MD, Smith J, Centeno CJ, Al-Sayegh H, Freeman MD, Smith J, Murrell WD, Bubnov R (2016) A multi-center analysis of adverse events among two thousand, three hundred and seventy two adult patients undergoing adult autologous stem cell therapy for orthopaedic conditions. Int Orthop 40:1755–1765. https://doi.org/10.1007/s00264-016-3162-y

234. De Morais PRS, Sousa ALL, Jardim T de SV, et al. Correlation of insulin resistance with anthropometric measures and blood pressure in adolescents. Arq Bras Cardiol 2016;106(4):319–326. doi:https://doi.org/10.5935/abc.20160041

235. Sasaki R, Yano Y, Yasuma T, Onishi Y, Suzuki T, Maruyama-Furuta N, Gabazza EC, Sumida Y, Takei Y (2016) Association of Waist Circumference and Body fat Weight with insulin resistance in male subjects with Normal body mass index and Normal glucose tolerance. Intern Med 55(11):1425–1432. https://doi.org/10.2169/internalmedicine.55.4100

236. Maenhaut N, Van de Voorde J (2011) Regulation of vascular tone by adipocytes. BMC Med 9:25

237. Boydens C, Maenhaut N, Pauwels B, Decaluwé K, Van de Voorde J (2012) Curr Hypertens Rep 14(3):270–278
238. van Dam AD, Boon MR, Berbée JFP, Rensen PCN, van Harmelen V (2017) Targeting white, brown and perivascular adipose tissue in atherosclerosis development. Eur J Pharmacol 816:82–92. https://doi.org/10.1016/j.ejphar.2017.03.051
239. Omar A, Chatterjee TK, Tang Y, Hui DY, Weintraub NL (2014) Proinflammatory phenotype of perivascular adipocytes. Arterioscler Thromb Vasc Biol 34(8):1631–1636. https://doi.org/10.1161/ATVBAHA.114.303030
240. Chatterjee TK, Stoll LL, Denning GM et al (2009) Pro-inflammatory phenotype of perivascular adipocytes: influence of high fat feeding. Circ Res 104(4):541–549. https://doi.org/10.1161/CIRCRESAHA.108.182998
241. Meyer MR, Fredette NC, Barton M, Prossnitz ER (2013) Regulation of vascular smooth muscle tone by adipose-derived contracting factor. PLoS One 8(11):e79245
242. Toda N, Okamura T (2013) Obesity impairs vasodilatation and blood flow increase mediated by endothelial nitric oxide: an overview. J Clin Pharmacol 53(12):1228–1239. https://doi.org/10.1002/jcph.179
243. Jonk AM, Houben AJ, Schaper NC, de Leeuw PW, Serné EH, Smulders YM (2011) Stehouwer CD obesity is associated with impaired endothelial function in the postprandial state. Microvasc Res 82(3):423–429
244. Chatsuriyawong S, Gozal D, Kheirandish-Gozal L, Bhattacharjee R, Khalyfa AA, Wang Y et al (2013) Genetic variance in nitric oxide synthase and endothelin genes among children with and without endothelial dysfunction. J Transl Med 11:227
245. Leung TF, Liu EK, Tang NL, Ko FW, Li CY, Lam CW et al (2005) Nitric oxide synthase polymorphisms and asthma phenotypes in Chinese children. Clin Exp Allergy 35(10):1288–1294
246. Pasarín M, Abraldes JG, Liguori E, Kok B, La Mura V (2017) Intrahepatic vascular changes in non-alcoholic fatty liver disease: potential role of insulin-resistance and endothelial dysfunction. World J Gastroenterol 23(37):6777–6787. https://doi.org/10.3748/wjg.v23.i37.6777
247. Karagiannides I, Pothoulakis C (2008) Neuropeptides, mesenteric fat, and intestinal inflammation. Ann N Y Acad Sci 1144:127–135. https://doi.org/10.1196/annals.1418.009
248. Cui X, Ye L, Li J, Jin L, Wang W, Li S, Bao M, Wu S, Li L, Geng B, Zhou X, Zhang J, Cai J (2018) Metagenomic and metabolomic analyses unveil dysbiosis of gut microbiota in chronic heart failure patients. Sci Rep 8(1):635. https://doi.org/10.1038/s41598-017-18756-2
249. Bubnov RV (2011) Ultrasonography diagnostic capability for mesenteric vascular disorders. Gut 60(Suppl 3):A104
250. Reginelli A, Genovese E, Cappabianca S et al (2013) Intestinal ischemia: US-CT findings correlations. Crit Ultrasound J 5(Suppl 1):S7. https://doi.org/10.1186/2036-7902-5-S1-S7
251. Glover LE, Lee JS, Colgan SP (2016) Oxygen metabolism and barrier regulation in the intestinal mucosa. J Clin Invest 126(10):3680–3688. https://doi.org/10.1172/JCI84429
252. Faber F, Bäumler AJ (2014) The impact of intestinal inflammation on the nutritional environment of the gut microbiota. Immunol Lett 162(0):48–53. https://doi.org/10.1016/j.imlet.2014.04.014
253. Bubnov RV, Moiseyenko YV (2017) Spivak MYa, NASC of Ukraine. The influence of environmental factors and stress on human health and chronic diseases: PPPM lessons from Antarctica, in EPMAWorld congress: traditional forum in predictive, preventive and personalised medicine for multi-professional consideration and consolidation. EPMA J 8(Suppl 1):S22–S23
254. Neis EP, van Eijk HM, Lenaerts K, Olde Damink SW, Blaak EE, Dejong CH, Rensen SS (2018. pii: gutjnl-2018-316161) Distal versus proximal intestinal short-chain fatty acid release in man. Gut. https://doi.org/10.1136/gutjnl-2018-316161
255. Van Hul M, Lijnen H (2012) Matrix metalloproteinase inhibition affects adipose tissue mass in obese mice. Clin Exp Pharmacol Physiol 39:544
256. Konieczka K, Ritch R, Traverso CE, Kim DM, Kook MS, Gallino A et al (2014) Flammer syndrome. EPMA J 5:11

257. Yeghiazaryan K, Flammer J, Golubnitschaja O (2010) Predictive molecular profiling in blood of healthy vasospastic individuals: clue to targeted prevention as personalised medicine to effective costs. EPMA J 1(2):263–272. https://doi.org/10.1007/s13167-010-0032-3

258. Bubnov R, Polivka J Jr, Zubor P, Koniczka K, Golubnitschaja O (2017) Pre-metastatic niches in breast cancer: are they created by or prior to the tumour onset? "Flammer syndrome" relevance to address the question. EPMA J 8:141–157. https://doi.org/10.1007/s13167-017-0092-8

259. Rashid SK, Khodja NI, Auger C et al (2014) Probiotics (VSL#3) Prevent Endothelial Dysfunction in Rats with Portal Hypertension: Role of the Angiotensin System. Peiró C, ed. PLoS One 9(5):e97458. https://doi.org/10.1371/journal.pone.0097458

260. Grech G, Zhan X, Yoo BC, Bubnov R, Hagan S, Danesi R, Vittadini G, Desiderio DM (2015) EPMA position paper in cancer: current overview and future perspectives. EPMA J 6(1):9. https://doi.org/10.1186/s13167-015-0030-6

261. Cox TR, Rumney RMH, Schoof EM et al (2015) The hypoxic cancer secretome induces pre-metastatic bone lesions through lysyl oxidase. Nature 522(7554):106–110. https://doi.org/10.1038/nature14492

262. Wang T, Cai G, Qiu Y, Fei N, Zhang M, Pang X, Jia W, Cai S, Zhao L (2012) Structural segregation of gut microbiota between colorectal cancer patients and healthy volunteers. ISME J 6:320–329

263. Roy S, Trinchieri G (2017) Microbiota: a key orchestrator of cancer therapy. Nat Rev Cancer 17(5):271–285. https://doi.org/10.1038/nrc.2017.13

264. York A (2018) Microbiome: gut microbiota sways response to cancer immunotherapy. Nat Rev Microbiol 16(3):121. https://doi.org/10.1038/nrmicro.2018.12

265. Ciorba MA, Riehl TE, Rao MS, Moon C, Ee X, Nava GM, Walker MR, Marinshaw JM, Stappenbeck TS, Stenson WF (2012) Lactobacillus probiotic protects intestinal epithelium from radiation injury in a TLR-2/ cyclo-oxygenase-2-dependent manner. Gut 61:829–838

266. Sivan A, Corrales L, Hubert N, Williams JB, Aquino-Michaels K, Earley ZM, Benyamin FW, Lei YM, Jabri B, Alegre ML, Chang EB, Gajewski TF (2015) Commensal Bifidobacterium promotes antitumor immunity and facilitates anti-PD-L1 efficacy. Science 350:1084–1089

267. Shapira N (2013) Women's higher health risks in the obesogenic environment: a gender nutrition approach to metabolic dimorphism with predictive, preventive, and personalised medicine. EPMA J 4(1):1

268. Haro C, Rangel-Zúñiga OA, Alcalá-Díaz JF, Gómez-Delgado F, Pérez-Martínez P, Delgado-Lista J et al (2016) Intestinal microbiota is influenced by gender and body mass index. PLoS One 11(5):e0154090. https://doi.org/10.1371/journal.pone.0154090

269. Reijnders D, Goossens GH, Hermes GD, Neis EP, van der Beek CM, Most J, Holst JJ, Lenaerts K, Kootte RS, Nieuwdorp M, Groen AK, Olde Damink SW, Boekschoten MV, Smidt H, Zoetendal EG, Dejong CH, Blaak EE (2016) Effects of gut microbiota manipulation by antibiotics on host metabolism in obese humans: A randomized double-blind placebo-controlled trial. Cell Metab 24(1):63–74. https://doi.org/10.1016/j.cmet.2016.06.016

270. Org E, Mehrabian M, Parks BW et al (2016) Sex differences and hormonal effects on gut microbiota composition in mice. Gut Microbes 7(4):313–322. https://doi.org/10.1080/19490976.2016.1203502

271. Rubinow KB (2017) Chapter 24: Estrogens and body weight regulation in men. Adv Exp Med Biol 1043:285–313. https://doi.org/10.1007/978-3-319-70178-3_14

272. Dakin RS, Walker BR, Seckl JR, Hadoke PW, Drake AJ (2015) Estrogens protect male mice from obesity complications and influence glucocorticoid metabolism. Int J Obes 39(10):1539–1547. Epub 2015 Jun 2

273. Medrikova D, Jilkova ZM, Bardova K, Janovska P, Rossmeisl M, Kopecky J (2012) Sex differences during the course of diet-induced obesity in mice: adipose tissue expandability and glycemic control. Int J Obes 36(2):262–272. Epub 2011 May 3

274. Codner E, Soto N, Lopez P, Trejo L, Avila A, Eyzaguirre FC, Iniguez G, Cassorla F (2006) Diagnostic criteria for polycystic ovary syndrome and ovarian morphology in women with type 1 diabetes mellitus. J Clin Endocrinol Metab 91(6):2250–2256
275. Lindheim L, Bashir M, Münzker J et al (2017) Alterations in Gut microbiome composition and barrier function are associated with reproductive and metabolic defects in women with Polycystic Ovary Syndrome (PCOS): A Pilot Study. Yu Y, ed. PLoS One 12(1):e0168390. https://doi.org/10.1371/journal.pone.0168390
276. Torres PJ, Siakowska M, Banaszewska B, Pawelczyk L, Duleba AJ, Kelley ST, Thackray VG (2018) Gut microbial diversity in women with polycystic ovary syndrome correlates with Hyperandrogenism. J Clin Endocrinol Metab. https://doi.org/10.1210/jc.2017-02153
277. Karamali M, Eghbalpour S, Rajabi S, Jamilian M, Bahmani F, Tajabadi-Ebrahimi M, Keneshlou F, Mirhashemi SM, Chamani M, Hashem Gelougerdi S, Asemi Z (2018) Effects of probiotic supplementation on hormonal profiles, biomarkers of inflammation and oxidative stress in women with polycystic ovary syndrome: a randomized, double-blind, placebo-controlled trial. Arch Iran Med 21(1):1–7
278. Metchnikoff E (1907) Lactic acid as inhibiting intestinal putrefaction. The prolongation of life: optimistic studies. William Heinemann, London, pp 161–183
279. Anukam KC, Reid G (2008) Probiotics: 100 years (1907–2007) after Elie Metchnikoff's observations. In: Mendez-Vilas A (ed) Communicating current research and educational topics and trends in applied microbiology, 2007th edn, pp 466–474
280. Hamilton-Miller J (2004) Probiotics and prebiotics in the elderly. Postgrad Med J 80(946):447–451. https://doi.org/10.1136/pgmj.2003.015339
281. Claesson MJ, Jeffery IB, Conde S, Power SE, O'Connor EM, Cusack S, Harris HM, Coakley M, Lakshminarayanan B, O'Sullivan O et al (2012) Gut microbiota composition correlates with diet and health in the elderly. Nature 488:178–184. https://doi.org/10.1038/nature11319
282. Vieira AT, Castelo PM, Ribeiro DA, Ferreira CM (2017) Influence of Oral and gut microbiota in the health of menopausal women. Front Microbiol 8:1884. https://doi.org/10.3389/fmicb.2017.01884
283. Wellons MF, Matthews JJ, Kim C (2017) Ovarian aging in women with diabetes: an overview. Maturitas 96:109–113. https://doi.org/10.1016/j.maturitas.2016.11.019
284. Landete JM, Gaya P, Rodríguez E, Langa S, Peirotén Á, Medina M, Arqués JL (2017) Probiotic bacteria for healthier aging: immunomodulation and metabolism of phytoestrogens. Biomed Res Int 2017:5939818. https://doi.org/10.1155/2017/5939818
285. Gupta VK, Paul S, Dutta C (2017) Geography, ethnicity or subsistence-specific variations in human microbiome composition and diversity. Front Microbiol 8:1162. https://doi.org/10.3389/fmicb.2017.01162
286. Rothschild D, Weissbrod O, Barkan E, Kurilshikov A, Korem T, Zeevi D, Costea PI, Godneva A et al (2018) Environment dominates over host genetics in shaping human gut microbiota. Nature 555(7695):210–215. https://doi.org/10.1038/nature25973
287. Marco ML, Tachon S (2013) Environmental factors influencing the efficacy of probiotic bacteria. Curr Opin Biotechnol 24(2):207–213. https://doi.org/10.1016/j.copbio.2012.10.002
288. Adams RI, Bateman AC, Bik HM, Meadow JF (2015) Microbiota of the indoor environment: a meta-analysis. Microbiome 3:49. https://doi.org/10.1186/s40168-015-0108-3
289. Zuo T, Kamm MA, Colombel JF, Ng SC (2018) Urbanization and the gut microbiota in health and inflammatory bowel disease. Nat Rev Gastroenterol Hepatol. https://doi.org/10.1038/s41575-018-0003-z
290. Ouwehand AC (2017) A review of dose-responses of probiotics in human studies. Benef Microb 8(2):143–151. https://doi.org/10.3920/BM2016.0140. Epub 2016 Dec 23
291. Larsen CN, Nielsen S, Kaestel P, Brockmann E, Bennedsen M, Christensen HR, Eskesen DC, Jacobsen BL, Michaelsen KF (2006) Dose-response study of probiotic bacteria Bifidobacterium animalis subsp lactis BB-12 and lactobacillus paracasei subsp paracasei CRL-341 in healthy young adults. Eur J Clin Nutr 60(11):1284–1293. Epub 2006 May 24

292. Chen J, He X, Huang J (2014) Diet effects in gut microbiome and obesity. J Food Sci 79(4):R442–R451. https://doi.org/10.1111/1750-3841.12397
293. Larsen N (2018) The effect of pectins on survival of probiotic lactobacillus spp. in gastrointestinal juices is related to their structure and physical properties. Food Microbiol 74:11e20. https://doi.org/10.1016/j.fm.2018.02.015
294. Wei M, Wang S, Gu P et al (2018) Comparison of physicochemical indexes, amino acids, phenolic compounds and volatile compounds in bog bilberry juice fermented by lactobacillus plantarumunder different pH conditions. J Food Sci Technol. https://doi.org/10.1007/s13197-018-3141-y
295. Putignani L, Dallapiccola B (2016 Sep) Foodomics as part of the host-microbiota-exposome interplay. J Proteome 16(147):3–20. https://doi.org/10.1016/j.jprot.2016.04.033
296. Bomba A, Petrov VO, Drobnych VG, Bubnov RV, Boyko NV (2016) Cells, animal, SHIME and in silico models for detection and verification of specific biomarkers of non-communicab.e chronic diseases. EPMA J 7(Suppl 1):A8
297. Petschow B, Doré J, Hibberd P, Dinan T, Reid G, Blaser M, Cani P, Degnan F, Foster J, Gibson G, Hutton J, Klaenhammer TR, Ley R, Nieuwdorp M, Pot B, Relma D, Serazin A, Sanders ME (2013) Probiotics, prebiotics, and the host microbiome: the science of translation. Ann N Y Acad Sci 1306:1–17
298. Shane AL, Cabana M, Vidry S, Merenstein D, Hummelen R, Ellis CL, Heimbach JT, Hempel S, Lynch S, Sanders ME, Tancredi DJ (2010) Guide to designing, conducting, publishing, and communicating results of clinical studies involving probiotic applications in human participants. Gut Microbes 1:243–253
299. Reid G, Gaudier E, Guarner F, Huffnagle GB, Macklaim JM, Munoz AM, Martini M, Ringel-Kulka T, Sartor B, Unal R, Verbeke K, Walter J (2010) International scientific Association for Probiotics and Prebiotics. Responders and non-responders to probiotic interventions: how can we improve the odds? Gut Microbes 1(3):200–204. https://doi.org/10.4161/gmic.1.3.12013
300. Zmora N, Zeevi D, Korem T, Segal E, Elinav E (2016) Taking it personally: personalized utilization of the human microbiome in health and disease. Cell Host Microb 19(1):12–20. https://doi.org/10.1016/j.chom.2015.12.016. Review
301. Gilbert JA, Quinn RA, Debelius J, Xu ZZ, Morton J, Garg N, Jansson JK, Dorrestein PC, Knight R (2016) Microbiome-wide association studies link dynamic microbial consort.a to disease. Nature 535(7610):94–103. https://doi.org/10.1038/nature18850. Review
302. Wu L, Ma D, Walton-Moss B, He Z (2014) Effects of low-fat diet on serum lipids in premenopausal and postmenopausal women: a meta-analysis of randomized controlled trials. Menopause 21(1):89–99
303. Jost L (2007) Partitioning diversity into independent alpha and beta components. Ecology 88(10):2427–2439. Erratum in: Ecology. 2009 Dec;90(12):3593
304. WGO updates guidelines on probiotics and prebiotics. http://www.worldgastroenterology.org/UserF.les/file/guidelines/Probiotics-and-prebiotics-English2017.pdf. Accessed 28 June 2017
305. Reid G (2011) Quo vadis – EFSA? Benef Microb 2(3):177–181. https://doi.org/10.3920/BM2011.0026
306. World Medical Association. Declaration of Helsinki. https://www.wma.net/policies-post/wma-declaration-of-helsinki-ethical-principles-for-medical-research-involving-human-subjects/. Accessed 29 Aug 2016
307. Rhodes R (2016) Rhodes Ethical issues in microbiome research and medicine. BMC Med 14:156. https://doi.org/10.1186/s12916-016-0702-7
308. Lima-Ojeda JM, Rupprecht R, Baghai TC (2017) "I am I and my bacterial circumstances": linking gut microbiome, neurodevelopment, and depression. Front Psych 8:153. https://doi.org/10.3389/fpsyt.2017.00153
309. Chuong KH, Hwang DM, Tullis DE et al (2017) Navigating social and ethical challenges of biobanking for human microbiome research. BMC Med Ethics 18:1. https://doi.org/10.1186/s12910-016-0150-y

310. Ma Y, Chen H, Lei R et al (2017) Biobanking for human microbiome research: promise, risks, and ethics. ABR 9:311. https://doi.org/10.1007/s41649-017-0033-9
311. Lewis ZT, Shani G, Masarweh CF, Popovic M, Frese SA, Sela DA, Underwood MA, Mills DA (2016) Validating bifidobacterial species and subspecies identity in commercial probiotic products. Pediatr Res 79(3):445–452. https://doi.org/10.1038/pr.2015.244
312. Akhmetov I, Bubnov RV (2017) Innovative payer engagement strategies: will the convergence lead to better value creation in personalized medicine? EPMA J 8:1. https://doi.org/10.1007/s13167-017-0078-6

Chapter 7
Selection of Prebiotic Substances for Individual Prescription

Oleksandra Pallah and Nadiya Boyko

Abstract Prebiotics are substances, more often than not ineffectively metabolized polysaccharides and oligosaccharides, that cannot be ingested successfully by the creature. A prebiotic may be a nonviable nourishment component that confers a well-being advantage on the have related to the balance of the microbiota. They invigorate the development of intestinal probiotic microbes, which can utilize these carbohydrates, in this manner advancing the health of the living being. A prebiotic must make strides development of bifidobacteria and lactic corrosive microbes and can increment the antimicrobial movement of probiotics. As a result, it can be said that probiotic microscopic organisms utilize for the treatment of pathogenic microscopic organisms with anti-microbials or their prebiotics and they repress the expression of destructiveness qualities by diverse instruments and metabolites. This review provides an insight on the current knowledge about the potential sources of plant-based prebiotics used in medicine.

Keywords Predictive · Preventive · Personalized medicine · Biologically active components · Prebiotics · Edible plants' extracts

O. Pallah (✉)
RDE Center of Molecular Microbiology and Mucosal Immunology, Department of Clinical Diagnostics and Pharmacology, Uzhhorod National University, Uzhhorod, Ukraine

N. Boyko
RDE Center of Molecular Microbiology and Mucosal Immunology, Uzhhorod National University, Uzhhorod, Ukraine

Department of Clinical Laboratory Diagnostics and Pharmacology, Uzhhorod National University, Uzhhorod, Ukraine

Ediens LLC, Uzhhorod, Ukraine

© The Author(s), under exclusive license to Springer Nature Switzerland AG 2023
N. Boyko, O. Golubnitschaja (eds.), *Microbiome in 3P Medicine Strategies*, Advances in Predictive, Preventive and Personalised Medicine 16, https://doi.org/10.1007/978-3-031-19564-8_7

7.1 Introduction

Healthy diet and nutrition are the very focus of the European Association for Predictive, Preventive and Personalised Medicine (EPMA), the main promoter of predictive, preventive, and personalized medicine (PPPM), and belong to the prioritized medical fields for long-term strategy of created multidisciplinary platform for progressing from "disease care" to "health care": "advancing participatory medicine", "well-being" concepts, and integrated approach [1, 2].

Numerous studies demonstrated a bunch of data on the possibility of using probiotics and prebiotics in medical practice (Table 7.1) [3].

Studies conducted *in vitro* and *in vivo*, including probiotics mechanism of function, intestine microbiota composition ecology, and metabolomic inquires about in respects to screening strains for clinical application are required to actualize personalized probiotic treatment in the clinical care and set important plans of clinical trials of specific strains [4, 5].

Clarification terms personalized/individualized for probiotics use is an important question [6] and also to establish correlations and associations between both approaches.

Probiotics could be useful in prevention of dental caries, respiratory tract infections, inflammatory bowel disease, and necrotizing enterocolitis [7]. Additional health benefits were reported to improve growth in healthy and malnourished persons. However, the health promoting effects are dependent on strain type, dosing regimen, and patient's individual responses [8].

The authors shows [9] that, on August 1, 2019, the ClinicalTrials.gov database contained 1341 studies which could be retrieved using the search term "probiotics". For comparison, during this same period, "microbiota" yielded 2151 studies and "prebiotics" 342 studies. Registrations of clinical studies with probiotics registered in ClinicalTrials.gov has been stable with around 100 studies annually since 2010 with a tendency to increase in numbers over the most recent years.

Identification of components of the microbiota and explenation of the molecular mechanisms of their action to induce pathological changes or exert beneficial, disease-protective activities could aid in our ability to influence the composition of the microbiota and to find bacterial strains and components (e.g., probiotics and prebiotics) the use of which can help in the prevention and treatment of diseases. The use of animal models to study the role of microbiota in human diseases allows the development and maintenance of chronic diseases [10, 11].

The prebiotic concept reflects follow-up developments built up on a broad probiotic acceptance and achievements in the area [12]. Further, prebiotics became included in the concept of functional and fermented foods synergistically demonstrating greater benefits compared with their individual microbial, nutritive, or bioactive components. Their relevance for the field of human nutrition and urgent necessity for inclusion into national and international dietary guidelines has been proposed [13]. The dedicated International Scientific Association for Probiotics and Prebiotics (ISAPP) expert group has proposed the definition of prebiotic as Ba

Table 7.1 Studies of the effects of probiotics on indicators of obesity

References	Subject of study	Study population	Duration	Results
Wang et al. (2019b)	Effect of probiotics on body weight and glycaemic control	Meta-analysis of 12 studies (416 subjects given placebo and 405 given probiotics)	8–24 weeks	Significant ↓ in BW, BMI, FM (%), and insulin levels in the probiotic group
Çelik and Ünlü Söğüt (2019)	Impact of probiotic supplementation on chemerin level, inflammation, and metabolic syndrome parameters	3 groups: Control, obese rats fed high-fat diet, and obese intervention given probiotics post obesity induction	16 weeks	↓ weight gain, and beneficial effects on insulin, fasting blood glucose, inflammatory markers, leptin, and chemerin levels
Qian et al. (2019)	Dietary therapy may alter the function and composition of microbes	4 groups: HFD, DI low-fat diet, HFD with probiotic (*Lactobacillus acidophilus*, *Bifidobacterium longum*, and *Enterococcus faecalis*), and DI with probiotic groups	4 months	In the group probiotics with DI: ↑ 2 Butyrate producing families (*Ruminococcacece* and *Lachnospiraceae*)
Krumbeck et al. (2018)	*Bifidobacterium* and galacto-oligosaccharides	114 subjects	3 weeks	↓ inflammation and improvement in intestinal permeability
Park et al. (2017)	*Lactobacillus plantarum*	Mice	12 weeks	↓ adiposity through lipid oxidation
Liu et al. (2017)	*Bacteroides thetaiotaomicron*	Mice and human obese and control groups	7 weeks	↓ total fat mass and weight gain in HFD with probiotic
Brooks et al. (2016)	Fermentable carbohydrate inulin	Mice	14 weeks	Suppression of peptide YY by 87%
Dao et al. (2016)	*Akkermansia muciniphila*	49 obese adults	6 weeks	Improvement of insulin resistance and metabolism
Schneeberger et al. (2015)	*Akkermansia muciniphila*	Mice	16 weeks	↓ inflammation and adiposity
Osterberg et al. (2015)	Mixture of *lactobacilli*, *streptococcus*, and *Bifidobacteria*	20 healthy males in 2 groups: HFD versus HFD with probiotics	4 weeks	Less weight and fat mass gain in HFD with probiotics
Plaza-Diaz et al. (2014)	*Lactobacillus paracasei*, *Bifidobacterium breve*, and *Lactobacillus rhamnosus* or mixture	Mice	30 days	↓ triacylglycerol liver content (mixture with *Lactobacillus rhamnosus*, *Bifidobacterium breve*), ↓ serum LPS levels

↓ decrease, ↑ increase, *BW* body weight, *BMI* body mass index, *DI* dietary intervention, *FM* fat mass, *HFD* high-fat diet, *LPS* serum levels of lipopolysaccharide

substrate that is selectively utilized by the host microorganisms conferring a health benefit [14] that opens an opportunity to test the prebiotic properties and corresponding health benefits for previously unconsidered broad spectrum of substances. As potential candidates, for example, fenugreek [15], gold-based nanomaterials [16].

The authors [17] described the possibility of using biologically active substances of plant origin (flavonoids, phenols) to effectively combat COVID-19-associated complications in primary, secondary, and tertiary care in the context of 3 PM Plant phenols are well-known antibacterial agents while the suggested antibacterial mechanisms enhance attenuation of pathogenicity and inhibition of: nucleic acid synthesis, cytoplasmic membrane function, attachment and biofilm formation, cell membrane porin, energy metabolism, and membrane permeability [18–20].

7.2 New Era of Prebiotics

Recent advances in microbiome science have opened new frontiers in the study of probiotics and prebiotics. New types, new mechanisms and new applications that are currently being studied can change the scientific understanding and application of these measures in nutrition and health.

Mechanisms of action of probiotics and prebiotics are complex, diverse, heterogeneous, and often strain- and compound-specific. Although much has been described, much remains to be understood, especially the structural and functional interpretations of the observed health effects and long-term consequences [21–23]. Currently, a narrow range of confirmed prebiotic substances exists, with galactans and fructans (e.g., inulin) dominating the market. The desire to stimulate a wider symbiotic biota has led to the development of new candidates for prebiotic compounds [24]. Some polyphenols have been shown to have prebiotic potential, such as cranberry-rich extracts stimulating *A. muciniphila* [25], or to provide antimicrobial action against pathogens [26].

In the future, prebiotics will likely be isolated from novel sources as focus on sustainability, cost, and scale emerges [27]. Future prebiotic compounds may also be chemically or structurally modified by the application of sonication, high pressure, acid, enzyme and oxidation treatments, in order to modify functionality. Further, unique combinations of prebiotics in optimized mixtures may provide the ability to create new profiles of benefits [28].

As of December 2020, there were 245 registered clinical trials (ClinicalTrials. gov) which have completed evaluation of prebiotics (alone or in combination with probiotics) on aging, autism, bariatric surgery, colic, colon cancer, atopic dermatitis, constipation, diarrhea, infant growth, irritable bowel syndrome (IBS), obesity, and other conditions. The number of studies and investigational targets are suggestive of significant investment in the development of prebiotics as bioactive ingredients or supplements for a range of potential applications.

During the most recent decade, several papers have proposed alternative definitions of prebiotics, with broader scope in order to better integrate emerging microbiome-modulating compounds [29, 30]. In particular, the requirement of "selective use" by a limited number of species or genera has been questioned (despite the general consensus on the contrary in the literature). The reason is that according to recent microbiome studies, the decisive factor is not the ability of pre- and probiotics to stimulate certain species of microorganisms, but their ability to normalize the ratio of the main species and genus.

7.3   NANO Prebiotics

Over the last few years, the application of nanotechnology to nutraceuticals has been rapidly growing due to its ability to enhance the bioavailability of the loaded active ingredients, resulting in improved therapeutic/nutraceutical outcomes.

Within the different definitions of nanomaterials, these can be described as the products of nanotechnology, characterized by at least one dimension within the size range below 100 nanometers [31–33].

There is a significant and growing interest in nanoparticles in both scientific and non-scientific circles since the former are used in food and non-food products. The versatility of new nanoparticle applications makes them potentially harmful in the food industry, healthcare, and environment, and thus requires the development of nanonutraceuticals from such nutrients as antioxidants, vitamins, fatty acids, fibers, probiotics, and prebiotics. The article [34] reveals the antimicrobial effect of silver and titanium dioxide nanoparticles on the growth of opportunistic bacteria and their interaction with the probiotic strains of *L. casei* ATCC 39392, *L. plantarum* ATCC 8014, and *L. fermentum* ATCC 9338.

In another study, a prebiotic formulation comprising *Pediococcus acidilactici* and phthalyl dextran nanoparticles with antimicrobial potentials was developed [35]. The investigators conjugated phthalic anhydride with dextran in the process and establish that phthalyl dextran nanoparticles were internalized by probiotics predicated on time, temperature, and glucose transporters, enhancing antimicrobial peptides product and antimicrobial exertion while adding good bacteria in mice through self- defense process compared with probiotics themselves.

Although little are the studies conducted purely on nano prebiotics, yet development of whey protein isolate/inulin nano complexes with prebiotic effects, and as a delivery system for various probiotics in food products [36], and the use of chitosan as a nanoencapsulation of *L. acidophilus* for enhanced viability and survival against gastrointestinal conditions are all current applications of nanotechnologies to prebiotics.

The authors [37] investigated the effectiveness of nanocerium compounds as a new prebiotic. The article studied cerium dioxide compounds in the composition with prebiotic strains and their effect on the intestinal microbiota in obese mice. The study demonstrated a positive effect of the *Lactobacillus* and *Bifidobacterium* strains (probiotics), if they contain nanocerium (a potential prebiotic), on lowering cholesterol and restoring intestinal microbiota in obese mice. The results presented provide a new understanding of food additives mechanisms and open new perspectives for the use of probiotics in combination with the substances that demonstrate prebiotic properties benefiting the health of the host.

7.4   Prebiotics of Plant Origin

Previously, prebiotics were considered an indigestible dietary fiber having some biological effects, including selective stimulation of growth and biological activity of beneficial microorganisms present in or introduced into the intestine therapeutically. Herewith, probiotics in functional foods are mainly found in dairy products. However, there is a need to develop non-dairy probiotics and fiber-free prebiotics.

According to the new definition of prebiotics, polyphenols and fatty acids may be included in this group together with some peptides catabolized by bacteria to active ingredients. Even inorganic materials (namely trace elements needed for bacterial growth) that are used externally and internally can be considered prebiotics.

The traditional prebiotics can be divided into three major groups different as regard their chemical composition, the constituent units and the glycosidic bonds and involving oligosaccharides (fructooligosaccharides (FOS), galactooligosaccharides (GOS), xylooligosaccharides (XOS), isomaltooligosaccharides (IMO), raffinose oligosaccharides (RFOS) man-mannooligosaccharides (MOS), isomaltulose, inulin, etc.) fibers (β-glucans, pectins, cellulose, dextrins, etc.) and polyols (xylitol, mannitol, lactitol lactulose) [38].

The function of a prebiotic ingredient is particular stimulation of certain intestinal bacteria, particularly lactic acid producing microorganisms similar as Bifidobacteria and Lactobacilli, occupant in the gut rather than introducing an exogenous species. The prebiotic composites can be classified grounded on their chemical nature, chain length or degree of polymerization(DP), mode of operation, etc. Depending on the chemical nature, prebiotic composites are classified into three types saccharide derivations(disaccharides, oligosaccharides and polysaccharides), proteins or peptides, and lipids [39]. Most of the prebiotics used as food adjuncts are saccharide derivatives and mainly from plants. This family of compounds includes several oligosaccharides (namely fructo-, gluco-, galacto-, isomalto-, xylo-, and soy-oligosaccharides), inulin, lactulose, lactosucrose, guar gum, resistant starch, pectin and chitosan. Cereals and legume crops like barley, wheat, chickpea and lentils are potential plant sources for prebiotic carbohydrates; vegetables like chicory, Jerusalem artichoke, onion, garlic, okra, and leek; and fruits like dragon fruit, jackfruit, palm fruit, nectarine and mushroom [40–47].

Today, there is evidence of the possibility of using not only inulin and chitosan as new generation prebiotics, but also other oligosaccharide compounds that are sources of carbon. For example, Table 7.2 presents promising biologically active compounds that can be used as prebiotics [48].

Table 7.2 Potential prebiotic carbohydrates

Soybean oligosaccharides	Mannan oligosaccharides (yeast cell wall)
Gluco oligosaccharides	Lactose
Cyclodextrins	Resistant starch and derivatives
Gentiooligosaccharides	Oligosaccharides from melobiose
Germinated barley foodstuffs	N-acetylchitooligosaccharides
Oligodextrans	Polydextrose
Glucuronic acid	Sugar alcohols
Gentiooligosaccharides	Konjac glucomannan
Pectic oligosaccharides	Whole grains

There are even noncarbohydrate food ingredients that have potential to be classified as prebiotics, including lactoferrin, phenolic compounds (e.g., flavonoids), and glutamine. To date, however, little research has been done on any of those compounds.

Concurrently, the presence of prebiotics and their intake helps the growth of beneficial bacteria, consequently limiting the pathogenic bacteria diffusion and colonization, so that they in turn join forces for the prevention of infections.

7.5 Pro-and Antibacterial Properties of Polyphenols and Anthocyanins In Vitro

Based on the determined gross content of anthocyanins and polyphenols and quantitative anthocyanins content in extracts, their ability to inhibit the growth of opportunistic bacteria of *E. faecium Ke-01, S. aureus Ks-01, K. pneumoniae K-01, P. aeruginosa Kp-01, E. coli EPEC,* and *E. cloacae Kc-01* clinical isolates was studied. The ability of fruit/berry extracts to inhibit opportunistic bacteria growth was studied in dynamics, that is with different duration (4, 14, and 24 h) of cultivation in the studied extracts.

The strains of *K. pneumoniae K-01, E. cloacae Kc-01, P. aeruginosa Kp-01, S. aureus Ks-01,* and *E coli Ksh-01* were sensitive to the impact of the cherry plum extract already after 4 h of cultivation: their number significantly decreased compared to the control sample. After 14 h of incubation, no growth was observed in all tested cultures except *K. pneumoniae K-01,* though the number of its living cells also decreased. After cultivation of the selected bacterial strains in the medium containing the extracts for 24 h, there were no viable cells of *K. pneumoniae K-01, E. cloacae Kc-01, P. aeruginosa Kp-01, S. aureus Ks-01,* and *E. coli Ksh-01.*

The *K. pneumoniae K-01, E coli Ksh-01,* and *P. aeruginosa Kp-01* strains were insensitive to the impact of the cherry plum extract after their cultivation for 4 h. The studied extract demonstrated an antibacterial effect on the strains of *E. cloacae Kc-01* and *S. aureus Ks-01.*

After cultivation for 14 h in the presence of the sweet cherry extract, there was an inhibition of growth of all tested strains and no growth of *E. cloacae Kc-01* and

S. aureus Ks-01 strains. After 24 h of incubation, there was no growth of all microorganisms.

Only the *P. aeruginosa Kp-01* strain was insensitive to the plum extract impact after 4 h of cultivation. The remaining tested strains were sensitive to the extract. There was an absence of growth of the *S. aureus Ks-01, E. cloacae Kc-01,* and *K. pneumoniae K-01* strains observed already after their 14-h cultivation. The most resistant to the plum extract impact was the *E coli Ksh-01* strain, which stopped growing only after cultivation for 24 h.

The strains of *S. aureus Ks-01, E. cloacae Kc-01,* and *K. pneumoniae K-01* were sensitive to the impact of the jostaberry extract. Their growth inhibition was observed after 4 h of their joint cultivation while after cultivation for 14 h, there was no growth of the studied strains. The jostaberry extract turned to be less effective against the strains of *P. aeruginosa Kp-01* and *E coli Ksh-01*: its antibacterial effect could be observed only after cultivation for 24 h.

The red currant extract had a weak antibacterial effect on *E. cloacae Kc-01* and *K. pneumoniae K-01.* After 14 h of incubation, the number of viable cells of *P. aeruginosa Kr-01, K. pneumoniae K-01,* and *E. cloacae Ks-01* significantly decreased.

The strains of *P. aeruginosa Kp-01, K. pneumoniae K-01, S. aureus Ks-01,* and *E coli Ksh-01* were sensitive to the impact of the blueberry extract; there was an inhibition of their growth observed after 4 h of incubation. The *E. cloacae Kc-01* strain was resistant to this extract. The number of bacteria did not change after 14 and 24 h of incubation, but was still lower than that after the 4-h incubation; thus, there was a bacteriostatic effect.

The study of antibacterial properties of berry extracts against the strains of *K. pneumoniae Rk-01, S. aureus Rs-01, P. aeruginosa Rp-01, E. cloacae Re-01, Y. pseudotuberculosis Ry-01, S. odorifera Rs-01, S. marcescens Rm-01,* and *P. agglomerans Ra-01* isolated from the surface of edible plants and fruits was a separate stage.

Analysis of the results (Table 7.3) of identification of berry extracts' antibacterial properties revealed the fact that they have a strong inhibitory effect on the bacteria isolated from the surface of edible plants. The results of the antibacterial effect evaluation are presented in the form of a logarithmic ratio.

The strains of *K. pneumoniae Rk-01, S. aureus Rs-01, P. aeruginosa Rp-01, E. cloacae Re-01, Y. pseudotuberculosis Ry-01, S. odorifera Rs-01, S. marcescens Rm-01,* and *P. agglomerans Ra-01* were most sensitive to the impact of the cherry plum, sweet cherry, and black currant extracts. Inhibition of their growth was observed already after 14 h of cultivation in the medium containing the above extracts. After 24 h of cultivation, there was no growth of all the above strains. The strains of *K. pneumoniae Rk-01, E. cloacae Re-01,* and *Y. pseudotuberculosis Ry-01* were more resistant to the extracts of jostaberry, red currant, plum, and blueberry compared to the extracts of cherry plum, sweet cherry, and black currant. After cultivation for 24 h, the number of viable cells decreased by six orders of magnitude, but the extracts could not completely inhibit the above bacteria's growth [49].

Table 7.3 Antibacterial impact of fruit/berry extracts on clinical isolates (lgN_t/N_0)

Time of cultivation, hours	K. pneumoniae K-01			P. aeruginosa Kp-01			S. aureus Ks-01			E. cloacae Kc-01			E. coli Ksh-01		
	24 h	14 h	4 h	24 h	14 h	4 h	24 h	14 h	4 h	24 h	14 h	4 h	24 h	14 h	4 h
Prunus cerasifera	–	–6.5 ± 0.29**	–5.2 ± 0.29**	–	–	–2.5 ± 0.38***	–	–	–3.13 ± 0.55**	–	–	–4.17 ± 0.58**	–	–6.5 ± 0.29**	–2.63 ± 0.29**
Ribes x nidigrolaria	–6.5 ± 0.5	–4.17 ± 0.5	–2.5 ± 0.14	*	–6.5 ± 0.23	–0.5 ± 0.29	–	–	–4.17 ± 0.27	–	–	–4.17 ± 0.29	–6.5 ± 0.27	–2.7 ± 0.27	–1.48 ± 0.27
Prunus avium	–	–4.47 ± 0.14	–2.47 ± 0.29	–	–6.47 ± 0.29	–0.47 ± 0.14	–	–	–4 ± 0.38	–	–	–	–	–5.48 ± 0.29	–2.47 ± 0.29
Ribes nigrum	–	–4.47 ± 0.29	–4 ± 0.76	–	–6.47 ± 0.29	–0.47 ± 0.29	–	–	–2 ± 0.25	–	–4.47 ± 0.29	–4 ± 0.76	–6.5 ± 0.3	–5.63 ± 0.29	–0.43 ± 0.2
Ribes rubrum	–	–	–2.47 ± 0.76	–	–	–4.47 ± 0.29	–	–6.47 ± 0.14	–5.17 ± 0.29	–3.17 ± 0.25	–2.17 ± 0.29	–2.17 ± 0.5	–6.5 ± 0.1	–5.48 ± 0.3	–3.48 ± 0.19
Vaccinium myrtillus	–	–4.47 ± 0.29	–3.87 ± 0.58	–6.47 ± 0.29	–6.47 ± 0.25	–6.47 ± 0.29	–	–6.47 ± 0.14	–4.36 ± 0.76	–3.17 ± 0.25	–2.17 ± 0.3	–2.17 ± 0.25	–	–	–4.48 ± 0.19
Prunus domestica	–	–	–4 ± 0.25**	–	–4.47 ± 0.29**	–0.47 ± 0.29**	–	–	–6.47 ± 0.29	–	–	–5.17 ± 0.29	–	–6.5 ± 0.29	–2.47 ± 0.25

Note: lgN_t/N_0 is the ratio of the number of bacteria that remain viable after cultivation (N_t) to the number of bacteria (N_0) in control samples (without extracts addition)

**p < 0.05 – the difference is reliable under the impact of the cherry plum extract on the strains of K. pneumoniae K-01, P. aeruginosa Kp-01, S. aureus Ks-01, E. coli Ksh-01, and E. cloacae Kc-01

Table 7.4 Lactobacilli viability after cultivation with various extracts for 14 and 24 h

Extracts	L. acidophilus C-01		L. acidophilus C-02		L. acidophilus C-03		L. acidophilus C-04	
	14 h	24 h	14 h	24 h	14 h	24 h	14 h	24 h
Prunus cerasifera	−1.04	−1.04	−0.76	−1.04	−1.04	−1.04	−0.4	−1.04
	±0.5*	±0.5*	±1.4	±0.79	±1.7	±0.19*	±0.25*	±0.17*
Ribes x nidigrolaria	−0.58	−0.46	−0.63	−0.01	−0.4	+0.01	−0.4	+0.05
	±1.1	±1.1	±0.5	±0.5	±0.35	±0.3	±0.2*	±0.3
Prunus avium	−0.09	−0.27	−0.24	−0.09	−0.57	+0.01	−0.34	−0.09
	±0.79	±0.5	±1.1	±0.26	±0.3*	±0.29	±0.42	±0.4
Ribes nigrum	−0.63	−0.43	−0.43	−0.01	−0.01	+0.05	−0.4	−0.34
	±0.5	±0.76	±0.5	±0.5	±0.35	±0.28	±0.38	±0.57
Ribes rubrum	−0.61	−0.65	−0.09	+0.02	−0.82	−0.75	−0.09	−1.04
	±0.5	±1.2	±0.26	±0.5	±0.5*	±0.17*	±0.38	±0.45*
Vaccinium myrtillus	−0.4	+0.05	−0.52	−0.09	−0.4	+0.05	−0.4	+0.06
	±0.83	±0.5	±0.29*	±0.29	±0.35	±0.5	±1.0	±0.92
Prunus domestica	−0.18	+0.05	−0.06	−0.09	−0.09	−0.06	−0.06	−0.09
	±0.2	±0.5	±0.26	±0.29	±0.5	±0.5	±0.78	±0.38

Notes: *There is a statistically significant confirmation of the inhibitory effect of the extract ($p < 0.05$)

lg (N_t/N_0) for strains L. acidophilus C-01, L. acidophilus C-02, L. acidophilus C-03, L. acidophilus C-04, 14, 24 h of cultivation

The ability of fruit extracts to stimulate lactobacilli growth was studied on four *Lactobacillus acidophilus* strains *(L. acidophilus C-01, L. acidophilus C-02, L. acidophilus C-03, L. acidophilus C-04)*, which were previously isolated from a healthy person's intestine and stored in the Museum of Cultures of the Scientific Research and Educational Center of Molecular Microbiology and the Immunology of Mucous Membranes, Uzhhorod National University [50].

The ability of cherry plums, sweet cherries, black and red currants, plums, and jostaberries extracts was studied via cultivation of the *L. acidophilus* strains in the studied extracts with different exposure times: 14 and 24 h (Table 7.4).

The ability to stimulate the *L. acidophilus C-01* strain's growth was observed in all extracts except the cherry plum extract. In addition, the jostaberry extract stimulated the growth of two more strains: *L. acidophilus C-04* and *L. acidophilus C-02*.

The blackcurrant extract stimulated *L. acidophilus C-03* growth. Red currant berries demonstrated probacterial properties against *L. acidophilus C-02*. The sweet cherry extract had an ability to stimulate *L. acidophilus C-03*, *L. acidophilus C-02*, and *L. acidophilus C-04* growth.

The plum extract had the best probacterial properties. It stimulated the growth of all tested lactobacilli strains: *L. acidophilus C-03*, *L. acidophilus C-04*, *L. acidophilus C-02*, and *L. acidophilus C-01*.

The quantitative data on extract composition demonstrated that the spectrum of anthocyanin and polyphenol compounds is narrower in fruit/berry extracts of the *Prúnus* genus plants than in the *Ribes* genus plant extracts. However, according to the experimental data, plum and sweet cherry extracts can better stimulate the growth of probiotic bacterial strains.

We have also noted blueberries' ability to stimulate the growth of all selected LAB strains; nevertheless, we had previously found that they inhibit the growth of human commensal microbiota.

Interestingly, the extract of cherry plum, which has significant antibacterial properties against opportunistic bacteria, is unable to stimulate the growth of the *L. acidophilus C-01*, *L. acidophilus C-02*, *L. acidophilus C-03*, and *L. acidophilus C-04* strains.

7.6 Determination of Minimum Inhibitory Concentrations of Edible Plant Extracts on Bacterial Strains Isolated from Medical Equipment Surfaces and Bacterial Strains Isolated from Plant Surfaces

Taking into account the results of the study of antibacterial properties of extracts, we determined the minimum inhibitory concentrations of extracts of plums, red and black currants, blueberries, cherries, and yoshta, which were observed inhibition of growth of the studied strains.

The study found that growth inhibition of *Y. pseudotuberculosis Ry-01* was observed at a concentration in the medium of yoshta extract of 0.0002 mg/ml. The value of the minimum inhibitory concentration of yoshta extract at which it had an antibacterial effect against *K. pneumoniae Rk-01*, *S. aureus Rs-01*, *P. aeruginosa Rp-01*, *E. cloacae Re-01*, *S. odorifera Rs-01*, *S marcescens Rm-01*, *P. agglomerans Ra-01* was 0.02 mg/ml. Minimum inhibitory concentrations of the extract against strains of *E. faecium Ke-01*, *S. aureus Ks-01*, *K. pneumoniae K-01*, *P. aeruginosa Kp-01*, *E. cloacae Ks-01*, *E. coli Ksh-01*, were in in the range from 0.02 to 0.2 mg/ml.

The minimum inhibitory concentration of plum extract for all strains isolated from plant surface, except *Y. pseudotuberculosis* Ry-01, was 0.0002 mg/ml, indicating the high ability of such an extract to inhibit the growth of this strain. The antimicrobial effect of the extract on the strain *Y. pseudotuberculosis* Ry-01, was observed at its concentration of 0.04 mg/ml. Analyzing the inhibitory properties of plum extract against clinical strains of *E. faecium Ke-01*, *S. aureus Ks-01*, *K. pneumoniae K-01*, *P. aeruginosa Kp-01*, *E. cloacae Ks-01*. *E. coli Ksh-01*, we found that the value of the minimum inhibitory concentration of plum extract was 0.002 mg/ml, which also indicates a significant antibacterial effect.

Under the action of black currant extract on the microorganisms we tested, both on clinical strains and bacteria isolated from the surface of edible plants, the MIC was 0.02 mg/ml.

Under the action of the cherry extract on test microorganisms, an inhibitory effect was observed on the growth of *P. aeruginosa Rp-01* (0.002 mg/ml), S. aureus Rs-01 (0.02 mg/ml), which were isolated from the plant surface. For clinical strains of *K. pneumoniae K-01*, the minimum inhibitory concentration was 0.04 mg/ml. For the remaining strains, the MIC was 0.2 mg/ml.

Under the action of the plum extract on the tested bacterial strains, the inhibitory effect was observed against clinical strains of *P. aeruginosa Kr-01* (0.0002 mg/ml), *S. aureus Ks-01* (0.02 mg/ml), *E. cloacae Ks- 01* (0.04 mg/ml), *K. pneumoniae K-01* (0.2 mg/ml). In relation to bacteria isolated from the plant surface, the extract was effective at maximum dilution, namely, 0.0002 mg/ml.

Under the action of red currant extract on the microorganisms we tested, an inhibitory effect was observed against *P. aeruginosa Kr-01* (0.002 mg/ml), *S. aureus Ks-01* (0.002 mg/ml). In relation to the remaining tested bacteria, the inhibitory effect of the extract was manifested at a concentration of 0.2 mg/ml.

The minimum inhibitory concentration of blueberry extract for all clinical isolates was 0.02 mg/ml. The MIC of the extract for bacteria isolated from the surface of edible plants and fruits was 0.0002 mg/ml.

We showed that only blueberry extract had a different inhibitory effect on all test strains that were isolated from different biological niches. That is, such an extract is more effective in eliminating bacteria isolated from food.

7.7 Application of Plum Extract as a Prebiotic Component of Biopreparations

The influence of the plum' (*Prunus domestica*) extract on the human cells culture had been investigated since its ability to inhibit the biofilm formation by various bacterial agents of opportunistic infections.

Further explotation of these beneficial properties are possible if case the absence of the toxicity. Thus the direct injection of the extract to the culture fluid of newborns dermal fibroblasts had been performed to detect their toxicity level. The explosion time was correspondingly 0 (control), 24 and 48 h [51].

Figure 7.1 is demonstrating screens of systems "plum' extract – cells culture" in the concentration 0.002, 0.02 mg/ml, and the control (CarlZeiss, Germany).

Fig. 7.1 The results of determining the minimum inhibitory concentration of plum extract on the growth of bacteria isolated from the plant surface

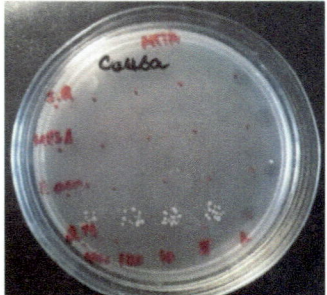

Concentration, mg/ml	0 h	24 h	48 h
0,002			
0,02			
Control			

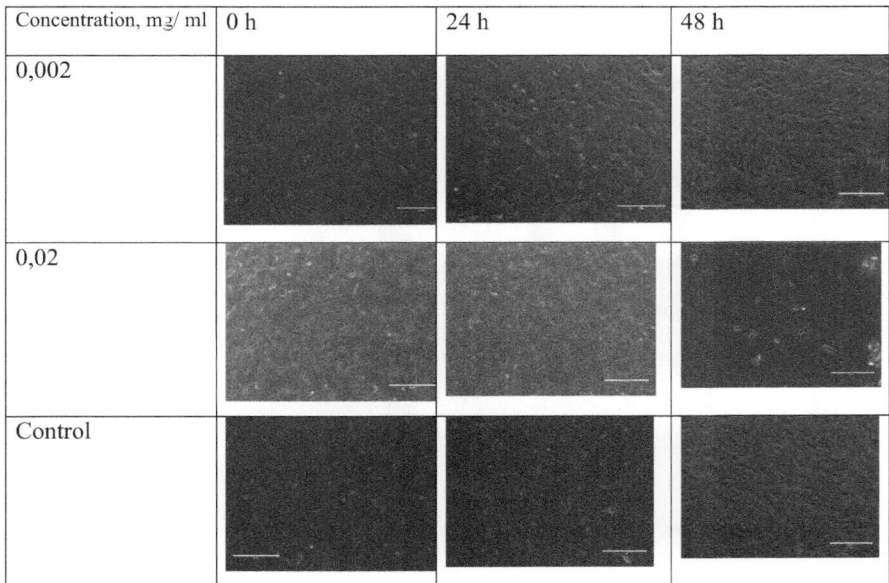

Fig. 7.2 Photographs of the system "sample plum extract - cell culture"at the beginning of the experiment, after 24 and 48 h, absolute length 400 μm

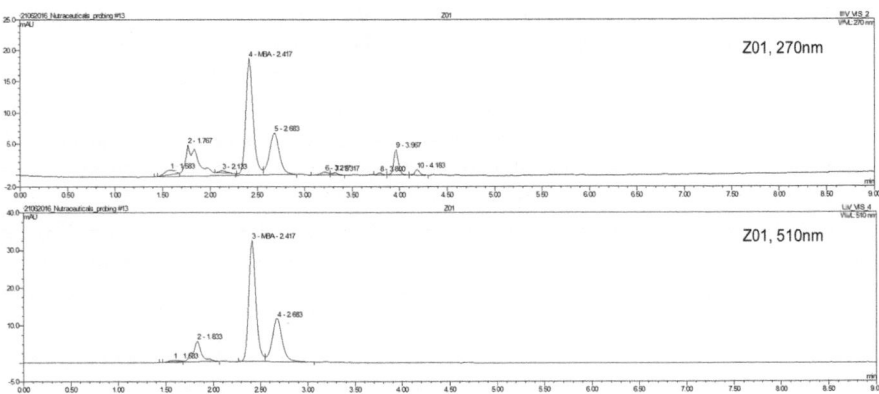

Fig. 7.3 Chromatogram of plum extract at two wavelengths, (λ = 270 and 510 nm)

Analysis of all obtained images showed that at a concentration of 0.002 mg/ml plum extract did not show toxic effects on dermal fibroblasts of newborns. However, the plum extract in the concentration of 0.02 mg/ml, the effect of this extract on dermal fibroblasts of newborns led to their death after 48 h of cultivation (Fig. 7.2).

Furthermore the culture fluids (plum extract and cell culture medium containing the extract) at two different concentrations – of 0.002 and 0.02 mg/ml, was investigated by high performance liquid chromatography (Fig. 7.3).

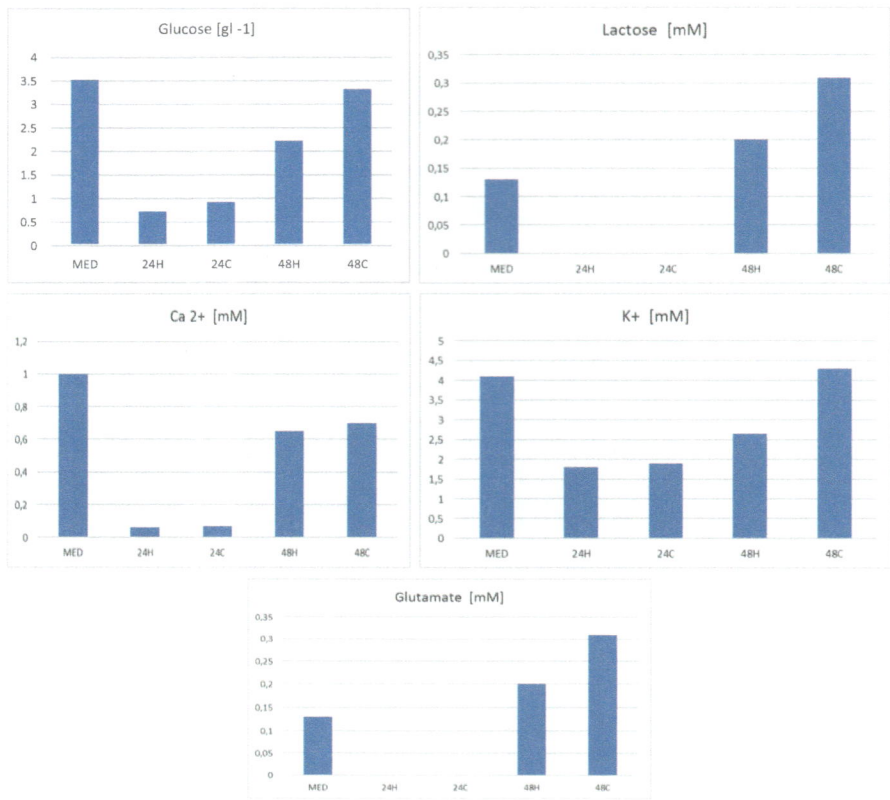

Fig. 7.4 Component content of culture fluid and ANOVA analysis. Note: MED – control of the experiment, H – extract concentration 0.002 mg/ml, C – extract concentration 0.02 mg/ml

According to the obtained results, the chromatogram at defined length $\lambda = 270$ nm was more informative and revealed the component composition of plum extract (Fig. 7.4). ANOVA analysis had been performed to demonstrate the statistically proved differences.

As you can see on the Fig. 7.3 the significant difference between the control and test samples in the content of lactose and glutamate had been registered.

No lactose or galactose content was detected after 24 h of cultivation. A rapid increase in lactose and glutamate was observed after 48 h of cultivation, but their level was significantly lower than in control samples.

The experimental values of the levels of glucose, calcium and potassium ions did not differ much from the values obtained in the control samples. Under the action of plum extract on fibroblast cultures when diluting the extract 0.02 mg/ml for 24 h of cultivation, the density of the monolayer is higher than in the control, but after 48 h of cultivation there is a sharp decrease in fibroblasts (or monolayer density), indicating inexpediency plum extract in this dilution.

In particular, a sharp increase in the level of glutamate was recorded, which is logical, because glutamate synthesizes proline, the part of the main connective tissue protein – collagen. Accordingly, the increase in glutamate in the culture fluid indicates their deficiency in the fibroblasts. In general, this experiment allows us to assess the safe for cells concentrations of plum extract and the time limits of its usage [52].

Principles of designing the modern pharmabiotics for the individual application.

In the pharmaceutical market the wide range of probiotic, prebiotic, synbiotics and a variety of biologically active additives available. One of the main problems of their sufficient exploitation is controversial data for their efficacy no or limited cohort testing available [53].

According to the principles of modern (3P) medicine, individualized selection of any preparations and in particular newly created biological products prescribed for the prevention and treatment of human NCD are required.

Based on the systemic biology approach, the same disease is differ in symptoms and recovering ratio in various patients. It is known that the combination of probiotic and also the prebiotic biologically active substance as the bio-preparations' component significantly modulate its multifaceted action. Therefore, the issue of development of modern pharmaceuticals on the basis of 3P medicine is relevant tool in order to increase the efficacy of individual biopreparations' application.

Taking into account our experimental data on the strain-specific pro-bacterial or antibacterial properties of the edible plants' (extracts) combine with clinically proved ability to inhibit biofilm formation, we first time demonstrate that they can be served as a novel prebiotic component of the modern pharmaceuticals [54].

This idea had been partially tested by using some known biopreparations used for the normalization of gut microbiota namely Enterogermina® (*Bacillus clausii*), Lactovit forte® (*Bacillus coagulans*), Enterol® (*Saccharomyces boulardii*). The selection of biopreparations aimed to achieve the difference of the microbiota compositions (in the content of the active bases).

The changes of antibacterial efficacy against the tested strains of clinically isolated opportunistic pathogens had been investigated under condition of modulated experimental induction of their "proficiency" with studied plant extract.

Table 7.5 shows that these biological additives (preparations) are capable selectively inhibit the growth of the bacteria – the agents of potential intestinal disorders.

Lakely, Enterogermina, Lactovit and Enterol 250 do not inhibit the growth of *K. pneumoniae* and *S. aureus Z03*, and Enterogermina does do not inhibit the growth of *P. aeruginosa*. Lactovit and Enterol 250 also do not inhibit the growth of *S. aureus, E. coli lac-*.

Since plum extract showed the strongest ability to inhibit biofilm activity with no any toxicity at a concentration of 0.002 mg/ml we used it to induce/redesign a new biological product with stronger antibacterial properties since its possible synergy.

The Table 7.6 shows the data obtained confirming the increase in the effectiveness of the composition of Enterogermina® together with anthocyanins extracted from plum berries.

Table 7.5 Activity spectrum of the biologically active additives – probiotics: Enterogermina®, Lactovit forte®, Enterol®

Clinical isolates	Enterogermina® (*Bacillus clausii*)	Lactovit forte® (*Bacillus coagulans*)	Enterol® (*Saccharomyces boulardii*)
S. aureus Z03	–	–	–
K. pneumoniae Z02	–	–	–
P. aeruginosa Z01	–	+	+
E. coli lac⁻ Z07	+	–	–
E. cloacae Z04	+	+	+

Note: Tested concentration $1 \cdot 10^8$, "–" – inhibition activity $\sim 10^7$ CUO/ml, "+" – inhibition 10^1–10^3

Table 7.6 Changed spectrum of action of plum extract combined with Enterogermina® and their compositions

Bacterial strains, agents of opportunistic infections, clinical isolates	Plum extract, 0.002 mg/ml	Enterogermina®	Composition of plum extract and Enterogermina®
S. aureus Z03	+	–	+
K. pneumoniae Z02	+	–	+
P. aeruginosa Z01	+	–	+
E. coli lac⁻ Z07	+	+	+
E. cloacae Z04	–	+	+

Note: Tested concentration $1 \cdot 10^8$, "–" – inhibition activity $\sim 10^7$ CUO/ml, "+" – inhibition 10^1–10^3

It was shown that plum extract was effective against strains of *S. aureus Z03, K. pneumoniae Z02, P. aeruginosa Z01, E. coli lac-Z07*, and the Enterogermina® – against *K. pneumoniae Z02, E. cloacae Z04*, while their composition consisting of plum extract at a concentration of 0.002 mg/ml and the Enterogermina® were active against all the tested clinical strains – *S. aureus Z03, K. pneumoniae Z02, P. aeruginosa Z01, E. coli lac- Z07, E. cloacae Z04*.

Given that plum extract is able to enhance the spectrum of antibacterial properties of the industrial drug "Enterogermina", we have developed new biological products based on the author's strains of lactobacilli and plum extract.

The design of the biological product was performed in collaboration with Albert Kristanov, Plovdiv, Bulgaria and the company "SELUR Pharma" (http://www.selur-pharma.com/ru/about.html). Deposited, authorial strains of lactobacilli were selected as the probiotic basis: *L. plantarum A, L rhamnosus S25, L. delbrueckii subsp. bulgaricus S19*, provided by SELUR Pharma, and strains of lactobacilli whose genome was sequenced, followed by depositing strains in the depository of the Institute of Microbiology and Virology named after D.K. Zabolotny National Academy of Sciences of Ukraine: *L. paracasei IMB B-7483, L. plantarum IIB B-7414, L. casei IMV B-7412, L. plantarum IMV B-7413*. These strains are included in the registered pharmabiotics **BioMe**₍combi₎ **10 + 1**™.

In the design of new biological products in which plum extract is combined with different strains of lactobacilli, the selection of the probiotic component was based on the results of previous studies [55]. Thus, we have previously shown that lactobacilli *L. paracasei IMB B-7483*, *L. casei IIB B-7412* and *L. plantarum IMB B-7414* have hypocholesterolemic activity, reduce glucose, low-density lipoprotein in serum in experimental obesity in rats. We also selected three strains of lactobacilli by the method of co-cultivation, which were characterized by high selective antagonistic activity against streptococcal and staphylococcal strains and significantly inhibitory efficacy against opportunistic enterobacteria and Candida – the dominant pathogen. Therefore, in the course of the study we developed three types of composite biological products for the treatment of caries, periodontal inflammation and chronic inflammation (obesity and type 2 diabetes). The possible combinations of lactobacilli strains in the amount of $1 \cdot 10^8$ CFU/ml of each strain (probiotic component) and plum extract (prebiotic component). To facilitate the treatment of caries, the most effective is the composition of strains of *L. paracasei IMB B-7483*, *L. plantarum IMB B-7414*, *L. casei IMV B-7412*, *L. plantarum IMV B-7413* and plum extract at a concentration of 0.002 mg/ml; in inflammation of periodontal tissues will be effective use of the composition of strains of *L. plantarum IMB B-7414*, *L. plantarum IMB B-7413*, *L. casei IMB B-7412* and plum extract at a concentration of 0.002 mg/ml, and in obesity and type 2 diabetes the most effective is the composition of strains of *L. casei IMB B-7412*, *L. plantarum A*, *L. rhamnosus S 25*, *L. delbrueckii subsp. bulgaricus S 19* and plum extract at a concentration of 0.002 mg/ml.

7.8 PPPM Strategies in the Field

Recent advances in predictive, preventive, and personalized medicine (PPPM) are expanding the use of prebiotics as a component of modern pharmaceuticals. Based on the data presented in this article, extracts of edible plants have great potential for use as a prebiotic component of modern pharmaceuticals. Our individual compositions of plant extracts and proprietary probiotic strains of Lactobacillus have been found to be effective against certain dental diseases, as well as chronic inflammations associated with common noncommunicable diseases – obesity and type 2 diabetes.

7.9 Expert Recommendations in the Frame-Work of 3 PM to the Practical Medicine

Biologically active compounds extracted from the edible plants have no harmful influence and that's why they can be used in medicine. The ability of biologically active substances contained in extracts of edible plants not to inhibit the growth of lactobacilli and intestinal microbiome has been proven, indicating the possibility of their use as a prebiotic component of biological products.

Taking into account our data on the effect of the plum extract on the culture of dermal fibroblast cells in newborns, only at a concentration of plum extract of 0.002 mg/ml harmful effect on dermal fibroblast cells in newborns is not observed. This suggests the need for cohort testing of any prebiotic components to clinically prove the effectiveness of each component. When choosing probiotics and prebiotics, it should be understood that there are clinical benefits in different situations, but the level of benefit in each case is individual and very variable. Therefore, the importance of probiotics, and therefore prebiotics can not be exaggerated. Therefore, it is necessary to develop approaches to the individual selection of probiotics, and hence prebiotics, to obtain the most effective result.

References

1. Golubnitschaja O, Costigliola V (2012) General report & recommendations in predictive, preventive and personalised medicine 2012: white paper of the European Association for predictive, preventive and personalised medicine. EPMA J 3(1):14. https://doi.org/10.1186/1878-5085-3-14
2. Golubnitschaja O, Costigliola V (2015) EPMA summit 2014 under the auspices of the presidency of Italy in the EU: professional statements. EPMA J 6(1):1–11. https://doi.org/10.1186/s13167-015-0026-2
3. Bubnov RV, Babenko LP, Lazarenko LM et al (2017) Comparative study of probiotic effects of lactobacillus and bifidobacteria strains on cholesterol levels, liver morphology and the gut microbiota in obese mice. EPMA J 8(4):357–376. https://doi.org/10.1007/s13167-017-0117-3
4. Fijan S (2014) Microorganisms with claimed probiotic properties: an overview of recent literature. Int J Environ Res Public Health 11(5):4745–4767. https://doi.org/10.3390/ijerph110504745
5. Golubnitschaja O, Baban B, Boniolo G (2016) Medicine in the early twenty-first century: paradigm and anticipation-EPMA position paper 2016. EPMA J 7(1):1–13. https://doi.org/10.1186/s13167-016-0072-4
6. Minocha A (2009) Probiotics for preventive health. Nutr Clin Pract 24(2):227–241. https://doi.org/10.1177/0884533608331177
7. Abdelhamid AG, El-Masry SS, El-Dougdoug NK (2019) Probiotic lactobacillus and Bifidobacterium strains possess safety characteristics, antiviral activities and host adherence factors revealed by genome mining. EPMA J 10(4):337–350. https://doi.org/10.1007/s13167-019-00184-z
8. Dronkers TM, Ouwehand AC, Rijkers GT (2020) Global analysis of clinical trials with probiotics. Heliyon 6(7):e04467. https://doi.org/10.1016/j.heliyon.2020.e04467
9. Grech G, Zhan X, Yoo BC et al (2015) EPMA position paper in cancer: current overview and future perspectives. EPMA J 6(1):1–31. https://doi.org/10.1186/s13167-015-0030-6
10. Gibson GR, Hutkins R, Sanders ME et al (2017) Expert consensus document: the International Scientific Association for Probiotics and Prebiotics (ISAPP) consensus statement on the definition and scope of prebiotics. Nat Rev Gastroenterol Hepatol 14(8):491–502. https://doi.org/10.1038/nrgastro.2017.75
11. Bubnov RV, Spivak MY, Lazarenko LM, Bomba A, Boyko NV (2015) Probiotics and immunity: provisional role for personalized diets and disease prevention. EPMA J 6(1):1–11

12. Marco ML, Heeney D, Binda S et al (2017) Health benefits of fermented foods: microbiota and beyond. Curr Opin Biotechnol 44:94–102. https://doi.org/10.1016/j.copbio.2016.11.010

13. Golubnitscha a O (ed) (2019) Flammer syndrome: from phenotype to associated pathologies, prediction, prevention and personalisation, vol 11. Springer. https://doi.org/10.1007/978-3-030-13550-8

14. Konopelniuk VV, Goloborodko II, Ishchuk TV et al (2017) Efficacy of fenugreek-based bionanocomposite on renal dysfunction and endogenous intoxication in high-calorie diet-induced obesity rat model – comparative study. EPMA J 8(4):377–390. https://doi.org/10.1007/s13167-017-0098-2

15. Bubnov RV (2012) Evidence-based pain management: is the concept of integrative medicine applicable? EPMA J 3(1):1–17. https://doi.org/10.1186/1878-5085-3-13

16. Liskova A, Koklesova L, Samec M et al (2021) Targeting phytoprotection in the COVID-19-induced lung damage and associated systemic effects – the evidence-based 3PM proposition to mitigate individual risks. EPMA J 12(3):325–347. https://doi.org/10.1007/s13167-021-00249-y

17. Ranilla LG, Huamán-Alvino C, Flores-Báez O et al (2019) Evaluation of phenolic antioxidant-linked in vitro bioactivity of Peruvian corn (Zea mays L.) diversity targeting for potential management of hyperglycemia and obesity. J Food Sci Technol 56(6):2909–2924. https://doi.org/10.1111/1750-3841.13973

18. Zaïri A, Nouir S, M'hamdi N et al (2018) Antioxidant, antimicrobial and the phenolic content of infusion, decoction and methanolic extracts of thyme and Rosmarinus species. Curr Pharm Biotechnol 19(7):590–599. https://doi.org/10.2174/1389201019666180817141512

19. Meleshko T, Pallah O, Petrov V et al (2020) Extracts of pomegranate, persimmon, nettle, dill, kale and Sideritis specifically modulate gut microbiota and local cytokines production: in vivo study. ScienceRise Biol Sci 2(23):4–14

20. Kleerebezem M, Binda S, Bron PA et al (2019) Understanding mode of action can drive the translational pipeline towards more reliable health benefits for probiotics. Curr Opin Biotechnol 56:55–60. https://doi.org/10.1016/j.copbio.2018.09.007

21. Guimarães JT, Balthazar CF, Silva R et al (2020) Impact of probiotics and prebiotics on food texture. Curr Opin Food Sci 33:38–44. https://doi.org/10.1016/j.cofs.2019.12.002

22. Monteagudo-Mera A, Rastall RA, Gibson GR et al (2019) Adhesion mechanisms mediated by probiotics and prebiotics and their potential impact on human health. Appl Microbiol Biotechnol 103(16):6463–6472. https://doi.org/10.1007/s00253-019-09978-7

23. Cardona F, Andrés-Lacueva C, Tulipani S et al (2013) Benefits of polyphenols on gut microbiota and implications in human health. J Nutr Biochem 24(8):1415–1422. https://doi.org/10.1016/j.jnutbio.2013.05.001

24. Anhê FF, Roy D, Pilon G et al (2015) A polyphenol-rich cranberry extract protects from diet-induced obesity, insulin resistance and intestinal inflammation in association with increased Akkermansia spp. population in the gut microbiota of mice. Gut 64(6):872–883. https://doi.org/10.1136/gutjnl-2014-307142

25. Kumar Singh A, Cabral C, Kumar R et al (2019) Beneficial effects of dietary polyphenols on gut microbiota and strategies to improve delivery efficiency. Nutrients 11(9):2216. https://doi.org/10.3390/nu11092216

26. Mano MCR, Neri-Numa IA, da Silva JB et al (2018) Oligosaccharide biotechnology: an approach of prebiotic revolution on the industry. Appl Microbiol Biotechnol 102(1):17–37. https://doi.org/10.1007/s00253-017-8564-2

27. Lam KL, Cheung PCK (2019) Carbohydrate-based prebiotics in targeted modulation of gut microbiome. J Agric Food Chem 67(45):12335–12340. https://doi.org/10.1021/acs.jafc.9b04811

28. Bird A, Conlon M, Christophersen C et al (2010) Resistant starch, large bowel fermentation and a broader perspective of prebiotics and probiotics. Benefic Microbes 1(4):423–431. https://doi.org/10.3920/BM2010.0041

29. Bindels LB, Delzenne NM, Cani PD et al (2015) Towards a more comprehensive concept for prebiotics. Nat Rev Gastroenterol Hepatol 12(5):303–310. https://doi.org/10.1038/nrgastro.2015.47

30. De Jong WH, Borm PJ (2008) Drug delivery and nanoparticles: applications and hazards. Int J Nanomedicine 3(2):133–149. https://doi.org/10.2147/ijn.s596

31. Jeevanandam J, Barhoum A, Chan YS et al (2018) Review on nanoparticles and nanostructured materials: history, sources, toxicity and regulations. Beilstein J Nanotechnol 9(1):1050–1074. https://doi.org/10.3762/bjnano.9.98

32. Auffan M, Rose BJY et al (2009) Towards a definition of inorganic nanoparticles from an environmental, health and safety perspective. Nat Nanotechnol 4(10):634–641. https://doi.org/10.1038/nnano.2009.242

33. Rezaee A, Rangkooy H, Khavanin A et al (2014) High photocatalytic decomposition of the air pollutant formaldehyde using nano-ZnO on bone char. Environ Chem Lett 12(2):353–357

34. Kim WS, Han GG, Hong L et al (2019) Novel production of natural bacteriocin via internalization of dextran nanoparticles into probiotics. Biomaterials 218:119360. https://doi.org/10.1016/j.biomaterials.2019.119360

35. Krithika B, Preetha R (2019) Formulation of protein based inulin incorporated synbiotic nanoemulsion for enhanced stability of probiotic. Mater Res Express 6(11):114003. https://doi.org/10.1088/2053-1591/ab4d1a

36. Bubnov R, Babenko L, Lazarenko L et al (2019) Can tailored nanoceria act as a prebiotic? Report on improved lipid profile and gut microbiota in obese mice. EPMA J 10(4):317–335. https://doi.org/10.1007/s13167-019-00190-1

37. Lu J, Ma KL, Ruan XZ (2019) Dysbiosis of gut microbiota contributes to the development of diabetes mellitus. Infect Microbes Dis 1(2):43–48

38. Prabhasankar P (2014) Prebiotics: application in bakery and pasta products. Crit Rev Food Sci Nutr 54(4):511–522. https://doi.org/10.1080/10408398.2011.590244

39. Fric P (2007) Probiotics and prebiotics – renaissance of a therapeutic principle. Open Med 2(3):237–270. https://doi.org/10.2478/s11536-007-0031-5

40. Stowell J (2006) Calorie control and weight management. In: Sweeteners and sugar alternatives in food technology, pp 54–61

41. Bañuelos O, Fernández L, Corral JM et al (2008) Metabolism of prebiotic products containing β (2-1) fructan mixtures by two lactobacillus strains. Anaerobe 14(3):184–189. https://doi.org/10.1016/j.anaerobe.2008.02.002

42. Pompei A, Cordisco L, Raimondi S et al (2008) In vitro comparison of the prebiotic effects of two inulin-type fructans. Anaerobe 14(5):280–286. https://doi.org/10.1016/j.anaerobe.2008.07.002

43. Bruzzese E, Volpicelli M, Squeglia V et al (2009) A formula containing galacto-and fructooligosaccharides prevents intestinal and extra-intestinal infections: an observational study. Clin Nutr 28(2):156–161. https://doi.org/10.1016/j.clnu.2009.01.008

44. Falony G, Calmeyn T, Leroy F et al (2009) Coculture fermentations of Bifidobacterium species and Bacteroides thetaiotaomicron reveal a mechanistic insight into the prebiotic effect of inulin-type fructans. Appl Environ Microbiol 75(8):2312–2319. https://doi.org/10.1128/AEM.02649-08

45. Hernot DC, Boileau TW, Bauer LL et al (2009) In vitro fermentation profiles, gas production rates, and microbiota modulation as affected by certain fructans, galactooligosaccharides, and polydextrose. J Agric Food Chem 57(4):1354–1361. https://doi.org/10.1021/jf802484j

46. Kelly G (2009) Inulin-type prebiotics: a review (part 2). Altern Med Rev 14(1):36–55

47. Pan W, Sunayama Y, Nagata Y et al (2009) Cloning of a cDNA encoding the sucrose: sucrose 1-fructosyltransferase (1-SST) from yacon and its expression in transgenic rice. Biotechnol Biotechnol Equip 23(4):1479–1484

48. Boler BMV, Fahey GC (2012) Prebiotics of plant and microbial origin. In: Direct-fed microbials and prebiotics for animals. Springer, New York, NY, pp 13–26

49. Pallah O, Meleshko T, Tymoshchuk S et al (2019) How to escape'the ESKAPE patho-
    gens' using plant extracts. ScienceRise Biol Sci 5-6 (20-21):30–37. https://doi.
    org/10.15587/2519-8025.2019.193155
50. Pallah OV, Meleshko TV, Bati VV et al (2019) Extracts of edible plants stimulators for benefi-
    cial microorganisms. Biotechnol Acta 12(3):67–74. https://doi.org/10.15407/biotech12.C3.067
51. Woodcock ME, Hollands WJ, Konic-Ristic A et al (2013) Bioac-tive-rich extracts of persim-
    mon, but not nettle,Sideritis, dill or kale, increase eNOS activation and NO bioavailability
    and decrease endothelin-1 secretion by human vascular endothelial cells. J Sci Food Agric
    93(14):3574–3580. https://doi.org/10.1002/jsfa.6251
52. Aranda Hernandez J, Heuer C, Bahnemann J, Szita N (2021) Microfluidic devices as process
    development tools for cellular therapy manufacturing. In: Advances in biochemical engineer-
    ing/biotechnology. Springer, Berlin/Heidelberg. https://doi.org/10.1007/10_2021_169
53. Pallah O, Bati V, Boyko N (2019, March 1–2) Antimicrobial properties of berry extracts as
    a prebiotic component of modern probiotic drugs. In: 8th international dental conference of
    students and young scientists, Uzhhorod
54. Sarvash O, Bati V, Markush N, Levchuk O, Melnik V, Mizernytskyy O, Boyko N (2013)
    Novel antimicrobials of complex origin. In: XIII congress of the society of microbiologists of
    Ukraine named after SM Vinogradsky: abstracts, Yalta, p 220
55. Bati VV, Meleshko TV, Pallah OV, Zayachuk IP, Boyko NV (2021) Personalised diet improve
    intestine microbiota and metabolism of obese rats. Ukr Biochem J

Chapter 8
Probiotic Administration for the Prevention and Treatment of Gastrointestinal, Metabolic and Neurological Disorders

Nicole Bozzi Cionci, Marta Reggio, Loredana Baffoni, and Diana Di Gioia

Abstract The interest in the intestinal microbiota of humans has increased in the last 20 years and significant advances have been achieved with regard to its structure and functions. Since gut microbiota is involved in a range of complex interactions with the host, its manipulation for promoting human health has been the object of several studies. Probiotics, with their long history of safety and effectiveness against harmful microorganisms, are a strategy to not only maintain or restore the correct balance in the microbial population of the intestinal tract, but also to prevent or treat a range of disease conditions. Personalized administration of selected probiotic strains is necessary to ensure the success of their application and is critical to implement their clinical use as a strategy to support preventive, personalized, and predictive medicine (PPPM).

The aim of this review is to explore preclinical and clinical studies focused on the use of probiotics for the prevention and treatment of gastrointestinal, metabolic and neurological diseases. *In vitro* and *in vivo* studies on probiotic administrations allow the screening of strains for clinical application and the development of personalized probiotic treatments, in accordance with the PPPM perspective. The results achieved in this field showed the positive effects of probiotic supplementation, especially for *Lactobacillus* and *Bifidobacterium* genera, but new strains belonging to different genera such as *Akkermansia* are gaining interest. The examined studies support the concept of "precision probiotics" that represent the future of probiotic-based intervention, according to which individuals will be recommended specific diets and probiotics administration tailored to their unique gut microbiome structure.

N. Bozzi Cionci (✉) · M. Reggio · L. Baffoni · D. Di Gioia
Department of Agricultural and Food Sciences, University of Bologna, Bologna, Italy
e-mail: nicole.bozzicionci@unibo.it; marta.reggio2@unibo.it; loredana.baffoni@unibo.it; diana.digioia@unibo.it

© The Author(s), under exclusive license to Springer Nature
Switzerland AG 2023
N. Boyko, O. Golubnitschaja (eds.), *Microbiome in 3P Medicine Strategies*,
Advances in Predictive, Preventive and Personalised Medicine 16,
https://doi.org/10.1007/978-3-031-19564-8_8

Keywords PPPM · Gut microbiota · Probiotics · Targeted microorganisms · Health · Diseases

8.1 Introduction

Personalization of care is a priority goal of contemporary medicine. In the last few years, fundamental innovations in the fields of molecular biology, biotechnology, genetics, and informatics have promoted the paradigm change from reactive to predictive, preventive, and personalized medicine (PPPM). Personalized medicine is defined as a comprehensive approach to the prevention, diagnosis, treatment and monitoring of disease based on the individual characteristics of a target person [1].

In this sense, a fundamental role for the application of personalized medicine is played by the maintenance of a healthy microbiota. In fact, it is now well established that a healthy microbiota protects the host from various diseases by influencing and regulating metabolic and immune functions. On the contrary, an altered composition of the ecosystem, which can potentially lead to a condition of intestinal dysbiosis, can promote the onset of a wide range of diseases. A healthy microbiota is also a prerequisite for the success of therapies, both nutritional and pharmacological, allowing a complete recovery of health status thanks to the largely known microbiota interactions with the immune system, the neuroendocrine system, and the central nervous system. The investigation and the comprehension of the gut microbiota structure and functions of each individual can constitute a resource to recovery, improve and preserve not only the intestinal homeostasis, but the overall host's healthy profile through appropriate dietary, probiotic administration and lifestyle changes, avoiding serious clinical implications in the near future [2].

Probiotics have enormous potential to promote and maintain a healthy microbiota and, therefore, they can be exploited for the advancement of personalized medicine in several disease conditions. Selected probiotic strains can be used for applications in personalized medicine and nutrition not only for disease prevention but also as adjuvants to specific drug treatments in order to prevent microbiota imbalance, counteracting potential pathogens. In addition, probiotics have a relevant role as biomarkers for predictive diagnosis based on their role in the gut-brain axis or gut microbiota [3, 4].

8.1.1 Probiotic Administration

The term "probiotic" means "for life" and it is currently used to define bacteria that provide beneficial effects to humans and animals. In 2001 they were described by the Food and Agriculture Organization of the United Nations (FAO)/World Health Organization (WHO) as "live microorganism which, when administered in adequate amounts confer a health benefit on the host". This definition has been revised in

2014 by the International Scientific Association for Probiotics and Prebiotics, including in the term probiotic "microorganism for which there are scientific evidence of safety and efficacy" and excluding "live cultures associated with fermented foods for which there is no evidence of a health benefit" [5]. Considering the close symbiotic relationship between the host and the gut microbiota that is crucial for the preservation of the host's health, the administration of beneficial microorganisms may represent a key determinant of diseases susceptibility and the host's health condition.

Species belonging to *Lactobacillus*, *Bifidobacterium*, *Streptococcus* and *Saccharomyces* genera have been largely used as probiotics in human studies. Probiotics exert their efficacy on the intestinal environment, having effects on the entire host's organism, with different mechanisms (Fig. 8.1), originally presented by Fuller [6]:

- antagonism against pathogens by producing antibacterial molecules and competing for nutrients and for the adhesion to specific receptors on intestinal epithelium;
- promotion of microbial metabolism by increasing the activity of useful enzymes and decreasing the activity of some enzymes that can be dangerous for the host;
- stimulation of the immune system by increasing antibodies and anti-inflammatory cytokines levels and inducing macrophages and natural killer cells [7].

A probiotic formulation can contain not only one strain, but also a mixture of different strains of the same species or it can be a multi-strain and multi-species mixture that, acting with a synergic effect, may enhance the effectiveness of each single strain [8, 9].

It is well established that imbalances in the gut microbiota, known as dysbiosis, have been associated with inflammation and chronic disease ranging from gastrointestinal inflammatory and metabolic conditions to neurological diseases [2] (Fig. 8.2).

Therefore, probiotics can be exploited for their health benefits in the treatment and prevention of many disease conditions. Recent advances in probiotic-human interaction research suggest their application in disease prevention, the development of personalized treatments and their use as biomarkers for predictive diagnosis, furthering the shift toward preventive, personalized and predictive medicine (PPPM) [4].

8.1.2 Gastrointestinal Disorders

A growing interest in studying and using probiotics has been observed for the treatment of gastrointestinal diseases and the improvement of human health, with strong potential for applications in personalized medicine and nutrition.

Inflammatory bowel diseases (IBDs), including Crohn's disease and ulcerative colitis, are chronic inflammatory disorders of the gastrointestinal tract, and represent the second most common chronic inflammatory diseases worldwide [10]. Gut

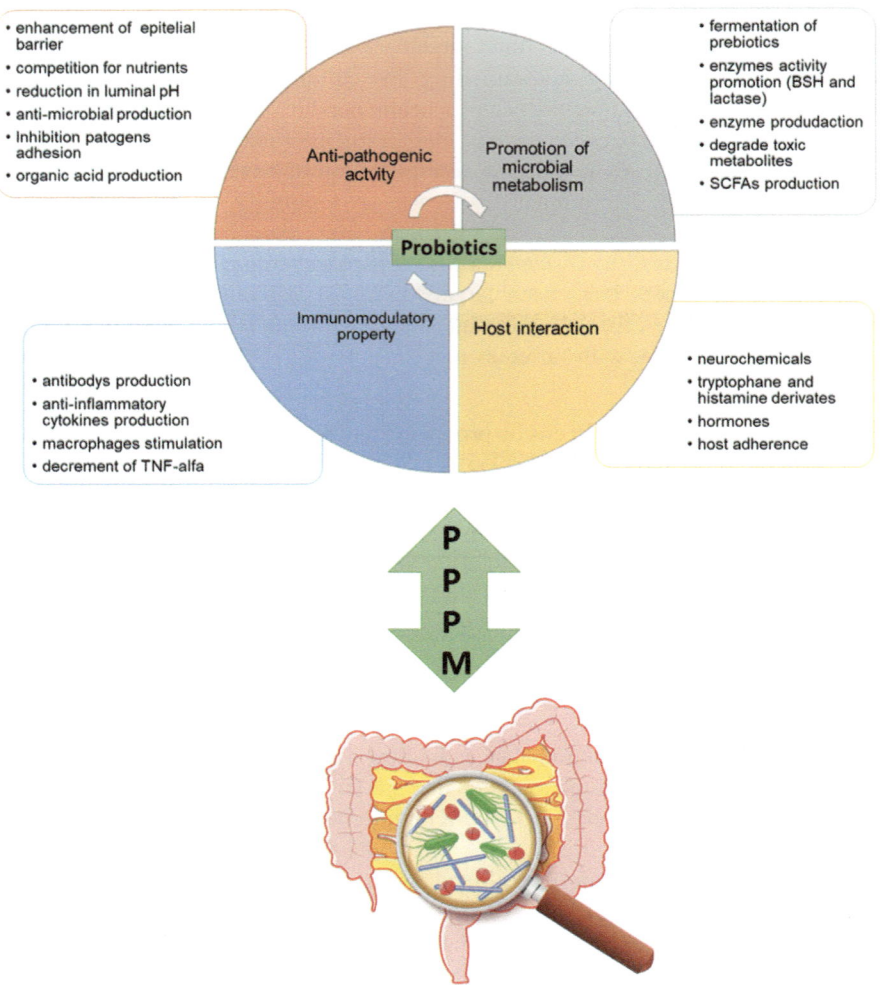

Fig. 8.1 Beneficial effect of probiotics on the gut microbiota and health, in compliance with PPPM strategies

microbiota and its interactions with the host immune system play critical roles in pathogenesis of IBDs together with genetic background and immunological factors. Recent studies with Next Generation Sequencing (NGS) have revealed a reduced diversity of the gut microbiota in IBD patients [11–14], characterized by a reduction of bacterial members of the phyla Firmicutes and Bacteroidetes, and an increase of Actinobacteria and Proteobacteria [15]. In addition, Sokol et al. [16] demonstrated that a member of the Firmicutes phylum, *Faecalibacterium prausnitzii*, is reduced in the stool of patients with Crohn's disease and has been associated with a higher risk of postoperative recurrence of ileal compliances in CD patients.

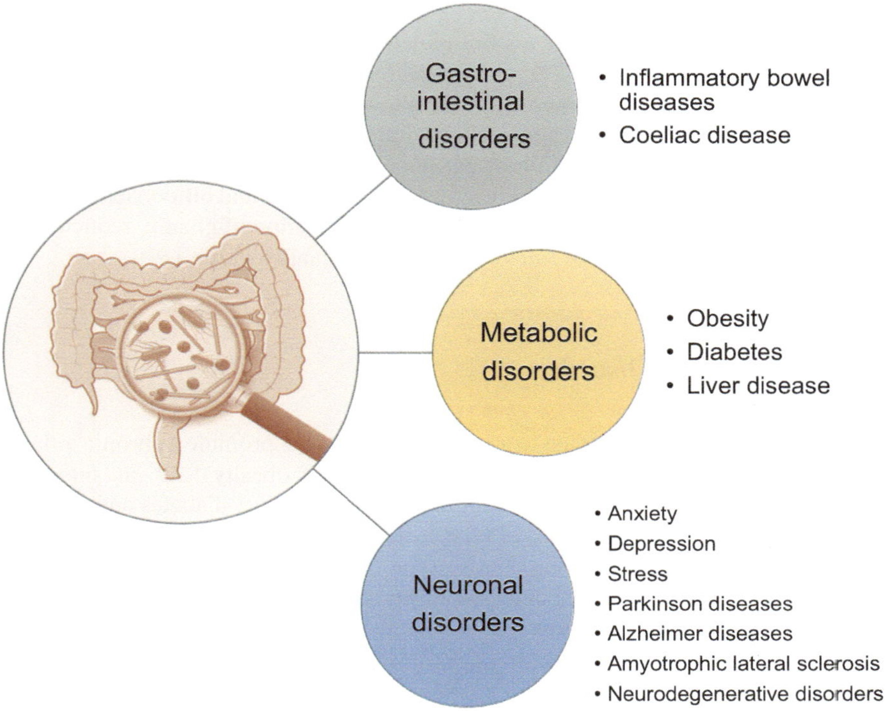

Fig. 8.2 Inflammation and chronic diseases associated to an altered and/or compromised intestinal microbiota

Concerning this, probiotics, have been recommended as preventive as well as therapeutic measures for IBDs and for the establishment of a healthy composition and function of the gut microbiome, for their ability to increase the integrity of the intestinal barrier, induce immune tolerance, hinder pathogenic bacterial translocation, and reduce gastrointestinal infections.

Furthermore, probiotics supplementations are expected to induce beneficial effects to counteract cancer, by reducing the systemic inflammatory status and restoring immune balance. In particular, the anticancer activity of probiotics, specifically *Lactobacillus* strains, occurs via various mechanisms, among which immune modulatory effects, apoptosis, antioxidant activity, epigenetic mechanisms and DNA damage prevention [17]. Following this approach, the administration of selected probiotics can be applied for the development of a personalized therapy.

Coeliac disease (CD) is a chronic autoimmune enteropathy causing damage to the small intestine, hypothetically linked to dietary exposure to gluten in genetically predisposed individuals. Evidence shows that the gut microbiota is somehow involved in the pathogenesis, progression, and clinical presentation of CD. In particular, several studies have reported the presence of an intestinal dysbiosis in patients with CD characterized by an increased abundance of *Bacteroides* spp. and a decrease of *Bifidobacterium* and *Lactobacillus* spp. Furthermore, the microbial imbalances often persist despite a strict gluten-free diet (GFD) and, consequently, coeliac subjects suffering from persistent gastrointestinal symptoms may present a compromised microbiota composition. Probiotic administration in CD patients' diets can be proposed as a new personalized therapy for reducing CD symptoms and achieving an effective restauration of gut microbiota [18, 19]. Indeed, the integration of CD patient's diet with probiotics such as lactobacilli and bifidobacteria could help in restoring gut microbiota composition, detoxification of gliadin, reduction of immune activation and improvement of the general quality of life [19].

8.1.3 Metabolic Disorders

A dysbiotic condition associated to a hyperlipidic diet can promote a chronic inflammatory response contributing to the pathogenesis of obesity [20]; an intestinal microbial imbalance can also have a certain impact on type 1 diabetes and it is recognized as the greatest risk factor of type 2 diabetes [21, 22]. Indeed, many studies evidenced that diseases including obesity, type II diabetes, liver disease, and hypercholesterolemia have been documented to be strongly associated with an imbalanced Firmicutes/Bacteroidetes ratio in the gut microbiota, marked by decreased butyrate-producing bacteria, such as *Roseburia intestinalis* and *F. prausnitzii* [22, 23].

Probiotics supplements can potentially contribute to the restoration of dysbiotic microbiota and prevention of metabolic disorders, thanks to their anti-obesity and anti-inflammatory effects [24, 25].

8.1.4 Neurological Disorders

It is largely known that there is an intimate relationship between the gastro-intestinal tract and the central nervous system. The gut microbiota is inserted in this interconnection, the gut-brain axis, which constitutes a complex bidirectional communication system, including different mechanisms, such as the alteration in circulating levels of inflammatory cytokines or the production of neurotoxins [26]. Alteration in the microbiota composition may lead to failures in this interaction that may be implicated in neurological diseases. It is known that enteric infections can cause anxiety, depression, and cognitive dysfunction, and it has recently been observed that the inflammatory bowel response induces closure of the cerebral choroid plexus and this appears to be associated with increased anxiety- and depression-like behaviors and alterations in the limbic system [27].

Probiotic administration can improve the production and the delivery of neuroactive compounds acting on the brain-gut axis and restoring the stability on nervous system. Many investigations assessed the potential of probiotics in preventing symptoms or alleviating comorbidities against neurological diseases, although the possible role of probiotics in some other neurological disorders, whose onset is obscure, as amyotrophic lateral sclerosis (ALS), has to be still evaluated.

8.2   Evidences Supporting Preclinical Studies

8.2.1   In Vitro Tests

Although clinical studies are of fundamental importance to understand the effect of targeted strains in humans, *in vitro* tests are essential to evaluate the probiotic potential and for the screening among a large number of candidates of the most suitable strain(s). This is particularly important within the contest of personalized medicine. *In vitro* tests may include the inhibition of targeted microbial species, anti-oxidant potential and studies on cell models. The beneficial role of probiotics in the targeted prevention or treatment of enteric disorder has been proven in several *in vitro* studies. Within the *Bifidobacterium* genus, *Bifidobacterium breve* and *Bifidobacterium longum* strains, mainly B632 (DSM 24706), B2274 (DSM 24707), B7840 (DSM 24708), and *B. longum* subsp. *longum* B1975 (DSM 24709), have been identified as potential probiotics for the treatment of enteric disorders in newborns, such as colics or as preventive agents for diarrhea of bacterial origin, for their *in vitro* ability to inhibit the growth of pathogens typical of the infant gastrointestinal tract and coliforms isolated from colicky infants. Particularly, *B. breve* B632 was also able to stimulate the activity of mitochondrial dehydrogenases of macrophages and the production of IL-6, related to a considerable activation of macrophages and endothelial cells in inflammatory condition [28, 29].

An *in vitro* study proved that treatment with *Lactobacillus plantarum* AN1 led to anti-inflammatory effects against IBD, by stimulating the production of nitric oxide (NO) in macrophage RAW264.7 cells and protecting RAW264.7 cells against hydrogen peroxide toxicity [30].

It has been suggested that the probiotic mixture VSL#3, probiotic formulation of four strains belonging to the species *Lactobacillus paracasei*, *Lactobacillus plantarum*, *Lactobacillus acidophilus* and *Lactobacillus delbrueckii* subsp. *bulgaricus*, three *Bifidobacterium* strains belonging to the species *Bifidobacterium longum*, *Bifidobacterium infantis* and *B. breve* and one strain of *Streptococcus salivarius* subsp. *thermophilus*, exerts its anticolitic effects by altering the morphology of polarized (M1/M2) and unpolarized macrophages but also modifying the secretion of cytokines and chemokines in these cells in a polarization-dependent manner [31]. Lammers et al. [32] also showed that stimulation of peripheral blood mononuclear cells (PBMC) with pure DNA extracted from the VSL#3 probiotic mixture and with total bacterial DNA from human feces modulated the immune response by an increase of IL-1 beta and of IL-10 cytokines, mitigating the IBD symptoms. In addition, immunomodulatory effects of *B. breve* on T cell polarization towards Th2 and Treg cell-associated responses has been proven *in vitro*, using PBMC [33].

Many researches have elucidated the protective effects of *Lactobacillus* and *Bifidobacterium* genera on epithelial cells against gliadin-derived toxic peptides (GDPs) and their therapeutic impact on patients suffering CD. In particular, an *in vitro* study performed by Lindforf et al. (2008) [34] has shown that two strains of *Lactobacillus fermentum* and *Bifidobacterium lactis* are able to reduce the toxic effects of gluten-derived peptides in intestinal CaCo2 cells by inhibiting the gliadin-induced intestinal permeability. Furthermore, the use of *Lactobacillus rhamnosus* GG inhibits *in vitro* the translocation of gliadin peptides into the cells, preventing the barrier dysfunction of the gut [35]. Other bifidobacterial strains are able to inhibit the inflammatory response induced by gliadins in CD, reducing inflammation, as demonstrated by Laparra et al. [36]. VSL#3 is also able to hydrolyze gliadin polypeptides, decreasing the toxic effects of wheat flour during the prolonged fermentation process. In addition, it has also been reported that the ability of VSL#3 to degrade gliadin is provided by the combination of probiotic strains and not by the individual strain [37].

The administration of probiotics has also a potential role in prevention and reduction of symptoms related to neurological disorders. A neuroprotective effect exerted by the probiotic formulation SLAB51, composed by a mixture of bifidobacteria and lactobacilli has been found in *in vitro* Parkinson's disease models. In particular, the probiotics can counteract the detrimental effect of 6-OHDA, modulating the brain-derived neurotrophic factor (BNDF) pathway, increasing neuroprotective protein levels and decreasing the neuronal death proteins in an *in vitro* 6-OHDA model of Parkinson's disease (PD) using dopaminergic-like SH-SY5Y neuro-blastoma cells [38].

8.2.2 Animals Experiments

The strong evidence of probiotic supplementation beneficial effects in patients with enteric disease is supported by a huge range of studies using animal models.

Alagón Fernández del Campo et al. [39] reviewed that the administration of *Lactobacillus* and *Bifidobacterium* strains was able to mitigate ulcerative colitis (UC) in mice decreasing the production of pro-inflammatory cytokines. In particular, *B. infantis* 35624 has been observed to reduce the secretion of pro-inflammatory IL-12, transforming growth factor-beta (TGF-β), interferon-gamma (IFN-γ) and tumor necrosis factor -alpha (TNF-α), in mice knock out for IL-10 [40].

Strains of *L. plantarum* are one of the most commonly used probiotics against IBD. Several studies with IBD murine models showed that administration of *L. plantarum* AN1, *L. plantarum* NCIMB8826 exhibited gut modulatory and anti-inflammatory properties, by reducing pro-inflammatory cytokines and by also producing bacteriocins [30, 41]. The combination of two different probiotics (*L. plantarum* ZDY2013 and *Bifidobacterium bifidum* WBIN03) reduced UC in mice through modification of gut microbiota and reduction of inflammation and oxidative stress [42].

Other lactobacilli, such as *L. reuteri* DSM 17938, have also been proven to be effective in modulating dysbiosis and decreasing inflammation in mice models [43]. Moreover, the co-administration of the *L. fermentum* KBL374 and *L. fermentum* KBL375 in mice with DSS (dextran sodium sulfate)-induced colitis corrected gut dysbiosis, decreased the proportions of *Bacteroides* and *Mucispirillium* and increased the proportion of *Lactobacillus*, as well as having an anti-inflammatory effect [44].

The effectiveness of the probiotic mixture VSL#3 to reduces colitis severity, colonic macrophage infiltration, and serum cytokine level has been examined in rats with acute trinitrobenzene sulfonic acid (TNBS)-induced colitis [31].

The administration of the probiotic *Escherichia coli* Nissle 1917 (EcN) in the DSS model of mouse colitis revealed an intestinal anti-inflammatory effect correlate with an amelioration of the altered gut microbiome. Moreover, this probiotic has shown an ability to modulate expression levels of miRNAs and different mediators of the immune response involved in gut inflammation [45].

Five strains of lactobacilli and bifidobacteria (*B. longum* CMUL CXL 001, *B. bifidum* IPL A7.31, *L. rhamnosus* A2.2, *L. gasseri* IPL A6.33, and *L. acidophilus* CMUL067) selected *in vitro* for their anti-inflammatory activity and ability to improve intestinal barrier function were found to prevent TNBS-induced acute colitis in BALB/c mice, thus having a high potential for the management of IBD [46].

Intervention with *B. breve* has been reported to attenuate the severity of DSS-induced colitis in mice by improving the immune balance [33]. Furthermore, a recent study conducted by Celiberto et al. [47] assessed the efficacy of a personalized probiotic strategy for the treatment of intestinal diseases DDS- induced colitis mice.

As reviewed by Serena et al. [48] most of the evidence on the possible therapeutic effect of probiotic bacteria on CD comes from animal models. D'arienzo et al. [49] research has shown that the administration of *L. casei* to immunized DQ8 mice was able to modulate both innate and adaptive immunity, to reduce gliadin-induced inflammation and to recovery the gut-associated lymphoid tissue (GALT) homeostasis, resembling healthy mucosal structure. Investigations performed in a gluten-sensitive model (BALB/c mice) have shown that *Saccharomyces boulardii* strain KK1 is able to hydrolyze toxic gliadin peptides, ameliorate enteropathy, and decrease histological damages and pro-inflammatory cytokine production [50]. Laparra et al. [36] studied the effect of *B. longum* CECT7347 in an animal model of gliadin-induced enteropathy and showed that administration of this strain reduced pro-inflammatory cytokine production and mediated immune response.

Recent *in vivo* animal studies demonstrated the probiotic anti-obesity effects and the ability to improve glycemic profile and modulate serum lipids associated to targeted microorganisms.

Stojanov et al. [51] reviewed that probiotics can have anti-obesity effects via F/B ratio modulation. Particularly, the administration of *L. rhamnosus* GG and *Lactobacillus sakei* NR28 has been proven to decrease the F/B ratio in obese mice.

Furthermore, the two probiotic strains reduced epididymal fat mass, acetyl-CoA carboxylase, fatty acid synthase, and stearoyl-CoA desaturase-1 in the liver [52]. In another study, *L. rhamnosus* GG consumed with a high-fat diet prevented weight gain and decreased the F/B ratio in a C57BL/6 J murine model [53]. Fifty-six days treatment with *L. rhamnosus* hsryfm1301 or its fermented milk reduced serum lipid levels in hyperlipidemic rats, induced a recovery of intestinal dysbiosis, increased Bacteroidetes and decreased in the abundance of Firmicutes by approximately 5% [54]. Oral administration of the probiotic *Lactobacillus paracasei* HII01 in a formulation also containing xyloligosaccharides for 12 weeks in diet-induced obese Wistar rats was able to reverse the effects induced by the high-fat diet, by improving insulin sensitivity, decreasing low-density lipoprotein cholesterol levels, reducing body weight and F/B ratio, thus reversing the dysbiosis [55].

In addition, other bacteriocin-producing bacteria, namely *Lactobacillus salivarius* UCC118 and *Bacillus amyloliquefaciens* SC06, also decreased the F/B ratio, body weight, and liver steatosis in mice consuming a high-fat diet [56, 57].

Two different studies evidenced that oral administration of *Saccharomyces boulardii* to leptin-resistant obese and type 2 diabetic mice increased Bacterioidetes by 37%, decreased Firmicutes by 30%, and reduced body weight, fat mass, hepatic steatosis, and inflammatory tone [58, 59]. The beneficial effects of *L. casei* IMV B-7280, *Bifidobacterium animalis* VKB, and *B. animalis* VKL combined with nanoceria, which is a potential prebiotic, to reduce cholesterol levels and restore gut microbiota has been observed in obese mice [60].

As reviewed by Bozzi Cionci et al. [2] *B. breve* was also involved in protective mechanisms against obesity; the orally administration of *B. breve* B-3 in a mouse model with diet-induced obesity could suppress the increase of body weight and epididymal fat, with improved serum levels of total cholesterol, fasting glucose and insulin and act by regulating gene expression pathways involved in lipid metabolism and response to stress in the liver [61, 62].

The positive effects of probiotics as adjuvants to standard metformin therapy against diabetes and colorectal cancer has been demonstrated in diabetic and CRC model mice [63].

Administration of *B. longum* has been shown to normalize the levels of the plasma LPS, IL-1b, intestinal myeloperoxidase, as well as intestinal inflammatory activity index in a rat model of metabolic syndrome induced by a high-fat; it significantly up-regulated the expression RegI protein, that is widely involved in inflammatory bowel disease and diabetes [64].

As reviewed by Sun et al. [20] different studies in animal models showed that probiotics restrict the development of type 2 diabetes, by improving glucose tolerance and insulin sensitivity. *L. reuteri* GMNL-263 supplementation in high-fructose-fed rat conferred a delaying effect on the development of type 2 diabetes by regulating the expression of peroxisome proliferator-activated receptor alpha (PPAR-c) and GLUT4 and down-regulating the expression of lipogenic genes, such as Srebp-1c, FAS, and Elvol6 [65]. Moreover, the same strain *L. rhamnosus* GG (LGG) has anti-inflammatory effects and decreases the expression of inflammatory cytokines, which are also involved in the development of diabetes [66]. In addition,

L. fermentum MTCC5689 resisted development of type 2 diabetes in mice and beneficially modulated all the biochemical and molecular alterations [67].

Several preclinical studies in animal models have shown that probiotics can affect brain activity by reducing neuroinflammation and, consequently, increasing neurogenesis.

As reviewed by Leta et al. [68], treatment with probiotics increased glucose metabolism, reduced peripheral and central inflammation, acting on IL-6 and TNF-a levels, reduced peripheral and central oxidative stress, decreased neurodegeneration, acting on tyrosine hydroxylase dopaminergic neurons and levels of brain derived neurotrophic factor, increased motor function and non-motor function.

Specifically, treatment with *L. plantarum* PS128, *Clostridium butyricum* WZMC1016, *L. fermentum* U-21 and mixtures of probiotics belonging to the Firmicutes or Actinobacteria prevented the decrease in tyrosine hydroxylase positive (TH +) dopaminergic neurons in the substantia nigra and the decrease in brain-derived neurotrophic factor (BDNF) and nerve growth factor (NGF) levels in the basal ganglia, in 6-OHDA, MPTP and rotenone models of PD of an MPTP model of PD [38, 69–73]. Moreover, it has been found that the same probiotics can decrease peripheral levels of inflammatory cytokines, such as interleukin-1beta (IL-1β), IL-6 and TNF-a [69–76], and ultimately prevent neuroinflammation [77]. Probiotics exert also beneficial effects on motor function preventing loss of dopaminergic neurons in the substantia nigra, which was observed in PD after the bacterial supplementation, and may also prevent cognitive-related deficits in a PD model in rodents [38, 69–73, 78, 79].

The original probiotic De Simone Formulation (DSF) has been shown to control the expression of several genes in the cerebral cortex of aging animals, attenuating inflammation and improving neuronal performance [80]. More recently, a novel probiotic formulation SLAB51 was studied in a mouse model of Alzheimer's disease (AD), exhibiting attenuation of cognitive impairment, reduction of Aβ aggregates and brain lesions, and partial restoration of the altered neuronal proteolytic pathway [81]. The same research group indicated that SLAB51 treatment leads to a strong oxidative reduction in mouse AD brain, through SIRT1-dependent mechanisms [82]. Positive effects of SLAB51 formulation were also noticed in PD. In particular, administration of the probiotic in *in vivo* study was able to protect dopaminergic neurons and improve behavioral alterations. This new probiotic formulation was also able to counteract neuroinflammation and oxidative stress, characteristics of PD, by reversing some molecular pathways underlying control conditions [38]. Ueda et al. [83], proved that the administration of specific *F. prausnitzii* strains improves cognitive impairment in an AD mouse model, showing their rule as possible candidates for gut microbiome-based intervention in Alzheimer's-type dementia.

A recent study also observed that individual supplementation of *Akkermansia muciniphila* in antibiotic-treated Sod1-Tg mice led to an improvement in ALS symptoms; furthermore, the treated mice were found to accumulate AM-associated nicotinamide in the central nervous system, and systemic nicotinamide supplementation improves motor symptoms and gene expression patterns in the spinal cord of Sod1-Tg mice [84].

8.3 Clinical Data

8.3.1 Clinical Trials

8.3.1.1 Probiotics in Gastrointestinal Disorders

Probiotic administration has a key role in the treatment likewise prevention of some disorders since early age, influencing the gut microbiota establishment [85].

Several studies have shown the efficacy of *L. reuteri,* in particular strains deriving from human milk, for the prevention and treatment of infant gastrointestinal disorders, such as colics, regurgitation, vomit, constipation [85–87]. Administration led to the improvement of symptoms and the reduction of anaerobic Gram-negative bacteria, Enterobacteriaceae and enterococci in colicky infants [88, 89]. A 3-months treatment with *B. breve* strains, i.e. B632 and BR03, in healthy breast-fed newborns helped to prevent functional gastrointestinal disorders, in particular by reducing daily vomit frequency, daily evacuation, and by improving the stool consistency [90]. The same administration led to a significant reduction of members of the *B. fragilis* group in bottle-fed infants. The use of the strains of *B. breve,* which is the dominant species in the gut of breast-fed infants and it has also been isolated from human milk, for the prevention and treatment of paediatric diseases has been largely explored [2]. Particularly, the target pathologies reported range from widespread gut diseases, including diarrhoea and infant colics, to celiac disease, obesity, allergic and neurological disorders; in addition, *B. breve* strains are used for the prevention of side infections in preterm newborns and during antibiotic treatments or chemotherapy.

L. reuteri ATCC 55730 efficacy has been proven in children with distal active UC improving mucosal inflammation and modulating mucosal expression levels of cytokines involved in the bowel inflammation [91]. The effectiveness of some *Bifidobacterium* strains for the treatment of acute rotavirus diarrhoea was evidenced in hospitalized children [92, 93] and in preterm and low birth weight infants, notably for the treatment of NEC [94, 95]. Among this genus, *B. breve* YIT4010 revealed the capability to improve abdominal symptoms and weight gain, favoring a correct microbial colonization in preterm infants [96, 97].

The activity of probiotics in inducing or prolonging a remission condition in IBD patients has been evidenced in several studies. The strains investigated were mainly belonging to *Lactobacillus* and *Saccharomyces* [98]. Several studies focused on VSL#3, a probiotic formulation, whose composition has been explained in the previous sections. Bibiloni et al. [99] showed that 6 weeks administration with this probiotic mixture improved UC remission and response in patients not responding to traditional therapy. VSL#3 efficacy was also tested in a study conducted on children with newly diagnosed UC [100]. The same probiotic mixture has also been approved for the prevention and the maintenance of remission of pouchitis and the efficacy is stated also in referral European guidelines [101, 102].

The efficacy of targeted probiotics was also highlighted for IBS [103]. Particularly, two different studies demonstrated the effectiveness of a mixture of *L. plantarum* and *B. breve* and VSL#3 in reducing IBS-associated pain [104, 105].

Although data deriving from *in vitro* and animal studies are promising [18], few clinical studies on the administration of probiotics in patients with coeliac disease have been carried out. *B. infantis* supplementation revealed an amelioration of gastrointestinal symptoms and the inflammatory state of coeliac subjects [106, 107]. A recent clinical investigation carried out by Francavilla et al. [108] demonstrated that the supplementation with a mixture bifidobacteria and lactobacilli led to an improvement of gastrointestinal symptoms combined with a significant increase of bifidobacteria in coeliac subjects with IBS symptoms. A pilot study conducted on coeliac children and adolescents evidenced that 3-month administration of *B. breve* strains helped in restoring the healthy percentage of main microbial components [109]. Particularly, the work compared coeliac children with a healthy control group and investigated on the effects deriving from the probiotic supplementation. The study revealed primarily an alteration in the intestinal microbial composition of coeliacs mainly characterized by a reduction of the Firmicutes/Bacteroidetes ratio, of Actinobacteria and Euryarchaeota compared to healthy controls [109]. Starting from these differences in gut microbiota composition, the probiotic administration resulted in an increase of Firmicutes, particularly Lactobacillaceae, and in a reestablishment of the physiological Firmicutes/Bacteroidetes ratio. A decrease of the proinflammatory cytokine TNF-alpha has also been observed within individuals administered with *B. breve* compared to placebo ones, evidencing the anti-inflammatory effects of the probiotic treatment [110].

Another study focusing on the same clinical trial highlighted, besides already documented microbiota changes in the Firmicutes phylum, that additional phyla, such as Verrucomicrobia, Parcubacteria and some yet unknown phyla belonging to Bacteria and Archaea Kingdom, may also play an important role in celiac disease-related pathology [111]. Moreover, the attention has been also shifted to previously unexplored phyla, particularly Synergistetes, which negatively correlated to acetic acid and total SCFAs, suggesting a potential role in microbiota restoration [111].

8.3.1.2 Probiotics in Metabolic Disorders

Although several *in vitro* and *in vivo* studies, as mentioned above, evidenced the anti-obesity effects of targeted probiotics and their ability to improve glycemic and lipidic profile, only a limited number of clinical studies focused on probiotic administration and metabolic diseases. Among these, Minami et al. [112] demonstrated that the administration of *B. breve* B-3 induced a significant decrease of the fat mass and an improvement of blood parameters in adults with a tendency for obesity. Subjects with high body mass index (BMI) and abdominal visceral fat area supplemented with *Lactobacillus gasseri* 2055 showed a decrease of abdominal adiposity and body weight [113]. VSL#3 administration in overweight subjects contributed to the improvement in lipid profile, insulin sensitivity and the decrease in C-reactive protein (CRP), which is a general marker for inflammation and infection [114]. Asemi et al. [115] evidenced that multispecies probiotic mixture, including *L. acidophilus, L. casei, L. rhamnosus, L. bulgaricus, B. breve, B. longum, S. thermophilus*, in type 2 diabetic patients led to an improvement of some blood, in particular preventing an increase of fasting plasma glucose, decreasing CRP and increasing the anti-oxidant plasma total glutathione. Ejtahed et al. [116] showed that probiotic yogurt containing *L. acidophilus* La5 and *Bifidobacterium lactis* Bb12 ameliorated the health condition of type 2 diabetic patients, by reducing the fasting blood glucose together with the glycated haemoglobin and favouring some antioxidant parameters. Other subjects with this pathology showed an improvement of the inflammatory status after a treatment with a multi-strain probiotic mixture comprising strains belonging to *Bifidobacterium, Lactobacillus, Lactococcus* and *Propionibacterium* genera [117].

Several studies evidenced the influence of probiotic supplementation on the lipidic profile. Particularly, works reviewed by Kobyliak et al. [118] showed that dietary intervention with yogurt containing probiotics (*Enterococcus faecium, S. thermophiles, L. acidophilus, B. longum, L. plantarum* and/or *B. lactis*) significantly reduced total serum cholesterol and LDL-cholesterol and improved the LDL:HDL cholesterol ratio. In addition, *L. gasseri* supplementation in hypercholesterolemic patients was efficient in reducing total and LDL-cholesterol and inulin [119].

A cross-over randomized clinical trial focused on obese children and adolescents showed the positive effects of probiotic *B. breve* strains (B632 and BR03) supplementation [120]. All subjects improved metabolic parameters, and decreased weight and intestinal *E. coli* counts. Probiotics improved some insulin parameters, i.e. sensitivity at fasting and during OGTT. An interesting outcome regarded fecal SCFAs, for which acetic acid and acetic acid pentyl-ester relative abundance remained stable in the treated subjects, while increased in the placebo group. A signature of five butanoic esters identified three clusters, one of them had better glucose responses during probiotic administration.

8.3.1.3 Probiotics in Neurological Disorders

Probiotic administration can improve the production and the delivery of neuroactive compounds acting on the brain-gut axis and improve the stability on the nervous system. Many investigations evidenced the potential of probiotics in preventing symptoms or attenuating comorbidities in neurological diseases, although the possible role of probiotics in some other neurological disorders, whose onset is unknown, as ALS, has to be still clarified.

The administration of *L. casei* Shirota in patients suffering from chronic fatigue syndrome led to a reduction of anxiety symptoms, besides an increase of lactobacilli and bifidobacteria in the gut microbial population [121]. Moreover, *Lactobacillus brevis* SBC8803 supplementation in males with poor quality of sleep was found to restore the sleep tone [122].

The efficacy of probiotics has also been highlighted in neuroinflammatory and autoimmune disorders. The supplementation with a probiotic mixture consisting of *L. acidophilus*, *L. casei*, *B. bifidum*, *L. fermentum* was associated to an improvement of cognitive functions in AD patients [123]. The same formulation was successful also in multiple sclerosis patients in ameliorating mental health condition [124].

In the scenery of neurological pathologies, probiotics can also assume the role of "psychobiotics". The first definition of psychobiotic was given by Dinan et al. [125]: "a live microorganism that, when ingested in adequate amounts, produces a health benefit in patients suffering from psychiatric hilliness". A probiotic formulation containing *L. acidophilus*, *L. casei* and *B. bifidum* administered in patients with major depressive disorders has been shown a general improvement of clinical picture [126]. In addition, it has been evidenced that *L. rhamnosus* HN001 treatment was effective in lowering post-partum depression and anxiety scores in women, receiving the probiotic supplementation from 14–16 weeks gestation to 6 months post-partum [127].

A pioneer study has been recently conducted on ALS patients and matched healthy controls [128, 129]. It showed, besides substantial differences in gut microbiota composition between the two groups involving potentially protective microbial groups, such as Bacteroidetes, and other with potential neurotoxic or pro-inflammatory activity, such as Cyanobacteria, the influence on microbiota composition of 6-month probiotic treatment on diseased subjects. The probiotic formulation was a mixture of five lactic acid bacteria: *S. thermophilus* ST10–DSM 25246, *L. fermentum* LF10–DSM 19187, *L. delbrueckii* subsp. *delbrueckii* LDD01–DSM 22106, *L. plantarum* LP01–LMG P-21021, and *L. salivarius* LS03–DSM 22776. This investigation represents a preliminary clinic probiotic application in ALS patients, a field of study that relies only on very few works mainly performed in animal models. The authors evidenced that some effects can be obtained in contrasting potentially pathogenic microbial groups and, thus, poses the bases for a microbial and personalized strategy to ameliorate the health status of the gut. The approach used can be applied in larger clinical studies incorporating different genetic and phenotypical disease variants and also patients at different stages of the disease in order to characterize the microbiota changes as a novel biomarker of the disease and, consequently, to design personalized therapies. In this contest, a study considering the application of a normocaloric ketogenic diet in ALS patients aimed at reducing hyperexcitability levels and modulating neuroinflammation has been recently proposed [130]. This study will also monitor the effect of the diet on the gut microbiota of the enrolled patients posing the basis for a combined personalized approach considering diet plus targeted strain administration.

A summary of the probiotic administration in clinical trials of the considered disease and their mode of action with is enlisted in Table 8.1.

Table 8.1 Summary of the beneficial role of probiotics in clinical trials

Diseases	Probiotics	Mode of action	References
Infants Gastroinstestinal disorders	*L. reuteri*	Reduction of pathogens and symptom improvement in colicky infants	[86–89]
	B. breve	Reduction of daily vomit frequency, daily evacuation, improved stool consistency, protection against developing metabolic disturbance in healthy infants	[90]
	L. reuteri	Modulation of mucosal inflammation and modulation of cytokines production in children with distal active ulcerative colitis	[91]
Necrotizing enterocolitis	*B. breve*	Reduction of abdominal symptoms, improvement weight gain and establishment of beneficial gut microbiota in preterm infants	[96, 97]
Inflammatory bowel diseases	VSL#3	Manifestation of high rate of remission and enhanced response to the therapy in UC children	[99, 100]
Inflammatory bowel diseases Coeliac disease	VSL#3 *L. plantarum* and *B. breve* and VSL#3	Prevention and the maintenance of remission of pouchitis	[101, 102]
		Reduction of IBS-associated pain	[104, 105]
	B. infantis	Improvement of gastrointestinal symptoms and inflammatory state	[106, 107]
Coeliac disease Obesity	*B. breve*	Restoration of the healthy percentage of main gut microbial components in celiac children	[109]
	B. breve *B. breve*	Reduction of pro-inflammatory TNF-α in celiac children	[110]
		Increment of novel phyla of the microbiota and modulation of faecal SCFAs profile	[111]
		Decrement of the fat mass and an improvement of blood parameters in adults with a tendency for obesity	[111]

(continued)

Table 8.1 (continued)

Diseases	Probiotics	Mode of action	References
Obesity Type II of diabetes	L. gasseri	Decrease of abdominal adiposity and body weight	[113]
	VSL#3	Improvement in lipid profile, insulin sensitivity and the decrease in CRP	[114]
	B. breve	Improvement of glucose metabolism, decrement of weight and E. coli in obese children	[120]
	L. acidophilus, L. casei, L. rhamnosus, L. bulgaricus, B. breve, B. longum and S. thermophilus	Reduction of blood glucose level, decrease of CRP and increase of the anti-oxidant plasma total glutathione	[116]
Type II of diabetes Hypercholesterolemia	L. acidophilus and B. lactis	Reduction of the fasting blood glucose, the glycated haemoglobin and promotion of some antioxidant parameters	[116]
	Enterococcus faecium, S. thermophiles, L. acidophilus, B. longum, L. plantarum and/or B. lactis	Redaction of total serum cholesterol and LDL-cholesterol and improvement of the LDL:HDL cholesterol ratio	[118]
	L. gasseri	Reduction of total and LDL-cholesterol and inulin	[119]
Neurological disease	L. casei Shirota	Reduction of anxiety and increment of lactobacilli and bifidobacterial in patients with chronic fatigue syndrome	[121]
Neurological disease	Lactobacillus brevis	Restoration of sleep quality	[122]
	L. acidophilus, L. casei, B. bifidum, L. fermentum	Improvement of cognitive functions in AD patients	[125]
	L. acidophilus, L. casei, B. bifidum, L. fermentum, L. acidophilus, L. casei and B. bifidum	Amelioration of the health condition in multiple sclerosis patients	[124]
		Improvement of depressive disorders	[126]
	L. rhamnosus	Melioration of post-partum depression	[127]
	S. thermophilus, L. fermentum, L. delbrueckii, L. plantarum and L. salivarius	Improvement of a heathy gut microbiota composition and counteracting pathogens	[128, 129]

8.3.2 Mechanisms of Action of Probiotics

The beneficial effect of probiotics on the gut microbiota and health o can be ascribed to a large array of actions (Fig. 8.1): reduction in luminal pH, competition with potential pathogens for nutritional sources, production of bacteriocin or bacteriocin-like substances and functions related to probiotic metabolites [131]. Specifically, some of these, such as organic acids, bacteriocins, hydrogen peroxide, and amines have been reported to interact with multiple targets in some metabolic pathways that regulate cellular proliferation, differentiation, apoptosis, inflammation, angiogenesis, and metastasis [132].

The positive effects deriving from the administration of bifidobacteria, as evidenced by Quagliariello et al. [109] and Primec et al. [111], can be related to their high ability to deep influence gut microbiota composition, by enhancing the blooming of some species and antagonizing others probably by the effect of the production of metabolites such as acetic acid [97]. In particular, there are evidences that *Bifidobacterium* strains supports Lactobacillaceae development [133]. Moreover, it is highly probable that the decrease of TNF-alfa observed in administered individuals can be attributed to the increase of lactobacilli, whose anti-inflammatory properties are largely known, promoted by the administration of *B. breve* [134].

The reduced amount of acetic acid and acetic acid pentyl-ester, among fecal SCFAs, detected in adolescents and children supplemented with *B. breve* [120], which is an acetate producer, can allow the speculation that acetic acid and acetic acid pentyl-ester were used by other acetate dependent species involved in metabolic diseases, through complex cross-feeding mechanisms [135, 136]. Noteworthy, acetate is used as fuel by peripheral tissues, including the liver, muscle, and pancreas, indicating an increased colonic absorption and transition to the systemic circulation to use it as an alternative source of energy during extended calorie restrictions apart from a regulatory action on glucose metabolism, insulin sensitivity and secretion, muscle function, adipose tissue metabolism and inflammation, and satiety [137]. In this scenario, it is important to highlight the ability of probiotics to produce bile-salt hydrolase (BSH), which plays a role in deconjugation of biliary salts and cholesterol absorption [138]. Therefore, this activity can be linked to the hypocholesterolemic effect deriving from probiotic intake [138, 139], which can explain the use of these beneficial microorganisms for treating and preventing metabolic disorders, like obesity. In fact, considering the beneficial effects of BSH-producing bacteria, BSH activity has been included in FAO/WHO guidelines for the evaluation of probiotics for food use [140].

The same *B. breve* treatment in healthy infants was associated with a concomitant lower weight gain in the population at higher risk of metabolic disturbances in later life [90]. Other formulations were not successful in inducing changes in weight in neonates [141, 142]. This inconsistency may be attributed to the strains used or, more probably, to the timing of the treatment being the protocol of Aloisio et al. [90] designed on 3 months, differently from the majority of the studies which followed infants for 1 month. Specifically, a lower weight gain has been evidenced in supplemented infants born by cesarean section; since epidemiological data suggest that children born by cesarean section have an increased risk to develop obesity later in life [143, 144], it is possible to hypothesize that *B. breve* may have a protective role for infants at risk of developing metabolic disorders.

Di Gioia et al. [129] demonstrated that 6 months probiotic administration in ALS patients may have contrasted some microbial groups potentially harmful for the host, such as *E. coli*, *Clostridium* cluster I, and Enterobacteriaceae. This can be ascribed to the ability of some lactobacilli and bifidobacteria to produce antimicrobial peptides known as bacteriocins, which prevent the proliferation of selected pathogens [145].

8.4 Remarks and Conclusion

The reported preclinical and clinical works showed mostly that the intestinal microbiota structure is altered in the disorders considered with respect to a healthy condition and that targeted probiotic administration can help in restoring the gut dysbiosis and in improving the symptoms associated to the considered pathologies.

Alterations in the gut microbiota between healthy and diseased subjects have been observed in two diseases targeting different organ systems, such as coeliac disease and ALS.

Quagliariello et al. [109] and Primec et al. [111] highlighted a clear separation of gut microbiota profiles between healthy and coeliac subjects, mostly attributable to Verrucomicrobia, Parcubacteria and some other unknown bacterial phyla. The absence of a serious dysbiosis in coeliac patients was probably attributed to the adherence to a gluten free diet. The study of Di Gioia et al. [129] is focused on a neurological disorder, whose relationship with the intestinal microbiota has been scarcely explored in clinical trials. The high abundance of Cyanobacteria observed in ALS patients with respect to controls, has given a renewed strength to the hypothesis formulated in 2000s and then abandoned on the role of Cyanobacteria in the pathogenesis of neurodegenerative diseases [146, 147]. Moreover, the study shed light on the microbiota composition in ALS posing the basis for a tailored probiotic approach as PPPM strategy for these patients.

Among probiotics, bifidobacteria have the potential for a targeted use in a wide range of diseases both in paediatric subjects and in adults. This has been demonstrated by the positive outcomes described in the reported trials. In this context, three works focused on the effects of *B.breve* based formulation (B632 and BR03) in preventing or treating paediatric can be discussed.

A 3-months treatment with the two *B. breve* strains in healthy breast-fed newborns [90] outlined that children born by caesarean section, who have a high risk to develop obesity later in life, showed a lower catch-up growth in weight. This allows the speculation that the administered formulation has a role, not only during infanthood but also in childhood, by reducing the risk of metabolic disturbances. Based on these evidences, it is possible to speculate that *B. breve* administration can represent a PPPM approach for infants with a predisposition to metabolic disorders.

The probiotic properties of *B. breve* strains in modulating gut microbiota rely on the effects of the administered strains not only on the bifidobacteria population but also on other microbial groups, acting as a "trigger" element for the increase of Firmicutes and the restoration of the physiological Firmicutes/Bacteroidetes ratio [109].

The Firmicutes family mostly influenced by the probiotic treatment was the Lactobacillaceae one that reached almost the values of the healthy subjects. The increase of Firmicutes was correlated to a decrease of pro-inflammatory TNF-alpha, confirming the role of gut microbiota in modulating the host's inflammatory condition [111]. These studies also reported new evidence about the potential anti-inflammatory role of Synergistetes, which negatively correlated with pro-inflammatory acetic acid in CD children after the probiotic administration. Therefore, *B. breve* supplementation for the increase of Lactobacillaceae and Synergistetes can constitute a PPPM strategy for CD patients, with the aim to reduce inflammatory condition.

Considering ALS, the 6-months probiotic treatment with a mixture of *Lactobacillus* strains influenced in different ways the gut microbial composition; however, none of the probiotic interventions brought the biodiversity of intestinal microbiota of patients closer to that of healthy subjects [129]. Although the suggested probiotic formulation only influenced few microbial groups not affecting the main clinical parameters measured in ALS, it was possible to observe some changes in the gut microbial environment in relation to the disease progression, confirming the specific role of gut microbiota also in pathologies involving districts relatively far from the intestinal tract. This is the first study that clearly showed the modifications of gut microbiota composition in ALS patients by applying novel and rigorous methodologies. This approach based on microorganisms' administration can be further improved with the design of a new formulation and applied in larger clinical studies in order to characterize the microbiota changes as a novel biomarker of the disease, following the principles of PPPM.

We would like to conclude this chapter with a general consideration. Being the gut microbiota a complex ecosystem, it possesses a certain resilience [148]. This property determines whether a particular perturbation will permanently shift its stable state or whether it will return to its initial homeostatic state after a perturbation. Dysbiosis can occur when resilience of the original community fails and this condition can lead to the acquisition of an unhealthy microbiota that can become resilient in turn. In this regard, a probiotic administration is necessary:

1. to generally maintain the resilience of a healthy microbiota in order to face any perturbation;
2. to protect the microbiota's resilience in a vulnerable period, such as during the first years of life when the microbiota is establishing, or during a period of stress or reduced activity of the immune system;
3. to contrast the establishment of an unhealthy resilient microbiota;
4. to reduce at the least the consequences associated to a permanently altered microbiota

The reviewed studies collected in this chapter confirmed that the gut microbiota represents a key determinant of the overall health status and disease susceptibility of humans. This study collection and elaboration will further help to decide on the appropriate treatment strategy for the considered diseases, in compliance with the perspective of PPPM. The use of tailored microorganisms can be an effective approach to reduce symptoms related to the disease, to maintain the host's health condition and to re-establish the beneficial microbiota in the gut.

8.5 PPPM Strategies in the Field of Probiotics Administration

The reported studies in this chapter evidenced that the intestinal microbial community acts as a fingerprint of the individual during the treatment of diseases. Thus, the gut microbial profile could be considered as a valuable biomarker for predictive diagnosis and targeted treatment strategy. Hence, a personalized diet and the use of specific probiotics to restore the gut microbiota composition and functions is crucial for the development of personalized medicine, in order to preserve and protect the individual's health status, in accordance with the PPPM perspective.

To this purpose, a comprehensive clinical study of the patient and screening for the strains to be used for such therapy are necessary to implement probiotic treatment tailored to clinical care and set up the relevant clinical trial designs of particular strains. Probiotic supplements are not homogenous beneficial microorganisms, rather, the reported studies show recognition of probiotic efficacy as both strain- and indication-specific. Person-specific factors also contribute to heterogeneity in the outcome of probiotic supplementations, including diet, age and the individual's gut microbiome. Therefore, the concept of "precision probiotics" has been developed in recent years [149] and will even gain greater popularity once the general population will have their microbiome sequenced. Even though nowadays this process is still not achievable, it is possible to state that it can be pursued in the following decades, serving as the basis for microbiome-centered precision nutrition and preventive medicine, including precision probiotics. In this future, individuals will be recommended diets, foods and precision probiotics tailored to their unique human–microbiome symbiosis, in compliance with the PPPM strategies (Fig. 8.1).

8.6 Expert Recommendations in the Framework of 3P Medicine

The use of probiotics in preventive and personalized medicine has a growing interest due to the considerable evidences of their beneficial effects on the treatment and prevention of many disease conditions. It is important to emphasize that their health benefits depend on the strain type, dosing regimen and the patient's individual responses. In the scenario of clinical application, matching the appropriate probiotic strain to patients who suffer from a certain disease can represent a challenging task. In this regard, McFarland et al. [150] demonstrated the importance of considering both probiotic strain and disease specificity. Individualized probiotic supplementation with selected strains should be based on the disease, site of infection, clinical situation or symptoms, phenotype and metabolic response of patients in order to maximize their efficacy and reduce treatment time [4]. Co-administration of prebiotics and probiotics is recommended to potentiate the effect and growth of probiotics, so the combination with an appropriate diet is very crucial. For this purpose, the application of probiotics in personalized therapy needs further investigation to reveal their specific proprieties and to ensure product quality, safety, viability and efficacy in animal and clinical settings. As already mentioned, the concept that one strain fits all diseases and all patients is now overcome by the need of a tailored strain administration. Based on this evidence, a technology that is able to guarantee a clear discrimination among probiotic strains is essential for the design of clinical trials focused on the prevention or treatment of diseases. DNA-based techniques, such as MLST, PFGE and whole genome sequencing, are required for the full characterization of the probiotic strains, the identification of genes associated with healthy features and host interaction and the screening of the strains that could be tested first *in vitro* and then *in vivo*, followed finally by clinical trials. In addition, gut microbiome profiling is useful for detecting alterations in microbial composition that may be associated with various human diseases; in this regard probiotics could serve as biomarkers for predictive diagnosis, especially with an integrated metabolic analysis. However, these types of analyses require the availability of specific high throughput technologies and cost. In this context, a very recent study conducted by Deidda et al. [151] suggested FTIR spectroscopy, as a quick, reliable and efficient method, for typing probiotic bacteria in a targeted probiotic formulation.

Probiotic products are unique in their properties to confer a health benefit, and they can present different challenges in design, development, scale-up, manufacturing, commercialization, and life cycle management [152]. It is necessary to consider that the quality and safety assessment of probiotic food and supplements are responsibilities of the industry and specific guidelines and legislative rules for approval and commercialization of the products need to be followed.

This comprehensive approach is necessary to select the most suitable probiotic strains for the formulation of personalized diets and treatments according to the PPPM approach.

References

1. Golubnitschaja O, Baban B, Boniolo G, Wang W, Bubnov R, Kapalla M, Krapfenbauer K, Mozaffari MS, Costigliola V (2016) Medicine in the early twenty-first century: paradigm and anticipation-EPMA position paper 2016. EPMA J 7:1–3. https://doi.org/10.1186/s13167-016-0072-4

2. Bozzi Cionc N, Baffoni L, Gaggìa F, Di Gioia D (2018) Therapeutic microbiology: the role of Bifidobacterium breve as food supplement for the prevention/treatment of paediatric diseases. Nutrients 10:1723. https://doi.org/10.3390/nu10111723

3. Bubnov RV, Babenko LP, Lazarenko LM, Mokrozub VV, Spivak MY (2018) Specific properties of probiotic strains: relevance and benefits for the host. EPMA J 9:205–223. https://doi.org/10.1007/s13167-018-0132-z

4. Abdelhamid AG, El-Masry SS, El-Dougdoug NK (2019) Probiotic Lactobacillus and Bifidobacterium strains possess safety characteristics, antiviral activities and host adherence factors revealed by genome mining. EPMA J 10:337–350. https://doi.org/10.007/s13167-019-00184-z

5. Hill C, Guarner F, Reid G, Gibson GR, Merenstein DJ, Pot B, Morelli L, Canani RB, Flint HJ, Salminen S, Calder PC (2014) The international scientific association for probiotics and prebiotics consensus statement on the scope and appropriate use of the term probiotic. Nat Rev Gastroenterol Hepatol 11:506–514. https://doi.org/10.1038/nrgastro.2014.66

6. Fuller R (1991) Probiotics in human medicine. Gut 32:439. https://doi.org/10.1136/gut.32.4.439

7. MacDonald TT, Monteleone G (2005) Immunity, inflammation, and allergy in the gut. Science 307:1920–1925. https://doi.org/10.1126/science.1106442

8. Chapman CM, Gibson GR, Rowland I (2011) Health benefits of probiotics: are mixtures more effective than single strains? Eur J Nutr 50:1–7. https://doi.org/10.1007/s00394-010-0166-z

9. Timmerman HM, Koning CJM, Mulder L, Rombouts FM, Beynen AC (2004) Monostrain, multistrain and multispecies probiotics – a comparison of functionality and efficacy. Int J Food Microbiol 96:219–233. https://doi.org/10.1016/j.ijfoodmicro.2004.05.012

10. Rohr M, Narasimhulu CA, Sharma D, Doomra M, Riad A, Naser S, Parthasarathy S (2018) Inflammatory diseases of the gut. J Med Food 1(21):113–126. https://doi.org/10.1089/jmf.2017.0138

11. Lloyd-Price J, Arze C, Ananthakrishnan AN, Schirmer M, Avila-Pacheco J, Poon TW, Andrews E, Ajami NJ, Bonham KS, Brislawn CJ, Casero D (2019) Multi-omics of the gut microbial ecosystem in inflammatory bowel diseases. Nature 569:655–662. https://doi.org/10.1038/s41586-019-1237-9

12. Vila AV, Imhann F, Collij V, Jankipersadsing SA, Gurry T, Mujagic Z, Kurilshikov A, Bonder MJ, Jiang X, Tigchelaar EF, Dekens J (2018) Gut microbiota composition and functional changes in inflammatory bowel disease and irritable bowel syndrome. Sci Transl Med 10:472. https://doi.org/10.1126/scitranslmed.aap8914

13. Pittayanon R, Lau JT, Leontiadis GI, Tse F, Yuan Y, Surette M, Moayyedi P (2020) Differences in gut microbiota in patients with vs without inflammatory bowel diseases: a systematic review. Gastroenterology 158:930–946. https://doi.org/10.1053/j.gastro.2019.11.294

14. Dalal SR, Chang EB (2014) The microbial basis of inflammatory bowel diseases. J Clin Invest 124:4190–4196. https://doi.org/10.1172/JCI72330

15. Hansen J, Gulati A, Sartor RB (2010) The role of mucosal immunity and host genetics in defining intestinal commensal bacteria. Curr Opin Gastroenterol 26:564. https://doi.org/10.1097/MOG.0b013e32833f1195

16. Sokol H, Pigneur B, Watterlot L, Lakhdari O, Bermúdez-Humarán LG, Gratadoux JJ, Blugeon S, Bridonneau C, Furet JP, Corthier G, Grangette C (2008) Faecalibacterium prausnitzii is an anti-inflammatory commensal bacterium identified by gut microbiota analysis of Crohn disease patients. Proc Natl Acad Sci U S A 105:16731–16736. https://doi.org/10.1073/pnas.0804812105

17. Monika K, Malik T, Gehlot R, Rekha K, Kumari A, Sindhu R, Rohilla P (2021) Antimicrobial property of probiotics. Environ Conserv J 22:33–48. https://doi.org/10.36953/ECJ.2021.SE.2204
18. Cristofori F, Indrio F, Miniello VL, De Angelis M, Francavilla R (2018) Probiotics in celiac disease. Nutrients 10:1824. https://doi.org/10.3390/nu10121824
19. Norouzbeigi S, Vahid-Dastjerdi L, Yekta R, Sohrabvandi S, Zendeboodi F, Mortazavian AM (2020) Celiac therapy by administration of probiotics in food products: a review. Curr Opin Food Sci 32:58–66. https://doi.org/10.1016/j.cofs.2020.01.005
20. Sun Z, Sun X, Li J, Li Z, Hu Q, Li L, Hao X, Song M, Li C (2020) Using probiotics for type 2 diabetes mellitus intervention: advances, questions, and potential. Crit Rev Food Sci Nutr 60:670–683. https://doi.org/10.1080/10408398.2018.1547268
21. Giongo A, Gano KA, Crabb DB, Mukherjee N, Novelo LL, Casella G, Drew JC, Ilonen J, Knip M, Hyöty H, Veijola R (2011) Toward defining the autoimmune microbiome for type 1 diabetes. ISME J 5:82–91. https://doi.org/10.1038/ismej.2010.92
22. Qin J, Li Y, Cai Z, Li S, Zhu J, Zhang F, Liang S, Zhang W, Guan Y, Shen D, Peng Y (2012) A metagenome-wide association study of gut microbiota in type 2 diabetes. Nature 490:55–60. https://doi.org/10.1038/nature11450
23. Stojanov S, Berlec A, Štrukelj B (2020) The influence of probiotics on the Firmicutes/Bacteroidetes ratio in the treatment of obesity and inflammatory bowel disease. Microorganisms 8:1715. https://doi.org/10.3390/microorganisms8111715
24. Hemarajata P, Versalovic J (2013) Effects of probiotics on gut microbiota: mechanisms of intestinal immunomodulation and neuromodulation. Ther Adv Gastroenterol 6:39–51. https://doi.org/10.1177/1756283X12459294
25. Abenavoli L, Scarpellini E, Colica C, Boccuto L, Salehi B, Sharifi-Rad J, Aiello V, Romano B, De Lorenzo A, Izzo AA, Capasso R (2019) Gut microbiota and obesity: a role for probiotics. Nutrients 11:2690. https://doi.org/10.3390/nu11112690
26. Rhee SH, Pothoulakis C, Mayer EA (2009) Principles and clinical implications of the brain–gut–enteric microbiota axis. Nat Rev Gastroenterol Hepatol 6:306–314. https://doi.org/10.1038/nrgastro.2009.35
27. Carloni S, Bertocchi A, Mancinelli S, Bellini M, Erreni M, Borreca A, Braga D, Giugliano S, Mozzarelli AM, Manganaro D, Fernandez PD (2021) Identification of a choroid plexus vascular barrier closing during intestinal inflammation. Science 374:439–448. https://doi.org/10.1126/science.abc6108
28. Aloisio I, Santini C, Biavati B, Dinelli G, Cenčič A, Chingwaru W, Mogna L, Di Gioia D (2012) Characterization of Bifidobacterium spp. strains for the treatment of enteric disorders in newborns. Appl Microbiol Biotechnol 96:1561–1576. https://doi.org/10.1007/s00253-012-4138-5
29. Simone M, Gozzoli C, Quartieri A, Mazzola G, Di Gioia D, Amaretti A, Raimondi S, Rossi M (2014) The probiotic Bifidobacterium breve B632 inhibited the growth of Enterobacteriaceae within colicky infant microbiota cultures. Biomed Res Int. https://doi.org/10.1155/2014/301053
30. Yokota Y, Shikano A, Kuda T, Takei M, Takahashi H, Kimura B (2018) Lactobacillus plantarum AN1 cells increase caecal L. reuteri in an ICR mouse model of dextran sodium sulphate-induced inflammatory bowel disease. Int Immunopharmacol 56:119–127. https://doi.org/10.1016/j.intimp.2018.01.020
31. Isidro RA, Lopez A, Cruz ML, Gonzalez Torres MI, Chompre G, Isidro AA, Appleyard CB (2017) The probiotic VSL# 3 modulates colonic macrophages, inflammation, and microflora in acute trinitrobenzene sulfonic acid colitis. J Histochem Cytochem 65:445–461. https://doi.org/10.1369/0022155417718542
32. Lammers KM, Brigidi P, Vitali B, Gionchetti P, Rizzello F, Caramelli E, Matteuzzi D, Campieri M (2003) Immunomodulatory effects of probiotic bacteria DNA: IL-1 and IL-10 response in human peripheral blood mononuclear cells. FEMS Immunol Med Microbiol 38:165–172. https://doi.org/10.1016/S0928-8244(03)00144-5

33. Zheng B, van Bergenhenegouwen J, Overbeek S, van de Kant HJ, Garssen J, Folkerts G, Vos P, Morgan ME, Kraneveld AD (2014) Bifidobacterium breve attenuates murine dextran sodium sulfate-induced colitis and increases regulatory T cell responses. PLoS One 9:e95441. https://doi.org/10.1371/journal.pone.0095441

34. Lindfors K, Blomqvist T, Juuti-Uusitalo K, Stenman S, Venäläinen J, Mäki M, Kaukinen K (2008) Live probiotic Bifidobacterium lactis bacteria inhibit the toxic effects induced by wheat gliadin in epithelial cell culture. Clin Exp Immunol 152:552–558. https://doi.org/10.1111/j.1365-2249.2008.03635.x

35. Orlando A, Linsalata M, Notarnicola M, Tutino V, Russo F (2014) Lactobacillus GG restoration of the gliadin induced epithelial barrier disruption: the role of cellular polyamines. BMC Microbiol 14:1–2. https://doi.org/10.1186/1471-2180-14-19

36. Laparra JM, Olivares M, Gallina O, Sanz Y (2012) Bifidobacterium longum CECT 7347 modulates immune responses in a gliadin-induced enteropathy animal model. PLoS One 7:e30744. https://doi.org/10.1371/journal.pone.0030744

37. De Angelis M, Rizzello CG, Fasano A, Clemente MG, De Simone C, Silano M, De Vincenzi M, Losito I, Gobbetti M (2006) VSL# 3 probiotic preparation has the capacity to hydrolyze gliadin polypeptides responsible for celiac sprue probiotics and gluten intolerance. Biochim Biophys Acta Mol basis Dis 1762:80–93. https://doi.org/10.1016/j.bbadis.2005.09.008

38. Castelli V, d'Angelo M, Lombardi F, Alfonsetti M, Antonosante A, Catanesi M, Benedetti E, Palumbo F, Cifone MG, Giordano A, Desideri G (2020) Effects of the probiotic formulation SLAB51 in vitro and in vivo Parkinson's disease models. Aging 12:4641. https://doi.org/10.18632/aging.102927

39. Alagón Fernández del Campo P, De Orta PA, Straface JI, López Vega JR, Toledo Plata D, Niezen Lugo SF, Alvarez Hernández D, Barrientos Fortes T, Gutiérrez-Kobeh L, Solano-Gálvez SG, Vázquez-López R (2019) The use of probiotic therapy to modulate the gut microbiota and dendritic cell responses in inflammatory bowel diseases. Med Sci 7:33. https://doi.org/10.3390/medsci7020033

40. McCarthy J, O'mahony L, O'callaghan L, Sheil B, Vaughan EE, Fitzsimons N, Fitzgibbon J, O'sullivan GC, Kiely B, Collins JK, Shanahan F (2003) Double blind, placebo controlled trial of two probiotic strains in interleukin 10 knockout mice and mechanistic link with cytokine balance. Gut 52:975–980. https://doi.org/10.1136/gut.52.7.975

41. Yin X, Heeney D, Srisengfa Y, Golomb B, Griffey S, Marco M (2018) Bacteriocin biosynthesis contributes to the anti-inflammatory capacities of probiotic lactobacillus plantarum. Benefic Micrcbes 9:333–344. https://doi.org/10.3920/BM2017.0096

42. Wang Y, Guo Y, Chen H, Wei H, Wan C (2018) Potential of lactobacillus plantarum ZDY2013 and Bifidobacterium bifidum WBIN03 in relieving colitis by gut microbiota, immune, and anti-oxidative stress. Can J Microbiol 64:327–337. https://doi.org/10.1016/j.nut.2020.110995

43. Liu Y, Tian X, He B, Hoang TK, Taylor CM, Blanchard E, Freeborn J, Park S, Luo M, Couturier J, Tran DQ (2019) Lactobacillus reuteri DSM 17938 feeding of healthy newborn mice regulates immune responses while modulating gut microbiota and boosting beneficial metabolites. Am J Physiol Gastrointest Liver Physiol 317(6):G824–G838. https://doi.org/10.1152/ajpgi.00107.2019

44. Jang YJ, Kim WK, Han DH, Lee K, Ko G (2019) Lactobacillus fermentum species ameliorate dextran sulfate sodium-induced colitis by regulating the immune response and altering gut microbiota. Gut Microbes 10:696–711. https://doi.org/10.1080/19490976.2019.1589281

45. Rodríguez-Nogales A, Algieri F, Garrido-Mesa J, Vezza T, Utrilla MP, Chueca N, Fernández-Caballero JA, García F, Rodríguez-Cabezas ME, Gálvez J (2018) The administration of Escherichia coli Nissle 1917 ameliorates development of DSS-induced colitis in mice. Front Pharmacol 9:468. https://doi.org/10.3389/fphar.2018.00468

46. Zaylaa M, Al Kassaa I, Alard J, Peucelle V, Boutillier D, Desramaut J, Dabboussi F, Pot B, Grangette C (2018) Probiotics in IBD: combining in vitro and in vivo models for selecting strains with both anti-inflammatory potential as well as a capacity to restore the gut epithelial barrier. J Funct Foods 47:304–315. https://doi.org/10.1016/j.jff.2018.05.029

47. Celiberto LS, Pinto RA, Rossi EA, Vallance BA, Cavallini DC (2018) Isolation and characterization of potentially probiotic bacterial strains from mice: proof of concept for personalized probiotics. Nutrients 10:1684. https://doi.org/10.3390/nu10111684

48. Serena G, D'Avino P, Fasano A (2020) Celiac disease and non-celiac wheat sensitivity: state of art of non-dietary therapies. Front Nutr 7:152. https://doi.org/10.3389/fnut.2020.00152

49. D'Arienzo R, Maurano F, Luongo D, Mazzarella G, Stefanile R, Troncone R, Auricchio S, Ricca E, David C, Rossi M (2008) Adjuvant effect of lactobacillus casei in a mouse model of gluten sensitivity. Immunol Lett 119:78–83. https://doi.org/10.1016/j.imlet.2008.04.006

50. Papista C, Gerakopoulos V, Kourelis A et al (2012) Gluten induces coeliac-like disease in sensitised mice involving IgA, CD71 and transglutaminase 2 interactions that are prevented by probiotics. Lab Investig 92:625–635. https://doi.org/10.1038/labinvest.2012.13

51. Stojanov S, Kreft S (2020) Gut microbiota and the metabolism of phytoestrogens. Rev Bras Farm 30:145–154. https://doi.org/10.1007/s43450-020-00049-x

52. Ji Y, Kim H, Park H, Lee J, Yeo S, Yang J, Park S, Yoon H, Cho G, Franz CM, Bomba A (2012) Modulation of the murine microbiome with a concomitant anti-obesity effect by lactobacillus rhamnosus GG and lactobacillus sakei NR28. Benefic Microbes 3:13–22. https://doi.org/10.3920/BM2011.0046

53. Ji Y, Park S, Park H, Hwang E, Shin H, Pot B, Holzapfel WH (2018) Modulation of active gut microbiota by lactobacillus rhamnosus GG in a diet induced obesity murine model. Front Microbiol 9:710. https://doi.org/10.3389/fmicb.2018.00710

54. Chen D, Yang Z, Chen X, Huang Y, Yin B, Guo F, Zhao H, Zhao T, Qu H, Huang J, Wu Y (2014) The effect of lactobacillus rhamnosus hsryfm 1301 on the intestinal microbiota of a hyperlipidemic rat model. BMC Complement Altern Med 14:1–9. https://doi.org/10.118 6/1472-6882-14-386

55. Thiennimitr P, Yasom S, Tunapong W, Chunchai T, Wanchai K, Pongchaidecha A, Lungkaphin A, Sirilun S, Chaiyasut C, Chattipakorn N, Chattipakorn SC (2018) Lactobacillus paracasei HII01, xylooligosaccharides, and synbiotics reduce gut disturbance in obese rats. Nutrition 54:40–47. https://doi.org/10.1016/j.nut.2018.03.005

56. Murphy EF, Cotter PD, Hogan A, O'Sullivan O, Joyce A, Fouhy F, Clarke SF, Marques TM, O'Toole PW, Stanton C, Quigley EM (2013) Divergent metabolic outcomes arising from targeted manipulation of the gut microbiota in diet-induced obesity. Gut 62:220–226. https://doi.org/10.1136/gutjnl-2011-300705

57. Wang Y, Wu Y, Wang B, Xu H, Mei X, Xu X, Zhang X, Ni J, Li W (2019) Bacillus amyloliquefaciens SC06protects mice against high-fat diet-induced obesity and liver injury via regulating host metabolism and gut microbiota. Front Microbiol 10:1161. https://doi.org/10.3389/fmicb.2019.01161

58. Everard A, Matamoros S, Geurts L, Delzenne NM, Cani PD (2014) Saccharomyces boulardii administration changes gut microbiota and reduces hepatic steatosis, low-grade inflammation, and fat mass in obese and type 2 diabetic db/db mice. MBio 5:e01011–e01014. https://doi.org/10.1128/mBio.01011-14

59. Yu L, Zhao XK, Cheng ML, Yang GZ, Wang B, Liu HJ, Hu YX, Zhu LL, Zhang S, Xiao Z, Liu YM (2017) Saccharomyces boulardii administration changes gut microbiota and attenuates D-galactosamine-induced liver injury. Sci Rep 7:1–7. https://doi.org/10.1038/s41598-017-01271-9

60. Bubnov R, Babenko L, Lazarenko L, Kryvtsova M, Shcherbakov O, Zholobak N, Golubnitschaja O, Spivak M (2019) Can tailored nanoceria act as a prebiotic? Report on improved lipid profile and gut microbiota in obese mice. EPMA J 10:317–335. https://doi.org/10.1007/s13167-019-00190-1

61. Kondo S, Xiao JZ, Satoh T, Odamaki T, Takahashi S, Sugahara H, Yaeshima T, Iwatsuki K, Kamei A, Abe K (2010) Antiobesity effects of Bifidobacterium breve strain B-3 supplementation in a mouse model with high-fat diet-induced obesity. Biosci Biotechnol Biochem 74:1656–1661. https://doi.org/10.1271/bbb.100267

62. Kondo S, Kamei A, Xiao JZ, Iwatsuki K, Abe K (2013) Bifidobacterium breve B-3 exerts metabolic syndrome-suppressing effects in the liver of diet-induced obese mice: a DNA microarray analysis. Benefic Microbes 4:247–251. https://doi.org/10.3920/BM2012.0019

63. Al Kattar S, Jurjus R, Pinon A, Leger DY, Jurjus A, Boukarim C, Diab-Assaf M, Liagre B (2020) Metformin and probiotics in the crosstalk between colitis-associated colorectal cancer and diabetes in mice. Cancers 12:1857. https://doi.org/10.3390/cancers12071857

64. Chen JJ, Wang R, Li XF, Wang RL (2011) Bifidobacterium longum supplementation improved high-fat-fed-induced metabolic syndrome and promoted intestinal Reg I gene expression. Exp Biol Med 236:823–831. https://doi.org/10.1258/ebm.2011.010399

65. Hsieh FC, Lee CL, Chai CY, Chen WT, Lu YC, Wu CS (2013) Oral administration of lactobacillus reuteri GMNL-263 improves insulin resistance and ameliorates hepatic steatosis in high fructose-fed rats. Nutr Metab 10:1–4. https://doi.org/10.1186/1743-7075-10-35

66. Gao K, Wang C, Liu L, Dou X, Liu J, Yuan L, Zhang W, Wang H (2017) Immunomodulation and signaling mechanism of lactobacillus rhamnosus GG and its components on porcine intestinal epithelial cells stimulated by lipopolysaccharide. J Microbiol Immunol Infect 50:700–713. https://doi.org/10.1016/j.jmii.2015.05.002

67. Balakumar M, Prabhu D, Sathishkumar C, Prabu P, Rokana N, Kumar R, Raghavan S, Soundarajan A, Grover S, Batish VK, Mohan V (2018) Improvement in glucose tolerance and insulin sensitivity by probiotic strains of Indian gut origin in high-fat diet-fed C57BL/6J mice. Eur J Nutr 57:279–295. https://doi.org/10.1007/s00394-016-1317-7

68. Leta V, Chaudhuri KR, Milner O, Chung-Faye G, Metta V, Pariante CM, Borsini A (2021) Neurogenic and anti-inflammatory effects of probiotics in Parkinson's disease: a systematic review of preclinical and clinical evidence. Brain Behav Immun 98:59–73. https://doi.org/10.1016/j.bbi.2021.07.026

69. Liao J-F, Cheng Y-F, You S-T, Kuo W-C, Huang C-W, Chiou J-J, Hsu C-C, Hsieh-Li H-M, Wang S, Tsai Y-C (2020) Lactobacillus plantarum PS128 alleviates neurodegenerative progression in 1-methyl-4-phenyl-1,2,3,6-tetrahydropyridine-induced mouse models of Parkinson's disease. Brain Behav Immun 90:26–46. https://doi.org/10.1111/j.1442-200x.2004.01953.x

70. Sun J, Li H, Jin Y, Yu J, Mao S, Su KP, Ling Z, Liu J (2020) Probiotic clostridium butyricum ameliorated motor deficits in a mouse model of Parkinson's disease via gut microbiota-GLP-1 pathway. Brain Behav Immun 91:703–715. https://doi.org/10.1016/j.bbi.2020.10.014

71. Marsova M, Poluektova E, Odorskaya M, Ambaryan A, Revishchin A, Pavlova G, Danilenko V (2020) Protective effects of lactobacillus fermentum U-21 against paraquat-induced oxidative stress in Caenorhabditis elegans and mouse models. World J Microbiol Biotechnol 36:1–10. https://doi.org/10.1007/s11274-020-02879-2

72. Srivastav S, Neupane S, Bhurtel S, Katila N, Maharjan S, Choi H, Hong Tae J, Choi DY (2019) Probiotics mixture increases butyrate, and subsequently rescues the nigral dopaminergic neurons from MPTP and rotenone-induced neurotoxicity. J Nutr Biochem 69:73–86. https://doi.org/10.1016/j.jnutbio.2019.03.021

73. Alipour Nosrani E, Tamtaji OR, Alibolandi Z, Sarkar P, Ghazanfari M, Azami Tameh A, Taghizadeh M, Banikazemi Z, Hadavi R, Naderi TM (2021) Neuroprotective effects of probiotics bacteria on animal model of Parkinson's disease induced by 6-hydroxydopamine: a behavioral, biochemical, and histological study. J Immunoassay Immunochem 42:106–120. https://doi.org/10.1080/15321819.2020.1833917

74. Magistrelli L, Amoruso A, Mogna L, Graziano T, Cantello R, Pane M, Comi C (2019) Probiotics may have beneficial effects in Parkinson's disease: in vitro evidence. Front Immunol 10:969. https://doi.org/10.3389/fimmu.2019.00969

75. Mohammadi G, Dargahi L, Peymani A, Mirzanejad Y, Alizadeh SA, Naserpour T, Nassiri-Asl M (2019) The effects of probiotic formulation pretreatment (lactobacillus helveticus R0052 and Bifidobacterium longum R0175) on a lipopolysaccharide rat model. J Am Coll Nutr 38:209–217. https://doi.org/10.1080/07315724.2018.1487346

76. Xin J, Zeng D, Wang H, Sun N, Khalique A, Zhao Y, Wu L, Pan K, Jing B, Ni X (2020) Lactobacillus johnsonii BS15 improves intestinal environment against fluoride-induced

memory impairment in mice-a study based on the gut-brain axis hypothesis. PeerJ 8:e10125. https://doi.org/10.7717/peerj.10125/supp-1

77. Shahbazi R, Yasavoli-Sharahi H, Alsadi N, Ismail N, Matar C (2020) Probiotics in treatment of viral respiratory infections and neuroinflammatory disorders. Molecules 25:4891. https://doi.org/10.3390/molecules25214891

78. Hsieh TH, Kuo CW, Hsieh KH, Shieh MJ, Peng CW, Chen YC, Chang YL, Huang YZ, Chen CC, Chang PK, Chen KY (2020) Probiotics alleviate the progressive deterioration of motor functions in a mouse model of Parkinson's disease. Brain Sci 10:206. https://doi.org/10.3390/brainsci10040206

79. Xie C, Prasad AA (2020) Probiotics treatment improves hippocampal dependent cognition in a rodent model of Parkinson's disease. Microorganisms 8:1661. https://doi.org/10.3390/microorganisms8111661

80. Distrutti E, O'Reilly JA, McDonald C, Cipriani S, Renga B, Lynch MA, Fiorucci S (2014) Modulation of intestinal microbiota by the probiotic VSL# 3 resets brain gene expression and ameliorates the age-related deficit in LTP. PLoS One 9:e106503. https://doi.org/10.1371/journal.pone.0106503

81. Bonfili L, Cecarini V, Berardi S, Scarpona S, Suchodolski JS, Nasuti C, Fiorini D, Boarelli MC, Rossi G, Eleuteri AM (2017) Microbiota modulation counteracts Alzheimer's disease progression influencing neuronal proteolysis and gut hormones plasma levels. Sci Rep 7:1–21. https://doi.org/10.1038/s41598-017-02587-2

82. Bonfili L, Cecarini V, Cuccioloni M, Angeletti M, Berardi S, Scarpona S, Rossi G, Eleuteri AM (2018) SLAB51 probiotic formulation activates SIRT1 pathway promoting antioxidant and neuroprotective effects in an AD mouse model. Mol Neurobiol 55:7987–8000. https://doi.org/10.1007/s12035-018-0973-4

83. Ueda A, Shinkai S, Shiroma H, Taniguchi Y, Tsuchida S, Kariya T, Kawahara T, Kobayashi Y, Kohda N, Ushida K, Kitamura A (2021) Identification of Faecalibacterium prausnitzii strains for gut microbiome-based intervention in Alzheimer's-type dementia. Cell Rep Med 2:100398. https://doi.org/10.1016/j.xcrm.2021.100398

84. Blacher E, Bashiardes S, Shapiro H, Rothschild D, Mor U, Dori-Bachash M, Kleimeyer C, Moresi C, Harnik Y, Zur M, Zabari M (2019) Potential roles of gut microbiome and metabolites in modulating ALS in mice. Nature 572:474–480. https://doi.org/10.1038/s41586-019-1443-5

85. Moubareck CA (2021) Human milk microbiota and oligosaccharides: a glimpse into benefits, diversity, and correlations. Nutrients 13:1123. https://doi.org/10.3390/nu13041123

86. Indrio F, Di Mauro A, Riezzo G, Civardi E, Intini C, Corvaglia L, Ballardini E, Bisceglia M, Cinquetti M, Brazzoduro E, Del Vecchio A (2014) Prophylactic use of a probiotic in the prevention of colic, regurgitation, and functional constipation: a randomized clinical trial. JAMA Pediatr 168:228–233. https://doi.org/10.1001/jamapediatrics.2013.4367

87. Chau K, Lau E, Greenberg S, Jacobson S, Yazdani-Brojeni P, Verma N, Koren G (2015) Probiotics for infantile colic: a randomized, double-blind, placebo-controlled trial investigating lactobacillus reuteri DSM 17938. J Pediatr 166:74–78. https://doi.org/10.1016/j.jpeds.2014.09.020

88. Savino F, Cordisco L, Tarasco V, Palumeri E, Calabrese R, Oggero R, Roos S, Matteuzzi D (2010) Lactobacillus reuteri DSM 17938 in infantile colic: a randomized, double-blind, placebo-controlled trial. Pediatrics 126:e526–e533. https://doi.org/10.1542/peds.2010-0433

89. Savino F, Fornasero S, Ceratto S, De Marco A, Mandras N, Roana J, Tullio V, Amisano G (2015) Probiotics and gut health in infants: a preliminary case–control observational study about early treatment with lactobacillus reuteri DSM 17938. Clin Chim Acta 451:82–87. https://doi.org/10.1016/j.cca.2015.02.027

90. Aloisio I, Prodam F, Giglione E, Bozzi Cionci N, Solito A, Bellone S, Baffoni L, Mogna L, Pane M, Bona G, Di Gioia D (2018) Three-month feeding integration with bifidobacterium strains prevents gastrointestinal symptoms in healthy newborns. Front Nutr 5:39. https://doi.org/10.3389/fnut.2018.00039

91. Oliva S, Di Nardo G, Ferrari F, Mallardo S, Rossi P, Patrizi G, Cucchiara S, Stronati L (2012) Randomised clinical trial: the effectiveness of lactobacillus reuteri ATCC 55730 rectal enema in children with active distal ulcerative colitis. Aliment Pharmacol Ther 35:327–334. https://doi.org/10.1111/j.1365-2036.2011.04939.x

92. Grandy G, Medina M, Soria R, Terán CG, Araya M (2010) Probiotics in the treatment of acute rotavirus diarrhoea. A randomized, double-blind, controlled trial using two different probiotic preparations in Bolivian children. BMC Infect Dis 10:1–7. https://doi.org/10.1186/1471-2334-10-253

93. Vandenplas Y, De Hert SG, PROBIOTICAL-Study Group (2011) Randomised clinical trial: the synbiotic food supplement Probiotical vs. placebo for acute gastroenteritis in children. Aliment Pharmacol Ther 34:862–867. https://doi.org/10.1111/j.1365-2036.2011.04835.x

94. Khailova L, Dvorak K, Arganbright KM, Halpern MD, Kinouchi T, Yajima M, Dvorak B (2009) Bifidobacterium bifidum improves intestinal integrity in a rat model of necrotizing enterocolitis. Am J Physiol Gastrointest Liver Physiol 297:G940–G949. https://doi.org/10.1152/ajpgi.00141.2009

95. Underwood MA, Kananurak A, Coursodon CF, Adkins-Reick CK, Chu H, Bennett SH, Wehkamp J, Castillo PA, Leonard BC, Tancredi DJ, Sherman MP, Dvorak B, Bevins CL (2012) Bifidobacterium bifidum in a rat model of necrotizing enterocolitis: antimicrobial peptide and protein responses. Pediatr Res 71:546. https://doi.org/10.1038/pr.2012.11

96. Kitajima H, Sumida Y, Tanaka R, Yuki N, Takayama H, Fujimura M (1997) Early administration of Bifidobacterium breve to preterm infants: randomised controlled trial. Arch Dis Child Fetal Neonatal Ed 76:F101–F107. https://doi.org/10.1136/fn.76.2.F101

97. Li Y, Shimizu T, Hosaka A, Kaneko N, Ohtsuka Y, Yamashiro Y (2004) Effects of Bifidobacterium breve supplementation on intestinal flora of low birth weight infants. Pediatr Int 46:509–515. https://doi.org/10.1111/j.1442-200x.2004.01953.x

98. Scaldaferri F, Gerardi V, Lopetuso LR, Del Zompo F, Mangiola F, Boškoski I, Bruno G, Petito V, Laterza L, Cammarota G, Gaetani E (2013) Gut microbial flora, prebiotics, and probiotics in IBD: their current usage and utility. Biomed Res Int. https://doi.org/10.1155/2013/435268

99. Bibiloni R, Fedorak RN, Tannock GW, Madsen KL, Gionchetti P, Campieri M, De Simone C, Sartor RB (2005) VSL# 3 probiotic-mixture induces remission in patients with active ulcerative colitis. Am J Gastroenterol 100:1539–1546

100. Miele E, Pascarella F, Giannetti E, Quaglietta L, Baldassano RN, Staiano A (2009) Effect of a probiotic preparation (VSL# 3) on induction and maintenance of remission in children with ulcerative colitis. Am J Gastroenterol 104:437–443

101. Veerappan GR, Betteridge J, Young PE (2012) Probiotics for the treatment of inflammatory bowel disease. Curr Gastroenterol Rep 14:324–333. https://doi.org/10.1007/s11894-012-0265-5

102. Floch MH, Walker WA, Madsen K, Sanders ME, Macfarlane GT, Flint HJ, Dieleman LA, Ringel Y, Guandalini S, Kelly CP, Brandt LJ (2011) Recommendations for probiotic use – 2011 update. J Clin Gastroenterol 45:S168–S171. https://doi.org/10.1097/MCG.0b013e318230928b

103. McFarland LV, Dublin S (2008) Meta-analysis of probiotics for the treatment of irritable bowel syndrome. World J Gastroenterol 14:2650. https://doi.org/10.3748/wjg.14.2650

104. Saggioro A (2004) Probiotics in the treatment of irritable bowel syndrome. J Clin Gastroenterol 38:S104–S106. https://doi.org/10.1097/01.mcg.0000129271.98814.e2

105. Kim HJ, Camilleri M, McKinzie S, Lempke MB, Burton DD, Thomforde GM, Zinsmeister AR (2003) A randomized controlled trial of a probiotic, VSL# 3, on gut transit and symptoms in diarrhoea-predominant irritable bowel syndrome. Aliment Pharmacol Ther 17:895–904. https://doi.org/10.1046/j.1365-2036.2003.01543.x

106. Smecuol E, Hwang HJ, Sugai E, Corso L, Chernavsky AC, Bellavite FP et al (2013) Exploratory, randomized, double-blind, placebo-controlled study on the effects of Bifidobacterium infantis natren life start strain super strain in active celiac disease. J Clin Gastroenterol 47:139–147. https://doi.org/10.1097/MCG.0b013e31827759ac

107. Pinto-Sánchez MI, Smecuol EC, Temprano MP, Sugai E, González A, Moreno ML, Huang X, Bercik P, Cabanne A, Vázquez H, Niveloni S (2017) Bifidobacterium infantis NLS super strain reduces the expression of α-defensin-5, a marker of innate immunity, in the mucosa of active celiac disease patients. J Clin Gastroenterol 51:814–817. https://doi.org/10.1097/MCG.0000000000000687
108. Francavilla R, Piccolo M, Francavilla A, Polimeno L, Semeraro F, Cristofori F, Castellaneta S, Barone M, Indrio F, Gobbetti M, De Angelis M (2019) Clinical and microbiological effect of a multispecies probiotic supplementation in celiac patients with persistent IBS-type symptoms: a randomized, double-blind, placebo-controlled, multicenter trial. J Clin Gastroenterol 53:e117. https://doi.org/10.1097/MCG.0000000000001023
109. Quagliariello A, Aloisio I, Bozzi Cionci N, Luiselli D, D'Auria G, Martinez-Priego L, Pérez-Villarroya D, Langerholc T, Primec M, Mičetić-Turk D, Di Gioia D (2016) Effect of Bifidobacterium breve on the intestinal microbiota of coeliac children on a gluten free diet: a pilot study. Nutrients 8:660. https://doi.org/10.3390/nu8100660
110. Klemenak M, Dolinšek J, Langerholc T, Di Gioia D, Mičetić-Turk D (2015) Administration of Bifidobacterium breve decreases the production of TNF-α in children with celiac disease. Dig Dis Sci 60:3386–3392. https://doi.org/10.1007/s10620-015-3769-7
111. Primec M, Klemenak M, Di Gioia D, Aloisio I, Cionci NB, Quagliariello A, Gorenjak M, Mičetić-Turk D, Langerholc T (2019) Clinical intervention using Bifidobacterium strains in celiac disease children reveals novel microbial modulators of TNF-α and short-chain fatty acids. Clin Nutr 38:1373–1381. https://doi.org/10.1016/j.clnu.2018.06.931
112. Minami JI, Kondo S, Yanagisawa N, Odamaki T, Xiao JZ, Abe F, Nakajima S, Hamamoto Y, Saitoh S, Shimoda T (2015) Oral administration of Bifidobacterium breve B-3 modifies metabolic functions in adults with obese tendencies in a randomised controlled trial. J Nutr Sci 4. https://doi.org/10.1017/jns.2015.5
113. Kadooka Y, Sato M, Imaizumi K, Ogawa A, Ikuyama K, Akai Y, Okano M, Kagoshima M, Tsuchida T (2010) Regulation of abdominal adiposity by probiotics (lactobacillus gasseri SBT2055) in adults with obese tendencies in a randomized controlled trial. Eur J Clin Nutr 64:636–643. https://doi.org/10.1038/ejcn.2010.19
114. Rajkumar H, Mahmood N, Kumar M, Varikuti SR, Challa HR, Myakala SP (2014) Effect of probiotic (VSL# 3) and omega-3 on lipid profile, insulin sensitivity, inflammatory markers, and gut colonization in overweight adults: a randomized, controlled trial. Mediators Inflamm. https://doi.org/10.1155/2014/348959
115. Asemi Z, Zare Z, Shakeri H, Sabihi SS, Esmaillzadeh A (2013) Effect of multispecies probiotic supplements on metabolic profiles, hs-CRP, and oxidative stress in patients with type 2 diabetes. Ann Nutr Metab 63:1–9. https://doi.org/10.1159/000349922
116. Ejtahed HS, Mohtadi-Nia J, Homayouni-Rad A, Niafar M, Asghari-Jafarabadi M, Mofid V, Akbarian-Moghari A (2011) Effect of probiotic yogurt containing lactobacillus acidophilus and Bifidobacterium lactis on lipid profile in individuals with type 2 diabetes mellitus. J Dairy Sci 94:3288–3294. https://doi.org/10.3168/jds.2010-4128
117. Kobyliak N, Falalyeyeva T, Mykhalchyshyn G, Kyriienko D, Komissarenko I (2018) Effect of alive probiotic on insulin resistance in type 2 diabetes patients: randomized clinical trial. Diabetes Metab Syndr 12:617–624. https://doi.org/10.1016/j.dsx.2018.04.015
118. Kobyliak N, Conte C, Cammarota G, Haley AP, Styriak I, Gaspar L, Fusek J, Rodrigo L, Kruzliak P (2016) Probiotics in prevention and treatment of obesity: a critical view. Nutr Metab 13:1–3. https://doi.org/10.1186/s12986-016-0067-0
119. He M, Shi B (2017) Gut microbiota as a potential target of metabolic syndrome: the role of probiotics and prebiotics. Cell Biosci 7:1–4. https://doi.org/10.1186/s13578-017-0183-1
120. Solito A, Bozzi Cionci N, Calgaro M, Caputo M, Vannini L, Hasballa I, Archero F, Giglione E, Ricotti R, Walker GE, Petri A, Agosti E, Bellomo G, Aimaretti G, Bona G, Bellone S, Amoruso A, Pane M, Di Gioia D, Vitulo N, Prodam F (2021) Supplementation with Bifidobacterium breve BR03 and B632 strains improved insulin sensitivity in children and adolescents with obesity in a cross-over, randomized double-blind placebo-controlled trial. Clin Nutr 40:4585–4594. https://doi.org/10.1016/j.clnu.2021.06.002

121. Rao AV, Bested AC, Beaulne TM, Katzman MA, Iorio C, Berardi JM, Logan AC (2009) A randomized, double-blind, placebo-controlled pilot study of a probiotic in emotional symptoms of chronic fatigue syndrome. Gut Pathog 1:1–6. https://doi.org/10.1186/1757-4749-1-6

122. Nakakita Y, Tsuchimoto N, Takata Y, Nakamura T (2016) Effect of dietary heat-killed lactobacillus brevis SBC8803 (SBL88™) on sleep: a non-randomised, double blind, placebo-controlled, and crossover pilot study. Benefic Microbes 7:501–509. https://doi.org/10.3920/BM2015.0118

123. Akbari E, Asemi Z, Daneshvar Kakhaki R, Bahmani F, Kouchaki E, Tamtaji OR, Hamidi GA, Salami M (2016) Effect of probiotic supplementation on cognitive function and metabolic status in Alzheimer's disease: a randomized, double-blind and controlled trial. Front Aging Neurosci 10:256. https://doi.org/10.3389/fnagi.2016.00256

124. Kouchaki E, Tamtaji OR, Salami M, Bahmani F, Kakhaki RD, Akbari E, Tajabadi-Ebrahimi M, Jafari P, Asemi Z (2017) Clinical and metabolic response to probiotic supplementation in patients with multiple sclerosis: a randomized, double-blind, placebo-controlled trial. Clin Nutr 36:1245–1249. https://doi.org/10.1016/j.clnu.2016.08.015

125. Dinan TG, Stanton C, Cryan JF (2013) Psychobiotics: a novel class of psychotropic. Biol Psychiatry 74:720–726. https://doi.org/10.1016/j.biopsych.2013.05.001

126. Akkasheh G, Kashani-Poor Z, Tajabadi-Ebrahimi M, Jafari P, Akbari H, Taghizadeh M, Memarzadeh MR, Asemi Z, Esmaillzadeh A (2016) Clinical and metabolic response to probiotic administration in patients with major depressive disorder: a randomized, double-blind, placebo-controlled trial. Nutrition 32:315–320. https://doi.org/10.1016/j.nut.2015.09.003

127. Slykerman RF, Hood F, Wickens K, Thompson JM, Barthow C, Murphy R, Kang J, Rowden J, Stone P, Crane J, Stanley T (2017) Effect of lactobacillus rhamnosus HN001 in pregnancy on postpartum symptoms of depression and anxiety: a randomised double-blind placebo-controlled trial. EBioMedicine 24:159–165. https://doi.org/10.1016/j.ebiom.2017.09.013

128. Mazzini L, Mogna L, De Marchi F, Amoruso A, Pane M, Aloisio I, Bozzi Cionci N, Gaggìa F, Lucenti A, Bersano E, Cantello R (2018) Potential role of gut microbiota in ALS pathogenesis and possible novel therapeutic strategies. J Clin Gastroenterol 52:S68–S70. https://doi.org/10.1097/MCG.0000000000001042

129. Di Gioia D, Bozzi Cionci N, Baffoni L, Amoruso A, Pane M, Mogna L, Gaggìa F, Lucenti MA, Bersano E, Cantello R, De Marchi F (2020) A prospective longitudinal study on the microbiota composition in amyotrophic lateral sclerosis. BMC Med 18:1–9. https://doi.org/10.1186/s12916-020-01607-9

130. De Marchi F, Collo A, Scognamiglio A, Cavaletto M, Bozzi Cionci N, Biroli G, Di Gioia D, Riso S, Mazzini L (2022) Study protocol on the safety and feasibility of a normocaloric ketogenic diet in people with amyotrophic lateral sclerosis. Nutrition 94:111525. https://doi.org/10.1016/j.nut.2021.111525

131. Collado MC, Gueimonde M, Salminen S (2010) Probiotics in adhesion of pathogens: mechanisms of action. In: Bioactive foods in promoting health. Academic, pp 353–370. https://doi.org/10.1016/B978-0-12-374938-3.00023-2

132. Kumar M, Nagpal R, Verma V, Kumar A, Kaur N, Hemalatha R, Gautam SK, Singh B (2013) Probiotic metabolites as epigenetic targets in the prevention of colon cancer. Nutr Rev 71:23–34. https://doi.org/10.1111/j.1753-4887.2012.00542.x

133. Ohtsuka Y, Ikegami T, Izumi H, Namura M, Ikeda T, Ikuse T, Baba Y, Kudo T, Suzuki R, Shimizu T (2012) Effects of Bifidobacterium breve on inflammatory gene expression in neonatal and weaning rat intestine. Pediatr Res 71:46–53. https://doi.org/10.1038/pr.2011.11

134. Tien MT, Girardin SE, Regnault B, Le Bourhis L, Dillies MA, Coppée JY, Bourdet-Sicard R, Sansonetti PJ, Pédron T (2016) Anti-inflammatory effect of Lactobacillus casei on Shigella-infected human intestinal epithelial cells. J Immunol 176:1228–1237. https://doi.org/10.4049/jimmunol.176.2.1228

135. Zhao L, Zhang F, Ding X, Wu G, Lam YY, Wang X, Fu H, Xue X, Lu C, Ma J, Yu L (2018) Gut bacteria selectively promoted by dietary fibers alleviate type 2 diabetes. Science 359:1151–1156. https://doi.org/10.1126/science.aao5774

136. Depommier C, Everard A, Druart C, Plovier H, Van Hul M, Vieira-Silva S, Falony G, Raes J, Maiter D, Delzenne NM, de Barsy M (2019) Supplementation with Akkermansia muciniphila in overweight and obese human volunteers: a proof-of-concept exploratory study. Nat Med 25:1096–1103. https://doi.org/10.1038/s41591-019-0495-2

137. Sowah SA, Hirche F, Milanese A, Johnson TS, Grafetstätter M, Schübel R, Kirsten R, Ulrich CM, Kaaks R, Zeller G, Kühn T (2020) Changes in plasma short-chain fatty acid levels after dietary weight loss among overweight and obese adults over 50 weeks. Nutrients 12:452. https://doi.org/10.3390/nu12020452

138. Pavlović N, Stankov K, Mikov M (2012) Probiotics – interactions with bile acids and impact on cholesterol metabolism. Appl Biochem Biotechnol 168:1880–1895. https://doi.org/10.1007/s12010-012-9904-4

139. Kumar R, Grover S, Batish VK (2011) Hypocholesterolaemic effect of dietary inclusion of two putative probiotic bile salt hydrolase-producing Lactobacillus plantarum strains in Sprague–Dawley rats. Br J Nutr 105:561–573. https://doi.org/10.1017/S0007114510003740

140. Food and Agriculture Organization/World Health Organization (2001) Health and nutritional properties of probiotics in food including powder milk with live lactic acid bacteria. American Cordoba Park Hotel, Cordoba, Argentina. FAO/WHO, Rome, pp 1–2

141. Xu M, Wang J, Wang N, Sun F, Wang L, Liu X (2015) The efficacy and safety of the probiotic bacterium *Lactobacillus reuteri* DSM 17938 for infantile colic: a meta-analysis of randomized controlled trials. PLoS One 10:1–16. https://doi.org/10.1371/journal.pone.0141445

142. Baldassarre ME, Di Mauro A, Mastromarino P, Fanelli M, Martinelli D, Urbano F, Capobianco D, Laforgia N (2016) Administration of a multi-strain probiotic product to women in the perinatal period differentially affects the breast milk cytokine profile and may have beneficial effects on neonatal gastrointestinal functional symptoms. A randomized clinical trial. Nutrients 8:677. https://doi.org/10.3390/nu8110677

143. Kuhle S, Tong OS, Woolcott CG (2015) Association between caesarean section and childhood obesity: a systematic review and meta-analysis. Obes Rev 16:295–303. https://doi.org/10.1111/obr.12267

144. Magne F, Puchi Silva A, Carvajal B, Gotteland M (2017) The elevated rate of cesarean section and its contribution to non-communicable chronic diseases in Latin America: the growing involvement of the microbiota. Front Pediatr 5:192. https://doi.org/10.3389/fped.2017.00192

145. Plaza-Diaz J, Ruiz-Ojeda FJ, Gil-Campos M, Gil A (2019) Mechanisms of action of probiotics. Adv Nutr 10:S49–S66. https://doi.org/10.1093/advances/nmy063

146. Rao SD, Banack SA, Cox PA, Weiss JH (2006) BMAA selectively injures motor neurons via AMPA/kainate receptor activation. Exp Neurol 201:244–252. https://doi.org/10.1016/j.expneurol.2006.04.017

147. Buenz EJ, Howe CL (2007) Beta-methylamino-alanine (BMAA) injures hippocampal neurons in vivo. Neurotoxicology 28:702–704. https://doi.org/10.1016/j.neuro.2007.02.010

148. Sommer F, Anderson JM, Bharti R, Raes J, Rosenstiel P (2017) The resilience of the intestinal microbiota influences health and disease. Nat Rev Microbiol 15:630–638. https://doi.org/10.1038/nrmicro.2017.58

149. Veiga P, Suez J, Derrien M, Elinav E (2020) Moving from probiotics to precision probiotics. Nat Microbiol 5:878–880. https://doi.org/10.1038/s41564-020-0721-1

150. McFarland LV, Evans CT, Goldstein EJ (2018) Strain-specificity and disease-specificity of probiotic efficacy: a systematic review and meta-analysis. Front Med 5:124. https://doi.org/10.3389/fmed.2018.00124

151. Deidda F, Cionci NB, Cordovana M, Campedelli I, Fracchetti F, Di Gioia D, Ambretti S, Pane M (2021) Bifidobacteria strain typing by Fourier transform infrared spectroscopy. Front Microbiol 12:692975. https://doi.org/10.3389/fmicb.2021.692975

152. Jackson SA, Schoeni JL, Vegge C, Pane M, Stahl B, Bradley M, Goldman VS, Burguière P, Atwater JB, Sanders ME (2019) Improving end-user trust in the quality of commercial probiotic products. Front Microbiol 10:739. https://doi.org/10.3389/fmicb.2019.00739

Chapter 9
Microbial Therapy with Indigenous Bacteria: From Idea to Clinical Evidence

Elena Ermolenko, Irina Koroleva, and Alexander Suvorov

Abstract Recent studies clearly demonstrated the role of indigenous microbiota in the development and sustaining of the host metabolic activity and immune status. It was also shown that in dysbiotic condition pathologically assembled microbiota cause the development of the chronic low-grade inflammation in different loci of microbial colorization including the gut, skin or the oral cavity. Present chapter describes an approach based on using indigenous bacterial strains – autoprobiotics for the treatment of dysbiosis and provides the arguments in favor of clinical usage of autoprobiotics for several clinical conditions. Data, presented in this chapter summarize the clinical experience of the personalized microbial therapy including irritated bowel syndrome, metabolic syndrome, *Helicobacter pylori* infection, colorectal cancer, Parkinson's disease, and chronic periodontitis.

Keywords Autoprobiotic · Microbial therapy · Microbiota · Microbiome · Idigenous bacteria

Abbreviations

AE	Autoprobiotic enterococcus
CGP	Chronic generalized periodontitis
CRC	Colorectal cancer
FMT	Fecal microbiota transplantation
H.p.	Helicobacter pylori
HDL	High density lipids
HV	Healthy volunteers
IBS	Irritable bowel syndrome
IL	Interleukins

E. Ermolenko · I. Koroleva · A. Suvorov (✉)
Institute of Experimental Medicine, Saint-Petersburg, Russia

© The Author(s), under exclusive license to Springer Nature Switzerland AG 2023
N. Boyko, O. Golubnitschaja (eds.), *Microbiome in 3P Medicine Strategies*,
Advances in Predictive, Preventive and Personalised Medicine 16,
https://doi.org/10.1007/978-3-031-19564-8_9

LDL   Low density lipids
MS   Metabolic syndrome
NGS   New generation sequencing
PD   Parkinson's disease

9.1 Introduction

Novel technologies including New Generation Sequencing (NGS), metagenome analysis, transcriptomics and metabolomics clearly revealed the fundamental role of microorganisms inhabiting any multicellular organism in all physiological functions including physical activity or food consumption. Present concept of any living organism as a holobiont with hologenome describes any organism, including human being, as a balanced consortium of the multicellular organism together with bacteria, fungi, protozoa and viruses [47]. Multicellular organisms exist as meta-organisms comprised of both the macroscopic host and its symbiotic commensal microbiota. With an estimated composition of 100 trillion cells, microbes colonizing humans outnumber host cells and express more unique genes than their host's genome [30].

The rules by which this microbial consortium is functioning are not completely understood by the present scientific knowledge, however it is easy to assume that microbiota inhabits us not coincidently but reflects our genomic features, style of life and dietarian habits. For example, human gut provides nutrients and a breeding environment for intestinal microbiota which in turn, assists in carbohydrate and protein fermentation, synthesize vitamins and reduce intestinal permeability of the epithelial and mucosal barrier [3]. Commensals also play a fundamental role in both the training of the immune system and its functional tuning, thereby acting as adjuvants to the immune system as a whole [2].

Considering the numbers of microorganisms (counting only bacteria) in the gut exceeding number of the cells in human body, the impact of microbiota on the host immune system seems very important. Host immunity and microbiota turned to be tightly bound and both adaptive and mucosal immunity maturate under the influence of microenvironment [21].

Each bacterial or fungi strain may express more than the hundred proteins just on the microbial surface. To note, each of these proteins carry several immunodominant epitopes responsible for the appearance of the specific pool of the antibodies. Taken together just in theory 10^{15} gut bacteria might generate incredible number of specific antibodies directed to these epitopes which is very costly energetically. The only logical explanation for this biological puzzle is provided by hygiene theory which is stating that the balance between the immune system and microbiota is guarded by the mechanisms of immune tolerance [35]. Indeed, it is well known that maturation of the human immune system is a time-consuming process taking about 3 first years of life. During this time period the child need to interact with maximum number of different antigens provided by the environment together with significant pool of internally located microorganisms. Many clinical studies including groups of people with allergic diseases shows the differences in the frequency of allergies in the countries with different sanitary standards or eating habits [58, 67]. Children born by caesarian section deprived from maternal bacteria aspirated during birth also often experience allergies or some neurological conditions in the future [20, 24].

The balance between the host and microbiota seems to be fairly stable in the healthy condition and highly individual because of the differences in the genomes, diet and life style. Contemporary bioinformatic analysis of metagenomic data allows distinguishing all the human microbiota into three or more stable consortiums – enterotypes which are able to sustain biomolecular and immunological status of the holobiont which is necessary for the healthy condition of the organism [11].

Taken together metagenomic and metabolomic data analyzed bioinformatically we can assume that each individual selects a set of bacterial, fungal or may be viral strains involved in the complex metabolic and (or) signaling interactions which we even can't completely comprehend at present level of scientific knowledge. However, even without understanding how the Swiss watch is working we unarguably believe that all little things in this time measuring devise are needed for its proper functioning. This analogy describes the malfunctioning of microbiota losing the original microbial composition in case of dysbiosis.

Intestinal dysbiosis, which is often associated with low-grade inflammation and caused by the various factors including antibiotics, infection, chemotherapy or stress condition, is responsible for the various health problems including gastrointestinal, immunopathological, neuronal, cancer and cardio-vascular diseases. In this respect restoration of host-microbial balance by using microorganisms for the therapy seems as a reasonable solution. However, the present therapeutic approaches for restoration of the microbiota are hardly became canonical.

Usage of probiotics has been already shown as powerful preventive and therapeutic intervention against inflammation aiming to maintain intestinal homeostasis [1, 5, 22, 29, 59, 62]. What is important – probiotics may significantly the immune and metabolic functioning of microbiome [6, 44].

However, probiotics as factors aimed for improving microbiota composition, limited by numerous factors including variety of the health conditions and immune statuses of the patients, and the original microbiome composition of the recipient of probiotic [51]. Another factor is inability of the probiotic strains to proliferate in the host [15, 17, 19]. Several side effects after using probiotics in clinical practice had been monitored: dyspeptic symptoms, acidosis, induction of dysbiotic condition and even the development of infectious pathology in immune compromised patients [26].

Microbiota composition can be changed by the diet, by pharmaceutical agents, by taking probiotics or even by the fecal microbiota transplantation (FMT). None of these approaches can be considered personified therapy and can't be applied universally. Recently we have developed a method of using indigenous microorganisms for the treatment of dysbiosis named autoprobiotics. Present paper is mainly devoted to the clinical applications of autoprobiotic technology performed on patients with different pathologies [66].

9.2   The Concept and Methodology of Making Autoprobiotics

The idea of usage of personal bacteria for microbial manipulation which was originally suggested by Boris Shenderov is based on the assumption that the innate immune system recognizes individual microbiota by the mechanisms of immune tolerance which makes it safe to use personal bacteria as probiotics [45]. Technology of making autoprobiotics was later modified by Alexander Suvorov and Vladimir Simanenkov who included into procedure genetic analysis and selection of the bacterial strains grown on the artificial media and applied for the organism as probiotic food [53].

Autoprobiotics are the indigenous representatives of the normal microbiota of host organism, isolated from the organism and orally consumed by himself after cultivation in the same concentrations as probiotics. Their indisputable advantage is immunological tolerance and compatibility with the microbiota and the metabolic status of the host, allowing autoprobiotics to persist in the body for a long time [18, 50].

Autoprobiotic strain is obtained after bacteriological selection between various bacterial isolates (lactobacilli, enterococci, bifidobacteria, or the others) obtained from the different loci of the body: intestinal mucosa, oral cavity, vagina, skin or the feces which was found as useful source of microbiota in case of intestinal dysbiosis. Autoprobiotics can be prepared as monocultures or as mixture of indigenous bacteria grown in anaerobe condition (so called anaerobe consortium) [54].

Requirements for the provider of autoprobiotic clones include:

1. Material is obtained from the healthy person or patient during remission, before the surgery, chemotherapy, antibiotics, radiotherapy, or expected stress conditions.
2. Preparation for sampling includes diet restrictions in fermented foods or taking commercial probiotics
3. Storage material (feces and individual clones of indigenous bacteria) in the biobank for delayed and repeated usage of personalized autoprobiotic therapy.

The pure bacterial cultures are obtaining by growing on selective nutrient media. After selection of individual bacterial colonies, they are subjected for DNA isolation and studied by polymerase chain reaction (PCR) in order to determine the type of microorganism (species) and for the presence of potential virulence factors.

For example, in order to make autoprobiotic enterococci, only *Enterococcus faecium* and *E. hirae* strains free from the following virulence and antibiotic resistance genes in the genome (*gelE, esp, sprE, fsrB, asaI, cylA, cyl M, efaA, vanA, vanB*) were used.

Non-pathogenic enterococci, lactobacilli, bifidobacteria can be used to obtain starter culture, which is introduced into the organism at a concentration of 9 lg CFU per day. Alternatively, autoprobiotics can be used in the form of freeze-dried bacteria, placed in capsules.

Selection of the bacterial species for autoprobiotic technology required testing number of different variants (species) applicable for the autoprobiotic technology. As to the gut autoprobiotics, we supposed, that they should be easily cultivated, present in every organism, and able to withstand high acidity in the stomach and bile in the duodenum to sustain viability in the intestines. By these criteria, for the animal study we selected lactobacilli, enterococci, bifidobacteria, and their mixture and tested the on the animals (male Wistar rats) employing the model of antibiotic induced dysbiosis according to Ermolenko et al. [14].

Indigenous enterococci, lactobacilli, or bifidobacteria were isolated from feces of rats before the animals were treated with antibiotics and were grown separately. Analyses of stool samples before and after taking autoprobiotics and identification of individual clones of bacteria (*E. faecium, Lactobacillus spp.* or *Bifidobacterium spp.*) were performed bacteriologically using selective media or by PCR with the following 16S metagenome analysis [55].

It was shown that the administration of metronidazole and ampicillin to rats *per os* was characterized by an increase in the representation of gamma *Proteobacteria* phylum, the families of Enterobacteriaceae (genus *Proteus, Enterobacter, Klebsiella, Erwinia, Enterobacter*) and decrease of *Faecalibacterium, Dorea, Dialister, Clostridium, Blautis* and *Bacteroides* genera representativeness (Fig. 9.1).

D – rats with intestinal dybiosis, Healthy – healthy animals (before induction of dysbiosis by ampicillin and metronidazole).

When correcting experimental dysbiosis with different autoprobiotics, it was shown that indigenous lactobacilli were most active against *Enterobacter spp.*,

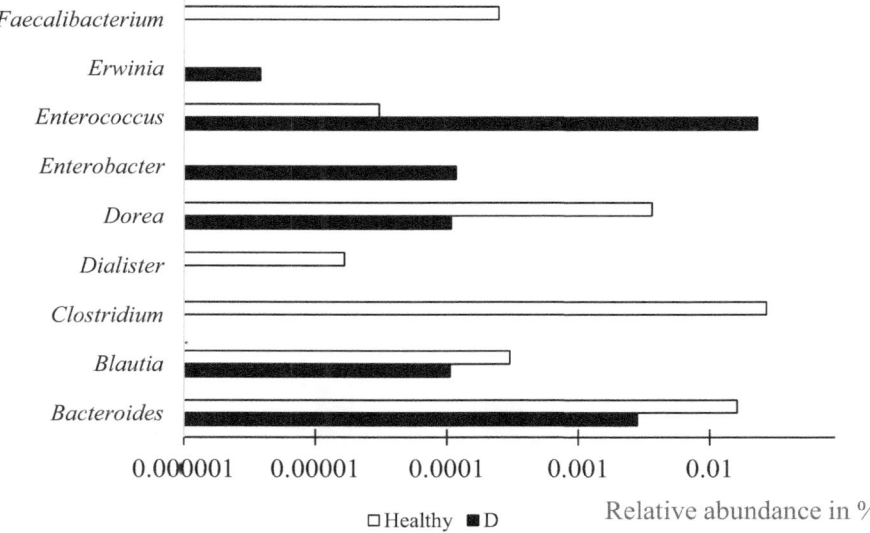

Fig. 9.1 Features of intestinal microbiome of rats with dysbiosis (4th day of experiment compared with healthy animals, relative abundance of different bacterial genera)

indigenous bifidobacteria – against *Klebsiella* spp. but the most activity against proteobacteria expressed autoprobiotic enterococci. Indigenous enterococci and bifidobacteria were most efficient in restoration of representatives of *Faecalibacterium* spp. Only autoprobiotic *Lactobacillus* spp. and *E. faecium* increased representation of *Prevotella* spp. Mixture of autoprobiotics along with a pronounced antagonistic activity against opportunistic enterobacteria: *Enterobacter, Klebsiella, Erwinia, Enterobacter*, inhibited the growth of obligate representatives of the microbiota: faecalibacteria and lactobacilli (Table 9.1). Usage of indigenous *E. faecium* was leading to an increase in the representation of *Ruminococcus* spp., *Bacteroides* spp. and *Bifidobacteria*, and decrease in the number of *Klebsiella* spp., *Staphylococcus* spp., *Streptococcus* spp. and *Lactobacillus* spp.

The level of regulatory and anti-inflammatory cytokine IL-10 in the blood serum remained unchanged in blood serum of rats from al the groups with mono-strain autoprobiotics and increased in all other experimental groups, however, after introduction of indigenous enterococci and lactobacilli we were monitoring an increase of this regulatory cytokine to a greater extent (Fig. 9.2).

Table 9.1 Changes in microbiome of rats with dysbiosis after autoprobiotic consumption[a]

Group of autoprobiotics	More than in rats with dysbiosis before therapy	Less than in rats with dysbiosis before therapy
Bifidobacteria	*Bacteroides, Blautia, Dorea, Ruminococcus, Faecalibacterium*	*Klebsiella, Proteus*
Enterococcus	*Bacteroides, Faecalibcterium, Ruminococcus, Bacteroides spp., Bifidobacterium*	*Klebsiella, lactobacillus, staphylococcus, Streptococcus*
Lactobacillus	*Bacteroides, Enterobacter, Klebsiella, Prevotella*	*Klebsiella*
Mixture of three autoprobiotics	*Bacteroides, Klebsiella*	*Klebsiella, Enterobacter, Erwinia, Faecalibacterium, Lactobacillus*

[a]The results of fecal samples metagenome study (16S rRNA and qPCR)

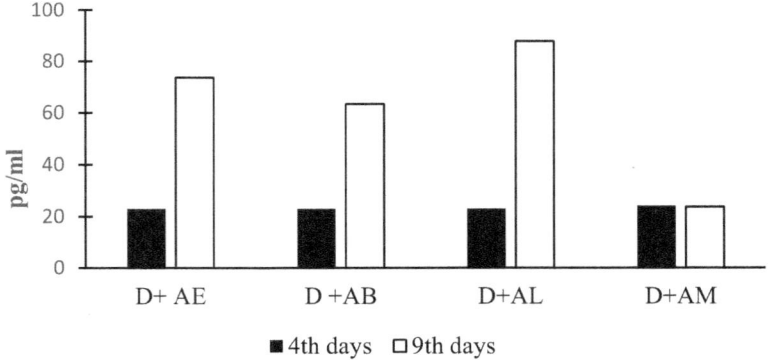

■4th days **□**9th days

Fig. 9.2 The IL-10 content in blood serum of rats with dysbiosis after autoprobiotic and their mixture consumption. AE-*Enterococcus* spp., AL-*Lactobacillus* spp., AB-*Bifidobacterium* spp. AM- mixture of three different autoprobiotic strains

Based on the results of the animal modeling we decided to focus our clinical studies on the mono-strain autoprobiotics with focus on the *Enterococcal faecium* strains without the virulence genes in the genome.

9.3 Usage of Autoprobiotic Enterococci in Medicine

Since 2004, on the basis of Institute of experimental medicine and clinic and outpatient clinics of Saint- Petersburg, clinical studies of autoprobiotics have been carried out on healthy volunteers and patients with irritable bowel syndrome (IBS), Parkinson's disease, metabolic syndrome, gastritis associated with *Helicobacter pylori* infection, surgical intervention in the treatment of colorectal molecular microbiologists, geneticists, immunologists, gastroenterologists, endocrinologists, oncologists and psychologists a comprehensive study of patients is being carried out, the results of which are currently being actively supplemented and subjected to statistical processing and comprehension. A large number of studies are associated with the clinical use of enterococci, which have been selected in experimental model and proved to be biotechnological when isolated and obtaining biomass of health beneficial food for administration.

9.4 Irritable Bowel Syndrome

Irritable bowel syndrome (IBS) is a functional bowel disorder in which recurrent abdominal pain is associated with bowel movements and changes in its rhythm. The pathophysiology of this disorder remains incompletely understood, medical treatment is empirical and is usually based on targeting the predominant symptoms [23]. In many publications regarding IBS, the main cause of it is considered as "low grade inflammation" [9, 56]. Microbiota of IBS patients demonstrated an increase in the representation of phylum *Proteobacteria*, changes in the ratio of individual representatives of phylum *Bacteroidetes* and *Firmicutes*, and decrease in biodiversity of microbiota [38]. Most often, low grade inflammation in patients with IBS is characterized by increased population of enterobacteria [40]. Also, unsaturated fatty acids are produced by other representatives of the intestinal microbiota, belonging to phylums *Bacteroidetes* and *Firmicutes* [27, 33, 48, 49]. A decrease in the representation of butyrate-producing bacteria and methanogens in IBS was also demonstrated, which may indicate the role of the deficiency of these functional groups of the phylometabolic nucleus of the microbiota in the violation of the intestinal barrier and excessive gas production due to insufficient utilization of hydrogen by methanogenes [38].

Clinical studies of autoprobiotics for the patients with IBS were performed in several medical facilities of Saint-Petersburg with positive clinical outcomes [50] however the microbiota previously was not studied by metagenome analysis.

In our study directed for evaluation of the effects of autoprobiotic treatment of IBS on microbiome 22 Patients fulfilling the Rome Criteria V for IBS with diarrhea were selected for clinical trial.

Metagenome analyses of the microbiota revealed several specific features of intestinal microbiota of patients with IBS: (1) an increase in the representation of *Actinobacteria*, including *Bifidobacterium spp., Firmicutes,* including representatives of *Streptococcaceae (Streptococcus), Lachnospiraceae (Dorea), Veillonellaceae (Dialister), Proteobacteria (Enterobacteriaceae* and *Desulfovibrionaceae)* and (2) decrease in the population of *Bacteroidetes,* including representatives of the families *Prevotellacea (Prevotella* spp), *Bacteroidaceae (Bacteroides spp.). Firmicutes* belonging to the families *Clostridiaceae* and *Ruminococcaceae (Faecalibacterium* spp) (Table 9.2).

After administration of autoprobiotics based on indigenous *E.faecium* (AE) we were able to monitor substantial changes in microbiota composition which included the changes in several microbial families (Table 9.3).

Analysis of the cytokines in the blood serum revealed, that the level of pro-inflammatory cytokines (IL-6, IL-8, IL-16), in blood serum of patients with IBS after autoprobiotic therapy was decreased. At the same time IL-10 content showed no dynamics (Fig. 9.3).

These data revealed that autoprobiotic enterococci were able to decrease «low grade inflammation» in patients with IBS.

During the course of the clinical study of autoprobiotic therapy, there was an improvement in all the monitored symptoms of IBS. Pain and flatulence decreased significantly; stool frequency returned to normal. Patients with diarrheal IBS noted a normalization of stool frequency and a change in its shape from 6-5 to 3-4 according to the Bristol Stool Scale. The figure (Fig. 9.4) shows the dynamics of changes in the clinical manifestations of IBS over time.

Table 9.2 Comparative analysis significant taxa in the feces of IBS patients before and after autoprobiotic therapy (Control – healthy volunteers)[a]

Name of the taxons	Healthy volunteers[a]	Patients with IBS[a]	Changes as compared to the volunteers
Faecalibacterium	2,5	0,8	↓
Bacteroides	39,1	21,2	↓
Dorea	0,00085	0,00211	↑
Bifidobacterium	0,00086	0,01475	↑
Desulfovibrio	0,000291	0,002832	↑
Streptococcus	0,00291	0,017398	↑
Dialister	0,079716	0,11,864	↑
Blautia	2,99	1,69	Tendency↓
Paraprevotella	0,65	0,2	Tendency↓
Prevotella	6,87	2,68	Tendency↓
Ruminococcus	1,97	3,94	Tendency↑

[a]Relative abundances of different bacterial genera (metagenome study 16S rRNA) in %

Table 9.3 Changes in the intestinal microbiome of IBS patients after consumption of autoprobiotics[a]

Increase of the relative abundancy after taking autoprobiotics	Decrease of the relative abundancy after taking autoprobiotics
Faecalibacterium	*Paraprevotella*
Blautia	*Enterobacter*
Bacteroides	
Coprococcus	
Enterococcus	

[a]The results of fecal samples metagenome study (16S rRNAand qPCR)

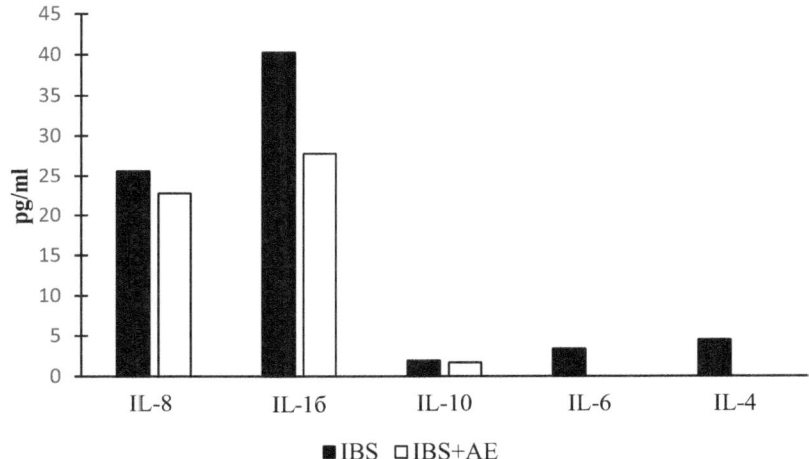

Fig. 9.3 The interleukin content in blood serum of patients with IBS before and after administration of autoprobiotic enterococcus (IBS+AE)

The graph on Fig. 9.4 shows almost complete disappearance of pain and flatulence by the 4th day of treatment. Improvement in general well-being occurred on days 5–6, reaching 2.5 points.

9.5   Parkinson's Disease

The quantitative and qualitative composition of the intestinal microbiota in PD is changed, there is an increase in the content of *Enterobacteriaceae*, a decrease in the number of *Lactobacillaceae* and *Prevotellaceae* families members just like *Faecalibacterium prausnitzii* [12, 13, 37, 42, 43].

All patients with PD had intestinal dysbiosis, which was partially corrected by taking the autoprobiotic. A bacteriological study revealed the ability of autoprobiotic enterococci to reduce the number of atypical *Escherichia coli* and

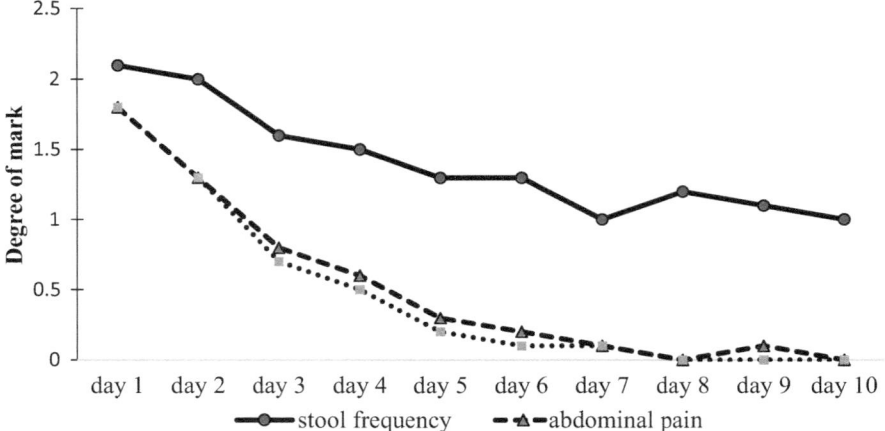

Fig. 9.4 Dynamics of the main clinical symptoms of IBS during therapy with autoprobiotics and at the end of treatment

staphylococci, as well as the tendency to crowd out the population of *Proteus spp.*, *Citrobacter spp.* and *Klebsiella. spp.* The decrease in the number of *Bacteroides fragilis, Fusobacterium spp.* and *Parvomonas spp.* and an increase in the number of bifidobacteria and fecalibacteria was demonstrated by metagenome study (16S rRNA). The peculiarities of intestinal dysbiosis in patients with PD and the possibility of its correction with autoprobiotic strain *E. faecium* were revealed (Fig. 9.5).

In our clinical study it was shown, that all patients with PD suffered from intestinal dysbiosis. That included the decrease of *Prevolella, Ruminococcus* and *Faecalibacterium* genera and an increase of *Lachnospira* spp. and opportunistic members *Enterobacteriaceae* family. Intestinal dysbiosis was partially corrected by taking the autoprobiotic enterococci during 20 days. A bacteriological and genetic study revealed the ability of autoprobiotic enterococci to reduce the number of atypical *Escherichia coli, Proteus* spp., *Citrobacter* spp. and *Klebsiella. spp., Bacteroides fragilis, Fusobacterium* spp. and *Parvomonas* spp. and to increase the number of *Bifidobacterium* spp. and *Faecalibacterium* sp. (Table 9.4).

In our clinical study it was shown, that all patients with PD suffered from intestinal dysbiosis. That included the decrease of *Prevolella, Ruminococcus* and *Faecalibacterium* genera and an increase of *Lachnospira spp.* and opportunistic members *Enterobacteriaceae*. Intestinal dysbiosis was partially corrected by taking the autoprobiotic enterococci during 20 days. A bacteriological and genetic study revealed the ability of autoprobiotic enterococci to reduce the number of atypical *Escherichia coli, Proteus* spp., *Citrobacter* spp. and *Klebsiella* spp., *Bacteroides fragilis, Fusobacterium* spp. and *Parvomonas* spp. and to increase the number of *Bifidobacterium* spp. and *Faecalibacterium* spp. (Table 9.4).

Somewhat unexpected was the increase in pro-inflammatory cytokines with the introduction of autoprobiotics in this group of patients (Fig. 9.6).

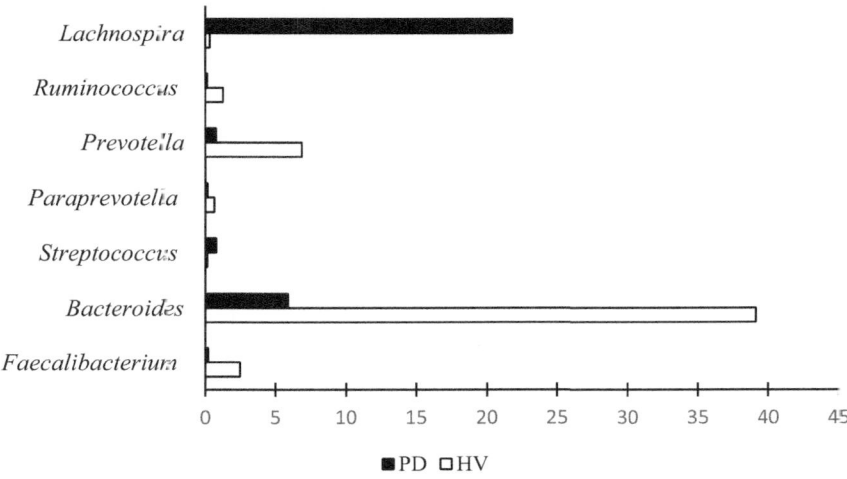

Fig. 9.5 Microbiota specific content of the patients with PD relatively to microbiota of healthy volunteers* (results of metagenome analysis 16S rRNA, in %)

Table 9.4 Changes in microbiome after autoprobiotic consumption for the therapy patients with PD[a]

Changes in microbiome after autoprobiotic consumption	More than in patients with PD before therapy	Less than in patients with PD before therapy
	Prevotella *Lachnospira* spp. *Lactobacillus* spp	*Fusobacteria* spp

[a]The results of fecal samples metagenome study (16S rRNA and qPCR)

The positive dynamics in the PD clinic were revealed. It was manifested in a decrease in neurological symptoms (Fig. 9.7) and a decrease in the severity of the constipation syndrome (Fig. 9.7). After introduction of autoprobiotic *E. faecium*, the frequency of defecation increased (from 1.8 to 2.6 times a week). The scores on the Bristol scale of the stool score increased from 2 to 4.

9.6 Metabolic Syndrome

Results of numerous preclinical and clinical studies have demonstrated the relationship between the qualitative and quantitative characteristics of the microbiota on the development of metabolic syndrome, obesity and type 2 diabetes mellitus [31, 39, 61, 64]. In the development of insulin resistance, an important role is played by an increase in the serum concentration of amino acids with a branched side chain, which is observed during the proliferation of bacteria *Prevotella copri* and

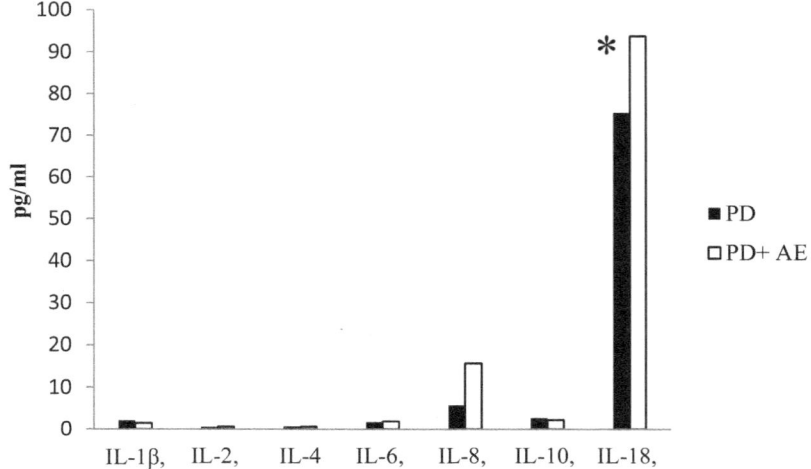

Fig. 9.6 Cytokines content before and after therapy BP with autoprobiotic enterococci

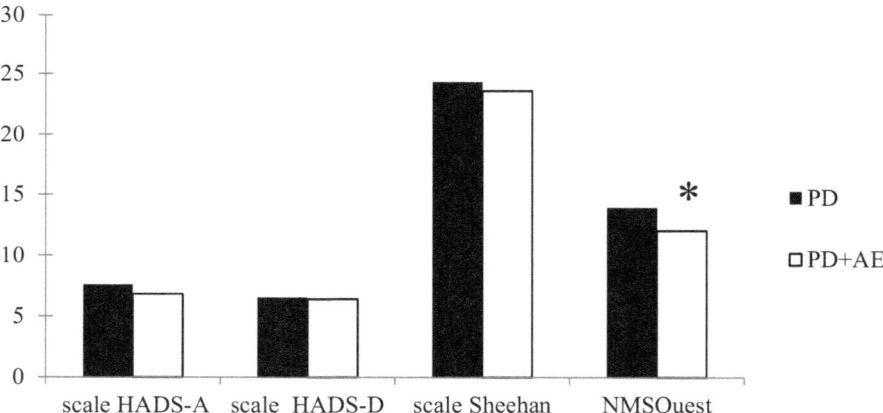

Fig. 9.7 Changes in the scale of nonmotor symptoms before and after therapy PD with autoprobiotic enterococci. *p < 0,05

Bacteroides vulgatus in the intestine [57]. In patients with prediabetes and type 2 diabetes, there is a decrease in the concentration of butyrate-producing bacteria, as well as the concentration of the *Akkermansia muciniphila* [10].

We have conducted a clinical study of patients with metabolic syndrome, the main symptom of which is a violation of carbohydrate metabolism. The intestinal microbiota was characterized by an increase in the content of *Prevotella, Roseburia, Streptococcus, Bacteroides*, and *Faecalibacterium* genera. The opportunistic enterobacteria were found in fecal samples of vast majority of the patients. After a course

Table 9.5 Changes in intestinal microbiota of patients with metabolic syndrome after autoprobiotic consumption[a]

More than in patients with MS before therapy	Less than in patients with MS before therapy
Escherichia, Prevotella Lachnospira, Lactobacillus	*Klebsiella, Acinetobacter, Staphylococcus, Streptococcus, Bacteroides fragilis, Faecalibacterium prausnitzii*

[a]The results of fecal samples metagenome study (16S rRNAand qPCR)

of autoprobiotics for 20 days, the following changes aimed at normalizing the microbiota were observed: an increase in the content of *Prevotella* spp., the disappearance or decrease the number of *Klebsiella* spp., *Acinetobacter, Streptococcus,* and *Staphylococcus* genera. At the same time, there was a decrease in butyrate-producing bacteria *Bacteroides fragilis, Faecalibacterium prausnitzii* (Table 9.5). A decrease in opportunistic enterobacteria and streptococci was also revealed.

Autoprobiotic therapy led to a significant improvement in the glycemic profile, which was expressed in a decrease in the level of glycated hemoglobin and fasting glycemia.

A trend was revealed for the beneficial effect of autoprobiotic therapy on the dynamics of the lipid spectrum: decrease of the LDL, total cholesterol, C-reactive protein and triglycerides and appropriate increase of HDL (Fig. 9.8). In addition anti-inflammatory effect of the autoprobiotics was characterized by decrease in the level of fecal calprotectin, decrease in the level of zonulin in blood serum and relative osmolarity of stool.

Immunological analysis of the blood serum for the presence of interleukins also revealed that this group of patients were characterized by low grade inflammation before autoprobiotic treatment with the autoprobiotic enterococci (Fig. 9.9).

Clinical evaluation of the patients with MC demonstrated significant improvement of the condition especially in the parameters of obesity. Autoprobiotic therapy caused a decrease in body weight, indexes of body weight and neck volume (Table 9.6).

9.7   Colorectal Cancer

Colorectal cancer (CRC) as one of the most common and severe types of cancer are is often affiliated with the changes in intestinal microbiome. Numerous bacterial species such as *Bacteroides fragilis, Fusobacterium nucleatum,* and *Peptostreptococcus stomatis* have been shown to be increased in case of colorectal cancer [65]. In most of studies CRC correlated with the decrease in biodiversity and with an increase in *Fusobacterium, Peptostreptococcus, Bacteroides, Eubacterium, Proteobacteria, Prevotella* and *Clostridium* spp. [25, 34, 60, 63].

In our study which included autoprobiotic therapy patients with colorectal cancer after the surgery were characterized by the presence of marker oncogenic

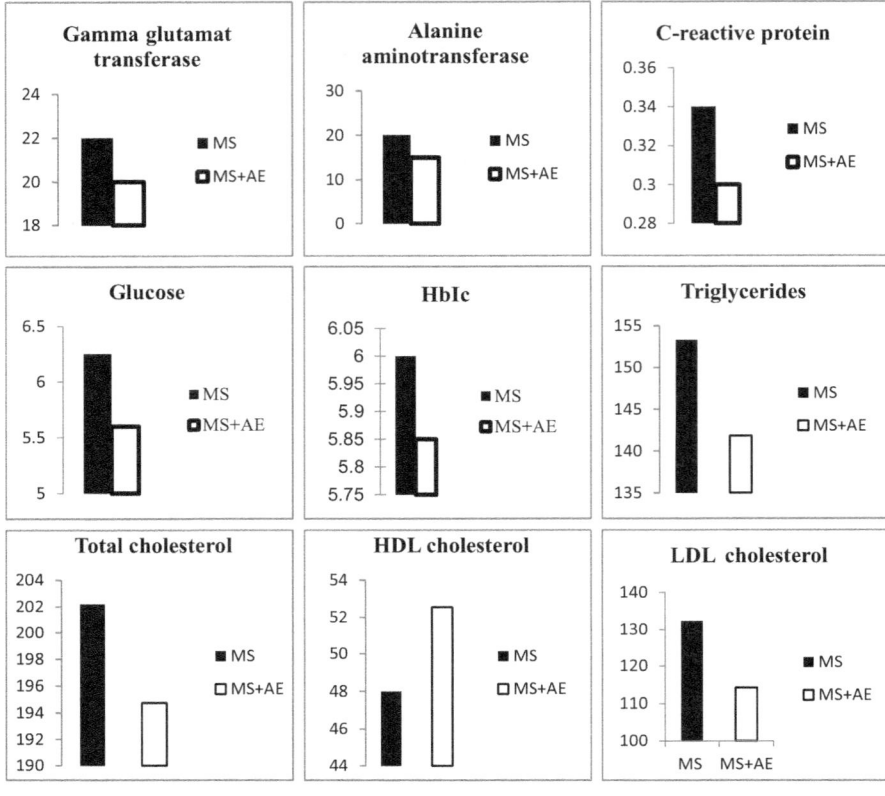

Fig. 9.8 Changes in biochemical parameters of blood serum of patients with MS after autoprobiotic therapy. Results of the biochemical analyses are presented in mg per deciliter

bacteria *Fusobactrium* spp., *Parvomonas* spp., and an increase in the ratio of *B. fragilis* and *Fecalibacterium prausnitzii*. After the introduction of autoprobiotic enterococci in the perioperative period for 10 days, we were able to monitor disappearance of marker oncogenic bacteria, and an increase in the representation of *Akkermansia* spp., *F. prausnitzii*. What seems especially important, we observed a decrease in the content of opportunistic representatives of the family *Enterobacteriacea, S.aureus, Streptococcus* spp., *C. difficile, Bifidobacterium* spp. (Table 9.7).

Against the background of a rapid normalization of the stool frequency, a gradual decrease in the severity of flatulence and a change in the properties of the stool were also observed from 6 to 4 after 3 days of taking autoprobiotic (Fig. 9.10).

What was particularly important, the category of patients with CRC with autoprobiotics experienced less number of complications after surgery which might be explained by the suppression of the pathogens such as *Clostridioides difficile*.

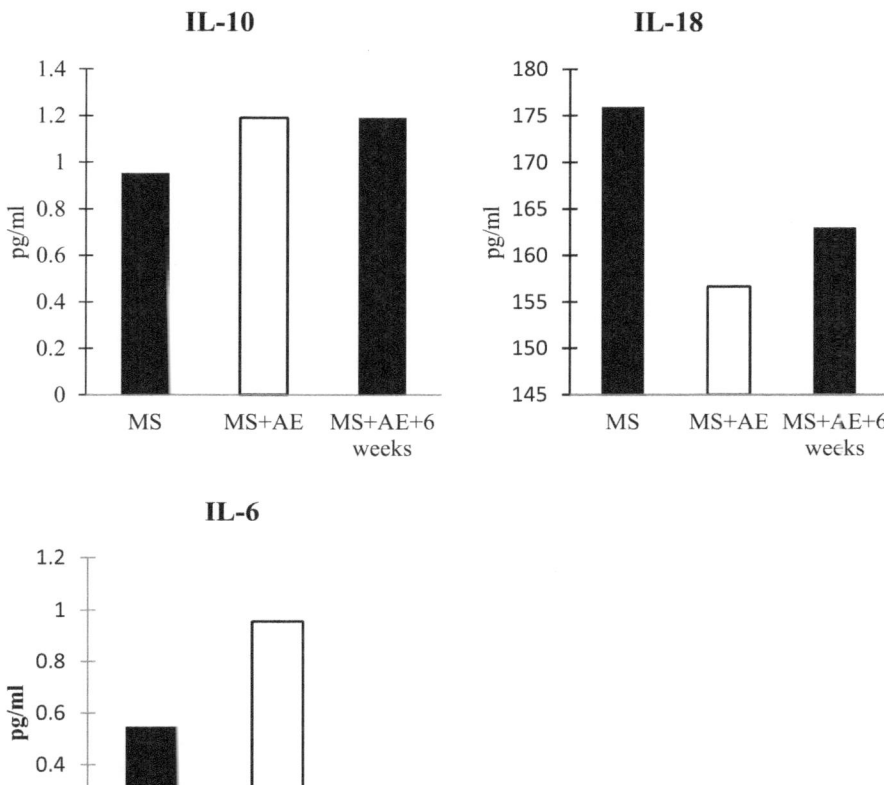

Fig. 9.9 Interleukin concentration in the blood serum before and after autoprobiotic treatment. (**a**) IL-10, (**b**) IL-18, (**c**) IL-6

Table 9.6 Statistically significant changes of anthropometric parameters of patients with metabolic syndrome after autoprobiotic therapy

Parameters	Before therapy	After therapy (MS+AE)
Body weight (in kg)	98	97
Index body weight	34,5	34,15
Neck volume (in cm)	39,5	39

Table 9.7 Changes in intestinal microbiome of patients with colorectal cancer after autoprobiotic consumption

More than in patients with CRC before therapy	Less than in patients with CRC before therapy
Akkermansia muciniphila, Faecalibacterium prausnitzii	*Klebsiella, Acinetobacter, Citrobacter, Staphylococcus aureus, Streptococcus, Bacteroides fragilis, Bifidobacterium, Clostridioides difficile*

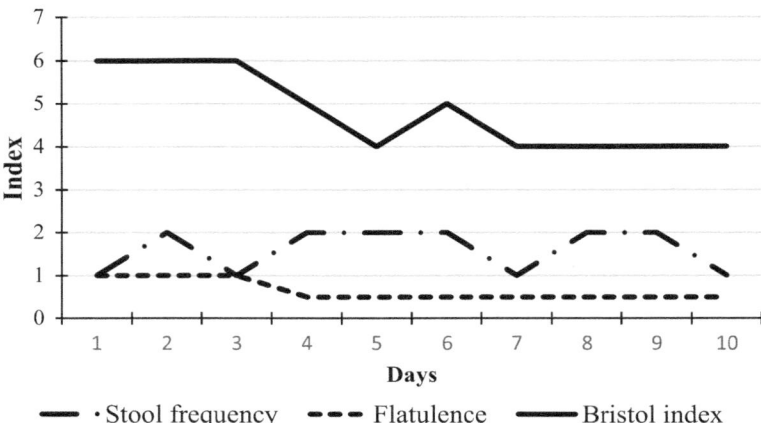

Fig. 9.10 Dynamics of the dyspeptic symptoms in patients with colorectal cancer during therapy with autoprobiotic. The graph shows that the almost complete disappearance of pain and flatulence is noted on the 4th day of treatment. Improvement in general well-being occurs on days 5–6, reaching 2.5 points or 1–7 Bristol scale

9.8 *Helicobacter pylori* Gastritis

According to present Maastricht V/Florence Consensus Report [32]. *H. pylori* gastritis is an infectious disease that leads to chronic active gastritis of varying severity in all infected subjects [52]. Cure of *H. pylori* infection heals the inflamed gastric mucosa can be achieved by antibiotics, however, usage of antibiotics for eradicaton of H.p. meets a lot of concerns in the medical community because of the rapid growth of antibiotic resistant strains and dysbiotic conditions at the end of the treatment.

Recent data on the effectiveness of monotherapy of H.p. infection with probiotic strains of Lactobacilli, Enterococci and Saccharomycetes provides the grounds for medical usage of the indigeneous bacteria for the treatment of H.p. gastritis. [4, 16, 41, 46].

In present study of the effectiveness of eradication therapy of *Helicobacter pylori* infection in patients with gastritis using autoprobiotic enterococci was carried out. The presence of *H.pylori* infection was confirmed by studying biopsy samples of the stomach and feces using bacteriological, biochemical (urease test) immuno-chromatographic methods, ELISA and PCR.

Autoprobiotic enterococci, originally isolated from feces and then prepared for the individual patients with gastritis as milk fermented product were administered for 20 days. Results of control study revealed the absence of H.p. after therapy in the vast majority of cases (Table 9.8).

Microbiota of the stomach and feces was analyzed before and after the treatment by qPCR and metagenome analysis 16S rRNA. Signs of intestinal dysbiosis such as a decrease in *Firmicutes* (*Lactobacillus, Enterococcus*), an increase in gamma-*Proteobacteria* (opportunistic representatives), of the family *Enterobacteriaceae*) were revealed in patients with H.p. gastritis before the therapy. A course of autoprobiotic monotherapy for 20 days led to the increase of bacteria belonging to the genera's *Lactobacillus, Enterococcus, Prevotella, Faecalibacterium prausnitzii, Enterococcus, Lachnobacterium* and to a decrease of *Klebsiella, Citrobacter, Burkcholderia, Acinetobacter, Staphylococcus aureus, Akkermansia muciniphila, Bacteroides fragilis, Bifidobacterium,* and *Ruminococcus* spp. (Table 9.9).

Gastrointestinal dysfunctions (belching, heartburn, flatulence, nausea, epigastric pain, poor appetite) disappeared after autoprobiotic usage (Table 9.10).

Table 9.8 The effect of autoprobiotic therapy on eradication of *H.pylori*

Material	Methods	Before therapy	After therapy
Fecal samples	PCR	12/12	2/12
	Immuno-chromatographic test	12/12	1/12
	Elisa	11/12	0/12
Biopsy samples	Urease test	11/12	0/12
	Bacteriological method and PCR	12/12	–

Table 9.9 Changes in intestinal microbiome after autoprobiotic consumption

More than in patients with HPG before therapy	Less than in patients with HPG before therapy
Lactobacillus, enterococcus, Prevotella, Faecalibacterium prausnitzii, Enterococcus, Lachnobacterium	*Klebsiella, Citrobacter, Burkcholderia Acinetobacter, Staphylococcus aureus, Akkermansia muciniphila, Bacteroides fragilis, Bifidobacterium, Ruminococcus*

Table 9.10 Changes in the gastrointestinal symptoms of patients with H.p. gastritis after autoprobiotic therapy

Symptoms	Before therapy	After therapy
Belching	4/12	0/12
Heartburn	6/12	0/12
Epigastric pain	10/12	0/12
Flatulence	4/12	0/12
Nausea	4/12	0/12

This pilot study revealed that autoprobiotic enterococci can be considered as candidates for an alternative method of eradication of H.p. infection in patients with gastritis.

9.9   Chronic Generalized Periodontitis

Inflammatory periodontal diseases including gingivitis and periodontitis are significant in terms of frequency and prevalence among dental diseases. Many authors have demonstrated the polyethological nature of periodontal diseases, with an important role in their development attributed to inflammatory reactions provoked by dysbiosis of the oral cavity, known as periodontal pathogens [36].

The importance of microbial therapy of the oral dysbiosis turns out to be significant in the context of the rapid recovery of the biological balance of microbiota when the environmental conditions for the development of periodontitis are eliminated. *Streptococcus salivarius* is an exclusively oral streptococcus species, which is often discovered in microbiota of the oral cavity with no negative effects on the human body. The use of *S.salivarius* as probiotic was evaluated in different clinical conditions including infections of the oral cavity and dental diseases [7, 8, 68]. However, usage of *S.salivarius* as autoprobiotics for the personalized treatment of the oral microbiota was never tested previously.

Studies on the oral application of *S.salivarius*-based autoprobiotic in the complex treatment of periodontitis [28]. A total of 37 patients aged 29–64 years with chronic generalized periodontitis (CGP) of mild severity without concomitant pathology were examined. Clinical examination of patients was carried out according to the generally accepted methodology: identification of complaints, collection of life history and medical history, external examination and examination of the oral cavity and X-ray examination. *S. salivarius* bacteria isolated from the individuals were analyzed bacteriologically, genetically and used to prepare autoprobiotics. For the monitoring of the detection of autoprobiotics in patients' periodontal pockets was performed by PCR during and after treatment.

It has been demonstrated that the use of *S. salivarius*-based autoprobiotics in the form of irrigation of periodontal pockets as well as in the form of oral baths resulted in faster normalization of clinical condition of periodontal tissues (OHI-S hygiene index, Plaque IndexPI index, Gingival Index PMA, Bleeding on probing Index BOP index, Periodontal Index CPITN) compared to the control group. At the same time, the patients of the main group underwent more accelerated elimination of periodontal pathogens (*P.gingivalis, T.forsythia, T.denticola, P.intermedia*) in periodontal pockets compared to the control group (Fig. 9.11).

The use of microorganisms of the individual's normal microbiota is an important component of the concept of personalized approach to the selection of a probiotic drug that guarantees the safety of their use based on their biocompatibility for the particular person. In this regard, it should be noted, that the above data confirm the validity of the use of *S. salivarius* as an oral autoprobiotic.

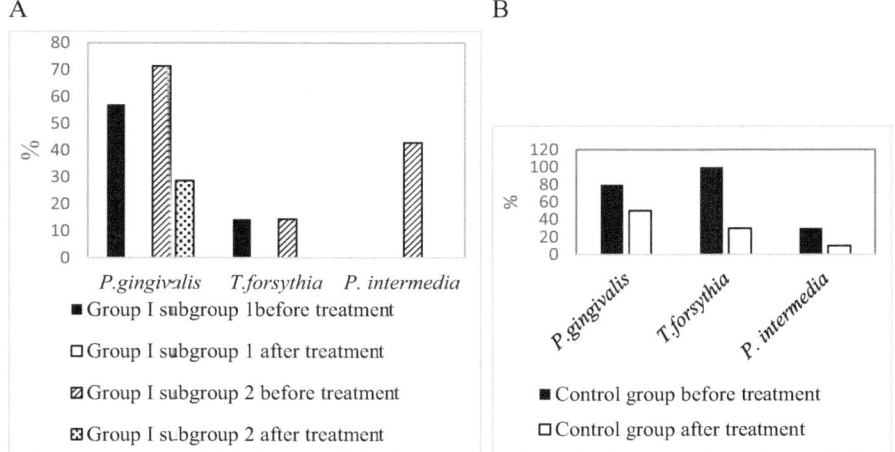

Fig. 9.11 (**a**) Change in the detection of periodontal pathogens in patients of the main group with mild CGP: Group I subgroup 1 – use of *S. salivarius*-based autoprobiotics in the form of irrigation of periodontal pockets in the complex treatment of mild CGP; Group I subgroup 2 – use of *S. salivarius*-based autoprobiotics in the form of oral baths in the complex treatment of mild CGP. (**b**) Change in the detection of periodontal pathogens in patients of the control group with the complex treatment of mild CGP

9.10 Summary

This chapter is devoted to the possible usage of indigenous bacteria as method of treatment intestinal dysbiosis. It summarizes some theoretical arguments regarding the role of microbial dysbiosis in the development of human diseases and the possible usage of autoprobiotic therapy. Chapter describes some animal studies and several examples when autoprobiotic enterococci were used for personified microbial therapy in the clinic. It is important to mention, that intestines are not the only loci of microbial colonization and the concept may be easily transferred for each loci including the skin, oral cavity and urogenital tract. Respectively, other kinds of bacteria should be selected for the treatment of dysbiotic condition such as *S.salivarius* for the treatment of periodontal diseases. It is obvious, that with accumulation of the knowledge of microbiome and it involvement in the pathological development of several chronic diseases it will be possible to improve and optimize selection of the best strategy of microbial therapy. However, based on the studies undertaken in several clinics of Russian Federation, it is possible conclude that autoprobiotic therapy did not cause any side effects and was quite efficient while curing several clinical conditions. The exact mechanisms of autoprobiotic therapy are still unknown. However, we were able to monitor some general clinical, immunological and laboratory features of autoprobiotic therapy of intestinal dysbiosis. Improvement of clinical condition mainly correlated with an improvement of the functioning of gastrointestinal tract and the decrease in the abundance of opportunistic bacteria predominantly by the members of *Enterobacteriaceae*. Immunological changes were characterized by the decrease of the low-grade inflammation. However, the cytokine level in blood serum seems to be not very stable clinical parameter. It looks like that host immune response follows the change of microbial consortium which is establishing after the course of microbial therapy. In any case more data on the subject of usage of autoprobiotics in dysbiosis conditions will be needed. An additional advantage of autoprobiotic technology is based on the present possibility for long term storage of individual microbiota and successfully used for microbial therapy autoprobiotic strains. Indigenous microbiota biobanks will open a new window of possibilities of microbial therapy in the future when more data on the role of microbiota in health and diseases will be collected.

Expert Recommendations for the Use of Autoprobiotics
1. For preventive purposes in conditions of sudden changes in environmental parameters.
2. For therapeutic purposes in patients with acute and chronic infectious processes, after antimicrobial therapy.
3. People who are exposed to frequent stress (police, military personnel, emergency workers, athletes, pilots, firefighters).
4. Patients with allergic, neuro-degenerative and immune diseases.
5. Patients after surgery.
6. Cancer patients after chemotherapy and radiation therapy.
7. Patients with metabolic syndrome.

Acknowledgements The work was supported by Scientific and Educational Center «Molecular bases of interaction of microorganisms and human» of the world-class research center «Center for personalized Medicine» FSBSI «IEM».

References

1. Azad MAK, Sarker M, Li T, Yin J (2018) Probiotic species in the modulation of gut microbiota: an overview. Biomed Res Int 2018:9478630. https://doi.org/10.1155/2018/9478630
2. Belkaid Y, Harrison OJ (2017) Homeostatic immunity and the microbiota. Immunity 46(4):562–576. https://doi.org/10.1016/j.immuni.2017.04.008
3. Berg D, Clemente JC, Colombel JF (2015) Can inflammatory bowel disease be permanently treated with short-term interventions on the microbiome? Expert Rev Gastroenterol Hepatol 9(6):781–795. https://doi.org/10.1586/17474124.2015.1013031
4. Boonyaritichaikij S, Kuwabara K, Nagano J, Kobayashi K, Koga Y (2009) Long-term administration of probiotics to asymptomatic pre-school children for either the eradication or the prevention of helicobacter pylori infection. Helicobacter 14(3):202–207. https://doi.org/10.1111/j.1523-5378.2009.00675.x
5. Bubnov RV, Babenko LP, Lazarenko LM et al (2017) Comparative study of probiotic effects of *lactobacillus* and *Bifidobacteria* strains on cholesterol levels, liver morphology and the gut microbiota in obese mice. EPMA J 8:357–376. https://doi.org/10.1007/s13167-017-0117-3

6. Bubnov RV, Spivak MY, Lazarenko LM et al (2015) Probiotics and immunity: provisional role for personalized diets and disease prevention. EPMA J 6:14. https://doi.org/10.1186/s13167-015-0036-0
7. Burton JP, Drummond BK, Chilcott CN, Tagg JR, Thomson WM, Hale JDF, Wescombe PA (2013) Influence of the probiotic streptococcus salivarius strain M18 on indices of dental health in children: a randomized double-blind, placebo-controlled trial. J Med Microbiol 62(Pt 6):875–884. https://doi.org/10.1099/jmm.0.056663-0
8. Chen TY, Hale JDF, Tagg JR, Jain R, Voss AL, Mills N, Best EJ, Stevenson DS, Bird PA, Walls T (2021) In vitro inhibition of clinical isolates of otitis media pathogens by the probiotic streptococcus salivarius BLIS K12. Probiotics Antimicrob Proteins 13(3):734–738. https://doi.org/10.1007/s12602-020-09719-7
9. Chong PP, Chin VK, Looi CY, Wong WF, Madhavan P, Yong VC (2019) The microbiome and irritable bowel syndrome – a review on the pathophysiology, current research and future therapy. Front Microbiol 10:1136. https://doi.org/10.3389/fmicb.2019.01136
10. Corb Aron RA, Abid A, Vesa CM, Nechifor AC, Behl T, Ghitea TC et al (2021) Recognizing the benefits of pre−/probiotics in metabolic syndrome and type 2 diabetes mellitus considering the influence of akkermansia muciniphila as a key gut bacterium. Microorganisms 9(3):618. https://doi.org/10.3390/microorganisms9030618
11. Costea PI, Hildebrand F, Arumugam M, Bäckhed F, Blaser MJ, Bushman FD et al (2018) Enterotypes in the landscape of gut microbial community composition. Nat Microbiol 3(1):8–16. https://doi.org/10.1038/s41564-017-0072-8
12. Dobbs SM, Dobbs RJ, Weller C, Charlett A, Augustin A, Taylor D, Ibrahim MA, Bjarnason I (2016) Peripheral aetiopathogenic drivers and mediators of Parkinson's disease and co-morbidities: role of gastrointestinal microbiota. J Neurovirol 22(1):22–32. https://doi.org/10.1007/s13365-015-0357-8
13. Ellis D, Dobbs RJ, Dobbs S, Curry A, Bjarnason I, Williams J, McCrossan MV, Weller C, Charlett A (2007) Duodenal enterocyte mitochondrial involvement and abnormal bowel function in idiopathic parkinsonism. In: Hanin I, Windisch M, Poewe W, Fisher A (eds) ADPD 2007 new trends in Alzheimer and Parkinson related disorders. Medimond S.r.l, Bologna, pp 269–272
14. Ermolenko E, Gromova L, Borschev Y, Voeikova A, Karaseva A, Ermolenko K, Gruzdkov A, Suvorov A (2013) Influence of different probiotic lactic acid bacteria on microbiota and metabolism of rats with dysbiosis. Biosci Microbiota Food Health 32(2):41–49. https://doi.org/10.12938/bmfh.32.41
15. Ermolenko E, Rybalchenko O, Borshev Y, Tarasova E, Kramskaya T, Leontieva G, Kotyleva M, Orlova O, Abdurasulova I, Suvorov A (2018) Influence of monostrain and multistrain probiotics on immunity, intestinal ultrastructure and microbiota in experimental dysbiosis. Benef Microbes 9(6):937–949. https://doi.org/10.3920/BM2017.0117
16. Ermolenko EI, Molostova AS, Gladyshev NS (2021) Eradication therapy of helicobacteriosis with probiotics, problems, and prospects. Exp Clin Gastroenterol 193:60–72. https://doi.org/10.31146/1682-8658-ecg-193-9-60-72
17. Ermolenko E (2011) Lactic acid bacteria. LAP Lambert Academic Publishing, Deutchland Labert. (In Russian)
18. Ermolenko EI, Donets VN, Dmitrieva YV, Ilyasov YY, Suvorova MA, Gromova LV (2009) Influence of probiotic enterococci on functional characteristics of rat bowel under dysbiosis induced by antibiotics. Bull St Petersburg State University 11:157–167. (In Russian)
19. Frese SA, Hutkins RW, Walter J (2012) Comparison of the colonization ability of autochthonous and allochthonous strains of Lactobacilli in the human gastrointestinal tract. Adv Microbiol 2(3):399–409. https://doi.org/10.4236/aim.2012.23051
20. Hällström M, Eerola E, Vuento R, Janas M, Tammela O (2004) Effects of mode of delivery and necrotising enterocolitis on the intestinal microflora in preterm infants. Eur J Clin Microbiol Infect Dis 23(6):463–470. https://doi.org/10.1007/s10096-004-1146-0

21. Hand TW (2016) The role of the microbiota in shaping infectious immunity. Trends Immunol 37(10):647–658. https://doi.org/10.1016/j.it.2016.08.007
22. Harmsen HJM, de Goffau MC (2016) The human gut microbiota. Adv Exp Med Biol 902:95–108. https://doi.org/10.1007/978-3-319-31248-4_7
23. Holtmann G, Shah A, Morrison M (2017) Pathophysiology of functional gastrointestinal disorders: a holistic overview. Dig Dis 35(Suppl 1):5–13. https://doi.org/10.1159/000485409
24. Hui-Beckman J, Kim BE, Leung DY (2022) Origin of allergy from *in utero* exposures to the postnatal environment. Allergy Asthma Immunol Res 14(1):8–20. https://doi.org/10.4168/aair.2022.14.1.8
25. Huipeng W, Lifeng G, Chuang G, Jiaying Z, Yuankun C (2014) The differences in colonic mucosal microbiota between normal individual and colon cancer patients by polymerase chain reaction-denaturing gradient gel electrophoresis. J Clin Gastroenterol 48(2):138–144. https://doi.org/10.1097/MCG.0b013e3182a26719
26. Kaibysheva V, Nikonov E (2019) Probiotics from the position of evidence-based medicine. Dokazatel'naya gastroenterologiya 8(3):45–54. https://doi.org/10.17116/dokgastro2019803145
27. Kassinen A, Krogius-Kurikka L, Mäkivuokko H, Rinttilä T, Paulin L, Corander J, Malinen E, Apajalahti J, Palva A (2007) The fecal microbiota of irritable bowel syndrome patients differs significantly from that of healthy subjects. Gastroenterology 133(1):24–33. https://doi.org/10.1053/j.gastro.2007.04.005
28. Koroleva IV, Mikhaylova ES, Kraeva LA, Suvorov AN (2021) Clinical and microbiological evaluation of the efficacy of autoprobiotics in the combination treatment of chronic generalized periodontitis. Revista Latinoamericana de Hipertension 16(1):15–27. http://www.revhipertension.com
29. Lee YK, Mazmanian SK (2010) Has the microbiota played a critical role in the evolution of the adaptive immune system? Science 330(6012):1768–1773. https://doi.org/10.1126/science.1195568
30. Ley RE, Peterson DA, Gordon JI (2006) Ecological and evolutionary forces shaping microbial diversity in the human intestine. Cell 124(4):837–848. https://doi.org/10.1016/j.cell.2006.02.017
31. Li J, Zhao F, Wang Y, Chen J, Tao J, Tian G et al (2017) Gut microbiota dysbiosis contributes to the development of hypertension. Microbiome 5(1):14. https://doi.org/10.1186/s40168-016-0222-x
32. Malfertheiner P, Megraud F, O'Morain CA, Gisbert JP, Kuipers EJ, Axon AT, Bazzoli F, Gasbarrini A, Atherton J, Graham DY, Hunt R, Moayyedi P, Rokkas T, Rugge M, Selgrad M, Suerbaum S, Sugano K, El-Omar EM, European Helicobacter and Microbiota Study Group and Consensus Panel (2017) Management of Helicobacter pylori infection-the Maastricht V/Florence Consensus Report. Gut 66(1):6–30. https://doi.org/10.1136/gutjnl-2016-312288
33. Mazmanian SK, Round JL, Kasper DL (2008) A microbial symbiosis factor prevents intestinal inflammatory disease. Nature 453(7195):620–625. https://doi.org/10.1038/nature07008
34. Nakatsu G, Li X, Zhou H, Sheng J, Wong SH, Wu WKK, Ng SC, Tsoi H et al (2015) Gut mucosal microbiome across stages of colorectal carcinogenesis. Nat Commun 6:8727. https://doi.org/10.1038/ncomms9727
35. Okada H, Kuhn C, Feillet H, Bach JF (2010) The 'hygiene hypothesis' for autoimmune and allergic diseases: an update. Clin Exp Immunol 160(1):1–9. https://doi.org/10.1111/j.1365-2249.2010.04139.x
36. Pei J, Li F, Xie Y et al (2020) Microbial and metabolomic analysis of gingival crevicular fluid in general chronic periodontitis patients: lessons for a predictive, preventive, and personalized medical approach. EPMA J 11:197–215. https://doi.org/10.1007/s13167-020-00202-5
37. Petrov VA, Saltykova IV, Zhukova IA, Alifirova VM, Zhukova NG, Dorofeeva YB, Tyakht AV, Kovarsky BA, Alekseev DG, Kostryukova ES, Mironova YS, Izhboldina OP, Nikitina MA, Perevozchikova TV, Fait EA, Babenko VV, Vakhitova MT, Govorun VM, Sazonov AE (2017) Analysis of gut microbiota in patients with Parkinson's disease. Bull Exp Biol Med 162(6):734–737. https://doi.org/10.1007/s10517-017-3700-7

38. Pozuelo M, Panda S, Santiago A, Mendez S, Accarino A, Santos J, Guarner F, Azpiroz F, Manichanh C (2015) Reduction of butyrate- and methane-producing microorganisms in patients with irritable bowel syndrome. Sci Rep 5:12693. https://doi.org/10.1038/srep12693
39. Razavi AC, Potts KS, Kelly TN, Bazanno LA (2019) Sex, gut microbiome and cardiovascular disease risk. Biol Sex Differ 10(1):29. https://doi.org/10.1186/s13293-019-0240-z
40. Rodes L, Khan A, Paul A, Coussa-Charley M, Marinescu D, Tomaro-Duchesneau C, Shao W, Kahouli I, Prakash S (2013) Effect of probiotics lactobacillus and Bifidobacterium on gut-derived lipopolysaccharides and inflammatory cytokines: an in vitro study using a human colonic microbiota model. J Microbiol Biotechnol 23(4):518–526. https://doi.org/10.4014/jmb.1205.05018
41. Sakamoto I, Igarashi M, Kimura K, Takagi A, Miwa T, Koga Y (2001) Suppressive effect of lactobacillus gasseri OLL 2716 (LG21) on helicobacter pylori infection in humans. J Antimicrob Chemother 47(5):709–710. https://doi.org/10.1093/jac/47.5.709
42. Sampson TR, Debelius JW, Thron T, Janssen S, Shastri GG, Ilhan ZE, Challis C, Schretter CE, Rocha S, Gradinaru V, Chesselet MF, Keshavarzian A, Shanno KM, Krajmalnik-Brown R, Wittung-Stafshede P, Knight R, Mazmanian SK (2016) Gut microbiota regulate motor deficits and Neuroinflammation in a model of Parkinson's disease. Cell 167(6):1469–1480.e12. https://doi.org/10.1016/j.cell.2016.11.018
43. Scher AI, Ross GW, Sigurdsson S, Garcia M, Gudmundsson LS, Sveinbjörnsdóttir S, Wagner AK, Gudnason V, Launer LJ (2014) Midlife migraine and late-life parkinsonism: AGES-Reykjavik study. Neurology 83(14):1246–1252. https://doi.org/10.1212/WNL.0000000000000840
44. Seong E, Bose S, Han SY et al (2021) Positive influence of gut microbiota on the effects of Korean red ginseng in metabolic syndrome: a randomized, double-blind, placebo-controlled clinical trial. EPMA J 12:177–197. https://doi.org/10.1007/s13167-021-00243-4
45. Shenderov BA (2003) Functional nutrition, cryogenic banks of microbiota and their role in health restoration. Bull Restorat Med 1:29–31. (In Russian)
46. Shimizu T, Haruna H, Hisada K, Yamashiro Y (2002) Effects of Lactobacillus gasseri OLL 2716 (LG21) on Helicobacter pylori infection in children. J Antimicrob Chemother 50(4):617–618. https://doi.org/10.1093/jac/dkf157
47. Simon JC, Marchesi JR, Mougel C, Selosse MA (2019) Host-microbiota interactions: from holobiont theory to analysis. Microbiome 7(1):5. https://doi.org/10.1186/s40168-019-0619-4
48. Sitkin SI, Tkachenko EI, Vakhitov TY (2016) Metabolic dysbiosis of the gut microbiota and its biomarkers. Eksp Klin Gastroenterol 12(12):6–29. (In English, Russian). PMID: 29889418
49. Smith PM, Howitt MR, Panikov N, Michaud M, Gallini CA, Bohlooly-Y M, Glickman JN, Garrett WS (2013) The microbial metabolites, short-chain fatty acids, regulate colonic Treg cell homeostasis. Science 341(6145):569–573. https://doi.org/10.1126/science.1241165
50. Soloviova O, Simanenkov V, Suvorov A (2017) The use of probiotics and autoprobiotics in the treatment of irritable bowel syndrome. Clin Exp Gastroenterol 7:115–120. URL: https://cyberleninka.ru/article/n/ispolzovanie-probiotikov-i-autoprobiotikov-v-lechenii-sindroma-razdrazhennoy-to
51. Song EJ, Han K, Lim TJ et al (2020) Effect of probiotics on obesity-related markers per enterotype: a double-blind, placebo-controlled, randomized clinical trial. EPMA J 11:31–51. https://doi.org/10.1007/s13167-020-00198-y
52. Sostres C, Carrera-Lasfuentes P, Benito R, Roncales P, Arruebo M, Arroyo MT, Bujanda L, García-Rodríguez LA, Lanas A (2015) Peptic ulcer bleeding risk. The role of helicobacter pylori infection in NSAID/low-dose aspirin users. Am J Gastroenterol 110(5):684–689. https://doi.org/10.1038/ajg.2015.98
53. Suvorov A (2013) Gut microbiota, probiotics, and human health. Biosci Microbiota Food Health 32(3):81–91. https://doi.org/10.12938/bmfh.32.81
54. Suvorov AN, Ermolenko EI, Kotyleva MP, Tsapieva AN (2020) Method for preparing autoprobiotic based on anaerobic bacteria consortium. Patent for invention 2734896 C2, 10/26/2020. Application No. 2018147697 dated 12/28/2018

55. Suvorov A, Karaseva A, Kotyleva M, Kondratenko Y, Lavrenova N, Korobeynikov A, Kozyrev P, Kramskaya T, Leontieva G, Kudryavtsev I, Guo D, Lapidus A, Ermolenko E (2018) Autoprobiotics as an approach for restoration of personalised microbiota. Front Microbiol 9:1869. https://doi.org/10.3389/fmicb.2018.01869

56. Tap J, Derrien M, Törnblom H, Brazeilles R, Cools-Portier S, Doré J, Störsrud S, Le Nevé B, Öhman L, Simrén M (2017) Identification of an intestinal microbiota signature associated with severity of irritable bowel syndrome. Gastroenterology 152(1):111–123.e8. https://doi.org/10.1053/j.gastro.2016.09.049

57. Tilg H, Moschen AR, Kaser A (2009) Obesity and the microbiota. Gastroenterology 136(5):1476–1483. https://doi.org/10.1053/j.gastro.2009.03.030

58. Vandegrift R, Bateman AC, Siemens KN, Nguyen M, Wilson HE, Green JL, Van Den Wymelenberg KG, Hickey RJ (2017) Cleanliness in context: reconciling hygiene with a modern microbial perspective. Microbiome 5(1):76. https://doi.org/10.1186/s40168-017-0294-2

59. Wang ZK, Yang YS, Chen Y, Yuan J, Sun G, Peng LH (2014) Intestinal microbiota pathogenesis and fecal microbiota transplantation for inflammatory bowel disease. World J Gastroenterol 20(40):14805–14820. https://doi.org/10.3748/wjg.v20.i40.14805

60. Warren RL, Freeman DJ, Pleasance S, Watson P, Moore RA, Cochrane K, Allen-Vercoe E, Holt RA (2013) Co-occurrence of anaerobic bacteria in colorectal carcinomas. Microbiome 1(1):16. https://doi.org/10.1186/2049-2618-1-16

61. Wiedermann CJ, Kiechl S, Dunzendorfer S, Schratzberger P, Egger G, Oberhollenzer F et al (1999) Association of endotoxemia with carotid atherosclerosis and cardiovascular disease: prospective results from the Bruneck study. J Am Coll Cardiol 34(7):1975–1981. https://doi.org/10.1016/S0735-1097(99)00448-9

62. Wilkins T, Sequoia J (2017) Probiotics for gastrointestinal conditions: a summary of the evidence. Am Fam Physician 96(3):170–178. PMID: 28762696

63. Wu N, Yang X, Zhang R, Li J, Xiao X, Hu Y, Chen Y, Yang F, Lu N, Wang Z, Luan C, Liu Y, Wang B, Xiang C, Wang Y, Zhao F, Gao GF, Wang S, Li L, Zhang H, Zhu B (2013) Dysbiosis signature of fecal microbiota in colorectal cancer patients. Microb Ecol 66(2):462–470. https://doi.org/10.1007/s00248-013-0245-9

64. Yan Q, Gu Y, Li X, Yang W, Jia L, Chen C et al (2017) Alterations of the gut microbiome in hypertension. Front Cell Infect Microbiol 7:381. https://doi.org/10.3389/fcimb.2017.00381

65. Yu J, Feng Q, Wong SH, Zhang D, Liang QY, Qin Y, Tang L, Zhao H, Stenvang J, Li Y, Wang X, Xu X, Chen N, Wu WK, Al-Aama J, Nielsen HJ, Kiilerich P, Jensen BA, Yau TO, Lan Z, Jia H, Li J, Xiao L, Lam TY, Ng SC, Cheng AS, Wong VW, Chan FK, Xu X, Yang H, Madsen L, Datz C, Tilg H, Wang J, Brünner N, Kristiansen K, Arumugam M, Sung JJ, Wang J (2017) Metagenomic analysis of faecal microbiome as a tool towards targeted non-invasive biomarkers for colorectal cancer. Gut 66(1):70–78. https://doi.org/10.1136/gutjnl-2015-309800

66. Yu JC, Khodadadi H, Baban B (2019) Innate immunity and oral microbiome: a personalized, predictive, and preventive approach to the management of oral diseases. EPMA J 10:43–50. https://doi.org/10.1007/s13167-019-00163-4

67. Zhong H, Penders J, Shi Z, Ren H, Cai K, Fang C, Ding Q, Thijs C, Blaak EE, Stehouwer CDA, Xu X, Yang H, Wang J, Wang J, Jonkers DMAE, Masclee AAM, Brix S, Li J, Arts ICW, Kristiansen K (2019) Impact of early events and lifestyle on the gut microbiota and metabolic phenotypes in young school-age children. Microbiome 7(1):2. https://doi.org/10.1186/s40168-018-0608-z

68. Zupancic K, Kriksic V, Kovacevic I, Kovacevic D (2017) Influence of Oral probiotic streptococcus salivarius K12 on ear and oral cavity health in humans. Probiotics Antimicrob Proteins 9(2):102–110. https://doi.org/10.1007/s12602-017-9261-2

Chapter 10
Fecal Microbiota Transplantation in Diseases Not Associated with *Clostridium difficile*: Current Status and Future Therapeutic Option

Sergii Tkach, Andrii Dorofeyev, Iurii Kuzenko, Nadiya Boyko, Tetyana Falalyeyeva, and Nazarii Kobyliak

Abstract Fecal microbiota transplantation (FMT) is one of several effective methods for modifying altered intestinal microbiota and treating certain gastrointestinal diseases. Currently, the only officially approved grounds for FMT is a recurrent infection of *Clostridium difficile*. Recently published data showed FMT success in paving the use of precision medicine in gastrointestinal disorders. Nonetheless, the effectiveness of FMT is presently being studied in treating other gastrointestinal and non-gastrointestinal pathologies not associated with recurrent infection of *Clostridium difficile*.

S. Tkach · I. Kuzenko
Ukrainian Research and Practical Centre of Endocrine Surgery, Transplantation of Endocrine Organs and Tissues of the Ministry of Health of Ukraine, Kyiv, Ukraine

A. Dorofeyev
Shupyk National Medical Academy of Postgraduate Education, Kyiv, Ukraine

N. Boyko
RDE Center of Molecular Microbiology and Mucosal Immunology, Uzhhorod National University, Uzhhorod, Ukraine

Department of Clinical Laboratory Diagnostics and Pharmacology, Uzhhorod National University, Uzhhorod, Ukraine

Ediens LLC, Uzhhorod, Ukraine

T. Falalyeyeva
Educational and Scientific Center "Institute of Biology and Medicine", Taras Shevchenko National University of Kyiv, Kyiv, Ukraine

Medical Laboratory CSD, Kyiv, Ukraine

N. Kobyliak (✉)
Medical Laboratory CSD, Kyiv, Ukraine

Department of Endocrinology, Bogomolets National Medical University, Kyiv, Ukraine

© The Author(s), under exclusive license to Springer Nature Switzerland AG 2023
N. Boyko, O. Golubnitschaja (eds.), *Microbiome in 3P Medicine Strategies*,
Advances in Predictive, Preventive and Personalised Medicine 16,
https://doi.org/10.1007/978-3-031-19564-8_10

Here, in the frame of Predictive, Preventive, and Personalized Medicine (PPPM), discovered a modern view of current research on the results of FMT use in patients with inflammatory bowel diseases, constipation, irritable bowel syndrome, antibiotic-associated diarrhea and hepatic encephalopathy, as well as such non-gastroenterological diseases as psoriasis, multiple sclerosis, autism, Parkinson's disease and metabolic syndrome. Based on a literature review, the authors conclude that FMT's potential effectiveness and feasibility in patients with ulcerative colitis, irritable bowel syndrome, antibiotic-associated diarrhea, hepatic encephalopathy, autism and metabolic syndrome. Research on Crohn's disease, psoriasis, multiple sclerosis and Parkinson's disease is still ongoing.

This part considers innovative concepts that an individual approach to each patient with appointment FMT and the basis of personalized medicine can improve clinical outcomes in patients with ulcerative colitis, irritable bowel syndrome, antibiotic-associated diarrhea, hepatic encephalopathy, autism and metabolic syndrome. So personalized FMT therapy in diseases not associated with *Clostridium difficile* would be a primary concrete practice for PPPM in the nearest future.

Keywords Predictive · Preventive · Personalized medicine · Fecal microbiota transplantation · Diseases not associated with *Clostridium difficile* · Ulcerative colitis · Pouchitis · Irritable bowel syndrome · Microscopic colitis · Functional constipation antibiotic-associated diarrhea · Gastro-intestinal cancers · Chronic liver disease · Acute pancreatitis · Psoriasis multiple sclerosis · Parkinson's disease · Autism spectrum disease · Epilepsy · Metabolic syndrome · Obesity

10.1   Introduction

The imbalance of the normal gut microbiota has been linked with many diseases [1–4]. This goes for both disorders of the digestive system (ulcerative colitis, irritable bowel syndrome, constipation, etc.) and diseases of another genesis such as cancer, metabolic syndrome, neuropsychiatric disorders, etc. [5–10]. Consequently, it is not surprising that various methods of modification of intestinal biocenosis have always attracted the close attention of the scientific community.

Fecal microbiota transplantation (FMT) is one such method, which consists of the administration of a solution of fecal matter from a healthy donor to an ill recipient to directly change the recipient's gut microbiota and confer a health benefit. The method of FTM was first described by the Chinese physician Ge Hong in the therapy of hard diarrhea as early as the fourth century AD [11]. In 1958, the current stage of FMT study started when American surgeon Ben Eisman published in the scientific literature four cases of antibiotic-associated diarrhea and clinical improvement after using enemas with fecal donor material [12]. However, this method wasn't officially acknowledged for over 50 years until the first controlled investigation discovered the elevated improvement of FMT in patients with recurrent *Clostridium difficile* [13]. Today, this disease is the only officially approved ground for FMT and is widely used worldwide [14].

Howsoever, the effectiveness of FMT in the therapy of other gastrointestinal (functional constipation antibiotic-associated diarrhea (AAD), pouchitis, irritable bowel syndrome (IBS), ulcerative colitis (UC), microscopic colitis (MC), gastrointestinal cancers, chronic liver disease, acute pancreatitis) and non-gastrointestinal pathologies (psoriasis multiple sclerosis, Parkinson's disease (PD), autism spectrum disease (ASD), epilepsy, metabolic syndrome/obesity) not associated with recurrent infection of *Clostridium difficile* is currently being studied [15–17].

10.2 FMT in Focus of Predictive Approach, Targeted Prevention and Personalisation of Medical Services

FTM is an innovative investigational treatment with administration in the colon of stool from a healthy donor to a recipient with a pathology connected with an unhealthy microbiota. FMT is successfully used for treating of recurrent *Clostridium difficile* infection [17]. Nonetheless, the effectiveness of FMT is currently being studied in treating other gastrointestinal and non-gastrointestinal pathologies not associated with recurrent infection of *Clostridium difficile*. The ever-increasing population of patients with this disease (irritable bowel syndrome, ulcerative colitis, antibiotic-associated diarrhea, hepatic encephalopathy, autism and metabolic syndrome), requires a change in the system of intervention measures to predictive, preventive and personalized medicine (PPPM). The current dramatic situation can be significantly improved by implementing appropriate early diagnostic, targeted preventive measures and therapies tailored to the person advancing healthcare and enhancing the quality of patient life.

10.3 Ulcerative Colitis (UC)

UC causes irritation and ulcers in the large intestine. It belongs to a group of conditions called IBD. Most of the studies devoted to the effectiveness of FMT in pathologies not associated with recurrent *Clostridium difficile* infections relate to inflammatory bowel diseases (primarily UC) due to significant violations of the qualitative and qualitative composition of the intestinal flora. For example, the diversity of gut microorganisms is lower in persons with UC than in healthy individuals, and there is a decrease in the relative quantity of *Bacteroidetes and Firmicutes*. It has been proven that with the development of UC, butyrate produced by *Fecalibacterium prausnitzii* decreases, and the number of *Proteobacteria* and *Actinobacteria* increases [18]. These changes in the intestine lead to a reduction of short-chain fatty acids (SCFAs) formation, especially butyrate, which is considered an essential source of sustenance for colonocytes and plays an important role in modulating immune and inflammatory responses [19].

There are three known randomized controlled trials (RCT) from six with the proven efficacy of FMT in patients with active UC. The study was conducted on 70 patients with active NVC, they participated in a single-blind RCT and received either allogeneic FMT by enema (main group) or water enema (control group) [17, 20]. The primary endpoints, including a reduction in total Mayo score to less than three and endoscopic healing (Mayo endoscopic score 0), were seen in 24% of FMT-treated patients and 5% of placebo-treated patients within 6 weeks of the start of the RCT. Notably, the general part of clinical cases showed high efficiency of obtaining FMT from 1 superdonor (39% compared to 10% with material from other donors), which, in turn, confirms the important role of donor selection. At the same time, it is emphasized that such data are not indicated in treating recurrent *Clostridium difficile* infection. The effectiveness of FMT was investigated in active NVC patients (n = 25) by conducting two duodenal infusions consisting of 500 ml of fresh allogeneic donor material. The Control group (n = 25) received their own fecal material in the same way [21]. The clinical and endoscopic responses were significantly better in UC subjects treated with allogeneic treatment. Paramsoti and others investigated the efficacy of FMT by colonoscopy in recipients with mild to moderate UC. Most patients received FMT by injecting donor material from 3–7 patients [17, 22]. Steroid-free remission and endoscopic response or remission were achieved in 11 of 41 (27%) recipients treated with active fecal material and in 3 of 40 (8%) patients treated with placebo (saline). An enhancement accompanied the clinical response in intestinal microbiota diversity and the lack of an effect was associated with a relative increase in *Fusobacterium*.

Also, the effectiveness of FMT was studied in recipients with mild to moderate UC through repeated receiving of frozen fecal material from several donors by enema [23]. When comparing the results with the previous study's data, remission was found in 9% of placebo-treated patients as opposed to remission in 32% of patients treated with fecal material. The LOTUS study, the first which used oral FMT as maintenance therapy in UC and assessed the long-term effectiveness of donor engraftment with clinical, endoscopic, and histological outcomes [24]. The primary outcome was corticosteroid-free clinical remission with endoscopic remission or response at week 8. At week 8, FMT responders were randomly assigned to either continue or withdraw FMT for a further 48 weeks. At week 8, 53% of patients in the FMT group achieved primary endpoint as compared to 15% in the placebo group (p = 0.027; OR 5.0, 95% CI 1.8–14.1) [24]. All patients who continued FMT in the open-label phase were in clinical, endoscopic, and histologic remission at week 56 compared with none of the patients who had FMT withdrawn [24].

A systematic review with the inclusion of 25 studies (2 RCT, 15 cohort studies, and 8 case studies) was designed to assess the safety and efficacy of FMT in UC. Of 234 patients with UC included for analysis, 65% (126/193) achieved clinical response and 42% (84/202) achieved clinical remission (CR). Among the cohort trials, the pooled estimate of cases that achieved CR and clinical response were 41% (95% CI 24.7–58.7%), and 66% (95% CI 43.7–83.0%) [25]. Another systematic meta-analysis was conducted to assess FMT as a treatment for active UC in 277 participants. FMT was connected with better remission between four RCTs than placebo [26]. A most recent meta-analysis, involving 6 RCT and 324 patients with UC demonstrated that compared with placebo, FMT has a significant benefit in inducing combined clinical and endoscopic remission (OR 4.11; 95% CI 2.19–7.72; p < 0.0001). Subgroup analyses of influencing factors showed no differences between fresh or frozen FMT (p = 0.35) and different routes or frequencies of delivery (p = 0.80 and p = 0.48, respectively) [27]. In contrast, a recent meta-analysis, with the inclusion of 14 RCT found that fresh (40.9%) as compared to frozen (32.2%) FMT can increase clinical remission rates in IBD patients, with no significant risk of study heterogeneity (I2 = 38%, p = 0.03) [28].

This way, in some people with UC, FMT demonstrates its obvious advantages over placebo. But its effect seems to connect with the method of introduction of the donor material and the donor material used (fresh or frozen material, from a normal healthy donor or a superdonor, from one or more donors), as well as the number of procedures (one-time FMT, repeated sessions), previous treatment and severity of UC. At the same time, it is still not known which microbes or microbial components ensure the effectiveness of FMT in patients with UC [17].

10.4 Pouchitis

After a patient has had surgery to remove the colon and rectum (large intestine), an ileal pouch-anal anastomosis is performed to create a structure or pouch that can store and eliminate stools. Pouchitis is a frequent complication in these patients, especially in individuals with UC. Pouchitis – the pouch inflammation – can be due to idiopathic or secondary causes. Chronic antibiotic-dependent pouchitis (CADP) and chronic antibiotic-resistant pouchitis (CARP) are the most difficult forms of chronic idiopathic pouchitis to treat [29]. The effectiveness of antibiotics in pouchitis determines the need to use tools that can modulate the life of the intestinal microbiota [30]. It is also emphasized that treatments such as FMT have good preliminary results, shown in small group studies, in the treatment of forms of chronic ileus. Thus, for 2 weeks, patients with CARP were administered a single FMT through a nasogastric tube; of the eight patients enrolled, none achieved clinical remission. However, two patients with previous resistance to ciprofloxacin showed sensitivity to the antibiotic, and a decrease in PDAI ≥ 3 after FMT was also observed [31]. Selvig et al. demonstrated the largest pilot, prospective, open-label study in 19 patients with chronic reservoir pouchitis, a single FMT by puchoscopy from healthy donors at screening. Although PDAI scores after FMT were unavailable for all patients, and some had PDAI scores ≤ 6 before FMT, endoscopic and histological findings did not significantly decrease [32]. Of note, there was a statistically significant improvement in bowel movements and a trend toward less abdominal pain after FMT [32]. Stallman et al. partially confirmed these findings and reported endoscopic response in all patients (n = 5) and endoscopic remission in one patient (20%) after FMT infusions (n = 1–7 instillations) [33]. Today, there is an urgent need for controlled research. Thus, in the only double-blind RCT known to date in which participants were randomized to one colonoscopically delivered FMT followed by oral capsules versus placebo, the study was stopped prematurely due to low donor FMT engraftment and lack of clinical response [34]. According to existing data, the use of antibiotics before FMT in UC patients promotes the establishment of beneficial xenomicrobiota, with improved clinical and histological responses [35]. Considering the fact that antibiotic therapy is the main method of treatment of pouchitis, the use of antibiotics before FMT is promising, provided that adequate antibiotic selection, dosage and duration of therapy are ensured [36]. Furthermore, successful multiple FMT in patients with UC and a reservoir is accompanied by an increase in *Firmicutes* and a decrease in *Proteobacteria* in the iliac bag [37]. In contrast to Selvig's study, microbiota profiling revealed no clear changes at the community level after FMT, although a small number of specific bacterial taxa differed significantly in relative abundance [32].

Data on the effectiveness of FMT in pouchitis were summarized (based on three studies) in a sub-analysis as part of a comprehensive meta-analysis of subtypes of IBD [38]. Furthermore, these studies were descriptive, with different FMT infusion regimens, endpoints, and debatable results. The frequency of clinical remission after FMT was 21.5% [38]. The most recent meta-analysis of pouchitis was based on nine studies (69 patients treated with FMT). Overall clinical response after FMT was reported in 14 (31.8%) of 44 evaluable patients at various time points after FMT, and clinical remission in 10 (22.7%) patients [39].

Accordingly, many questions need to be answered, namely: the optimal dosing regimen (frequency and interval of infusions), advantages of using multi-donor/ combined infusions, pre-selection of donors according to microbial profiles, etc. [17, 36].

10.5   Crohn's Disease (CD)

CD is a special type of IBD that causes chronic inflammation of the gastrointestinal tract. Given the positive results of FMT in patients with UC, its effectiveness was also studied in patients with CD. To date, the question of the effectiveness of FMT in patients with CD is still open. Studies with mixed results have been published, but there has been no RCT on the effectiveness of FMT in patients with CD [40, 41]. No significant improvement was found among the 6 patients with CD at week 8 following FMT via ileocolonoscopy. However, one patient experienced temporary clinical remission for 6 weeks. A recent study with an open-label design included 30 patients with refractory CD (Harvey-Bradshaw Index (HBI) score $\geq$ 7) for single FMT through mid-gut and assessed during follow-up. It was established that the rate of clinical improvement and remission in the first month was 87% and 77%, which was higher than other assessment points within a 15-month follow-up. A decreased abdominal pain associated with sustaining CD was also observed using FTM [41]. Nine patients aged 12–19 years with mild to moderate symptoms as defined by the Pediatric Crohn's Disease Activity Index (PCDAI 10–29) participated in an open-label study of nasogastric tube FMT with follow-up at 2, 6, and 12 weeks. The mean PCDAI score improved at 2 weeks by 67% and at 6 weeks by

56%. 7 of 9 patients were in remission at 2 weeks, and 5 of 9 patients who did not receive additional medical therapy were in remission at 6 and 12 weeks. Patients who did not receive a graft or whose microbiome was most similar to the donor microbiome showed no or moderate improvement [42]. Meta-analysis with the inclusion of 39 CD patients estimated the efficacy of FMT. It was found a pooled estimate of CR of 61% (95% CI 28.4–85.6%) for CD (P = 0.05; I^2 = 37%) [43]. Data from 2020 studies (for CD and for UC) were summarized based on the latest meta-analysis. In particular, it was found that frozen fecal material from universal donors may be associated with a higher CR rate, especially for CD. Pairwise meta-analyses of six controlled trials showed significant effectiveness of FMT compared with placebo. An overall CR of 37% and overall clinical response of 54% were established for the interventional investigations [44].

Sokol et al., initiated and conducted the first randomized, single-blind, controlled trial (NCT02097797) of FMT in adults with colon or ileocolonic CC. The primary endpoint was donor microbiota implantation at week 6 (Sorensen index >0.6). Importantly, no patient met the primary endpoint. Steroid-free clinical remission rates at 10 and 24 weeks were 44 and 33% in the sham transplant group and 88 and 50.0% in the FMT group. Endoscopic Crohn's Disease Severity Index decreased 6 weeks after FMT (p = 0.03) but not after sham transplantation (p = 0.8). In contrast, CRP levels increased 6 weeks after sham transplantation (p = 0.008) but not after FMT (p = 0.5) [45].

In patients with CD, fever is often a side effect several hours after FMT. A recent large-scale study evaluated long-term adverse event (AD) risk factors and short-term efficacy of FMT in patients with CD. Within 1 month of FMT, 14% of mild adverse events occurred, including increased frequency of defecation, fever, abdominal pain, flatulence, hematochezia, vomiting, abdominal distension, and herpes zoster. It is important that AE were not observed for more than 1 month. Thus, to determine the short-term and long-term side effects of FMT, a cut-off of 1 month can be proposed [46]. Other side effects include dyspepsia, gastroenteritis, *Escherichia coli* bacteremia, flares of UC, gastrointestinal tract bleeding, peritonitis, or enteritis, colon micro-perforation, pneumonia, transient sore throat and headache, hypotension, herpes zoster, significant weight gain and other [47].

FMT is a safe, highly effective therapy for CD. However, in addition to the already clear advantages, the issues of donor screening, stool preparation and application, the exact mechanism of microbial recovery, and long-term safety remain open [17]. Accordingly, in the FMT perspective are sealed frozen FMT agents, they can be administered orally, which significantly reduces risks and costs [47].

10.6 Irritable Bowel Syndrome (IBS)

IBS is a group of symptoms that occur together, including repeated pain in your abdomen and changes in your bowel movements, which may be diarrhea, constipation, or both. Patients with IBS exhibit a reduced diversity of intestinal microbiota, an increased number of *Enterobacteria* and a relatively low level of *Bifidobacteria* and *Lactobacilli* [48]. Underproduction of butyrate and overproduction of acetate and propionate in patients with IBS is usually accompanied by symptoms such as abdominal pain and distension. Currently, there are few controlled investigations about the effectiveness of FMT in the treatment of IBS. For example, in a single-blind RCT patients (n = 90) with IBS with diarrhea (IBS-D) or mixed IBS (IBS-M) received FMT or placebo. It was found significant difference in clinical improvement of the FMT group (mixed material from two donors) in the after 3 months [49]. Instead, Halkjær et al. reported a greater reduction in IBS-SSS (-125.71 ± 90.85) in the placebo group (n = 23) at 3 months compared to the FMT group (n = 22) who received FMT capsules for 12 days (-52.45 ± 97.72) [50]. Moreover, patients treated with FMT in capsules had greater microbial diversity, but symptomatic improvement was greater than patients in the placebo group [50]. Holvoet et al. conducted an examination among 64 patients with IBS without constipation [51]. Patients participated in single-blind, randomized trials in a ratio of 2:1. The aim of the studies was to obtain FMT by colonoscopy from 2 donors (experimental group) or FMT from own feces (control group). It is worth noting that in the experimental group there was a significant decrease in discomfort, abdominal pain and bloating compared to the control group. According to the microbiome analysis, it was found that patients who received effective FMT had a higher baseline concentration of *Streptococcus* as well as a greater enrichment of the whole gut microbiome [51]. A double-blind, placebo-controlled RCT on the efficacy of FMT capsules in patients with IBS-D was conducted by Aroniadis JC et al. in three centers in the USA [52]. Patients were allocated (1:1) in blocks of four by computer randomization sequence to receive 75 capsules of either donor stool (each containing 0.38 g) or 75 placebo capsules for 3 consecutive days (25 capsules per day). Interestingly, after 12 weeks, there was almost no difference in treatment efficacy between the two groups (as measured by the IBS Severity Index). Accordingly, it was concluded that the effectiveness of FMT in patients with IBS requires further thorough research [52].

Today, the results of the studies reviewed by us are combined in a meta-analysis. Importantly, a meta-analysis of four [53], five RCT [54] and seven [55, 56] found that FMT did not improve overall IBS symptoms compared with placebo. At the same time, the meta-analysis evaluated data from single-group studies (SAT) and RCTs separately. Thus, in SAT, 60% (95% CI 49.1–69.3) of patients with IBS showed significant improvement. In the RCT, there were no differences between FMT and controls in terms of improvement (HR = 0.93 (95% CI 0.50–1.75)) or change in IBS-SSS [57]. Delivery method is probably worth considering, as classic FMT is promising, but capsule delivery is not. In a subanalysis, advantages of single-dose FMT using colonoscopy and nasojejunal tubes were observed compared to autologous FMT for placebo treatment and reduced likelihood of multiple capsule improvement [53, 54]. In contrast to previous meta-analysis, two recent ones which were published in 2022, similar does not show significant global improvement in IBS patients. Howsoever, FMT operated by invasive routes, *via* colonoscopy or gastroscope, significantly improved global IBS symptoms [55, 56].

Benno et al., reviewed FMT in the form of anaerobically cultured human intestinal microbiota (ACHIM) via upper gastrointestinal endoscopy to achieve treatment response in 50 patients with IBS. Symptom reduction in the majority of patients, 32 with a 50-point IBS-SSS reduction and 21 with a 100-point IBS-SSS reduction, was observed independent of changes in SCFA baseline [58]. El-Salhy et al. published the first placebo-controlled RCT in IBS that used gastroscopy as a delivery route to the distal duodenum and compared the efficacy of two different doses of donor stool in ultra-high-density fecal solution (30 g FMT and 60 g FMT in 40 ml saline) [59]. The study included 165 patients with IBS, resulting in higher FMT response rates than in previous studies [60]. Response, defined as a reduction of 50 points or more in the IBS-SSS at 3 months, corresponding to a 10% improvement in symptoms, was achieved by 77% of patients in the 30 g FMT group, 89% of patients in the FMT group, 60 g FMT group, and 24% of patients in the placebo group. Symptom remission, defined as an improvement in IBS-SSS score of at least 175 points, was achieved in 35% of patients treated with 30 g stool, 47% of patients treated with 60 g stool, and 5% of patients treated with placebo [59]. Currently the same group, reported a 3-year follow-up of the patients in our previous clinical trial to clarify the long-term efficacy and possible adverse events of FMT [61]. The response rates were significantly higher in the 30-g and 60-g groups than in the placebo group at 2, 3 years after FMT. No long-term adverse events were recorded [61].

Despite the undeniable facts of the effectiveness of FMT in the treatment of IBS, we still consider it necessary to further develop this problem, as well as a detailed study of previous antibiotic therapy [62] and the profile of the intestinal microbiome before and after the procedure [17]. In addition, scientists should pay attention to which changes in the gut microbiota are associated with clinical improvement. The mechanisms for the cross talks between gut microbiota and colonic enteroendocrine cells remain also to be investigated. It was found, that colonic enteroendocrine cells densities significantly change in FMT responders [63].

10.7   Microscopic Colitis (MC)

MC is determined by the presence of microscopic modifications of the mucous membrane of the large intestine in patients without macroscopic lesions who had signs of chronic diarrhea. There are two types of MC lymphocytic colitis (LC) and collagenous colitis (CC). However, it is not entirely clear whether these are different manifestations of one unique disease or whether they are other conditions [64]. Note that 1–6% of patients with microscopic colitis are not amenable to medical treatment. Accordingly, the possibility of effectiveness of FMT in MC was considered [17]. Thus, in one of the recent reports, an assessment of the effect of repeated FMT was made in a patient with CC refractory to drug treatment, in particular, severe symptoms were observed, including profuse diarrhea and weight loss. After 3 courses of FMT, the patient remained in remission for 11 months [65].

Another study discussed the development of a new MC in a patient with UC after FMT [66]. Thus, there is a known case of studying the effect of three repeated FMT (day 0, 2 weeks, 4 weeks) in patients with CC (n = 9), where feces from two donors were studied. The primary endpoint (remission at 6 weeks, defined as <3 stools of which <1 watery stool per day) was achieved in two patients and in one at 8 weeks. Thus, FMT resulted in an increase in the number of lamina propria lymphocytes, which probably indicates the initial immune activation of the mucosa [67].

In the future, studies that will help characterize this subgroup and understand whether the stated results apply to all patients with MC.

10.8   Functional Constipation

Functional constipation is directly related to intestinal dysbacteriosis, but in this case there is much less evidence of disturbance of the intestinal microbiota. Thus, in a RCT (n = 60), the effectiveness of FMT in constipation with delayed intestinal transit was investigated [68]. Patients treated with FMT by nasogastric tube administration of donor material had significant improvement and acceleration of intestinal transit compared to the control group. Improvements in a clinical remission rate were also observed in other RCT studied in patients with chronic functional constipation after three courses of FMT by gastroscopy [69]. Also, FMT treatment promoted intestinal motility, increased levels of NO, 5-HT and amounts of *Prevotella* and *Acidaminococcus* [69]. According to the processed data, a decrease in the frequency of clinical recovery over time and a higher effectiveness of FMT in combination with the polysaccharide pectin were observed in patients with chronic constipation [70, 71]. However, it should be emphasized that the FMT treatment was quite intensive and invasive, as patients received up to 18 nasoduodenal FMTs within 3 months [71]. The outlined limitations can be solved with the help of FMT capsules. The effectiveness of faecal microbiota in FMT can also be considered due to the fact that glycerol has a relaxing effect. It is used to protect the gut biomaterial from freezing [72].

10.9   Antibiotic-Associated Diarrhea (AAD)

Antibiotics can cause dysbacteriosis, which is accompanied by diarrhea. In particular, there is one study devoted to the expediency of the use of FMT or probiotics in patients with AAD [73]. FMT has been shown to promote rapid recovery of IM and improve symptoms, whereas multi-probiotic treatment slows recovery of IM [17].

10.10   Immune Checkpoint Inhibitors Associated Colitis and Gastro-intestinal Cancers

Fatefully that the interest to FMT as a therapeutic option in the treatment of oncology has risen after the discovery of the impact of gut microbiota in determining the response to immune checkpoint inhibitors [74].

Thereby, immunotherapy significantly influenced the development of oncology as a whole. Treatments with immune checkpoint inhibitors (ICI) targeting cytotoxic T-lymphocyte-associated antigen 4 (CTLA-4), programmed cell death protein 1 (PD-1) and programmed cell death ligand 1 (PD-L1) is associated with increasing the number of T cells activation and effective anti-tumor immune responses [75]. At the same time, a number of side effects were observed, in particular colitis, which was caused by immune checkpoint inhibitors [76]. After FMT, there was a significant decrease in CD8+ T-cell density with a concomitant increase in mucosal CD4+ FoxP3+ in one participant, meaning that FMT may actually reverse immunotherapy-related toxicity [76, 77]. To date, there are several studies, most of which are pre-clinical, devoted to the study of FMT in gastro-intestinal cancer, including colorectal cancer [78], hepatocellular carcinoma [79], and pancreatic cancer [80].

10.11   Chronic Liver Disease

Chronic liver disease is the most common cause of morbidity in the world [81, 82]. Despite the achievements of modern medicine in the treatment of most chronic viral hepatitis, many causes of chronic liver disease remain insufficient. According to recent research data, the human gut microbiome plays an important role in the development and progression of chronic liver diseases. Accordingly, FMT may become an important segment in general therapy related to liver diseases [83].

Thus, alcoholic hepatitis has an extremely high short-term mortality rate of up to 50%, and currently there are no specific treatment methods other than steroids [84]. The only treatments are a liver transplant and abstinence from alcohol. FMT has also been considered as a potential treatment for severe alcoholic hepatitis (SAH) [85]. Philips et al. administered FMT for patients during 7 days, and eventually the researchers reported an improvement at 1 year in survival compared to the control group (88% vs. 33%) [86]. In particular, the improvement was related to the relative abundance of *Proteobacteria* and *Actinobacteria*, as well as a decrease in pathogenic *Klebsiella pneumoniae* species (from 10% to 1% in 1 year) and an increase in beneficial species (*Enterococcus villorum*, *Bifidobacterium longum* and *Megasphaera elsdenii*) [86]. Based on a study involving 51 patients, it was established that FMT is associated with a decrease in the frequency of hepatic encephalopathy (HE), respectively, there were higher survival rates after 3 months (survived 75% of patients with FMT versus 38, 29 and 30% in group of corticosteroids, nutrition and pentoxifylline) [87]. Preliminary results of the first RCT (NCT 03091010) comparing patients (n = 112) with SAH treated with steroids or FMT showed good results with improved 90-day survival in the FMT group [84]. Recent open-label study, in patients with SAH presenting as acute-on-chronic liver failure (ACLF), demonstrated single FMT administered as a freshly prepared stool suspension from pre-identified healthy family member, increase survival rate at 28 and 90 days in the FMT arm (100% versus 60%, p = 0.01; 53.84% versus 25%, p = 0.02) [88]. HE resolved in 100% versus 57.14% (FMT versus SOC, p = 0.11) patients, while ascites resolved in 100% versus 40% survivors (p = 0.04) [88]. Taking these data into account, we can conclude that further studies of FMT are promising [17].

Primary sclerosing cholangitis (PSC) is a chronic cholestatic liver disease, which in the majority of patients progresses to liver transplantation or death [89]. About 70–80% patients with PSC have a concomitant diagnosis of IBD, usually UC [90]. According to the conducted studies, the intestinal microbiota plays an important role in the etiopathogenesis of PSC [91]. Thus, according to one of the hypotheses, bacteria from the inflamed intestinal mucosa move through the portal vein system to the liver and stimulate an adverse immune response in the biliary system [92]. Studies of the fecal microbiota have highlighted enriched levels of certain species, including *Veillonella*, *Streptococcus* and *Enterococcus*, among others [93]. To date, no medical treatment has been proven to be of benefit. Several antibiotics, including vancomycin [94] and FMT, have shown good results [95]. In 2018, the first case report described the utility of weekly FMT in healthy donors for the treatment of recurrent acute bacterial cholangitis in PCS. After FMT, the patient's bilirubin and alkaline phosphatase (ALP) levels decreased, and remission lasted for more than a year [96]. One of the most recent pilot studies presented the results of using FMT in 10 patients with PSC and an increase in ALP more than 1.5 times the upper limit of normal and simultaneous IBD. As this was the first study of its kind, the primary endpoint was safety. There were no related adverse events. 30% (3/10) of patients had at least a 50% decrease in ALP, with the relative abundance of these bacteria present in the stool microbiota after FMT, but not before FMT, being negatively correlated with alkaline phosphatase levels [97]. Therefore, there is an urgent need for further research that would confirm the effectiveness of FMT in PSC.

Hepatitis B is one of the most common infectious diseases globally. It has been estimated that there are 350 million chronic hepatitis B virus (HBV) carriers worldwide [98]. New technologies have enabled systematic gut microbiota studies and provided more realistic information about its composition and pathological variance [99]. FMT may be a useful therapy for HBV-related diseases. Thus, in a study involving 18 participants with persistently positive hepatitis B e-antigen (HBeAg), it was found that 1 month of FMT treatment reduced HBeAg titers over time, and 2/5 in the treatment group achieved HBeAg clearance versus 0/13 in the control group [100]. They found that monthly FMT treatment reduced HBeAg titers over time, and 2/5 patients in the treatment arm achieved HBeAg clearance versus 0/13 in the control [100]. Most recently, HBeAg-positive patients, despite being on antiviral treatment for >1 year were given six cycles of FMT via gastroscope (nasoduodenal route) at 4 weekly intervals along with antiviral therapy (AVT) In the FMT arm, 16.7% (2/12) patients had HBeAg clearance in comparison to none in the AVT arm

(P = 0.188). None of the patients in either arm had HBsAg loss. The FMT was tolerated well and patients reported one or more minor adverse events [101]. Optimistically, it may allow patients with persistently positive HBeAg to come of otherwise lifelong antiviral therapy and thus warrants further investigation [102].

Non-alcoholic fatty liver disease (NAFLD) has become the most common chronic liver condition and described as a multifactorial complication due to genetic predisposition, metabolic functions, inflammatory, gut microbiota, and environmental factors [103, 104]. A major current research effort is ongoing to find potential strategies to treat NAFLD-non-alcoholic steatohepatitis (NASH), with special attention to the gut microbiota [105, 106]. In preclinical studies in mice fed a high-fat diet, FMT reduced weight gain, nonalcoholic fatty liver disease, and the accumulation of intrahepatic lipids and intrahepatic proinflammatory cytokines 8 weeks after FMT [107]. Similarly, mice transplanted with feces from human NASH exhibited increased hepatic steatosis and inflammatory cell infiltration compared to mice transplanted with feces from healthy humans [108]. Currently, human testing of FMT is limited to obese adults with metabolic syndrome (without defined NAFLD) [109]. To date, only one pilot RCT has investigated the effect of allogeneic (n = 15) or autologous (n = 6) FMT delivered to the distal duodenum in patients with NAFLD. However, FMT did not improve IR as measured by HOMA-IR or liver proton density fat fraction, but had the potential to reduce small bowel permeability in patients with NAFLD [110]. Disadvantages of this study are the different methods used to confirm NAFLD: biopsy, FibroScan, and MR elastography. In another RCT, allogenic FMT using lean vegan donors as compared to autologous, in individuals with hepatic steatosis shows an effect on intestinal microbiota composition [111]. Unfortunately, the research due to slow recruitment prematurely terminated.

NASH is the leading cause of liver cirrhosis and hepatocellular carcinoma (HCC). End-stage liver cirrhosis can lead to recurrent hepatic encephalopathy (HE), which is often associated with the development of intestinal dysbacteriosis [112]. Animal studies prove the beneficial effect of FMT for HE, in particular, the FMT study was conducted in rats. Thus, FMT prevented liver necrosis and intestinal mucosal barrier damage, leading to hepatic clearance of ammonia, decreased intestinal permeability, and decreased expression of TLR4 and TLR9 [113]. In one study involving 20 male outpatients with cirrhosis and hepatic encephalopathy, the participants were divided into 2 groups. In the first group (experimental), patients received broad-spectrum antibiotics for 5 days, and then FMT (enema) from one donor. The second group (control) received standard therapy [114]. Notably, hepatic encephalopathy recurred in 5 of 10 patients with standard treatment but in none of 10 patients treated with FMT. It is also important that patients in the experimental group showed improved cognitive function and a relative increase in *Lactobacillaceae* and *Bifidobacteriaceae*.

Furthermore, a recent study determined the long-term effects of FMT compared to standard of care (SOC) over 12 months on cognitive function, hospitalization and HE by extending the results of previous RCTs [114, 115]. SOC therapy for this condition includes intestinal laxatives (especially lactulose) and non-absorbable antibiotics (rifaximin). Overall, in those assigned to FMT, the intervention was well-tolerated and had a non-concerning long-term safety profile, without infections or need for new antibiotic initiation. There was a significantly higher number of hospitalizations and HE episodes in the SOC arm compared to the FMT arm during the long-term follow-up [115]. Therefore, many questions remain open, especially regarding the effectiveness and safety of FMT in HE, because the duration of treatment and the number of necessary transplants are still unknown. In the analyzed studies, all faecal transplants used a single donor selected on the basis of microbiome characteristics. Accordingly, it is not entirely clear whether positive results can be expected from another donor, and therefore how the composition of the donor's microbiome affects the results in general [83]. Several RCT active recruiting (NCT02862249, NCT03796598, NCT03439982) patients with HE and cirrhosis can help resolve the controversial issues mentioned above questions [17].

10.12   Acute Pancreatitis

Currently, only one case of the efficacy of FMT in an acute form of pancreatitis of moderate severity is known, which was complicated by a severe infection caused by *Clostridium difficile*. In particular, this is a 51-year-old patient from China who suffered from diarrhea during acute pancreatitis. Because the patient provided informed consent for FMT treatment, he was initially treated by FMT rather than metronidazole. Diarrhea resolved within 5 days post FMT. The patient remained asymptomatic, and the follow-up colonoscopy performed 40 days after discharge showed a complete recovery. Regarding these findings, the authors concluded that FMT might become a critical component of treating severe *Clostridium difficile* infection in acute pancreatitis patients [116]. Currently, the impact of FMT in pancreatitis patients is under active investigation and 3 RCT are registered (NCT03015467, NCT02318134, and NCT02318147) [17].

10.13   Non-gastroenterological Diseases

10.13.1   Systemic Lupus Erythematosus (SLE) and Sjögren's Syndrome (SS)

SLE and SS are two important conditions which show the properties of chronicity and autoinflammation. SLE is a complex condition which especially affects young women, is characterized with a chronic course of exacerbations and remissions and may lead to serious organ damages [117]. The first clinical trial of FMT in active SLE patients provide supportive evidence that FMT might be a feasible, safe, and potentially effective therapy in SLE patients by modifying the gut microbiome and its metabolic profile [118]. In this 12-week, single-arm pilot clinical trial oral encapsulated FMT once a week for three consecutive weeks along with standard treatment lead to significant reductions in the SLEDAI-2K scores and the level of serum anti-dsDNA antibody and IL-6 and CD4+ memory/naïve ratio, enrichment of SCFAs-producing bacterial taxa and decrease of inflammation-related bacterial taxa [118].

Sjögren's syndrome develops due to accumulation of lymphocytes in exocrine glands [117]. The most common clinical findings include xerostomia and immune-mediated dry eye (DE) which occur as a result of involvement of salivary glands and lacrimal glands. In open-label study, after 2 FMTs from a single healthy donor delivered via enema, 1 week apart, five individuals subjectively reported improved DE symptoms 3 months after FMT [119].

10.13.2   Psoriasis

Psoriasis is a widespread inflammatory skin disease with symptoms similar to IBD. It was established that patients with psoriasis have abnormalities of the intestinal microbiota, namely: a decrease in the relative abundance of *Akkermansia mucinophila* and a threefold increase in the ratio of *Firmicutes/Bacteroidetes* [120]. Therefore, the key to effective treatment of psoriasis is the improvement of intestinal microbiota. The first scientifically based evidence of the effectiveness of FMT in psoriasis involved a 36-year-old Chinese man who had suffered from severe plaque psoriasis for 10 years and IBS for 15 years. At the same time, he was administered FMT twice with the help of upper endoscopy and colonoscopy with an interval of 5 weeks After FMT, there were significant improvements in psoriasis area, dermatological quality of life index (DLQI), histological examination, intestinal symptoms and serum TNF-α level compared to baseline [121]. A 6-month, double-blind, placebo-controlled RCT of the efficacy and safety of FMT in patients with psoriatic arthritis is currently underway [122].

10.13.3   Multiple Sclerosis

Multiple sclerosis (MS) is a chronic autoimmune demyelinating disease that causes severe neurological changes, most of which have no effective treatment options. Most patients with multiple sclerosis have gastrointestinal symptoms and changes in the gut microbiota [17, 123]. It was observed that the inflammatory response in mice with autoimmune encephalomyelitis (AE) was reduced when strains of butyrate-producing bacteria were introduced [124]. But, faecal transplantation from MS patients can induce an MS-like autoimmune disease in mice [125]. Thus, Li et al. tested FMT on mice with experimental AE, a mouse model of MS. At the same time, it was established that FMT can improve the altered intestinal microbiota, reduce the level of activation of microglia and astrocytes, and therefore provide protection of the blood-brain barrier (BBB), myelin and axons in experimental AE [126]. The first case describes short-term improvement in neurological symptoms after FMT for constipation in three individuals with MS [127]. A possible long-term benefit of FMT on MS progression over 10 years has recently been proposed [128]. FMT can become a treatment method for relapsing MS. These results are associated with an increase in the relative abundance of "presumed" anti-inflammatory butyrate-producing bacteria in the subject's stool, particularly *Faecalibacterium prausnitzii*. During the FMT study, there was an increase in serum brain-derived neurotrophic factor (BDNF), subjective/objective signs of improvement in gait/ walking matrices (the MS patient's chief complaint). At the same time, patients had typical gastrointestinal symptoms, with no episodes of relapsing/remitting relapses/ exacerbations of MS symptoms during the 12-month follow-up period [129]. Despite the fact that only one patient participated in the research, the obtained findings can be the basis for the development of future RCTs evaluating FMT in patients with MS. Accordingly, it is quite logical to think that MS can be treated with FMT in the same way as UC. There are currently three active researches devoted to the effectiveness of FMT in patients with multiple sclerosis.

10.13.4   Parkinson's Disease (PD)

PD is an intractable neurodegenerative disease, usually associated with gastrointestinal disorders (constipation, IBD, and IBS). Also, patients with PD often have increased intestinal permeability [130] and bacterial overgrowth in the small intestine [131]. The gut microbiome of patients with PD (especially in complicated cases) is characterized by an excessive number of *Bacteroidetes*, *Faecalibacterium prausnitzii*, *Enterococci*, *Prevotella* and *Clostridium* [132, 133]. In general, patients with PD have more pro-inflammatory gut bacteria, particularly LPS-producing proteobacteria, and less anti-inflammatory butyrate-producing gut bacteria [132].

The transplantation of microbiota from patients with PD in a mouse model led to a worsening of neurological manifestations, whereas depletion of gut microbiota in the same model reduced neurological symptoms [134]. In another research, a mouse model of PD improved motility, increased striatal neurotransmitters, and reduced neuroinflammation after receiving feces from healthy mice. FMT of healthy mouse donors had neuroprotective effects in PD mice by suppressing neuro-inflammation and reducing TLR4/TNF-α signaling [135]. Zhou et al., observed less motor function decline and loss of dopaminergic neurons in the substantia nigra in PD mice that received a fasting mimicking diet (FMD) compared to ad-libitum-fed PD mice. Furthermore, FMT from normal mice with FMD treatment to antibiotic-pretreated PD mice increased dopamine levels in the recipient PD mice, suggesting that gut microbiota contributed to the neuroprotection [136].

The first case report, which assesses the effectiveness of FMT in PD was described in a 71-year-old male patient who presented with 7 years of resting tremors. FMT resulted in a reduction in leg tremor and other PD symptoms as early as

1 week after three FMTs. However, leg tremors returned to baseline 2 months after FMT, as did other PD symptoms, unfortunately. Instead, the constipation also decreased, which lasted until the end of the observation [137]. Xue et al., represented the data of the first pilot study of the efficacy and safety of FMT, in which 15 patients with PD participated. In particular, colonic-preferred FMT, rather than naso-intestinal, has been found to alleviate motor and non-motor symptoms.

Furthermore, 2 from 10 of patients the colonic FMT group achieved self-satisfying outcomes that last for more than 24 months. In contrast, no patients were satisfied with nasointestinal FMT for more than 3 months. Overall, it can be argued that FMT was safe, with only five mild side effects reported during the study [138]. Therefore, FMT may be considered a research prospect in treating PD, but further thorough experiments are needed [17, 139].

10.13.5 Autism Spectrum Disease (ASD)

ASD is a group of neurodevelopmental disorders characterized by altered social communication and interaction as well as repetitive, stereotyped behavior [17, 140]. ASD is usually characterized by constipation, bloating, diarrhea, and changes in the gut microbiome [141, 142]. Children with autism typically have a reduced *Bactero idetes/Firmicutes* ratio [143] and elevated levels of the genus *Clostridium*, which is known to produce potentially toxic metabolites [140]. Presumably, changes in the microbiome may interact with tryptophan metabolism and contribute to behavioral change, but strong evidence is currently lacking [144].

FMT from ASD children to germ-free wild-type mice was associated with the development of ASD-like symptoms and demonstrated alternative splicing of ASD-relevant genes in the offspring [145]. Another study demonstrated a reduction in cerebral oxidative stress after FMT with normal hamster feces in a hamster model of ASD. Importantly, this effect was more pronounced after the introduction of *Lactobacillus paracaseii* [146].

A significant improvement in ASD behavior indicators and decreased intestinal symptoms (abdominal bloating, constipation, diarrhea) were noted in a pilot open-label study on 18 children (aged between 7 and 16 years old) with autism after a 2-week course of antibiotic treatment. Such processes were observed for more than 8 weeks after the introduction of FMT. In addition, there was engraftment of donor stool microbiota with an increase in both overall bacterial α- diversity as well as the abundance of *Bifidobacteria, Prevotella* and *Desulfovibrio* which persisted for more than 8 weeks post-FMT [17, 147]. These same gains were maintained in participants for 2 years after FMT [148]. But, as already mentioned, the study was open-label, so there was no comparison group of patients who received placebo/autologous FMT. Also, unfortunately, diet or nutritional supplements were not controlled. And most importantly, there is no information about side effects during long-term follow-up. Zhao et al., in an open-label, randomized trial, found improvements in ASD symptoms and changes in gastrointestinal symptoms 2 months after two FMTs in 24 children with ASD compared with 24 control children with ASD. In seven patients with FMT, side effects, namely nausea, fever, and allergy, were found, which fortunately were mild [149]. At the same time, we emphasize that the established potential cause-and-effect relationship between the microbiome and ASD still requires additional arguments [17].

10.13.6 Epilepsy

Epilepsy is a chronic disease accompanied by sudden abnormal cerebral neuron discharge, leading to transient brain dysfunction [150]. Although the origin of most cases of epilepsy remains unknown, susceptibility to epilepsy is primarily linked to genetic and environmental factors [151]. Intestinal microbiota composition in patients with refractory epilepsy differs significantly from healthy control groups. Thus, some studies indicate an increase in the ratio of *Firmicutes/Bacteroides* and α-diversity, as well as the number of rare genera of bacteria, namely: *Ruminococcus, Akkermansia, Neisseria, Coprococcus, Methanobrevibacter*, and *Roseburia* [152, 153]. High levels of *Bifidobacterium* and *Lactobacillus* were found to be associated with fewer attacks per year [152], and a ketogenic diet reduced attack frequency by modulating the gut microbiota [154].

A recent animal study demonstrated that FMT from stressed donor rats increased the progression and duration of seizures (standing, with or without falling) in sham-stressed rats. Proepileptic effects were eliminated in stressed recipients of donor feces from sham-stressed rats [155]. Olson et al. established that transplantation of ketogenic microbiota or long-term administration of *Akkermansia muciniphila, Parabacteroides merdae*, and *Parabacteroides distasonis* reduced the number of seizures in mice at a higher threshold level [17, 156]. In addition, the researchers reported a case of generalized epilepsy and Crohn's disease in a 17-year-old patient who had worsening neurological and intestinal symptoms after three rounds of FMT. This antiepileptic therapy discontinued sodium valproate after 20 months, and no seizures were observed [17, 157].

10.13.7 Other Neurological Disorders

Vendrik et al. research and description of cases of FMT in neurological disorders in humans or animals. Thus, among 541 studies, 34 were taken into account in the analysis. There have been studies involving animal models for stroke, Alzheimer's disease, and Guillain-Barré syndrome. At the same time, the conducted studies confirmed the positive effect of FMT of healthy donors. Instead, one study in an animal model of stroke found increased mortality after FMT. For Guillain-Barre, only one study was identified. Therefore, we cannot confirm the positive results of animal research in treating human diseases. It should also be noted that several trials of

FMT are ongoing or planned for both the above-mentioned neurological disorders and for amyotrophic lateral sclerosis [139].

10.13.8 Metabolic Syndrome/Obesity/Hypertension

As a rule, the development of metabolic syndrome is associated with changes in the intestinal microbiome. Thus, until recent studies, the difference in bacterial flora in the human intestine is one of the main aspects of obesity [158, 159]. Metagenomic studies and analysis of 16S ribosomal DNA analysis revealed differences in the composition of the gut microbiota and most genes, including when comparing faecal subjects with obesity and those with a healthy weight [160]. There is a direct relation between dysbiosis and obesity because of diminished fasting-induced adipose factor (also known as an inhibitor of lipoprotein lipase) and AMP-activated protein kinase activity (known as a key enzyme that controls cellular energy status), decreased vitamin absorption, disabled production of SCFA, reduced inflammatory processes, and metabolic endotoxemia [17, 161].

Because the gut microbiota is so important for metabolism and a large number of research woks demonstrated that the microbiota changes under obesity, the question arises whether certain manipulations of intestinal bacteria can improve the composition of the gut microbiome in obese people to lose weight and improve metabolism [162, 163]. In the future, this may become a therapeutic strategy for treating obesity. In one pilot study, 18 patients were divided into two groups: allogeneic FMT from lean donors (n = 9) and obese patients who received autologous FMT (n = 9). The allogeneic FMT group demonstrated improved insulin sensitivity at 6 weeks [164]. In contrast, another, more robust study (n = 38) found that allogeneic FMT (n = 26) did not reduce insulin resistance compared with autologous FMT (n = 12) after 18 weeks and was directly related to the lack of overall changes in composition intestinal microbiome [17, 165]. Allegretti et al. in RCT investigated using oral capsules of FMT from lean donors in obese patients. It was determined that capsules with gut stool did not influence patients' body mass index (BMI) but led to a decrease in the concentration of taurocholic acid. Consequently, there was a change in the composition of the intestinal microbiota, which corresponded to the diversity of the lean patient. Both α and β-diversity were increased, as well as the number of genus *Faecalibacterium*. It can be concluded that FMT from lean donors may be effective for bacterial diversity improvement, but unfortunately, not for BMI reduction [166]. The same group reported a secondary analysis of a previous RCT with analysis of postprandial glucose and insulin levels. A significant change in glucose AUC was observed at week 12, and insulin AUC changed at week 6 [17, 167]. Similar placebo-controlled studies were conducted by Yu et al., who applied FMT from lean subjects to obese and IR patients. FMT was found to be neither associated with changes in BMI, adipose tissue, or insulin sensitivity nor with gut microbiome diversity [168]. Zhang et al. compared data on the use of FMT described in different studies. In particular, two studies reported improvements in peripheral insulin sensitivity with FMT. FMT increased the species of *Roseburia intestinalis*, *A. muciniphila* and *Clostridium spp.* [169]. In meta-analyses, based on 6 RCTs (n = 154), investigated the effectiveness of FMT from the frail donor(s) for the therapy of metabolic disorders in comparison with any form of placebo (placebo capsules or sham, saline, autologous FMT). There was no difference in obesity parameters 6 to 12 weeks after intervention. Nevertheless, HbA1c level was decreased and mean HDL cholesterol was up in the FMT group in 6 weeks of the study [170].

It has now been established that the response of patients with metabolic syndrome to a change in the intestinal microbiome depends on the initial state of the microbiome and diet. So, for example, Guirro et al. investigated the effect of a hypercaloric diet on the intestinal microbiota. Significantly, this diet changed the gut microbiota. In this way, the use of FMT, after antibiotic therapy, increased the number of *Bacteroidetes*, *Firmicutes* to the initial level observed before antibiotic treatment [171]. The largest as for nowadays RCT included 90 participants, which made it possible to assess the effectiveness and safety of autologous FMT (the DIRECT PLUS trial) [172]. Participants were randomly assigned to receive 100 capsules of their own frozen fecal microbiota or a placebo for up to 14 months. Thus, it has been found that FMT administered during the recovery phase can maintain weight loss and glycemic control and is directly associated with specific microbiome signatures. A diet based on consumption of green plants high in polyphenols optimizes the microbiome for FMT procedure. In details, FMT significantly attenuated weight regain in the green-Mediterranean group (FMT, 17.1%, vs placebo, 50%; P = .02), but not in the dietary guidelines or Mediterranean diet groups [173]. Recent studies [174, 175], also confirmed the suggestion that FMT followed with dietary and lifestyle interventions, associated with microbiota engraftment, improvement in lipid profile and liver stiffness in obese T2D patients.

A hypothesis of gut microbiota intervention for treating hypertension is now postulated [176]. Fan et al., reported the protocol for multicenter, placebo-controlled RCT in 120 grade 1 hypertensive patients for 3 months (NCT04406129). All recruited patients will be randomly assigned in a 1:1 ratio to take oral FMT capsules or placebo capsules. The primary outcome is the office systolic blood pressure change from baseline to day 30 [176].

Positive impact studies and meta-analyses provide further evidence for the benefits of FMT as an option for obese patients with metabolic syndrome. Most of the physiological changes that occur after FMT are antidiabetic in nature, namely improved glucose handling, increased basal metabolism, and reduced systemic inflammation [177]. However, determining the role of FMT in these patients requires high-quality RCTs with long-term follow-up [17].

10.14 Discussion/Conclusion/Obstacles and Limitation of Wide Implementation

Thus, despite advances in gut microbiome research, few controlled studies have investigated the effectiveness of therapeutic interventions that can genuinely modify it. The same applies to FMT. As we know, FMT was first used to treat patients with recurrent *Clostridium difficile*-associated colitis, which is now the only official reason for its use. At the same time, FMT rules and classifications have changed. Let us emphasize that the vectors of FMT research differ significantly in different countries (Fig. 10.1).

The Medicines and Healthcare Products Regulatory Agency (MHRA) in the UK and the FDA in the USA now define FMT as a medicinal product. In Europe, however, in 2014, the European Commission qualified FMT as a "combined substance" consisting of human and non-human components. Accordingly, FMT is not subject to the European Tissue and Cell Directive; therefore, its status must be determined locally [74].

The condition is under investigation and has active recruiting status according to the https://www.clinicaltrials.gov/ database on a beginning of 2022 presented in Fig. 10.2. Nevertheless, the effectiveness of FMT is currently being studied intensively or refined in treating other gastroenterological and non-gastroenterological diseases, some of which were discussed above (Table 10.1).

PPPM strategies in the field of well-designed clinical trials of FMT will help validate the microbiota's role in disease and identify the specific bacteria or metabolites responsible for this effect. We emphasize that the mentioned tests will determine the development vectors of targeted bacteriotherapy. Accordingly, with the increase in the number of FMT treatments and clinical trials in other indications, there is a need to develop standardized guidelines to ensure patient safety, and therefore to develop microbiota-based medicines that are safer [16].

So far, many questions that potentially have a great impact on the effectiveness of FMT remain unanswered. These questions are related to the origin of the disease, the initial state of the recipient patient's IM and the selection of a suitable donor (super donor?) or donors, as well as the means of introducing the material and the frequency of procedures. We hope that research in the next 5–10 years will be able to provide answers to all of the aforementioned questions and that causes need to be elucidated to efficiently enable PPPM.

10.15 Expert Recommendations in the Framework of 3PM to the Practical Medicine

Clinical research on FMT has dramatically increased in the last few years and is still ongoing. The results of existing clinical studies show that FMT treatment can be beneficial for a wide range of acute and chronic diseases and new findings are being

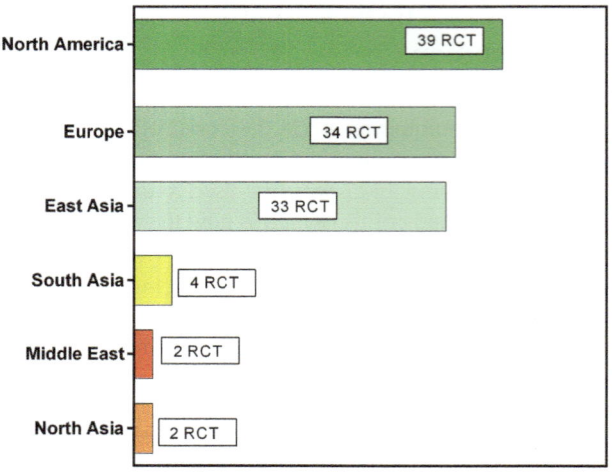

Worldwide distribution of FMT research across regions

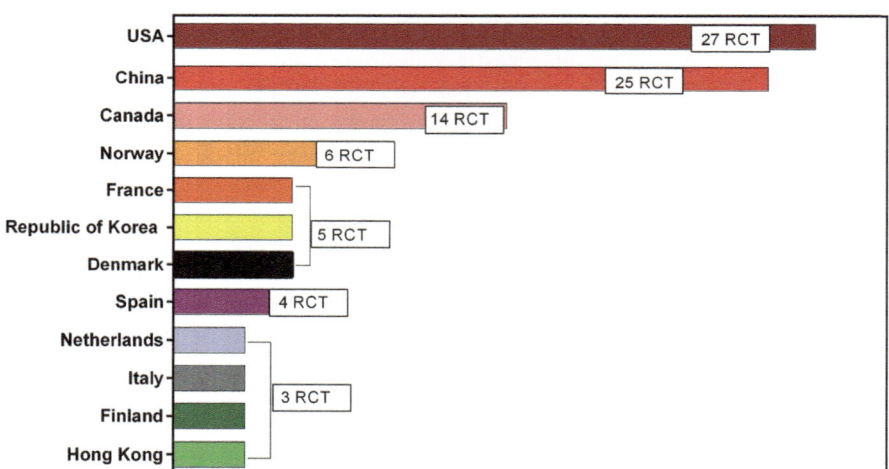

List of countries with most active contribution in FMT investigation

Fig. 10.1 The contribution for FMT research in different regions. (Published [17] and modified for this chapter)

published constantly. Future PPPM/3P-related studies should consider the proposals presented below:

– This article emphasizes that an individual approach to each patient with appointment FMT which is the basis of personalized medicine, can improve clinical

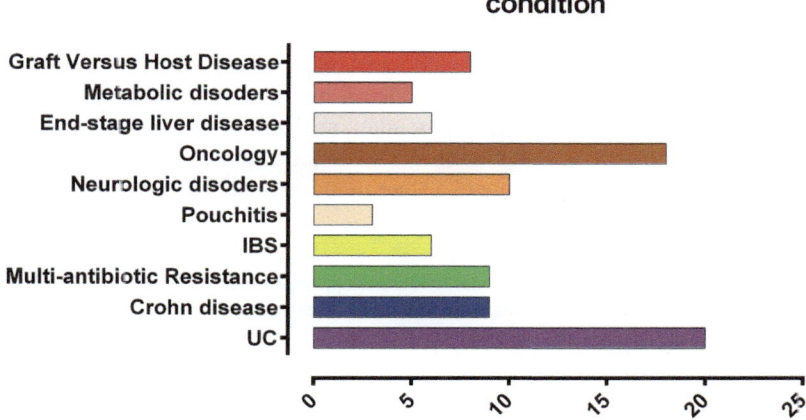

Fig. 10.2 The contribution for FMT research between different regions. (Published [17] and modified for this chapter)

Table 10.1 Diseases for which the effectiveness of FMT is being investigated

Disease	Study type	Assessment of FMT effect to date
Gatroenterological pathology		
NUC	RCT	Overall positive
Crohn's disease (CD)	Series of cases	Negative thus far
IBS	RCT	Assumed
AAD	Series of cases	Assumed
Constipation	RCT	Assumed
Hepatic encephalopathy	RCT	Assumed
Neurological diseases		
Multiple sclerosis	RCTs ongoing	Unknown
Parkinson's disease	RCTs planned	Unknown
Autism	Uncontrolled pilot study	Assumed
Psoriasis	RCTs ongoing	Unknown
Metabolic syndrome	Controlled study	Assumed

Note: RCTs are randomized clinical trials

outcomes in patients with ulcerative colitis, irritable bowel syndrome, antibiotic-associated diarrhea, hepatic encephalopathy, autism and metabolic syndrome.

- In this context, the role of PPPM/3P medicine is to introduce innovative predictive approaches to develop predictive diagnostics methods, targeted prevention, and personalized medical services for diseases not associated with *Clostridium difficile*.

- All donors need to undergo thorough and regular medical examinations that include testing for a wide variety of conditions, including infectious diseases such as HIV and hepatitis. During this process, the medical history and predispositions of the donors are also assessed and donors with an increased risk of developing certain conditions aren't eligible to donate stool. All donors must follow lifestyle and diet recommendations, ensuring their intestinal ecosystem is at an optimal, balanced level.

– In PPPM/3P medicine, doctors should focus on providing a comprehensive treatment protocol that aims to optimize intestinal microbiota, revitalize the intestinal mucosa, and offer lifestyle recommendations. During the patient stay at the clinic, they would be advised on how to take good care of your new gut microbiota even after returning home to ensure that you're able to maintain the profits of FMT treatment long-term.

Legalislation Issues Data/Results/Biomarkers/Other approaches recommended as a Preliminary Protocols for the Prevention and Predictive Personalised Patients treatment and Relevance to the Existent Medical Protocols (Country, National, etc.)

No declare.

Founding No declare.

Other Acknowledgments No declare.

References

1. Eslami M, Bahar A, Hemati M et al (2020) Dietary pattern, colonic microbiota and immunometabolism interaction: new frontiers for diabetes mellitus and related disorders. Diabet Med:e14415. https://doi.org/10.1111/dme.14415
2. Macchione IG, Lopetuso LR, Ianiro G et al (2019) Akkermansia muciniphila: key player in metabolic and gastrointestinal disorders. Eur Rev Med Pharmacol Sci 23:8075–8083. https://doi.org/10.26355/eurrev_201909_19024
3. Sivamaruthi BS, Kesika P, Suganthy N, Chaiyasut C (2019) A review on role of microbiome in obesity and antiobesity properties of probiotic supplements. Biomed Res Int 2019:1–20. https://doi.org/10.1155/2019/3291367
4. Kobyliak N, Abenavoli L, Falalyeyeva T et al (2021) Metabolic benefits of probiotic combination with absorbent smectite in type 2 diabetes patients a randomised controlled trial. Rev Recent Clin Trials 16:109–119. https://doi.org/10.2174/1574887115666200709141131
5. Mancabelli L, Milani C, Lugli GA et al (2017) Unveiling the gut microbiota composition and functionality associated with constipation through metagenomic analyses. Sci Rep 7:9879. https://doi.org/10.1038/s41598-017-10663-w
6. Eslami M, Sadrifar S, Karbalaei M et al (2020) Importance of the microbiota inhibitory mechanism on the Warburg effect in colorectal cancer cells. J Gastrointest Cancer 51:738–747. https://doi.org/10.1007/s12029-019-00329-3
7. Miyake S, Yamamura T (2019) Gut environmental factors and multiple sclerosis. J Neuroimmunol 329:20–23
8. Kobyliak N, Abenavoli L, Mykhalchyshyn G et al (2019) Probiotics and smectite absorbent gel formulation reduce liver stiffness, transaminase and cytokine levels in NAFLD associated with type 2 diabetes: a randomized clinical study. Clin Diabetol 8:205–214. https://doi.org/10.5603/dk.2019.0016
9. Golubnitschaja O, Liskova A, Koklesova L et al (2021) Caution, "normal" BMI: health risks associated with potentially masked individual underweight—EPMA Position Paper 2021. EPMA J 12:243–264. https://doi.org/10.1007/s13167-021-00251-4

10. Abdelhamid AG, El-Masry SS, El-Dougdoug NK (2019) Probiotic Lactobacillus and Bifidobacterium strains possess safety characteristics, antiviral activities and host adherence factors revealed by genome mining. EPMA J 10:337–350. https://doi.org/10.1007/s13167-019-00184-z

11. Zhang F, Luo W, Shi Y et al (2012) Should we standardize the 1,700-year-old fecal microbiota transplantation. Am J Gastroenterol 107:1755

12. Eiseman B, Silen W, Bascom GS, Kauvar AJ (1958) Fecal enema as an adjunct in the treatment of pseudomembranous. Surgery 44:854–859

13. van Nood E, Vrieze A, Nieuwdorp M et al (2013) Duodenal infusion of donor feces for recurrent Clostridium difficile. N Engl J Med 368:407–415. https://doi.org/10.1056/nejmoa1205037

14. Ott SJ, Waetzig GH, Rehman A et al (2017) Efficacy of sterile fecal filtrate transfer for treating patients with Clostridium difficile infection. Gastroenterology 152:799–811.e7. https://doi.org/10.1053/j.gastro.2016.11.010

15. D'Haens GR, Jobin C (2019) Fecal microbial transplantation for diseases beyond recurrent clostridium difficile infection. Gastroenterology 157:624–636

16. Giles EM, D'Adamo GL, Forster SC (2019) The future of faecal transplants. Nat Rev Microbiol 17:719–719. https://doi.org/10.1038/s41579-019-0271-9

17. Tkach S, Dorofeyev A, Kuzenko I et al (2022) Current status and future therapeutic options for fecal microbiota transplantation. Medicina (Kaunas) 58:84. https://doi.org/10.3390/MEDICINA58010084

18. Frank DN, St. Amand AL, Feldman RA et al (2007) Molecular-phylogenetic characterization of microbial community imbalances in human inflammatory bowel diseases. Proc Natl Acad Sci U S A 104:13780–13785. https://doi.org/10.1073/pnas.0706625104

19. Guillemot FÇ, Colombel JF, Neut C et al (1991) Treatment of diversion colitis by short-chain fatty acids – prospective and double-blind study. Dis Colon Rectum 34:861–864. https://doi.org/10.1007/BF02049697

20. Moayyedi P, Surette MG, Kim PT et al (2015) Fecal microbiota transplantation induces remission in patients with active ulcerative colitis in a randomized controlled trial. Gastroenterology 149:102–109 e6. https://doi.org/10.1053/j.gastro.2015.04.001

21. Rossen NG, Fuentes S, Van Der Spek MJ et al (2015) Findings from a randomized controlled trial of fecal transplantation for patients with ulcerative colitis. Gastroenterology 149:110–118.e4. https://doi.org/10.1053/j.gastro.2015.03.045

22. Paramsothy S, Kamm MA, Kaakoush NO et al (2017) Multidonor intensive faecal microbiota transplantation for active ulcerative colitis: a randomised placebo-controlled trial. Lancet 389:1218–1228. https://doi.org/10.1016/S0140-6736(17)30182-4

23. Costello SP, Waters O, Bryant RV et al (2017) Short duration, low intensity, pooled fecal microbiota transplantation induces remission in patients with mild-moderately active ulcerative colitis: a randomised controlled trial. Gastroenterology 152:S198–S199. https://doi.org/10.1016/s0016-5085(17)30969-1

24. Haifer C, Paramsothy S, Kaakoush NO et al (2022) Lyophilised oral faecal microbiota transplantation for ulcerative colitis (LOTUS): a randomised, double-blind, placebo-controlled trial. Lancet Gastroenterol Hepatol 7:141–151. https://doi.org/10.1016/S2468-1253(21)00400-3

25. Shi Y, Dong Y, Huang W et al (2016) Fecal microbiota transplantation for ulcerative colitis: a systematic review and meta-analysis. PLoS One 11:e0157259. https://doi.org/10.1371/journal.pone.0157259

26. Narula N, Kassam Z, Yuan Y et al (2017) Systematic review and meta-analysis: fecal microbiota transplantation for treatment of active ulcerative colitis. Inflamm Bowel Dis 23:1702–1709

27. El Hage CN, Ghoneim S, Shah S et al (2022) Efficacy of fecal microbiota transplantation in the treatment of active ulcerative colitis: a systematic review and meta-analysis of double-blind randomized controlled trials. Inflamm Bowel Dis. https://doi.org/10.1093/IBD/IZAC135

28. Tan X-Y, Xie Y-J, Liu X-L et al (2022) A systematic review and meta-analysis of randomized controlled trials of fecal microbiota transplantation for the treatment of inflammatory bowel disease. Evid Based Complement Alternat Med 2022:8266793. https://doi.org/10.1155/2022/8266793

29. Shah H, Zezos P (2020) Pouchitis: diagnosis and management. Curr Opin Gastroenterol 36:41–47

30. Segal JP, Ding NS, Worley G et al (2017) Systematic review with meta-analysis: the management of chronic refractory pouchitis with an evidence-based treatment algorithm. Aliment Pharmacol Ther 45:581–592

31. Landy J, Walker AW, Li JV et al (2015) Variable alterations of the microbiota, without metabolic or immunological change, following faecal microbiota transplantation in patients with chronic pouchitis. Sci Rep 5:12955. https://doi.org/10.1038/srep12955

32. Selvig D, Piceno Y, Terdiman J et al (2020) Fecal microbiota transplantation in Pouchitis: clinical, endoscopic, histologic, and microbiota results from a pilot study. Dig Dis Sci 65:1099–1106. https://doi.org/10.1007/s10620-019-05715-2

33. Stallmach A, Lange K, Buening J et al (2016) Fecal microbiota transfer in patients with chronic antibiotic-refractory pouchitis. Am J Gastroenterol 111:441–443

34. Herfarth H, Barnes EL, Long MD et al (2019) Combined endoscopic and oral fecal microbiota transplantation in patients with antibiotic-dependent pouchitis: low clinical efficacy due to low donor microbial engraftment. Inflamm Intest Dis 4:1–6. https://doi.org/10.1159/000497042

35. Keshteli AH, Millan B, Madsen KL (2017) Pretreatment with antibiotics may enhance the efficacy of fecal microbiota transplantation in ulcerative colitis: a meta-analysis. Mucosal Immunol 10:565–566

36. Castaño-Rodríguez N, Paramsothy S, Kaakoush NO (2020) Promise of fecal microbiota transplantation therapy in pouchitis. Dig Dis Sci 65:1107–1110

37. Tominaga K, Tsuchiya A, Yokoyama J, Terai S (2019) How do you treat this diversion ileitis and pouchitis? Gut 68:593–758

38. Paramsothy S, Paramsothy R, Rubin DT et al (2017) Faecal microbiota transplantation for inflammatory bowel disease: a systematic review and meta-analysis. J Crohns Colitis 11:1180–1199. https://doi.org/10.1093/ecco-jcc/jjx063

39. Cold F, Kousgaard SJ, Halkjaer SI et al (2020) Fecal microbiota transplantation in the treatment of chronic pouchitis: a systematic review. Microorganisms 8:1433

40. Vermeire S, Joossens M, Verbeke K et al (2016) Donor species richness determines faecal microbiota transplantation success in inflammatory bowel disease. J Crohns Colitis 10:387–394. https://doi.org/10.1093/ecco-jcc/jjv203

41. Cui B, Feng Q, Wang H et al (2015) Fecal microbiota transplantation through mid-gut for refractory Crohn's disease: safety, feasibility, and efficacy trial results. J Gastroenterol Hepatol 30:51–58. https://doi.org/10.1111/jgh.12727

42. Suskind DL, Brittnacher MJ, Wahbeh G et al (2015) Fecal microbial transplant effect on clinical outcomes and fecal microbiome in active Crohn's disease. Inflamm Bowel Dis 21:556–563. https://doi.org/10.1097/MIB.0000000000000307

43. Colman RJ, Rubin DT (2014) Fecal microbiota transplantation as therapy for inflammatory bowel disease: a systematic review and meta-analysis. J Crohns Colitis 8:1569–1581

44. de Fátima Caldeira L, Borba HH, Tonin FS et al (2020) Fecal microbiota transplantation in inflammatory bowel disease patients: a systematic review and meta-analysis. PLoS One 15:e0238910. https://doi.org/10.1371/journal.pone.0238910

45. Sokol H, Landman C, Seksik P et al (2020) Fecal microbiota transplantation to maintain remission in Crohn's disease: a pilot randomized controlled study. Microbiome 8:12. https://doi.org/10.1186/s40168-020-0792-5

46. Wang H, Cui B, Li Q et al (2018) The safety of fecal microbiota transplantation for Crohn's disease: findings from a long-term study. Adv Ther 35:1935–1944. https://doi.org/10.1007/s12325-018-0800-3

47. Shen Z, Zhu C, Quan Y et al (2017) Update on intestinal microbiota in Crohn's disease 2017: mechanisms, clinical application, adverse reactions, and outlook. J Gastroenterol Hepatol 32:1804–1812

48. Rajilić-Stojanović M, Biagi E, Heilig HGHJ et al (2011) Global and deep molecular analysis of microbiota signatures in fecal samples from patients with irritable bowel syndrome. Gastroenterology 141:1792–1801. https://doi.org/10.1053/j.gastro.2011.07.043

49. Johnsen PH, Hilpüsch F, Cavanagh JP et al (2018) Faecal microbiota transplantation versus placebo for moderate-to-severe irritable bowel syndrome: a double-blind, randomised, placebo-controlled, parallel-group, single-centre trial. Lancet Gastroenterol Hepatol 3:17–24. https://doi.org/10.1016/S2468-1253(17)30338-2

50. Halkjær SI, Christensen AH, Lo BZS et al (2018) Faecal microbiota transplantation alters gut microbiota in patients with irritable bowel syndrome: results from a randomised, double-blind placebo-controlled study. Gut 67:2107–2115. https://doi.org/10.1136/gutjnl-2018-316434

51. Holvoet T, Joossens M, Vázquez-Castellanos JF et al (2021) Fecal microbiota transplantation reduces symptoms in some patients with irritable bowel syndrome with predominant abdominal bloating: short- and Long-term results from a placebo-controlled randomized trial. Gastroenterology 160:145–157.e8. https://doi.org/10.1053/j.gastro.2020.07.013

52. Aroniadis OC, Brandt LJ, Oneto C et al (2019) Faecal microbiota transplantation for diarrhoea-predominant irritable bowel syndrome: a double-blind, randomised, placebo-controlled trial. Lancet Gastroenterol Hepatol 4:675–685. https://doi.org/10.1016/S2468-1253(19)30198-0

53. Xu D, Chen VL, Steiner CA et al (2019) Efficacy of fecal microbiota transplantation in irritable bowel syndrome: a systematic review and meta-analysis. Am J Gastroenterol 114:1043–1050

54. Ianiro G, Eusebi LH, Black CJ et al (2019) Systematic review with meta-analysis: efficacy of faecal microbiota transplantation for the treatment of irritable bowel syndrome. Aliment Pharmacol Ther 50:240–248

55. Wu J, Lv L, Wang C (2022) Efficacy of fecal microbiota transplantation in irritable bowel syndrome: a meta-analysis of randomized controlled trials. Front Cell Infect Microbiol 12:827395. https://doi.org/10.3389/FCIMB.2022.827395

56. Zhao H, Zhang X, Zhang N et al (2022) Fecal microbiota transplantation for patients with irritable bowel syndrome: a meta-analysis of randomized controlled trials. Front Nutr 9:890357. https://doi.org/10.3389/FNUT.2022.890357

57. Myneedu K, Deoker A, Schmulson MJ, Bashashati M (2019) Fecal microbiota transplantation in irritable bowel syndrome: a systematic review and meta-analysis. United Eur Gastroenterol J 7:1033–1041

58. Benno P, Norin E, Midtvedt T, Hellström PM (2019) Therapeutic potential of an anaerobic cultured human intestinal microbiota, ACHIM, for treatment of IBS. Best Pract Res Clin Gastroenterol 40–41:101607

59. El-Salhy M, Hatlebakk JG, Gilja OH et al (2020) Efficacy of faecal microbiota transplantation for patients with irritable bowel syndrome in a randomised, double-blind, placebo-controlled study. Gut 69:859–867. https://doi.org/10.1136/gutjnl-2019-319630

60. König J, Brummer RJ (2020) Faecal microbiota transplantation in IBS—new evidence for success? Nat Rev Gastroenterol Hepatol 17:199–200. https://doi.org/10.1038/s41575-020-0282-z

61. El-Salhy M, Winkel R, Casen C et al (2022) Efficacy of fecal microbiota transplantation for patients with irritable bowel syndrome at three years after transplantation. Gastroenterology. https://doi.org/10.1053/J.GASTRO.2022.06.020

62. Singh P, Alm EJ, Kelley JM et al (2022) Effect of antibiotic pretreatment on bacterial engraftment after fecal microbiota transplant (FMT) in IBS-D. Gut Microbes 14:2020067. https://doi.org/10.1080/19490976.2021.2020067

63. Mazzawi T, Hausken T, El-Salhy M (2022) Changes in colonic enteroendocrine cells of patients with irritable bowel syndrome following fecal microbiota transplantation. Scand J Gastroenterol 57:792–796. https://doi.org/10.1080/00365521.2022.2036809

64. Mosso E, Boano V, Grassini M et al (2019) Microscopic colitis: a narrative review with clinical approach. Minerva Gastroenterol Dietol 65:53–62
65. Günaltay S, Rademacher L, Hörnquist EH, Bohr J (2017) Clinical and immunologic effects of faecal microbiota transplantation in a patient with collagenous colitis. World J Gastroenterol 23:1319–1324. https://doi.org/10.3748/wjg.v23.i7.1319
66. Tariq R, Smyrk T, Pardi DS et al (2016) New-onset microscopic colitis in an ulcerative colitis patient after fecal microbiota transplantation. Am J Gastroenterol 111:751–752
67. Holster S, Rode J, Bohr J et al (2020) Faecal microbiota transfer in patients with microscopic colitis – a pilot study in collagenous colitis. Scand J Gastroenterol 55:1454–1466. https://doi.org/10.1080/00365521.2020.1839544
68. Tian H, Ge X, Nie Y et al (2017) Fecal microbiota transplantation in patients with slow-transit constipation: a randomized, clinical trial. PLoS One 12:e0171308. https://doi.org/10.1371/journal.pone.0171308
69. Tian Y, Zuo L, Guo Q et al (2020) Potential role of fecal microbiota in patients with constipation. Ther Adv Gastroenterol 13:1756284820968423. https://doi.org/10.1177/1756284820968423
70. Ding C, Fan W, Gu L et al (2018) Outcomes and prognostic factors of fecal microbiota transplantation in patients with slow transit constipation: results from a prospective study with long-term follow-up. Gastroenterol Rep 6:101–107. https://doi.org/10.1093/gastro/gox036
71. Zhang X, Tian H, Gu L et al (2018) Long-term follow-up of the effects of fecal microbiota transplantation in combination with soluble dietary fiber as a therapeutic regimen in slow transit constipation. Sci China Life Sci 61:779–786. https://doi.org/10.1007/s11427-017-9229-1
72. Wortelboer K, Nieuwdorp M, Herrema H (2019) Fecal microbiota transplantation beyond Clostridioides difficile infections. EBioMedicine 44:716–729
73. Suez J, Zmora N, Zilberman-Schapira G et al (2018) Post-antibiotic gut mucosal microbiome reconstitution is impaired by probiotics and improved by autologous FMT. Cell 174:1406–1423.e16. https://doi.org/10.1016/j.cell.2018.08.047
74. Ianiro G, Segal JP, Mullish BH et al (2020) Fecal microbiota transplantation in gastrointestinal and extraintestinal disorders. Future Microbiol 15:1173–1186
75. Michot JM, Bigenwald C, Champiat S et al (2016) Immune-related adverse events with immune checkpoint blockade: a comprehensive review. Eur J Cancer 54:139–148
76. Wang Y, Wiesnoski DH, Helmink BA et al (2018) Fecal microbiota transplantation for refractory immune checkpoint inhibitor-associated colitis. Nat Med 24:1804–1808. https://doi.org/10.1038/s41591-018-0238-9
77. Wardill HR, Secombe KR, Bryant RV et al (2019) Adjunctive fecal microbiota transplantation in supportive oncology: emerging indications and considerations in immunocompromised patients. EBioMedicine 44:730–740
78. Wong SH, Zhao L, Zhang X et al (2017) Gavage of fecal samples from patients with colorectal cancer promotes intestinal carcinogenesis in germ-free and conventional mice. Gastroenterology 153:1621–1633.e6. https://doi.org/10.1053/j.gastro.2017.08.022
79. Ma C, Han M, Heinrich B et al (2018) Gut microbiome–mediated bile acid metabolism regulates liver cancer via NKT cells. Science (80-) 360:eaan5931. https://doi.org/10.1126/science.aan5931
80. Pushalkar S, Hundeyin M, Daley D et al (2018) The pancreatic cancer microbiome promotes oncogenesis by induction of innate and adaptive immune suppression. Cancer Discov 8:403–416. https://doi.org/10.1158/2159-8290.CD-17-1134
81. Kobyliak N, Abenavoli L, Falalyeyeva T et al (2016) Prevention of NAFLD development in rats with obesity via the improvement of pro/antioxidant state by cerium dioxide nanoparticles. Clujul Med 89:229–235. https://doi.org/10.15386/cjmed-632
82. Mykhalchyshyn G, Kobyliak N, Bodnar P (2015) Diagnostic accuracy of acyl-ghrelin and it association with non-alcoholic fatty liver disease in type 2 diabetic patients. J Diabetes Metab Disord 14. https://doi.org/10.1186/s40200-015-0170-1
83. Lechner S, Yee M, Limketkai BN, Pham EA (2020) Fecal microbiota transplantation for chronic liver diseases: current understanding and future direction. Dig Dis Sci 65:897–905

84. Shasthry SM (2020) Fecal microbiota transplantation in alcohol related liver diseases. Clin Mol Hepatcl 26:294–301. https://doi.org/10.3350/cmh.2020.0057
85. Thursz MR, Richardson P, Allison M et al (2015) Prednisolone or pentoxifylline for alcoholic hepatitis. N Engl J Med 372:1619–1628. https://doi.org/10.1056/nejmoa1412278
86. Philips CA, Pande A, Shasthry SM et al (2017) Healthy donor fecal microbiota transplantation in steroid-ineligible severe alcoholic hepatitis: a pilot study. Clin Gastroenterol Hepatol 15:600–602
87. Philips CA, Phadke N, Ganesan K et al (2018) Corticosteroids, nutrition, pentoxifylline, or fecal microbiota transplantation for severe alcoholic hepatitis. Indian J Gastroenterol 37:215–225. https://doi.org/10.1007/s12664-018-0859-4
88. Sharma A, Roy A, Premkumar M et al (2022) Fecal microbiota transplantation in alcohol-associated acute-on-chronic liver failure: an open-label clinical trial. Hepatol Int 16:433–446. https://doi.org/10.1007/S12072-022-10312-Z
89. Williamson KD, Chapman RW (2016) New therapeutic strategies for primary sclerosing cholangitis. Semin Liver Dis 36:5–14. https://doi.org/10.1055/s-0035-1571274
90. Rodriguez EA, Carey EJ, Lindor KD (2017) Emerging treatments for primary sclerosing cholangitis. Expert Rev Gastroenterol Hepatol 11:451–459
91. Shah A, MacDonald GA, Morrison M, Holtmann G (2020) Targeting the gut microbiome as a treatment for primary sclerosing cholangitis: a conceptional framework. Am J Gastroenterol 115:814–822
92. Tripathi A, Debelius J, Brenner DA et al (2018) The gut-liver axis and the intersection with the microbiome. Nat Rev Gastroenterol Hepatol 15:397–411
93. Little R, Wine E, Kamath BM et al (2020) Gut microbiome in primary sclerosing cholangitis: a review. World J Gastroenterol 26:2768–2780
94. Rahimpour S, Nasiri-Toosi M, Khalili H et al (2016) A triple blinded, randomized, placebo-controlled clinical trial to evaluate the efficacy and safety of oral vancomycin in primary sclerosing cholangitis: a pilot study. J Gastrointest Liver Dis 25:457–464. https://doi.org/10.15403/jgld.2014.1121.254.rah
95. Ali AH, Carey EJ, Lindor KD (2016) The microbiome and primary sclerosing cholangitis. Semin Liver Dis 36:340–348. https://doi.org/10.1055/s-0036-1594007
96. Philips CA, Augustine P, Phadke N (2018) Healthy donor fecal microbiota transplantation for recurrent bacterial cholangitis in primary sclerosing cholangitis – a single case report. J Clin Transl Hepatcl 6:438–441. https://doi.org/10.14218/JCTH.2018.00033
97. Allegretti JR, Kassam Z, Carrellas M et al (2019) Fecal microbiota transplantation in patients with primary sclerosing cholangitis: a pilot clinical trial. Am J Gastroenterol 114:1071–1079. https://doi.org/10.14309/ajg.0000000000000115
98. Lavanchy D (2004) Hepatitis B virus epidemiology, disease burden, treatment, arid current and emerging prevention and control measures. J Viral Hepat 11:97–107
99. Kang Y, Cai Y (2017) Gut microbiota and hepatitis-B-virus-induced chronic liver disease: implications for faecal microbiota transplantation therapy. J Hosp Infect 96:342–348
100. Ren YD, Ye ZS, Yang LZ et al (2017) Fecal microbiota transplantation induces hepatitis B virus e-antigen (HBeAg) clearance in patients with positive HBeAg after long-term antiviral therapy. Hepatology 65:1765–1768. https://doi.org/10.1002/hep.29008
101. Chauhan A, Kumar R, Sharma S et al (2020) Fecal microbiota transplantation in hepatitis B e antigen-positive chronic hepatitis B patients: a pilot study. Dig Dis Sci. https://doi.org/10.1007/s10620-020-06246-x
102. Terrault NA, Bzowej NH, Chang KM et al (2016) AASLD guidelines for treatment of chronic hepatitis B. Hepatology 63:261–283. https://doi.org/10.1002/hep.28156
103. Ebrahimzadeh Leylabadlo H, Ghotaslou R, Samadi Kafil H et al (2020) Non-alcoholic fatty liver diseases: from role of gut microbiota to microbial-based therapies. Eur J Clin Microbiol Infect Dis 39:613–627
104. Falalyeyeva T, Komisarenko I, Yanchyshyn A et al (2021) Vitamin D in the prevention and treatment of type-2 diabetes and associated diseases: a critical view during

COVID-19 time. Minerva Biotechnol Biomol Res 33:65–75. https://doi.org/10.23736/S2724-542X.21.02766-X

105. Koopman N, Molinaro A, Nieuwdorp M, Holleboom AG (2019) Review article: can bugs be drugs? The potential of probiotics and prebiotics as treatment for non-alcoholic fatty liver disease. Aliment Pharmacol Ther 50:628–639

106. Abenavoli L, Falalyeyeva T, Pellicano R et al (2020) Next-generation of strain specific probiotics in diabetes treatment: the case of Prevotella copri. Minerva Endocrinol 45:277–279. https://doi.org/10.23736/S0391-1977.20.03376-3

107. Zhou D, Pan Q, Shen F et al (2017) Total fecal microbiota transplantation alleviates high-fat diet-induced steatohepatitis in mice via beneficial regulation of gut microbiota. Sci Rep 7:1529. https://doi.org/10.1038/s41598-017-01751-y

108. Chiu CC, Ching YH, Li YP et al (2017) Nonalcoholic fatty liver disease is exacerbated in high-fat diet-fed gnotobiotic mice by colonization with the gut microbiota from patients with nonalcoholic steatohepatitis. Nutrients 9:1220. https://doi.org/10.3390/nu9111220

109. Sharpton SR, Ajmera V, Loomba R (2019) Emerging role of the gut microbiome in non-alcoholic fatty liver disease: from composition to function. Clin Gastroenterol Hepatol 17:296–306

110. Craven L, Rahman A, Nair Parvathy S et al (2020) Allogenic fecal microbiota transplantation in patients with nonalcoholic fatty liver disease improves abnormal small intestinal permeability: a randomized control trial. Am J Gastroenterol 115:1055–1065. https://doi.org/10.14309/ajg.0000000000000661

111. Witjes JJ, Smits LP, Pekmez CT et al (2020) Donor fecal microbiota transplantation alters gut microbiota and metabolites in obese individuals with steatohepatitis. Hepatol Commun 4:1578–1590. https://doi.org/10.1002/hep4.1601

112. Manzhalii E, Moyseyenko V, Kondratiuk V et al (2022) Effect of a specific Escherichia coli Nissle 1917 strain on minimal/mild hepatic encephalopathy treatment. World J Hepatol 14:634–646. https://doi.org/10.4254/WJH.V14.I3.634

113. Wang WW, Zhang Y, Huang XB et al (2017) Fecal microbiota transplantation prevents hepatic encephalopathy in rats with carbon tetrachloride-induced acute hepatic dysfunction. World J Gastroenterol 23:6983–6994. https://doi.org/10.3748/wjg.v23.i38.6983

114. Bajaj JS, Kassam Z, Fagan A et al (2017) Fecal microbiota transplant from a rational stool donor improves hepatic encephalopathy: a randomized clinical trial. Hepatology 66:1727–1738. https://doi.org/10.1002/hep.29306

115. Bajaj JS, Fagan A, Gavis EA et al (2019) Long-term outcomes of fecal microbiota transplantation in patients with cirrhosis. Gastroenterology 156:1921–1923.e3. https://doi.org/10.1053/j.gastro.2019.01.033

116. Hu Y, Xiao HY, He C et al (2019) Fecal microbiota transplantation as an effective initial therapy for pancreatitis complicated with severe Clostridium difficile infection: a case report. World J Clin Cases 7:2597–2604. https://doi.org/10.12998/wjcc.v7.i17.2597

117. Taşdemir M, Hasan C, Ağbaş A et al (2016) Sjögren's syndrome associated with systemic lupus erythematosus. Turkish Arch Pediatr Pediatr Arşivi 51:166–168. https://doi.org/10.5152/TURKPEDIATRIARS.2016.2001

118. Huang C, Yi P, Zhu M et al (2022) Safety and efficacy of fecal microbiota transplantation for treatment of systemic lupus erythematosus: an EXPLORER trial. J Autoimmun 130:102844. https://doi.org/10.1016/J.JAUT.2022.102844

119. Watane A, Cavuoto KM, Rojas M et al (2022) Fecal microbial transplant in individuals with immune-mediated dry eye. Am J Ophthalmol 233:90–100. https://doi.org/10.1016/J.AJO.2021.06.022

120. Benhadou F, Mintoff D, Schnebert B, Thio H (2018) Psoriasis and microbiota: a systematic review. Diseases 6:47. https://doi.org/10.3390/diseases6020047

121. Yin G, Li JF, Sun YF et al (2019) Fecal microbiota transplantation as a novel therapy for severe psoriasis. Zhonghua Nei Ke Za Zhi 58:782–785. https://doi.org/10.3760/cma.j.issn.0578-1426.2019.10.011

122. Kragsnaes MS, Kjeldsen J, Horn HC et al (2018) Efficacy and safety of faecal microbiota transplantation in patients with psoriatic arthritis: protocol for a 6-month, double-blind, randomised, placebo-controlled trial. BMJ Open 8:e019231

123. Chen J, Chia N, Kalari KR et al (2016) Multiple sclerosis patients have a distinct gut microbiota compared to healthy controls. Sci Rep 6:28484. https://doi.org/10.1038/srep28484

124. Lee YK, Menezes JS, Umesaki Y, Mazmanian SK (2011) Proinflammatory T-cell responses to gut microbiota promote experimental autoimmune encephalomyelitis. Proc Natl Acad Sci U S A 108:4615–4622. https://doi.org/10.1073/pnas.1000082107

125. Berer K, Gerdes LA, Cekanaviciute E et al (2017) Gut microbiota from multiple sclerosis patients enables spontaneous autoimmune encephalomyelitis in mice. Proc Natl Acad Sci U S A 114:10719–10724. https://doi.org/10.1073/pnas.1711233114

126. Li K, Wei S, Hu L et al (2020) Protection of fecal microbiota transplantation in a mouse model of multiple sclerosis. Mediat Inflamm 2020:2058272. https://doi.org/10.1155/2020/2058272

127. Borody T, Leis S, Campbell J et al (2011) Fecal microbiota transplantation (FMT) in multiple sclerosis (MS). Am J Gastroenterol 106:S352. https://doi.org/10.14309/00000434-201110002-00942

128. Makkawi S, Camara-Lemarroy C, Metz L (2018) Fecal microbiota transplantation associated with 10 years of stability in a patient with SPMS. Neurol Neuroimmunol NeuroInflammation 5:e459. https://doi.org/10.1212/NXI.0000000000000459

129. Engen PA, Zaferiou A, Rasmussen H et al (2020) Single-arm, non-randomized, time series, single-subject study of fecal microbiota transplantation in multiple sclerosis. Front Neurol 11:978. https://doi.org/10.3389/fneur.2020.00978

130. Forsyth CB, Shannon KM, Kordower JH et al (2011) Increased intestinal permeability correlates with sigmoid mucosa alpha-synuclein staining and endotoxin exposure markers in early Parkinson's disease. PLoS One 6:e28032. https://doi.org/10.1371/journal.pone.0028032

131. Tan AH, Mahadeva S, Thalha AM et al (2014) Small intestinal bacterial overgrowth in Parkinson's disease. Park Relat Disord 20:535–540. https://doi.org/10.1016/j.parkreldis.2014.02.019

132. Keshavarzian A, Green SJ, Engen PA et al (2015) Colonic bacterial composition in Parkinson's disease. Mov Disord 30:1351–1360. https://doi.org/10.1002/mds.26307

133. Scheperjans F, Aho V, Pereira PAB et al (2015) Gut microbiota are related to Parkinson's disease and clinical phenotype. Mov Disord 30:350–358. https://doi.org/10.1002/mds.25069

134. Sampson TR, Debelius JW, Thron T et al (2016) Gut microbiota regulate motor deficits and neuroinflammation in a model of Parkinson's disease. Cell 167:1469–1480.e12. https://doi.org/10.1016/j.cell.2016.11.018

135. Sun MF, Zhu YL, Zhou ZL et al (2018) Neuroprotective effects of fecal microbiota transplantation on MPTP-induced Parkinson's disease mice: gut microbiota, glial reaction and TLR4/TNF-α signaling pathway. Brain Behav Immun 70:48–60. https://doi.org/10.1016/j.bbi.2018.02.005

136. Zhou ZL, Jia XB, Sun MF et al (2019) Neuroprotection of fasting mimicking diet on MPTP-induced Parkinson's disease mice via gut microbiota and metabolites. Neurotherapeutics 16:741–760. https://doi.org/10.1007/s13311-019-00719-2

137. Huang H, Xu H, Luo Q et al (2019) Fecal microbiota transplantation to treat Parkinson's disease with constipation: a case report. Medicine (Baltimore) 98:e16163. https://doi.org/10.1097/MD.0000000000016163

138. Xue LJ, Yang XZ, Tong Q et al (2020) Fecal microbiota transplantation therapy for Parkinson's disease: a preliminary study. Medicine (Baltimore) 99:e22035. https://doi.org/10.1097/MD.0000000000022035

139. Vendrik KEW, Ooijevaar RE, de Jong PRC et al (2020) Fecal microbiota transplantation in neurological disorders. Front Cell Infect Microbiol 10:98

140. Fattorusso A, Di Genova L, Dell'isola GB et al (2019) Autism spectrum disorders and the gut microbiota. Nutrients 11:521

141. Slykerman RF, Thompson J, Waldie KE et al (2017) Antibiotics in the first year of life and subsequent neurocognitive outcomes. Acta Paediatr Int J Paediatr 106:87–94. https://doi.org/10.1111/apa.13613

142. McElhanon BO, McCracken C, Karpen S, Sharp WG (2014) Gastrointestinal symptoms in autism spectrum disorder: a meta-analysis. Pediatrics 133:872–883. https://doi.org/10.1542/peds.2013-3995

143. Tomova A, Husarova V, Lakatosova S et al (2015) Gastrointestinal microbiota in children with autism in Slovakia. Physiol Behav 138:179–187. https://doi.org/10.1016/j.physbeh.2014.10.033

144. Hsiao EY, McBride SW, Hsien S et al (2013) Microbiota modulate behavioral and physiological abnormalities associated with neurodevelopmental disorders. Cell 155:1451–1463. https://doi.org/10.1016/j.cell.2013.11.024

145. Sharon G, Cruz NJ, Kang DW et al (2019) Human gut microbiota from autism spectrum disorder promote behavioral symptoms in mice. Cell 177:1600–1618.e17. https://doi.org/10.1016/j.cell.2019.05.004

146. Aabed K, Bhat RS, Moubayed N et al (2019) Ameliorative effect of probiotics (lactobacillus paracaseii and Protexin®) and prebiotics (propolis and bee pollen) on clindamycin and propionic acid-induced oxidative stress and altered gut microbiota in a rodent model of autism. Cell Mol Biol 65:1–7. https://doi.org/10.14715/cmb/2019.65.1.1

147. Kang DW, Adams JB, Gregory AC et al (2017) Microbiota transfer therapy alters gut ecosystem and improves gastrointestinal and autism symptoms: an open-label study. Microbiome 5:10. https://doi.org/10.1186/s40168-016-0225-7

148. Kang DW, Adams JB, Coleman DM et al (2019) Long-term benefit of microbiota transfer therapy on autism symptoms and gut microbiota. Sci Rep 9:5821. https://doi.org/10.1038/s41598-019-42183-0

149. Zhao H, Gao X, Xi L et al (2019) Mo1667 fecal microbiota transplantation for children with autism spectrum disorder. Gastrointest Endosc 89:AB512–AB513. https://doi.org/10.1016/j.gie.2019.03.857

150. Xu H-M, Huang H-L, Zhou Y-L et al (2021) Fecal microbiota transplantation: a new therapeutic attempt from the gut to the brain. Gastroenterol Res Pract 2021:6699268. https://doi.org/10.1155/2021/6699268

151. Lum GR, Olson CA, Hsiao EY (2020) Emerging roles for the intestinal microbiome in epilepsy. Neurobiol Dis 135:104576

152. Peng A, Qiu X, Lai W et al (2018) Altered composition of the gut microbiome in patients with drug-resistant epilepsy. Epilepsy Res 147:102–107. https://doi.org/10.1016/j.eplepsyres.2018.09.013

153. Lindefeldt M, Eng A, Darban H et al (2019) The ketogenic diet influences taxonomic and functional composition of the gut microbiota in children with severe epilepsy. npj Biofilms Microbiomes 5:5. https://doi.org/10.1038/s41522-018-0073-2

154. Dahlin M, Prast-Nielsen S (2019) The gut microbiome and epilepsy. EBioMedicine 44:741–746

155. Medel-Matus JS, Shin D, Dorfman E et al (2018) Facilitation of kindling epileptogenesis by chronic stress may be mediated by intestinal microbiome. Epilepsia Open 3:290–294. https://doi.org/10.1002/epi4.12114

156. Olson CA, Vuong HE, Yano JM et al (2018) The gut microbiota mediates the anti-seizure effects of the ketogenic diet. Cell 173:1728–1741.e13. https://doi.org/10.1016/j.cell.2018.04.027

157. He Z, Cui BT, Zhang T et al (2017) Fecal microbiota transplantation cured epilepsy in a case with Crohn's disease: the first report. World J Gastroenterol 23:3565–3568. https://doi.org/10.3748/wjg.v23.i19.3565

158. Castaner O, Goday A, Park Y-M et al (2018) The gut microbiome profile in obesity: a systematic review. Int J Endocrinol 2018:1–9. https://doi.org/10.1155/2018/4095789

159. Kobyliak N, Falalyeyeva T, Boyko N et al (2018) Probiotics and nutraceuticals as a new frontier in obesity prevention and management. Diabetes Res Clin Pract 141:190–199. https://doi.org/10.1016/j.diabres.2018.05.005
160. Pasolli E, Truong DT, Malik F et al (2016) Machine learning meta-analysis of large metagenomic datasets: tools and biological insights. PLoS Comput Biol 12:e1004977. https://doi.org/10.1371/journal.pcbi.1004977
161. Kyriachenko Y, Falalyeyeva T, Korotkyi O et al (2019) Crosstalk between gut microbiota and antidiabetic drug action. World J Diabetes 10:154–168. https://doi.org/10.4239/wjd.v10.i3.154
162. Kobyliak N, Falalyeyeva T, Tsyryuk O et al (2020) New insights on strain-specific impacts of probiotics on insulin resistance: evidence from animal study. J Diabetes Metab Disord 19:289–296. https://doi.org/10.1007/s40200-020-00506-3
163. Kobyliak N, Falalyeyeva T, Mykhalchyshyn G et al (2020) Probiotic and omega-3 polyunsaturated fatty acids supplementation reduces insulin resistance, improves glycemia and obesity parameters in individuals with type 2 diabetes: a randomised controlled trial. Obes Med 19:100248. https://doi.org/10.1016/j.obmed.2020.100248
164. Vrieze A, Van Nood E, Holleman F et al (2012) Transfer of intestinal microbiota from lean donors increases insulin sensitivity in individuals with metabolic syndrome. Gastroenterology 143:913–916. https://doi.org/10.1053/j.gastro.2012.06.031
165. Kootte RS, Levin E, Salojärvi J et al (2017) Improvement of insulin sensitivity after lean donor feces in metabolic syndrome is driven by baseline intestinal microbiota composition. Cell Metab 25:611–619.e6. https://doi.org/10.1016/j.cmet.2017.09.008
166. Allegretti JR, Kassam Z, Mullish BH et al (2020) Effects of fecal microbiota transplantation with oral capsules in obese patients. Clin Gastroenterol Hepatol 18:855–863.e2. https://doi.org/10.1016/j.cgh.2019.07.006
167. Allegretti JR, Kassam Z, Hurtado J et al (2021) Impact of fecal microbiota transplantation with capsules on the prevention of metabolic syndrome among patients with obesity. Hormones 20:209–211
168. Yu EW, Gao L, Stastka P et al (2020) Fecal microbiota transplantation for the improvement of metabolism in obesity: the FMT-TRIM double-blind placebo-controlled pilot trial. PLoS Med 17:e1003051. https://doi.org/10.1371/journal.pmed.1003051
169. Zhang Z, Mocanu V, Cai C et al (2019) Impact of fecal microbiota transplantation on obesity and metabolic syndrome—a systematic review. Nutrients 11:2291. https://doi.org/10.3390/nu11102291
170. Proença IM, Allegretti JR, Bernardo WM et al (2020) Fecal microbiota transplantation improves metabolic syndrome parameters: systematic review with meta-analysis based on randomized clinical trials. Nutr Res 83:1–14
171. Guirro M, Costa A, Gual-Grau A et al (2019) Effects from diet-induced gut microbiota dysbiosis and obesity can be ameliorated by fecal microbiota transplantation: a multiomics approach. PLoS One 14:e0218143. https://doi.org/10.1371/journal.pone.0218143
172. Tsaban G, Yaskolka Meir A, Rinott E et al (2020) The effect of green Mediterranean diet on cardiometabolic risk; a randomised controlled trial. Heart. https://doi.org/10.1136/heartjnl-2020-317802
173. Rinott E, Youngster I, Yaskolka Meir A et al (2021) Effects of diet-modulated autologous fecal microbiota transplantation on weight regain. Gastroenterology 160:158–173.e10. https://doi.org/10.1053/j.gastro.2020.08.041
174. Su L, Hong Z, Zhou T et al (2022) Health improvements of type 2 diabetic patients through diet and diet plus fecal microbiota transplantation. Sci Rep 12:1152. https://doi.org/10.1038/S41598-022-05127-9
175. Ng SC, Xu Z, Mak JWY et al (2022) Microbiota engraftment after faecal microbiota transplantation in obese subjects with type 2 diabetes: a 24-week, double-blind, randomised controlled trial. Gut 71:716–723. https://doi.org/10.1136/GUTJNL-2020-323617

176. Fan L, Ren J, Chen Y et al (2022) Effect of fecal microbiota transplantation on primary hypertension and the underlying mechanism of gut microbiome restoration: protocol of a randomized, blinded, placebo-controlled study. Trials 23:178. https://doi.org/10.1186/S13063-022-06086-2

177. Napolitano M, Covasa M (2020) Microbiota transplant in the treatment of obesity and diabetes: current and future perspectives. Front Microbiol 11:590370

Chapter 11
Personalized Microbiome Correction by Application of Individual Nutrition for Type 2 Diabetes Treatment

Tamara Meleshko and Nadiya Boyko

Abstract Type 2 diabetes is one of the most common noncommunicable diseases in the world regardless of their age and region due to the changes in lifestyles, genetics, and environmental factors, all of which together influence the disorder. Recent researches have demonstrated that the development of low-grade inflammation is a consequence of gut microbiota alteration, which is closely related to metabolic disorders such as obesity and type 2 diabetes. Gut microbiota-targeted therapy has gained significant importance in recent decades for its health-promoting role in the prevention and treatment of type 2 diabetes. However, applying all observations about personalized diet in practice taking into consideration patients' microbiome uniqueness is a challenge. The fundamental purpose of this review was to summarize our studies that have investigate the possibility of correction of lipid metabolism of patients with type 2 diabetes using a personalized diet based on the most important microbial, biochemical, and immunological biomarkers of chronic inflammation.

Keywords Predictive · Preventive · Personalized medicine · Human microbiota · Personalized diet · Prebiotic and probiotic components · Metabolism regulation · Prognostic correction

T. Meleshko (✉)
RDE Center of Molecular Microbiology and Mucosal Immunology, Uzhhorod National University, Uzhhorod, Ukraine

Department of Clinical Laboratory Diagnostics and Pharmacology, Uzhhorod National University, Uzhhorod, Ukraine

N. Boyko
RDE Center of Molecular Microbiology and Mucosal Immunology, Uzhhorod National University, Uzhhorod, Ukraine

Department of Clinical Laboratory Diagnostics and Pharmacology, Uzhhorod National University, Uzhhorod, Ukraine

Ediens LLC, Uzhhorod, Ukraine

© The Author(s), under exclusive license to Springer Nature
Switzerland AG 2023
N. Boyko, O. Golubnitschaja (eds.), *Microbiome in 3P Medicine Strategies*,
Advances in Predictive, Preventive and Personalised Medicine 16,
https://doi.org/10.1007/978-3-031-19564-8_11

11.1   Introduction

Type 2 diabetes (T2D) has gradually become an increasingly striking social health problem around the world. According to the International Diabetes Federation, there were about 425 million of diabetics worldwide in 2017, and this figure is expected to increase to 693 million by 2045 [1]. T2D and the increased incidence of its complications are among the leading causes of death and cause huge burden to patients, especially those living in underdeveloped or developing countries. Another problem is that the annual pool of new cases of individuals diseased on T2D becomes enriched by patients in the teenager-age. Also, particularly individuals diseased on diabetes early in life are particularly predisposed to a cascade of every diabetes-related complications with poor outcomes such as cardiovascular disease, several types of cancer and neurological disorders [2].

T2D is a group of metabolic syndrome characterized by absolute or relative insufficiency of insulin secretion and decreased sensitivity of target organs to insulin, followed by fat, protein, water, electrolytes and other metabolic disorders [3]. Typical clinical markers of T2D include glucose and glycosylated hemoglobin, increased cholesterol, triglycerides, low-density lipoprotein (LDL-cholesterol), very low-density lipoprotein (VLDL), and decreased high-density lipoprotein (HDL) [4]. T2D is a multi-factorial disease, the manifestation of which is regulated by both genetic and epigenetic components. Several exogenous and endogenous risk factors are synergistically involved in the disease predisposition and clinical manifestation [2]. Although the etiology and pathogenesis of T2D are still unclear, its occurrence is related to insufficient insulin secretion or insulin resistance (IR), and IR is often associated with obesity. Age, sex, family history, lifestyle related factors (e.g. disturbed dietary patterns, lack of exercise, smoking, stress) and anthropometric data (e.g. waist circumference, the Body Mass Index (BMI)) form non-laboratory based clinical risk assessment of diabetes [5–7]. Modern researches have shown that the development of low-grade inflammation is a consequence of gut microbiota alteration and plays a vital role in the occurrence and development of T2D [8, 9]. Host-microbial interaction and the possibility of targeted regulation of [gut] microbiota following to maintenance mucosal immune response is subject of great practical interest for prevention and treatment of noncommunicable diseases (NCD) initiated by low grade inflammation such as T2D, cardiovascular diseases (CVDs), metabolic syndrome, obesity [10].

In particular, in the majority of patients suffering from diabetes the levels of *Bifidobacterium* and *Lactobacillus* decrease, which leads to an increase in the levels of *Bacteroides, Prevotella, Peptococcus, Clostridium, Proteus, Staphylococcus,* and *Candida*. Importantly, T2D subjects have smaller amounts of butyrate producing bacteria, such as *Roseburia intestinalis* and *Faecalibacterium prausnitzii*, and a mucus-degrading bacterium *Akkermansia muciniphila* [11].

Studies conducted within the "Human Microbiome" project [12] demonstrated that intestinal microbiome can be dominated by different ratios of beneficial microorganisms and still perform identical functions. Thus, it is not only the species composition of the microbiome, but also its "function" that is important. Herewith, it is obvious that the microbiome of each individual is unique [13].

In the new era of predictive, preventive, and personalized medicine (PPPM), it is no longer enough to diagnose a disease and its accompanying complications and treat patients using the classical tools, but rather take the process to a higher level of developing platforms that can predict the onset of such events, prevent the disease or its complications, and administer treatment based on personalized variations between patients [14]. T2D can be prevented or delayed by lifestyle and/or pharmacological interventions [15, 16]. The White Paper of the "European Association for Predictive, Preventive and Personalised Medicine (PPPM)" (EPMA) [17] suggested that a central component of preventive strategies is identification of individuals at risk for development of T2D. It is known that today there are no specific methods of prediction, prevention, and treatment of T2D, and they are therefore still to be in the focus of clinical research [18].

Recently, numerous research studies have been conducted to find a relationship between nutrition and its impact on human health. The reason is, on the one hand, that people misunderstand (underestimate) the role of food as a source of essential balanced nutrients. On the other hand, there are huge amounts of data on "proper nutrition" (rational nutrition) available and they are often contradictory, scientifically unsubstantiated, and clinically unconfirmed. A new modern challenge is the use of P3 approaches, in particular personalized nutrition, in medical practice.

The diet-microbiome interplay is currently the basis for personalized nutrition introduction and microbiota composition is the key factor affecting responsiveness to nutritional interventions that will soon take into account initial stratification of individuals on the basis of microbiota [19].

The health benefits of adherence to the different diets, as well as the relationship between microbiota and its associated metabolome in people consuming varied diets ranging from vegan to omnivorous, are now evidence-based [20]. The most popular diets whose positive health effects on the human body are considered to be established include the Mediterranean diet, vegetarian/vegan diet, high-fiber diet, and high-protein diet.

The antioxidant and anti-inflammatory effects of the Mediterranean diet on the whole as well as the effects of this diet's individual components, in particular olive oil, fruits and vegetables, whole grains, and fish, have a beneficial impact on abdominal obesity, lipids levels, glucose metabolism, and blood pressure levels [21]. Gut microbiota in individuals following the Mediterranean diet is characterized by high levels of *Lactobacillus* spp., *Bifidobacterium* spp., and *Prevotella* spp. and low levels of *Clostridium* spp., which relates to weight loss, improvement of the lipid profile, and decreased inflammation [22].

For vegetarians and vegans, the most relevant risk factors for chronic disease, such as BMI, lipid variables, and fasting glucose, are significantly lower. People following a plant-based dietary pattern demonstrate significantly lower levels of BMI, total cholesterol, LDL-cholesterol, triglycerides, and blood glucose when vegetarians were compared to nonvegetarians, and lower levels of BMI, total cholesterol, and LDL-cholesterol when vegans were compared to nonvegans [23].

People following vegan and vegetarian diets rich in fermentable plant-based foods were reported to have a microbiota characterized by a lower abundance of *Bacteroides* spp. and *Bifidobacterium* spp. [24].

High fiber intake is associated with lower serum cholesterol concentrations, lower risk of coronary heart disease, reduced blood pressure, enhanced weight control, better glycemic control, reduced risk of certain forms of cancer, and improved gastrointestinal function [25]. One study revealed that three diets containing different fiber-rich whole grains (barley, brown rice, or a combination of both) increased microbial diversity, the Firmicutes/Bacteroidetes ratio, and the abundance of the genus *Blautia* in fecal samples [26].

High-protein diet decreases weight, fasting glucose, and insulin concentrations as well as total and abdominal fat. In addition, this diet significantly decreases LDL-cholesterol concentrations [27]. Dietary protein intake in humans has been associated with the *Bacteroides* enterotype [26].

However, applying all these diets in practice taking into consideration patients' microbiome uniqueness is a challenge.

Extremely promising in this regard are new approaches to the adjustment of microbiota, in particular those described by the authors [28]. Their research demonstrated the ability to predict the glycemic response to the consumption of certain foods (PPGR), which resulted in the possibility of making plans for personalized nutrition and adjusting the intestinal microbiota. Baseline data included: intestinal microbiome analysis, anthropometric data, blood test, food frequency questionnaire, lifestyle and health surveys. The authors also performed constant monitoring of blood glucose and exercise. Prediction of glycemic response was performed using machine learning approaches. Cohort studies have shown that such a personalized change in diet leads to consistent changes in the intestinal microbiota and a decrease in PPGR. According to the results of these studies, the company DAY TWO was created, which uses simplified approaches to personalized food. However, this approach has some drawbacks. In particular, it does not take into account a significant part of important immune markers. Also, the adjustment of the microbiota with the help of personalized food is based on a dietary approach and in no way takes into account the presence or absence of certain diseases in the patient. A similar approach is used by the established company Viome [29]. However, unlike previous authors, the use of Metatranscriptome Sequencing is suggested here. Intestinal microbiome analysis and a patient-completed questionnaire, as well as, in some cases, a blood sample taken from a finger, are used as baseline data. The obtained data are analyzed using artificial intelligence. As a result, the patient receives, in a convenient format, recommendations for lifestyle and diet. In our opinion, the information provided is not enough to assess the provability of this selection.

Taking into account the paradigm of PPPM/3P medicine [30], in our opinion, the most promising way of individual microbiome correction, as well as prognostic modulation of local immune response, is the use of complete personalized diets rather than individual components [13].

This chapter is devoted to investigation the possibility of creating or developing personalized (individual) approaches (diet plans) using traditional dishes (based on traditional dishes) of our region as a source of biologically active components (BAC) selected for their known biological effects on the microbiome and local immune response and that could be used to treat T2D in a controlled diet study [13, 31, 32].

To achieve this goal, it was necessary: (1) to determine the typical ratios of the main functional groups of representatives of the intestinal microbiota of patients with atherosclerosis (AT), obesity, T2D and CVD in Zakarpattia; (2) to investigate the biological properties of extracts of edible plants – potential prebiotic components; (3) to determine the effect of individually designed nutrition on the intestinal microbiota and metabolic parameters in *in vivo* studies; (4) to create personalized nutrition plans for target intestinal microbiota correction; (5) to prove clinically effectiveness of proposed targeted correction of the intestinal microbiota through the use of personalized diet plans in the comprehensive treatment of patients with diabetes mellitus-2.

Research on intestinal microbiome changes and its role in the occurrence of NCDs have become one of the main today's scientific topic [33]. We believe that among many NCDs it is necessary to identify a group of diseases that are directly related to changes in the microbiome, the main triggers of which are chronic inflammation, namely atherosclerosis, obesity, T2D, CVDs.

Chronic low-grade inflammation, caused by gut microbiota perturbation and changes in gut permeability, accompanies all stages of atherosclerotic disease [34]. In obese people, the increased number of capillaries in the small intestine epithelium [35] leading to more monosaccharides absorption. At the same time, the increase number of microorganisms capable of fermenting indigestible carbohydrates in the colon [36, 37], significantly increase the ability to obtain more energy from food. The major risk for the development of type 2 diabetes, which leads to the destruction of insulin receptors and causes resistance to insulin, is obesity. In turn, patients with diabetes have an extremely high cardiovascular risk due to development of comorbidities, such as hypertension and dyslipidemia, which further accelerates the atherosclerotic process [38].

The results obtained from limited clinical case study presents data of the typical ratios of the main functional groups of representatives of the intestinal microbiota of patients with atherosclerosis, obesity, T2D and CVD in Zakarpattia [39]. Also, the individual differences of the intestinal microbiota of patients with these nosologies are characterized.

The represented study showed the typical ratio of the main functional groups of the gut microbiota of persons with atherosclerosis, namely: an increase in the number of LPS-containing bacteria of the genera *Klebsiella, Proteus,* and *Enterobacter,* immunomodulatory bacteria of the genus *Enterococcus,* and commensal *Streptococcus* spp., slight excess of the norm for immunomodulatory *E. coli,* as well as a decrease in the number of neuroactive and lactate-producing bacteria of the genus *Bifidobacterium,* representatives of acetate-propionate-producing bacteria of the genus *Bacteroides,* butyrate-producing bacteria *F. prausnitzii,* and *R. intestinalis,* as well as the mucin-degrading *A. muciniphila* (Figs. 11.1 and 11.2).

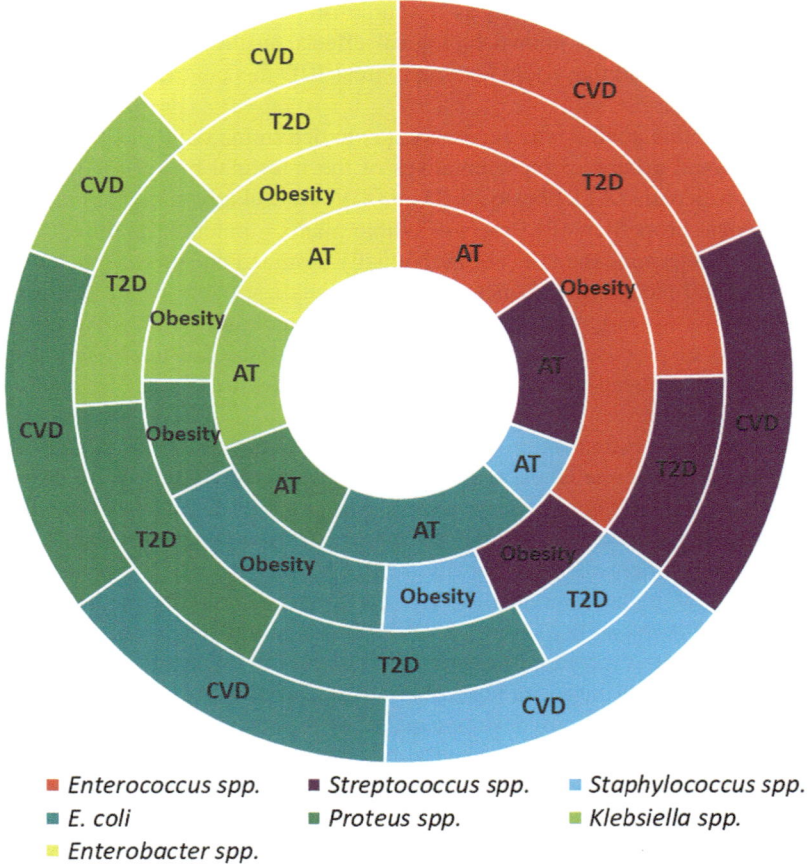

Fig. 11.1 The patients gut microbiota composition: coccal microorganisms and enterobacteriaceae

It should be noted that the decreased number of acetate-propionate-producing bacteria of the genus *Bacteroides,* butyrate-producing bacteria *R. intestinalis*, neuroactive and lactate-producing bacteria of the genus *Bifidobacterium*, mucin-degrading *A. muciniphila*, as well as immunomodulatory *E. coli* along with an increased number of neuroactive, lactate-producing bacteria *Lactobacillus* spp., and immunomodulatory bacteria of the genus *Enterococcus* can be considered typical for Zakarpattia ratio of the main functional groups of the intestinal microbiota of patients with obesity (Figs. 11.1 and 11.2).

It was found that the most characteristic ratio of the main functional groups of gut microbiota representatives of people with T2D was a decrease in the number of immunomodulatory *E. coli*, neuroactive, lactate-producing bacteria of the genus *Bifidobacterium*, acetate-propionate-producing bacteria of the genus *Bacteroides*,

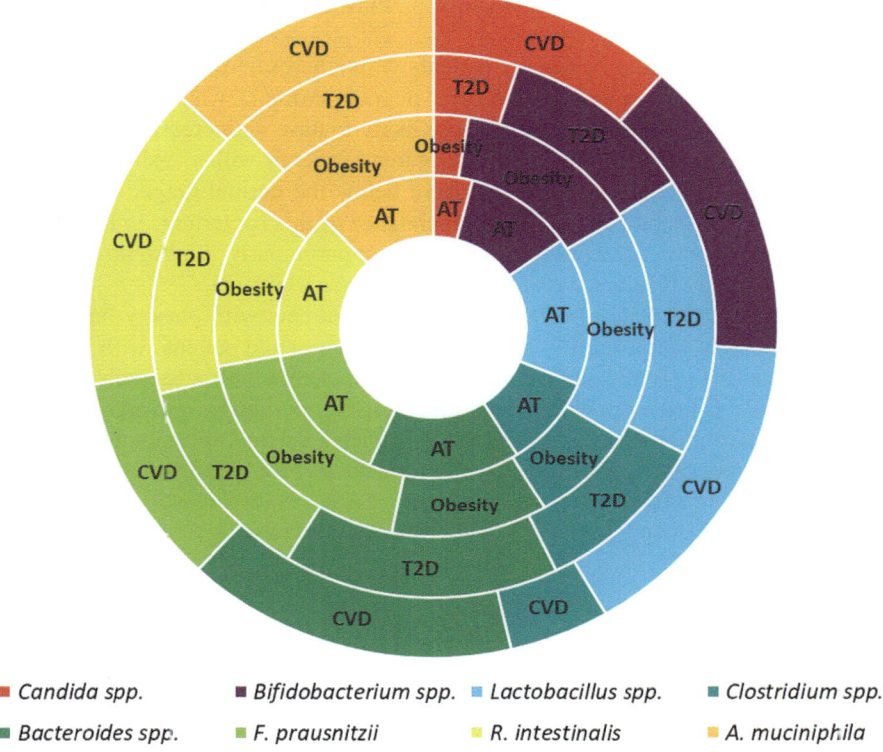

■ *Candida spp.*   ■ *Bifidobacterium spp.*   ■ *Lactobacillus spp.*   ■ *Clostridium spp.*

■ *Bacteroides spp.*   ■ *F. prausnitzii*   ■ *R. intestinalis*   ■ *A. muciniphila*

Fig. 11.2 The patients gut microbiota composition: anaerobes, facultative-anaerobes and obligate anaerobes

butyrate-producing strains *F. prausnitzii*, and *R. intestinalis*, and mucin-degrading bacteria *A. muciniphila* (Figs. 11.1 and 11.2).

Also, an increase in the number of immunomodulatory *Enterococcus* spp., commensal bacterial genera *Streptococcus* and *Staphylococcus*, LPS-containing bacteria of genera *Proteus* and *Enterobacter*, as well as *Candida* spp., along with decreased number of immunomodulatory strains of *E. coli*, neuroactive and lactate-producing *Bifidobacterium* spp., acetate-propionate-producing bacteria of the genus *Bacteroides*, butyrate-producing strains of *F. prausnitzii* and *R. intestinalis*, and mucin-degrading *A. muciniphila* can be considered as a typical ratio of the main functional groups of the gut microbiota of persons with CVDs (Figs. 11.1 and 11.2).

Analysis of the data gathered revealed that patients with CVDs were the only group to have a substantial rise in the staphylococci in their intestinal microbiota, but patients with atherosclerosis and CVDs also showed an increase in the streptococci. Gut microbiota of patients with obesity, atherosclerosis, and CVDs was characterised with high levels of enterococci. Therefore, an increase in the number of coccal microbial forms in the gut microbiota, including *Enterococcus* spp.,

Streptococcus spp., and *Staphylococcus* spp., may be a sign of atherosclerosis and CVDs development.

According to the study's results, patients with atherosclerosis alone were more likely to have an increase in *Klebsiella* spp. in gut microbiota, whereas patients with both atherosclerosis and CVDs were more likely to have an increase in *P. vulgaris* and *Enterobacter* spp. of intestine microbiota. Patients with obesity, T2D, and CVDs had *E. coli* concentrations that were below the normal range, but patients with atherosclerosis had a modest overabundance of this bacterium. In light of the aforementioned, an increase in enterobacteria, particularly *Klebsiella* spp. and *Enterobacter* spp., is indicative of the onset of atherosclerosis.

According to the data obtained in the study, patients with obesity had higher concentrations of lactobacilli in their gut microbiota, while patients with T2D had lower concentrations of *Lactobacillus* spp., and patients with atherosclerosis and CVDs had levels that were within the normal range. Individuals from all groups had normal concentrations of *Clostridium* spp. in their intestinal microbiota, but only patients with obesity had normal concentrations of *F. prausnitzii*. Patients from all nosological groups had lower concentrations of *Bifidobacterium* spp., *Bacteroides* spp., *R. intestinalis*, and *A. muciniphila* in their gut microbiota.

11.2   Effect of Edible Plant Extracts on Intestinal Microbiota: *In Vitro* Studies

Modern human nutrition needs to meet a lot of demands and provide not only safe and necessary composition of micro and microelements, but also be functional, which means enable the prevention of the development or even treatment of various diet-associated diseases, such as T2D, CVDs, etc. [32].

Whole grains, fruits, and vegetables as well as other traditional foods are the primary sources of BAC. It is well acknowledged that bioactive compounds – primarily vitamins, trace minerals, and antioxidants such polyphenols and anthocyanins [40] – are known to have an impact on a number of critical bodily functions, including gene expression, the synthesis of hormones and enzymes, the activation of receptors, immunological and antioxidant functions, etc. [41, 42]. The creation of contemporary preparations based on BAC, such as prebiotics and synbiotics, is a well-known trend of recent decades [43]. It is well recognized that BAC have long been utilized in practical medicine.

The edible plants' fruits which are characterized by high BAC contents and ability to stimulate the growth of beneficial microorganisms and inhibit the growth of conditionally pathogenic microorganisms could be perspective components for personalized nutrition. Additionally, it is known that the geographical location of plant food ingredients' growth affects the quantitative and qualitative composition of their BAC. Herewith, local plants are extremely important, because they are most often used for food in a particular region. Therefore, for our study, we selected the most typical edible plants for Zakarpattia region – *Ribes rubrum* (red currant),

Prunus avium (sweet cherry), *Prunus × domestica* (plum), *Ribes × nidigrolaria* (jostaberry), *Vaccinium myrtillus* (blueberry), *Ribes nigrum* (black currant), *Prunus cerasifera* (alycha) and *Cornus mas L.* (cornelian cherry). Our group has recently studied the gross polyphenols' content and total anthocyanins' amount of those edible plants [44].

In our previously published study, we investigated the antibacterial properties of the edible plants fruits in relation to the selected microorganisms such as *Escherichia coli, Enterobacter cloacae, Klebsiella pneumoniae, Klebsiella oxytoca, Proteus mirabilis, Pseudomonas aeruginosa, Streptococcus pyogenes, Staphylococcus aureus, Enterococcus faecalis, Candida albicans* by culturing them in extracts obtained from these edible plants' fruits [39].

The red currant extract completely inhibited the growth of *K. pneumoniae, K. oxytoca,* and *P. aeruginosa* on 48 h of co-cultivation, as well as *S. aureus* on 72 h of co-cultivation. After 48 h of co-cultivation with *S. aureus,* and after 72 h co-cultivation with *K. pneumoniae, K. oxytoca,* and *P. aeruginosa,* the black currant extract completely inhibited their growth (Figs. 11.3 and 11.4).

After 48 and 72 h of co-cultivation, the sweet cherry extract totally inhibited the growth of *E. cloacae, S. aureus, K. pneumoniae, K. oxytoca, P. aeruginosa, E.*

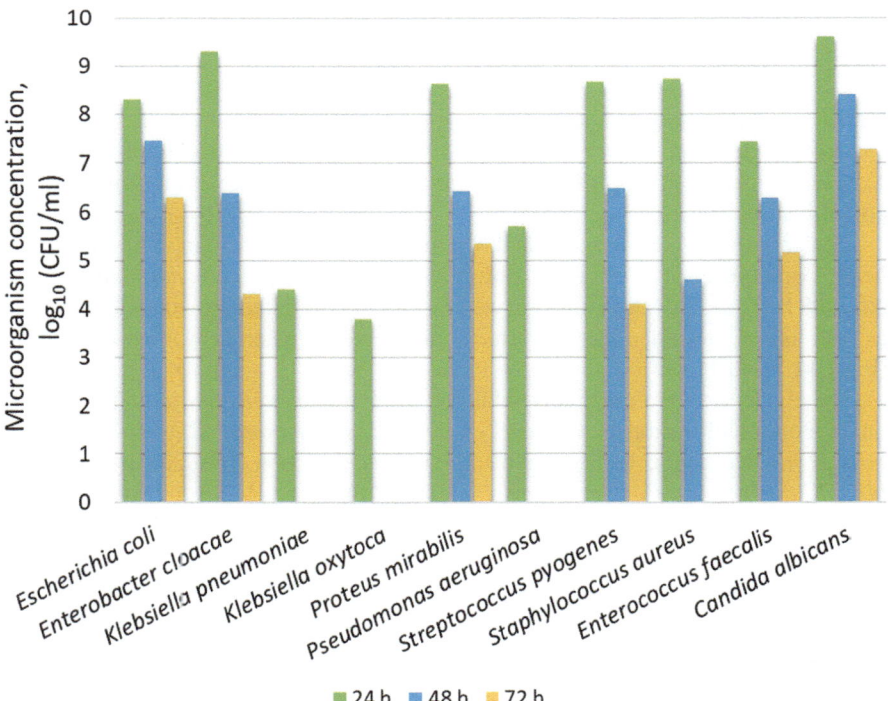

Fig. 11.3 Biological influence of *Ribes rubrum* extract on growth of selected microorganisms in dynamics

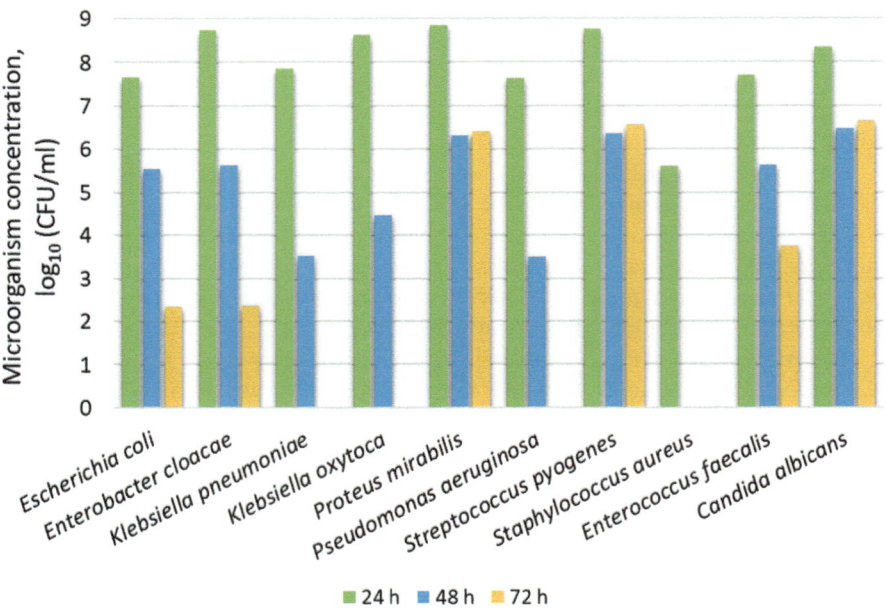

Fig. 11.4 Biological influence of *Ribes nigrum* extract on growth of selected microorganisms in dynamics

faecalis, and *P. mirabilis.* The plum extract completely inhibited growth of *K. pneumoniae, K. oxytoca, S. aureus*, and *C. albicans* after 48 h of co-cultivation and *E. coli, E. cloacae, P. mirabilis, P. aeruginosa*, and *S. pyogenes* strains after 72 h of co-cultivation (Figs. 11.5 and 11.6).

The growth of *P. aeruginosa* and *S. aureus* was completely inhibited by jostaberry extract after 48 h of co-cultivation while growth of *E. cloacae* – after 72 h. After 48 h of co-cultivation blueberry extract totally inhibited growth of *P. mirabilis* as well as s *K. pneumoniae, K. oxytoca*, and *S. aureus* after 72 h of co-cultivation (Figs. 11.7 and 11.8).

The alycha extract completely inhibited growth of *P. aeruginosa* and *S. aureus* on 48 h of co-cultivation as well as *E. cloacae, K. pneumoniae, K. oxytoca, P. mirabilis, E. faecalis,* and *C. albicans* after 72 h of co-cultivation. The growth of *P. mirabilis, S. aureus*, and *C. albicans* was totally inhibited after 48 h of co-cultivation with cornelian cherry extract (Figs. 11.9 and 11.10).

According to the data obtained in the study, it can be concluded that extracts of red currant and plum can be used as a growth inhibitors of *Klebsiella* spp.; the extract of sweet cherry can be used for inhibition of *Enterobacter* spp. growth; the extracts of blueberry and cornelian cherry are effective growth inhibitors of *Proteus* spp.; the extracts of plum and cornelian cherry can be used as *Candida* spp. growth inhibitors; the extracts of black currant, sweet cherry, plum, jostaberry, alycha and cornelian cherry are effective growth inhibitors of *Staphylococcus* spp.

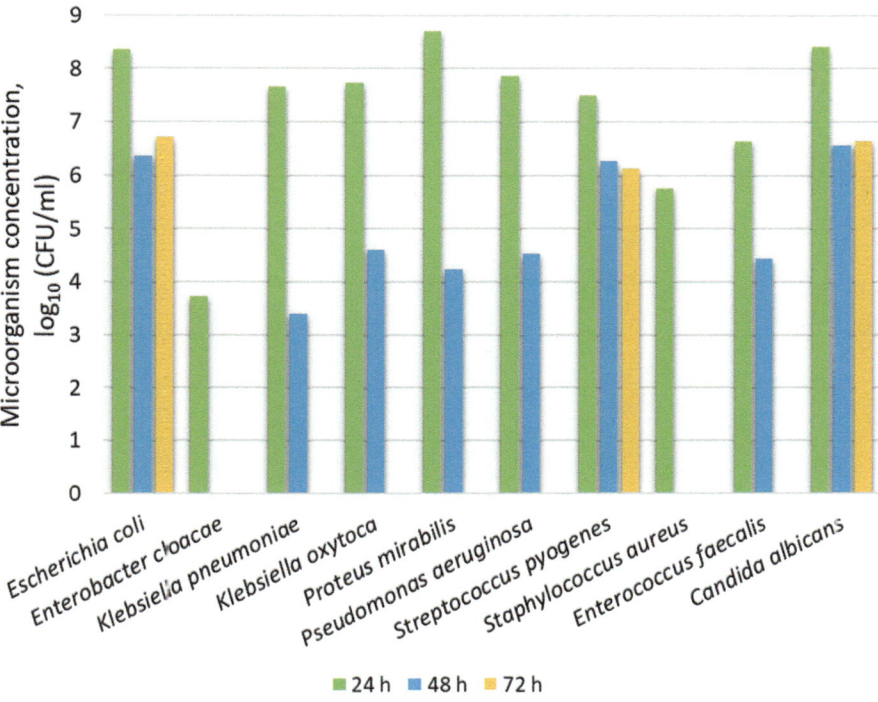

Fig. 11.5 Biological influence of *Prunus avium* extract on growth of selected microorganisms in dynamics

After analyzing the experimental data and considering the findings of our earlier research [32], we can recommend that specific extracts from edible plants and fruits be utilized as parts of individualized nutrition plans for the prevention and treatment of NCDs caused by chronic inflammation.

11.3 Investigation of the Impact of Individually Designed Nutrition on Gut Microbiota and Metabolism: *In Vivo* Studies

In order to confirm the interactions between modulation of microbiota by individually designed nutrition and intestinal beneficial effects our research group performed *in vivo* investigations [45]. The aim of the study was to determine the effect of individually designed nutrition on the intestinal microbiota and metabolic parameters of rats. Outbred laboratory rats with obesity were randomly divided into 9 groups (n = 12) depending on the type of food ingredients taken orally for 3 months. The

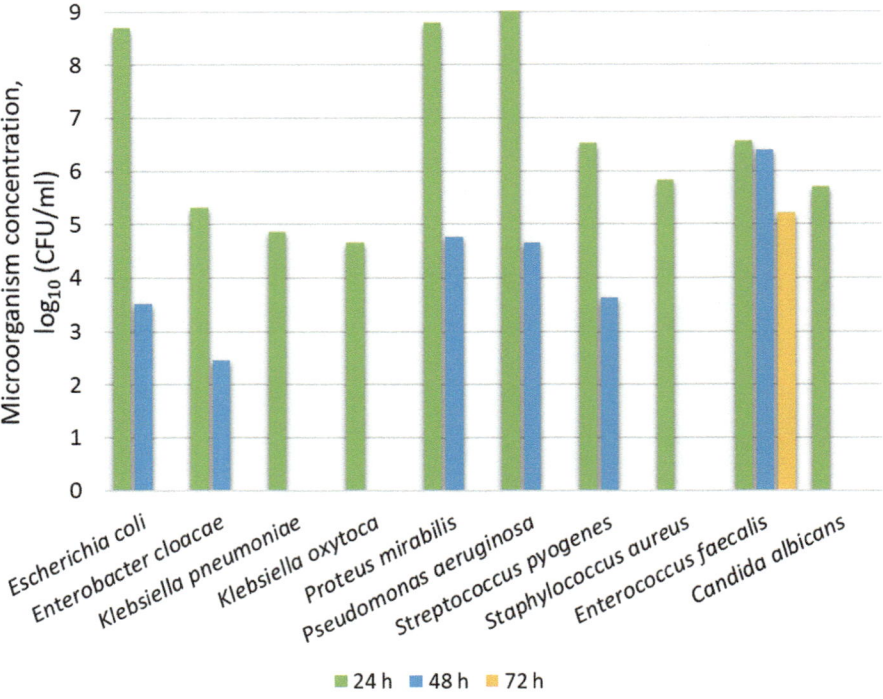

Fig. 11.6 Biological influence of *Prunus x domestica* extract on growth of selected microorganisms in dynamics

ratio of the gut commensal microorganisms' main groups, as well as the lipid profile and the content of glucose, urea, calcium in the serum of animals were determined.

In this *in vivo* study, all experimental animals were divided into nine groups depending on the orally consumed different ingredient(s): Group 1 – lactobacilli suspension (*Lactobacillus casei* IMB B-7412, *Lactobacillus plantarum* IMB B-7414, *Lactobacillus paracasei* IMB B-7483, and *Lactobacillus plantarum* KR-1); Group 2 – blueberry juice (*Vaccinium murtillus*); Group 3 – fermented milk drink with strains of lactobacilli (*L. paracasei* IMB B-7483, *L. casei* IMB B-7412, *L. plantarum* IMB B-7414, and *L. plantarum* KR-1) and blueberry juice (the ratio of the fermented drink and blueberry juice was 4:1); Group 4 – fermented milk drink with strains of lactobacilli (*L. paracasei* IMB B-7483, *L. casei* IMB B-7412, *L. plantarum* IMB B-7414, and *L. plantarum* KR-1) without plant components; Group 5 – sauerkraut juice with *L. casei* IMB B-7412 and *L. plantarum* IMB B-7414; Group 6 – persimmon juice (*Diospyros kaki*); Group 7 – lignin; Group 8 – pectin (15%); Group 9 – control group, standard vivarium diet food.

In the present study, we observed a significant decrease of body weight of all groups' experimental animals except the control one and the fourth group of rats under the influence of various components of the new generation functional foods.

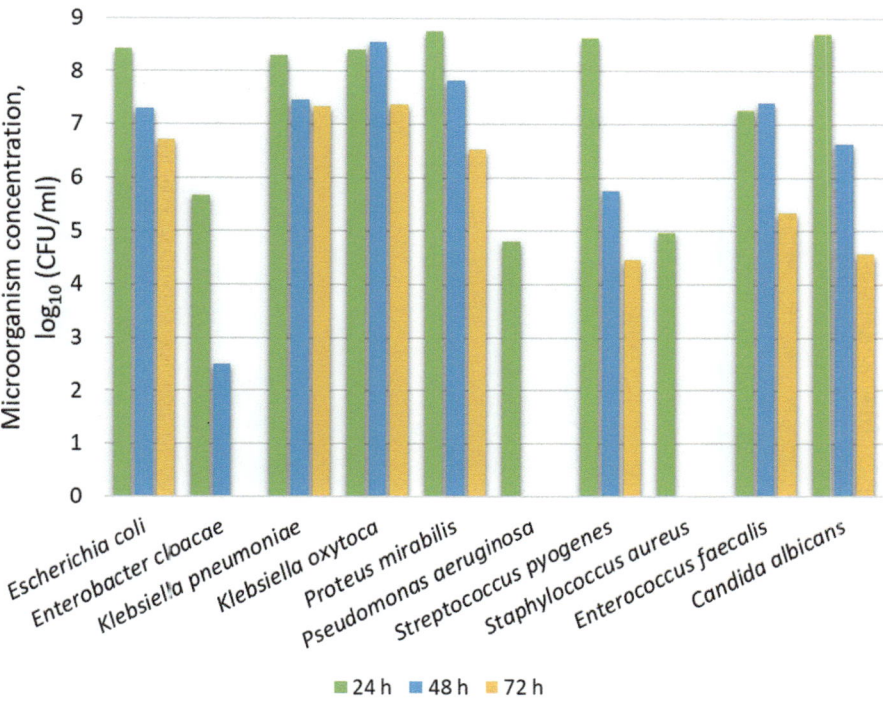

Fig. 11.7 Biological influence of *Ribes x nidigrolaria* extract on growth of selected microorganisms in dynamics

The body weight of the rats that followed the fermented milk drink diet without plant components (Group 4) increased on (15 ± 4) g. The body weight of the rats that followed the fermented milk drink diet without plant components (Group 4) increased on (15 ± 4) g, and in the control one – on (20 ± 4) g. It should be noted, that under the influence of lactobacilli, 15% apple pectin concentrate, lignins, and sauerkraut juice LDL-cholesterol significantly decreased ($p < 0.05$), while in the control group this parameter significantly increased (namely 2.5 times), and in other experimental groups there were no significant changes observed.

Under the influence of apple pectin and lignin, and also in the control group total lipid content significantly increased, while in all other groups total lipid level was significantly decreased. Interestingly, that triglyceride concentration decreased only in case of apple pectin consumption and reduction of cholesterol level occurred in all experimental groups of animals.

As a result of the introduction of the diets we developed, blood urea concentration decreased only in the group of animals following a diet enriched with probiotic bacterial strains. Under the effect of blueberry juice and fermented milk drink without plant components the glucose content decreased significantly, but not when this ingredients was used separately. Calcium level was increased under the

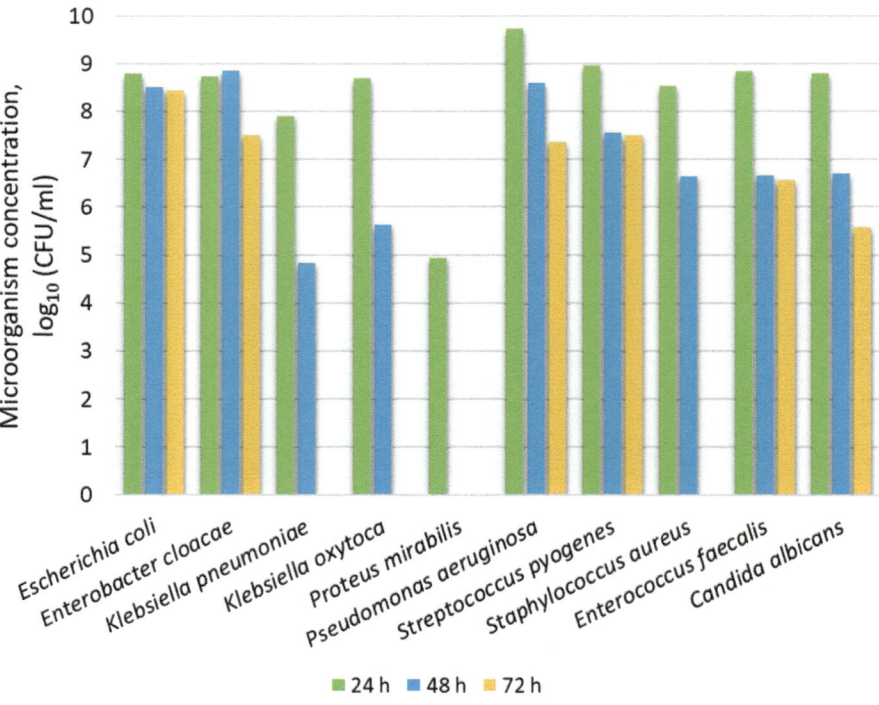

Fig. 11.8 Biological influence of *Vaccinium myrtillus* extract on growth of selected microorganisms in dynamics

consumption of fermented milk drink without plant components, while using all other diets, a decrease in calcium levels was observed.

Interesting, that despite the weight increasing of two animals groups, namely the control group and animals consuming fermented milk drink with lactobacilli without plant components, the other lipid profile indices of these groups significantly differed. Unlike the control group, an improvement in almost all registered biochemical parameters (including a slight decrease in total blood plasma lipids and a similar increase in calcium levels) was detected in animals consuming fermented milk drink without plant components.

It should be noted, that this study have shown that animals can be overweight not only because of high levels of opportunistic microorganisms such as *Staphylococcus nepalensis* and *E. faecalis*, but also because of the increasing level of the probiotic strain *Bifidobacterium breve*. The *in vivo* results proved the ability of developed fermented milk drink and blueberry juice to regulate blood glucose levels that in turn confirmed a prognostic value of these ingredients as recommended components of diet to type 2 diabetes patients.

No less important were changes in rats' intestinal microbiota throughout the experiment under the influence of different diets. In almost all experimental groups

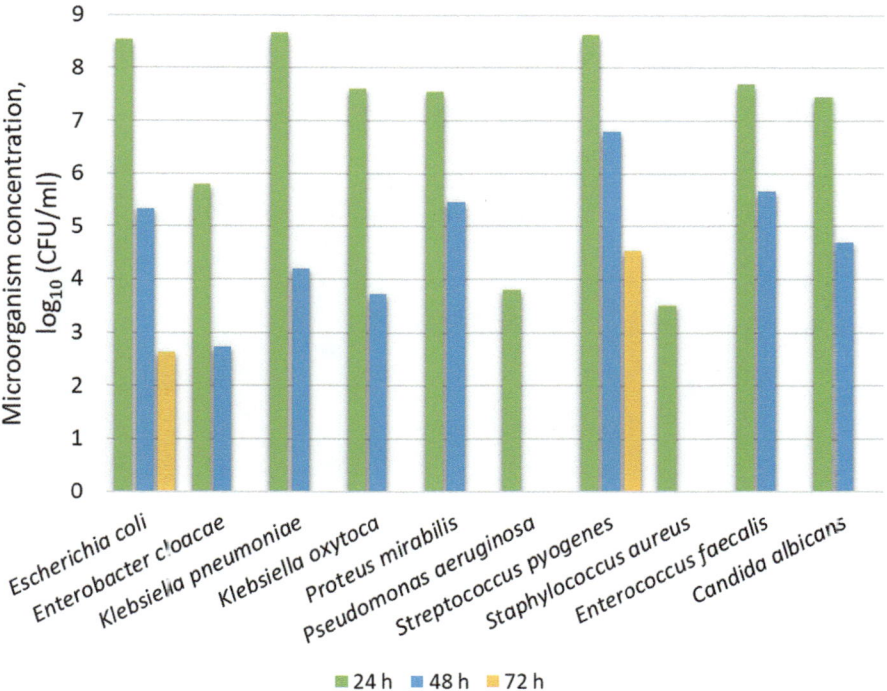

■ 24 h   ■ 48 h   ■ 72 h

Fig. 11.9 Biological influence of *Prunus cerasifera* extract on growth of selected microorganisms in dynamics

compared to the control group we observed a consistently high level of lactic acid and other beneficial bacteria and a decreasing level of opportunistic microorganisms.

The daily oral administration of almost all the developed food components (diets) leads to decrease in the level of *E. faecalis* and different *Staphylococcus* species compared to that in the control group. Consumption of blueberry juice caused a decrease in the concentration of *K. pneumoniae* and *Morganella morganii*. During the whole experiment under the influence of the test samples based on plant extracts, plant pectins, and lignins the level of *Bacillus subtilis* remained almost unchanged.

In obese rats consumption of lactobacillus strains suspension (*L. casei* IMB B-7412, *L. plantarum* IMB B-7414, *L. paracasei* IMB B-7483, and *L. plantarum* KR-1) caused an antagonistic activity against *S. aureus, Peptostreptococcus anaerobius,* and *E. faecalis* and led to a decrease in *E. coli* and *E. cloacae* levels. Under the influence of a blueberry juice-based diet the concentration of commensal *K. pneumoniae, M. morganii, E. coli, Actinomyces naeslundii,* and *Bacteroides* significantly decreased as well as the complete elimination of *Streptococcus parvulus* and increase in *E. faecalis* and staphylococci were detected.

During the whole study we observed a decreased concentration of *E. coli* and an increase in commensal lactobacilli, elimination of *Staphylococcus* spp. and *E.*

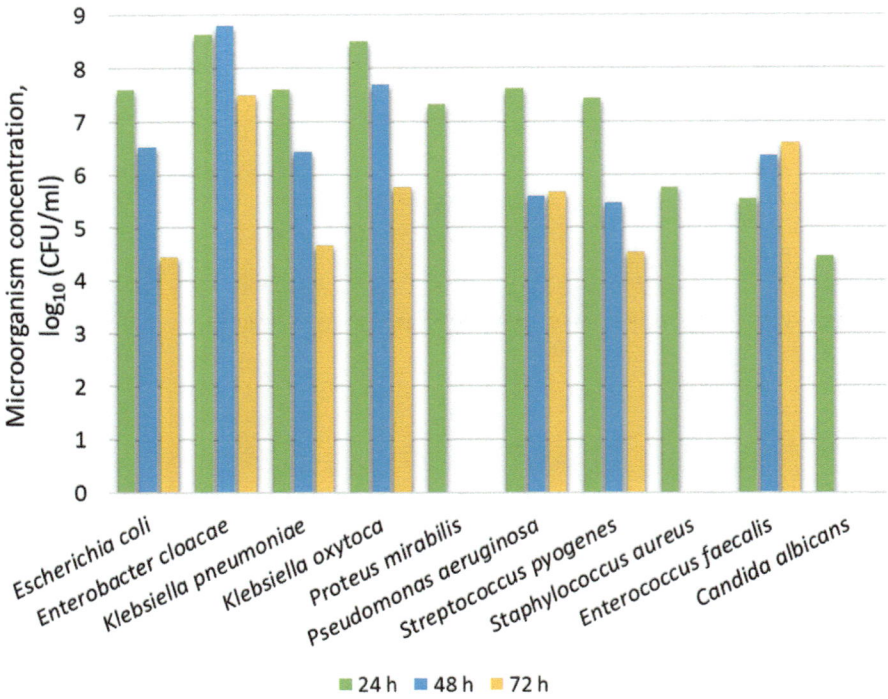

■ 24 h ■ 48 h ■ 72 h

Fig. 11.10 Biological influence of *Cornus mas. L.* extract on growth of selected microorganisms in dynamics

faecalis under the influence of the developed fermented milk product with blueberry juice compared to the control group.

Reduction in concentration of *E. coli* and *Actinomyces israelii* and elimination of *M. morganii* and *E. faecalis* were observed in obese rats after 2-weeks oral administration of fermented milk drink with strains of lactobacilli (*L. paracasei* IMB B-7483, *L. casei* IMB B-7412, *L. plantarum* IMB B-7414, and *L. plantarum* KR-1).

Under the influence of sauerkraut juice increased commensal lactobacilli concentration, partial elimination of *E. coli* and *B. subtilis*, as well as complete elimination of cocci were demonstrated in this study.

The observed result indicate that long-term oral administration of persimmon juice to animals caused a decrease in the levels of *E. coli, P. mirabilis, E. cloacae,* and *E. faecalis* and led to a significant antagonistic effect against *Staphylococcus cohnii.* Also, it should be noted that in case of persimmon juice consumption a consistently high levels of microorganisms, such as *Lactobacillus acidophilus, Bifidobacterium longum,* and *B. subtilis* were observed compared to the control group.

Commonly known, that dietary fiber (pectin, lignin, cellulose, and hemicellulose) can be found in most vegetables and fruits, and they are well known as a natural enterosorbent with can affect gut microbiota composition. Therefor we investigated the effect of apple pectin and lignin in a chronic experiment on obese rats.

Oral administration of lignin-based diet demonstrated reduction in enterococci, *B. subtilis*, *E. coli*, and *P. mirabilis* with a slight increase in commensal lactobacilli concentration. Complete elimination of *E. faecalis* and a decrease in the concentration of commensal *Bifidobacterium*, staphylococci, *E. coli*, and *B. subtilis* were observed under the influence of apple-pectin concentrate. In this study, an increase in the concentration of *P. mirabilis* as well as a decrease in the concentration of *E. faecalis* and *B. subtilis* were observed in the control group of animals. It should be noted that the decrease in the levels of *Lactobacillus* spp., *B. subtilis*, and *P. mirabilis* in the intestinal microbiome may serve as a biomarker of LDL-cholesterol and cholesterol reduction and triglyceride increase (and vice versa). The increase in *B. breve* and *B. subtilis* is a biomarker of decreased plasma calcium (and vice versa).

A direct correlation between the amount of *S. aureus* and *L. acidophilus* in the gut microbiota of rats using the tested products was found by us. A positive effects of the novel functional food itself, namely fermented milk drink with strains of lactobacilli and blueberry juice as well as separate components of this food were demonstrated in our study. Apparently, this is due to the fact that plant extracts and probiotic strains have the ability to adjust and stimulate the beneficial intestinal microbiota of the experimental animals. Additionally, probiotic strains of bifidobacteria and lactobacilli create unfavorable conditions for pathogens' colonization by decreasing their adhesive features.

Since the long-term oral administration of suspension of lactobacilli *L. paracasei* IMB B-7483, *L. casei* IMB B-7412, *L. plantarum* IMB B-7414, and *L. plantarum* KR-1 led to a decrease in cholesterol levels from 2.95 mmol/l to 0.84 mmol/l it can be argued that the suspension of these strains demonstrated the best hypocholesterolemic activity.

While analyzing the results of this study, we concluded that the particular components – blueberry juice, as well as fermented milk drink with strains of *L. casei* IMB B-7412, *L. plantarum* IMB B-7414, *L. paracasei* IMB B-7483, and *L. plantarum* KR-1 can be offered to patients with type 2 diabetes because of their ability to decrease the glucose level in the serum of obese animals.

Interestingly, that the decreasing in LDL-cholesterol levels were demonstrated under the oral administration of suspensions of lactobacilli *L. paracasei* IMB B-7483, *L. casei* IMB B-7412, *L. plantarum* IMB B-7414, and *L. plantarum* KR-1, sauerkraut juice (enriched with strains of *L. plantarum* IMB B-7414 and *L. casei* IMB B-7412), 15% apple pectin concentrate, and lignin. Since the long-term consumption of 15% apple pectin concentrate leads to a decrease in not only LDL-cholesterol, but also triglycerides and cholesterol this component has the highest hypolipidemic activity and can be used to prevent atherosclerosis and cardiovascular disease.

The correlation between the changes in key representatives of gut microbiota and values of biochemical indicators of the organism's state together with the results obtained during this study provide an opportunity to develop new generation functional food products capable of regulating the balance of intestinal microbiota and preventing the development of diet-related pathologies.

11.4 Personalized Nutrition for Microbiota Correction and Metabolism Restore in Type 2 Diabetes Mellitus Patients: Clinical Trail

Previous research has focused on fecal microbiota has provided evidence that both the composition and function of gut microbial communities were critical for maintaining physical health and also associated with metabolic diseases like diabetes and obesity. [46]. Recent studies suggest the possibility of prevention and treatment of T2D by using personalized approaches to nutrition [13].

In our previously published study [13], we conducted a randomized controlled trial on the effects of a personalized diet on 56 female patients in two parallel groups. Patients from experimental group followed an 18-day personalized diet, which included individually selected products rich in BAC and yogurts with unique microbial starters (sequenced strains of *L. casei* IMB B-7412, *L. plantarum* IMB B-7414, and *L. plantarum* IMB B-7413). Control group patients for 18 days ate usual for them food and additionally included to the ration berries and yogurt prepared without microbial starters. Biochemical, physical, and immunological parameters were measured by standard methods on days 1 and 18 of the experiment. Gut and oral microbiota detection were performed in dynamics on days 1, 7, 11, and 18 using selective culture mediums and real-time polymerase chain reaction. With the help of the developed information system, a personalized diet was developed for each participant of the experiment. Developed diets included products that contain functioning groups of biologically active substances such as polyphenols, anthocyanins, and flavonoids as well as unique microbial starters for fermentation. The selection of food products was based on WHO recommendations (https://www.who.int/nutrition/publications/nutrient/en/), taking into account individual wishes and contraindications, as well as when determining the portion size – individual characteristics of patients such as the level of physical activity, body mass index, etc. [13].

It was found out that the suggested correction of intestinal microbiota with help of personalized nutrition plans in complex treatment of T2D patients clinically has proven its effectiveness. It was shown that after compliance with personalized nutrition plans in patients of the experimental group the normalization of intestinal and oral microbiota was observed. In particular, the statistically significant changes in the numbers of representatives of the gut microbiota (*E. faecalis*, *E. coli* (lac +), *E. coli* (lac-), *Lactobacillus* spp., and *Candida* spp.) and microorganisms of oral cavity (*E. faecalis, Lactobacillus* spp., *P. aeruginosa,* and *Candida* spp.) were detected.

This indicates the normalization of intestinal microbiota, which, in turn, leads to metabolism improvement, including glucose and cholesterol metabolism.

The issue of diagnostics and treatment of T2D and other NCDs are still relevant because emergence of a number of new markers greatly simplifies and increases the accuracy of the disease diagnosis. In our work, we used a "classic" set of diabetes markers which still mostly used by medical personnel in particular because of their availability for analysis [47, 48] as well as indicators of gut and oral microbiota as they play a significant role in the development of type 2 diabetes and human health in general [49, 50]. During the experiment there were improvements in a number of markers, such as VLDL-cholesterol, glucose, creatinine, urea, magnesium, sodium, thymol test, and uric acid according to the results of indicators' change.

It should be noted that according to the results of our research, a decrease in LDL cholesterol indicates an improvement in lipid metabolism, a reduction in the risk of developing atherosclerosis and coronary heart disease [51]. In patients in the experimental group, a statistically significant drop in creatinine and urea levels and an increase in magnesium concentration within normal ranges may be signs that the renal excretory function has returned to normal [52–54]. The normalization of sodium level indicates the effectiveness of the proposed diet [13] as sufficient concentration of this indicator is extremely important for proper functioning of membrane transport, muscle contraction, nerve impulse transmission, and many other vital functions [55, 56].

A statistically significant decrease in thymol test levels within normal limits can indicate improvement of liver function [57]. A statistically significant increase in uric acid levels within normal limits can be explained by the increase in the consumption of foods containing fructose, such as apples, persimmons, blueberries, pears and dried fruits [13].

There was also a statistically significant decrease in the levels of immune parameters – secretory IgA (SIgA) and tumor necrosis factor alpha (TNF-α) which is considered one of the many risk factors in the development of T2D [3, 58].

Normalization of patients' gut and oral microbiota, above mentioned changes in biochemical and immunological parameters led to changes in physical parameters of all patients of the experimental group, namely a statistically significant decrease in body weight and the circumference of waist, hips, and upper thighs [13].

All this confirms that our proposed targeted correction of the intestinal microbiota through the use of personalized diet plans in the comprehensive treatment of patients with diabetes mellitus-2 has clinically proven its effectiveness.

11.5 Analysis of Experimental Data, Development and Application of Methods for Selecting Markers of the Process of Intestinal Microbiota Personalized Correction

For detection of the most promising clinical indicators as biomarkers the obtained data of the 62 indicators (indicators of the experiment participant' state - IEPS) of a randomized controlled clinical trial of the effectiveness of personalized correction of the intestinal microbiota for the treatment of patients with type 2 diabetes were analyzed. Basic statistical analysis of the obtained data was performed by known methods of descriptive statistics and statistical inference. Its results showed a statistically significant presence for the group of patients and the absence for the participants of the control group of a clear dynamics of changes in IEPS during the experiment (Fig. 11.11).

Peculiarities of this dynamics for a group of patients were identified and investigated by correlation analysis, Principal Component Analysis (PCA) and cluster analysis. One of the main established features is manifested in the presence of a direction (in the space IEPS), along which the set of points of experimental data is clearly extended by the variance (respectively, in other directions it is compressed). And this indicates the existence of clear patterns of development of the process of personalized correction of the intestinal microbiome (Fig. 11.12). Accordingly, the task of identifying informative markers of this process among IEPS becomes relevant.

A methodology for identifying such markers has been developed, which is reduced to quantifying each IEPS on: (1) formation of correlations with other indicators, (2) role in the formation of the first main components of PCA related to the above direction, (3) impact on cluster compactness experimental points in the space of the studied indicators and (4) importance in the "tree-based" model of classification of available data on IEPS (Fig. 11.13).

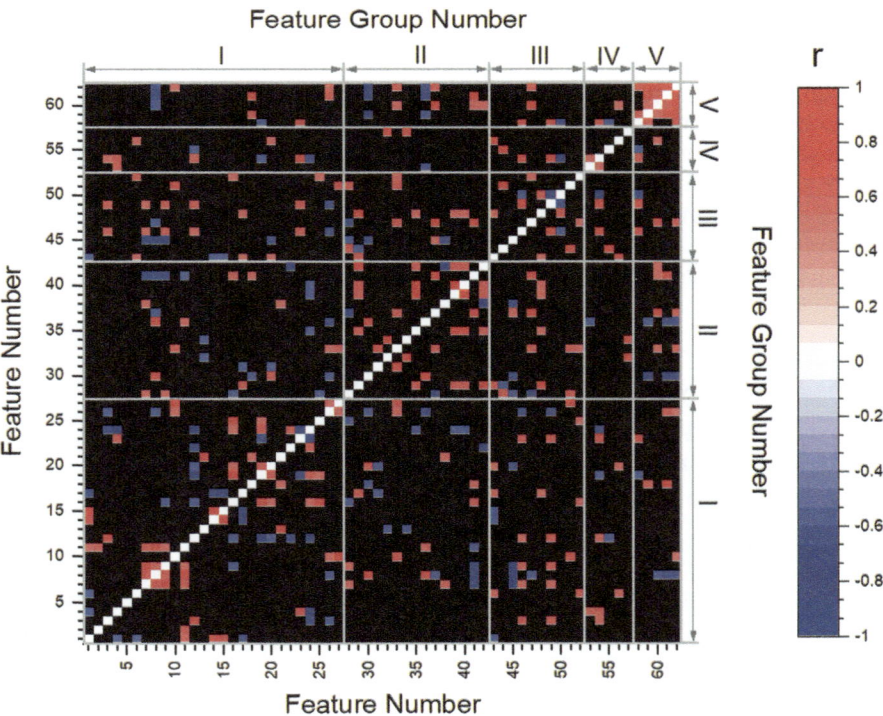

Fig. 11.11 Heat map matrix of dynamic correlation coefficients, which corresponds to the data set for the group of patients on the 1st and 18th days of the study

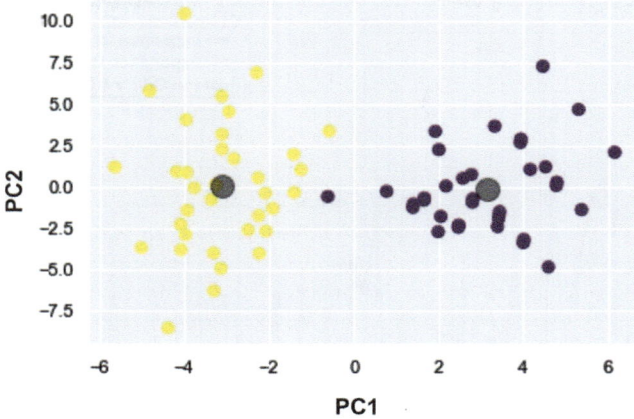

Fig. 11.12 Obtained within the framework of the studied model of k-means picture of clusters of experimental points of the 1st (yellow color) and 18th (violet color) days of the experiment. The centers of the clusters are marked with gray circles

On the basis of correlation, PCA and cluster analysis of experimental data and using a specially developed method, an orderly list of microbial, biochemical and immune markers of the process of personalized treatment of diabetes mellitus-2 by correcting the intestinal microbiota. The first 10 places in this list are occupied by *E. faecalis* (saliva), TNF-a, creatinine, urea, thymol test, Na, *Lactobacillus* spp. (saliva), *E. faecalis* (stool), *E. coli* (lac -) (stool), *Lactobacillus* spp. (stool). All this indicators should be used as a prioritized biomarkers of type 2 diabetes.

11.6   PPPM Strategies in the Field

The fact that co-evolutionary relationships between nutrition, gut microbiota, stress, lifestyle, and environmental factors lead to epigenetic influence on human health outcomes is already widely accepted [59]. Nowadays, it is obvious that changes in gut microbiota often act as a trigger of noncommunicable diseases connected to low-grade inflammation and metabolic disorders, such as obesity, T2D, CVDs, and others. One of the commonly known and most widely used approaches to correct human microbiota is to prescribe various biopharmaceuticals like pre-, pro-, and synbiotics, or pharmabiotics, if their efficacy is clinically proven [45]. There are numerous studies in animal models on the use of various probiotic microorganisms and prebiotic components such as biologically active substances for microbiota correction. However, there are very few clinical studies that confirm the effectiveness of such microbiota correction approaches to prevent NCDs. Therefore, to date, the question of individual selection of a composition of pro- and pre-biotic components in the form of a personalized nutritional plan remains open.

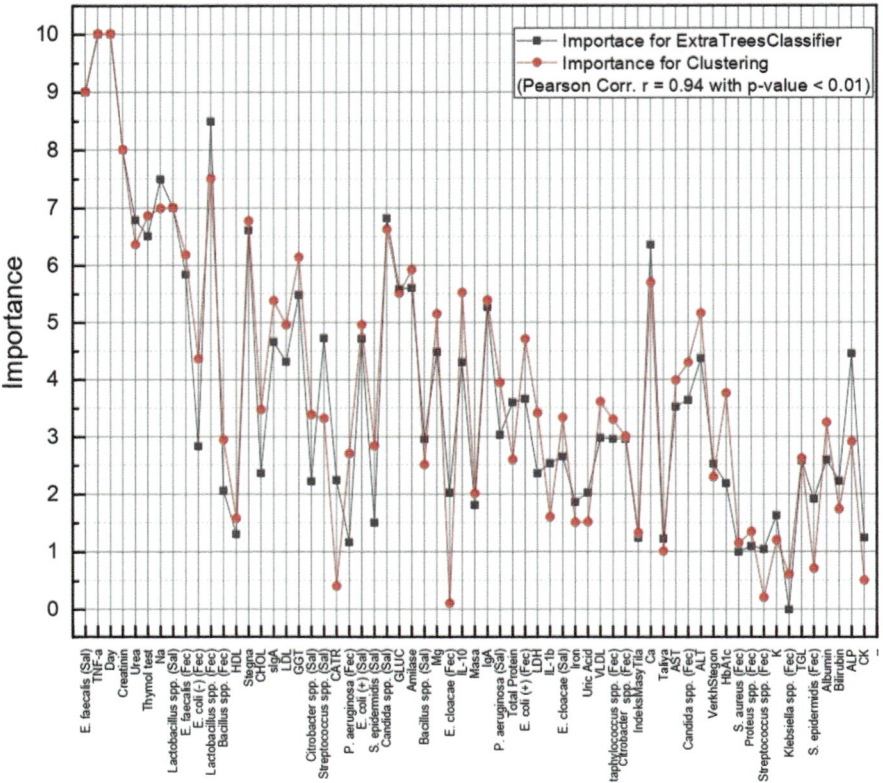

Fig. 11.13 Correlation of estimates calculated based on the results of cluster analysis and analysis of the importance of IEPS using the "tree-based" model of data classification

11.7   Expert Recommendations in the Frame-Work of 3PM to the Practical Medicine

- In diagnostics of different NCDs such as type 2 diabetes, cardiovascular diseases, atherosclerosis and etc. and more accurate interpretation of obtained results systemic approach should be used. In this case attention should be addressed not only to "classic" markers of diseases but also to new biomarkers which will indicate the general condition of the patient, including not only biochemical parameters, but the condition of the patient's microbiota and his immune parameters.
- For more precise personalized diagnostic of type 2 diabetes such biomarkers as levels of *E. faecalis* (saliva), TNF-a, creatinine, urea, thymol test, Na, *Lactobacillus* spp. (saliva), *E. faecalis* (stool), *E. coli* (lac -) (stool), *Lactobacillus* spp. (stool) should be used.

- Targeted correction of the intestinal microbiota through the use of personalized diet plans which include individually selected appropriate probiotics and components of food products of plant origin of prebiotic purpose, can be used for treatment of patients with type 2 diabetes.
- Taking into account P3 medicine's position on the correction of the human microbiome the prescription of individual nutrition should be made on an evidence-based basis. Such a necessary verification is the conduct of randomized clinical trials and cohort studies.
- Developed and applied in this work, the methodology of search and quantitative assessment of the quality of microbial, biochemical and immune markers of the process of personalized treatment of NCDs by correcting the intestinal microbiota has practical use as a component of a new method of implementing "feature selection" – optimal selection of data for machine learning, which will maximise an efficacy of the disease prevention and prognosis, and optimised treatment algorithms.

Founding This work was supported by Ministry of Education and Science, Topic: "The introduction of new approaches to the creation and use of modern pharmabiotics" Registration number 0117 U000379.

Other Acknowledgments No declare.

References

1. Cho NH, Shaw JE, Karuranga S et al (2018) IDF Diabetes Atlas: global estimates of diabetes prevalence for 2017 and projections for 2045. Diabetes Res Clin Pract 138:271–281. https://doi.org/10.1016/j.diabres.2018.02.023
2. Duarte AA, Mohsin S, Golubnitschaja O (2018) Diabetes care in figures: current pitfalls and future scenario. EPMA J 9(2):125–131. https://doi.org/10.1007/s13167-018-0133-y
3. Ma Q, Li Y, Li P et al (2019) Research progress in the relationship between type 2 diabetes mellitus and intestinal flora. Biomed Pharmacother 117:109138. https://doi.org/10.1016/j.biopha.2019.109138
4. Krauss RM (2004) Lipids and lipoproteins in patients with type 2 diabetes. Diabetes Care 27(6):1496–1504. https://doi.org/10.2337/diacare.27.6.1496
5. Kyrou I, Tsigos C, Mavrogianni C et al (2020) Sociodemographic and lifestyle-related risk factors for identifying vulnerable groups for type 2 diabetes: a narrative review with emphasis on data from Europe. BMC Endocr Disord 20(1):1–13. https://doi.org/10.1186/s12902-019-0463-3
6. Kraniotou C, Karadima V, Bellos G et al (2018) Predictive biomarkers for type 2 of diabetes mellitus: bridging the gap between systems research and personalized medicine. J Proteome 188:59–62. https://doi.org/10.1016/j.jprot.2018.03.004
7. Scirica BM (2017) Use of biomarkers in predicting the onset, monitoring the progression, and risk stratification for patients with type 2 diabetes mellitus. Clin Chem 63(1):186–195. https://doi.org/10.1373/clinchem.2016.255539
8. Song S, Lee JE (2018) Dietary patterns related to triglyceride and high-density lipoprotein cholesterol and the incidence of type 2 diabetes in Korean men and women. Nutrients 11(1):8. https://doi.org/10.3390/nu11010008

9. Jeon J, Jang J, Park K (2018) Effects of consuming calcium-rich foods on the incidence of type 2 diabetes mellitus. Nutrients 11(1):31. https://doi.org/10.3390/nu11010031

10. Meleshko T, Pallah O, Petrov V et al (2020) Extracts of pomegranate, persimmon, nettle, dill, kale and Sideritis specifically modulate gut microbiota and local cytokines production: in vivo study. ScienceRise Biol Sci 2(23):4–14

11. Tilg H, Moschen AR (2014) Microbiota and diabetes: an evolving relationship. Gut 63(9):1513–1521. https://doi.org/10.1136/gutjnl-2014-306928

12. Peterson J, Garges S, Giovanni M et al (2009) The NIH human microbiome project. Genome Res 19(12):2317–2323. http://www.genome.org/cgi/doi/10.1101/gr.096651.109

13. Meleshko T, Rukavchuk R, Levchuk O et al (2021) Personalized nutrition for microbiota correction and metabolism restore in type 2 diabetes mellitus patients. Adv Exp Med Biol. https://doi.org/10.1007/5584_2021_621

14. Golubnitschaja O, Baban B, Boniolo G et al (2016) Medicine in the early twenty-first century: paradigm and anticipation – EPMA position paper 2016. EPMA J 7(1):23. https://doi.org/10.1186/s13167-016-0072-4

15. le Roux CW, Astrup A, Fujioka K et al (2017) 3 years of liraglutide versus placebo for type 2 diabetes risk reduction and weight management in individuals with prediabetes: a randomised, double-blind trial. Lancet 389(10077):1399–1409. https://doi.org/10.1016/S0140-6736(17)30069-7

16. Ley SH, Hamdy O, Mohan V et al (2014) Prevention and management of type 2 diabetes: dietary components and nutritional strategies. Lancet 383(9933):1999–2007. https://doi.org/10.1016/S0140-6736(14)60613-9

17. Golubnitschaja O, Costigliola V, EPMA (2012) General report & recommendations in predictive, preventive and personalised medicine 2012: white paper of the European Association for Predictive, Preventive and Personalised Medicine. EPMA J 3(1):14. https://doi.org/10.1186/1878-5085-3-14

18. Wang K, Gong M, Xie S et al (2019) Nomogram prediction for the 3-year risk of type 2 diabetes in healthy mainland China residents. EPMA J 10(3):227–237. https://doi.org/10.1007/s13167-019-00181-2

19. Ercolini D, Fogliano V (2018) Food design to feed the human gut microbiota. J Agric Food Chem 66(15):3754–3758. https://doi.org/10.1021/acs.jafc.8b00456

20. Shanahan F, van Sinderen D, O'Toole PW et al (2017) Feeding the microbiota: transducer of nutrient signals for the host. Gut 66(9):1709–1717. https://doi.org/10.1136/gutjnl-2017-313872

21. Kastorini CM, Milionis HJ, Esposito K et al (2011) The effect of Mediterranean diet on metabolic syndrome and its components: a meta-analysis of 50 studies and 534,906 individuals. J Am Coll Cardiol 57(11):1299–1313. https://doi.org/10.1016/j.jacc.2010.09.073

22. Singh RK, Chang HW, Yan D et al (2017) Influence of diet on the gut microbiome and implications for human health. J Transl Med 15(1):73. https://doi.org/10.1186/s12967-017-1175-y

23. Dinu M, Abbate R, Gensini GF et al (2017) Vegetarian, vegan diets and multiple health outcomes: a systematic review with meta-analysis of observational studies. Crit Rev Food Sci Nutr 57(17):3640–3649. https://doi.org/10.1080/10408398.2016.1138447

24. Wu GD, Compher C, Chen EZ et al (2016) Comparative metabolomics in vegans and omnivores reveal constraints on diet-dependent gut microbiota metabolite production. Gut 65(1):63–72. https://doi.org/10.1136/gutjnl-2014-308209

25. Anderson JW, Baird P, Davis RH Jr et al (2009) Health benefits of dietary fiber. Nutr Rev 67(4):188–205. https://doi.org/10.1111/j.1753-4887.2009.00189.x

26. Oriach CS, Robertson RC, Stanton C et al (2016) Food for thought: the role of nutrition in the microbiota-gut-brain axis. Clin Nutr Exp 6:25–38. https://doi.org/10.1016/j.yclnex.2016.01.003

27. Parker B, Noakes M, Luscombe N et al (2002) Effect of a high-protein, high-monounsaturated fat weight loss diet on glycemic control and lipid levels in type 2 diabetes. Diabetes Care 25(3):425–430. https://doi.org/10.2337/diacare.25.3.425

28. Zeevi D, Korem T, Zmora N et al (2015) Personalized nutrition by prediction of glycemic responses. Cell 163(5):1079–1094. https://doi.org/10.1016/j.cell.2015.11.001
29. Hatch A, Horne J, Toma R et al (2019) Robust metatranscriptomic technology for population-scale studies of diet, gut microbiome, and human health. Int J Genomics 2019:1718741–1718749. https://doi.org/10.1155/2019/1718741
30. Golubnitschaja O, Topolcan O, Kucera R et al (2020) 10th anniversary of the European Association for Predictive, Preventive and Personalised (3P) Medicine – EPMA world congress supplement 2020. EPMA J 11(Suppl 1):1–133. https://doi.org/10.1007/s13167-020-00206-1
31. Danesi F, Pasini F, Caboni MF et al (2013) Traditional foods for health: screening of the antioxidant capacity and phenolic content of selected Black Sea area local foods. J Sci Food Agric 93(14):3595–3603. https://doi.org/10.1002/jsfa.6339
32. Pallah O, Meleshko T, Bati V et al (2019) Extracts of edible plants stimulators for beneficial microorganisms. Biotechnologia Acta 12(3):67–74. https://doi.org/10.15407/biotech12.03.067
33. Lazar V, Ditu LM, Pircalabioru GG et al (2019) Gut microbiota, host organism, and diet trialogue in diabetes and obesity. Front Nutr 6:21. https://doi.org/10.3389/fnut.2019.00021
34. Cox AJ, West NP, Cripps AW (2015) Obesity, inflammation, and the gut microbiota. Lancet Diabetes Endocrinol 3(3):207–215. https://doi.org/10.1016/S2213-8587(14)70134-2
35. Tazoe H, Otomo Y, Kaji I et al (2008) A roles of short-chain fatty acids receptors, GPR41 and GPR43 on colonic functions. J Physiol Pharmacol 59(2):251–262
36. Boulangé CL, Neves AL, Chilloux J et al (2016) Impact of the gut microbiota on inflammation, obesity, and metabolic disease. Genome Med 8(1):42. https://doi.org/10.1186/s13073-016-0303-2
37. Remely M, Tesar I, Hippe B et al (2015) Gut microbiota composition correlates with changes in body fat content due to weight loss. Benef Microbes 6(4):431–439. https://doi.org/10.3920/BM2014.0104
38. Long AN, Dagogo-Jack S (2011) Comorbidities of diabetes and hypertension: mechanisms and approach to target organ protection. J Clin Hypertens (Greenwich) 13(4):244–251. https://doi.org/10.1111/j.1751-7176.2011.00434.x
39. Meleshko TV, Pallah OV, Rukavchuk RO et al (2020) Edible fruits extracts affect intestinal microbiota isolated from patients with noncommunicable diseases associated with chronic inflammation. Biotechnologia Acta 13(5):87–100. https://doi.org/10.15407/biotech13.05 087
40. Moyer R, Hummer K, Wrolstad RE et al (2002) Antioxidant compounds in diverse ribes and rubus germplasm. Acta Hortic 585:501–505. https://doi.org/10.17660/actahortic.2002.585.80
41. Correia RT, Borges KC, Medeiros MF et al (2012) Bioactive compounds and phenolic-linked functionality of powdered tropical fruit residues. Food Sci Technol Int 18(6):539–547. https://doi.org/10.1177/1082013211433077
42. Bene K, Varga Z, Petrov VO et al (2017) Gut microbiota species can provoke both inflammatory and tolerogenic immune responses in human dendritic cells mediated by retinoic acid receptor alpha ligation. Front Immunol 8:427. https://doi.org/10.3389/fimmu.2017.00427
43. Perez-Gregoric R, Simal-Gandara J (2017) A critical review of bioactive food components, and of their functional mechanisms, biological effects and health outcomes. Curr Pharm Des 23(19):2731–2741. https://doi.org/10.2174/1381612823666170317122913
44. Meleshko T, Rukavchuk R, Buhyna L et al (2021) Biologically active substance content in edible plants of Zakarpattia and their elemental composition model. Biol Trace Elem Res 199(6):2387–2398. https://doi.org/10.1007/s12011-020-02345-y
45. Bati VV, Meleshko TV, Pallah OV et al (2021) Personalised diet improve intestine microbiota and metabolism of obese rats. Ukr Biochem J 93(4):87–102. https://doi.org/10.15407/ubj93.04.077
46. Li Q, Chang Y, Zhang K et al (2020) Implication of the gut microbiome composition of type 2 diabetic patients from northern China. Sci Rep 10(1):1–8. https://doi.org/10.1038/s41598-020-62224-3

47. Grygoruk G, Myshchuk V, Tserpiak N (2019) Changes in gut microbiota and blood lipid profile in patients with irritable bowel syndrome in association with obesity. Bukovinian Med Her 23(1 (89)):32–38
48. DeFronzo RA, Ferrannini E, Groop L et al (2015) Type 2 diabetes mellitus. Nat Rev Dis Primers 1:15019. https://doi.org/10.1038/nrdp.2015.19
49. Gurung M, Li Z, You H et al (2020) Role of gut microbiota in type 2 diabetes pathophysiology. EBioMedicine 51:102590. https://doi.org/10.1016/j.ebiom.2019.11.051
50. Sharma S, Tripathi P (2019) Gut microbiome and type 2 diabetes: where we are and where to go? J Nutr Biochem 63:101–108. https://doi.org/10.1016/j.jnutbio.2018.10.003
51. Xie X, Zhang X, Xiang S et al (2017) Association of very low-density lipoprotein cholesterol with all-cause and cardiovascular mortality in peritoneal dialysis. Kidney Blood Press Res 42(1):52–61. https://doi.org/10.1159/000469714
52. Gowda S, Desai PB, Kulkarni SS et al (2010) Markers of renal function tests. N Am J Med Sci 2(4):170–173
53. Liamis G, Liberopoulos E, Barkas F et al (2014) Diabetes mellitus and electrolyte disorders. World J Clin Cases 2(10):488–496. https://doi.org/10.12998/wjcc.v2.i10.488
54. Kostov K (2019) Effects of magnesium deficiency on mechanisms of insulin resistance in type 2 diabetes: focusing on the processes of insulin secretion and signaling. Int J Mol Sci 20(6):1351. https://doi.org/10.3390/ijms20061351
55. Liamis G, Tsimihodimos V, Elisaf M (2015) Hyponatremia in diabetes mellitus: clues to diagnosis and treatment. J Diabetes Metab 6(5):559–561. https://doi.org/10.4172/2155-6156.1000560
56. Constantin M, Alexandru I (2011) The role of sodium in the body. Balneo Res J 2(1):70–74. https://doi.org/10.12680/balneo.2011.1015
57. Djiambou-Nganjeu H (2019) Relationship between portal HTN and cirrhosis as a cause for diabetes. J Transl Int Med 7(2):79–83. https://doi.org/10.2478/jtim-2019-0009
58. Akash MSH, Rehman K, Liaqat A (2018) Tumor necrosis factor-alpha: role in development of insulin resistance and pathogenesis of type 2 diabetes mellitus. J Cell Biochem 119(1):105–110. https://doi.org/10.1002/jcb.26174
59. D'Argenio V, Salvatore F (2005) The role of the gut microbiome in the healthy adult status. Clin Chim Acta 451:97–102. https://doi.org/10.1016/j.cca.2015.01.003

Chapter 12
Pro- Pre- and Synbiotic Supplementation and Oxalate Homeostasis in 3 PM Context: Focus on Microbiota Oxalate-Degrading Activity

Ganna Tolstanova, Iryna Akulenko, Tetiiana Serhiichuk, Taisa Dovbynchuk, and Natalia Stepanova

Abstract Hyperoxaluria has been proven to be the major risk factor for the formation of oxalate-calcium stones, which are the cause of 75% of all kidney stones. The oxalate-degrading bacteria (ODB) is a diverse group of bacteria consisting of the "narrow-profile oxalotrophs" that use oxalate as their sole source of carbon and energy (e.g. *Oxalobacter formigenes*) and the "broad-profile oxalotrophs" that use oxalate as an alternative or additional source of metabolism (e.g. *Lactobacillus spp.*, *Bifidobacterium spp.*, *Bacillus spp., etc*).

Taking into account the difficulties in assessing ODB due to their wide range of colonization density, which can reflect both diet and competing microorganisms, and considering the need for multiple tests to detect ODB colonization, we hypothesized that the total ability of intestinal ODB to degrade oxalate rather than the ODB number is a key determinant in oxalate homeostasis.

Therefore, in the present chapter, based on the results of our previous research and other published data, we present a new perspective on the translational relevance of oxalate-degrading activity assessing in the prediction of nephrolithiasis formation and individualized prescribing of probiotics or synbiotic in kidney stone disease patients which would allow shifting the reactive medicine paradigm to PPPM/3P model.

G. Tolstanova (✉) · I. Akulenko · T. Serhiichuk · T. Dovbynchuk
Taras Shevchenko National University of Kyiv, Kyiv, Ukraine
e-mail: ganna.tolstanova@knu.ua; iv_akulenko@knu.ua; stm1972@knu.ua

N. Stepanova
State Institution "Institute of Nephrology of the National Academy of Medical Sciences of Ukraine", Kyiv, Ukraine

N. Boyko, O. Golubnitschaja (eds.), *Microbiome in 3P Medicine Strategies*,
Advances in Predictive, Preventive and Personalised Medicine 16,
https://doi.org/10.1007/978-3-031-19564-8_12

Keywords PPPM /3P medicine · Ceftriaxone · Synbiotic · Probiotic · Oxalate homeostasis · Oxalic acid · Hyperoxaluria · Oxalate-degrading activity · Oxalate-degrading bacteria

12.1 Introduction

Oxalate is a potentially toxic anion of dicarboxylic acid ($C_2O_4H_2$), that is widespread in both plants and mammals [13, 35, 75]. In humans, the balance of body oxalate is achieved due to its renal (up to 90%) and intestinal (10%) excretion [10, 64]. The oxalate homeostasis is maintained by kidneys'(the bulk 90% to 95% of circulating oxalate) and intestinal (10%–15%) excretions [13, 37]. A high-oxalate diet and/or gut microbiota violation can result in hyperoxaluria, defined as urinary oxalate excretion greater than 44 mg per day [74], and calcium oxalate stone formation [10, 40]. In turn, kidney function decline in non-stone-formers leads to a decrease in renal oxalate clearance and an increase in plasma oxalic acid (POx) concentration corresponding to the progression of chronic kidney disease (CKD) [57]. The accumulation of oxalate has been postulated to be associated with oxidative stress [43], inflammation [22, 24] and a high risk for cardiovascular disease (CVD) in kidney stone disease patients [40].

The intestine plays a complex role in oxalate homeostasis via interconnected mechanisms of intestinal epithelium oxalate transport (both secretion and absorption) and microbiota-dependent degradation of intraluminal oxalate reducing its intestinal absorption [74].

The absorptive flux of oxalate across the intestine can occur via both the paracellular (predominantly passive) and transcellular pathways. The active transcellular movement of oxalate is provided by anion exchange proteins belonging to the multifunctional SLC26 gene family. Among them, Slc26a6 is highly expressed in the intestine and renal proximal tubule and plays an important role in the control of systemic oxalate metabolism [42]. Its pivotal role in oxalate homeostasis was demonstrated in Slc26a6$^{-/-}$ mice. These mice had a defect in intestinal oxalate secretion resulting in enhanced net absorption of oxalate accompanied by an increase in serum and urinary oxalate [39].

Despite the potential toxicity, humans lack the oxalate metabolizing enzymes and are therefore unable to degrade excessive amounts of oxalate [6]. The gut oxalate-degrading bacteria (ODB) play an important role in oxalate destruction expressing the catabolic enzymes formyl-CoA transferase (Frc) and oxalyl-CoA decarboxylase (Oxc). The genes encoding these proteins are clustered on the genomes and show strong phylogenetic relationships [1]. Humans' intestinal ODB that complement the missing oxalate metabolizing ability in a mammalian host are collectively identified as Oxalate Metabolizing Bacterial Species (OMBS) [71]. OMBS include "narrow-profile oxalotrophs"that use oxalate as their sole source of carbon and energy, for example, *Oxalobacter formigenes (O. formigenes)*, and "broad-profile oxalotrophs" that use oxalate as an alternative or additional source of metabolism, for example, *Lactobacillus spp., Bifidobacterium spp., Bacillus spp., E. faecalis, N. albigula* [8, 32].

O. formigenes described in 1985 and isolated from the rumen of a sheep, the cecum of a pig, and human feces were all similar Gram-negative, obligate anaerobic rods [8]. *O. formigenes* uses oxalate as its sole source of energy and carbon; it metabolizes oxalate into formate and CO_2. *O. formigenes* is the only specialized oxalate degrader in humans [8]. In a recent comparative study on the prevalence of *O. formigenes*, among the USA adults and infants (the first three years of life) with Venezuela and Tanzania ethnic groups, a higher prevalence of *O. formigenes* in both adults and children from the developed countries *vs.* the USA population have been found [56].

In addition to the oxalate-degrading activity (ODA), *O. formigenes* have been shown to produce small proteins that can directly induce oxalate transport through the oxalate transporter SLC26A6 –dependent pathway in intestinal Caco-2 cells, confirming a specific role of the gut microbiota in oxalate homeostasis [11].

The therapeutic use of antibiotics and other medications may contribute to the loss of ODB. *O. formigenes* and other ODB are susceptible to commonly used antibiotics [34, 44]. Patients with cystic fibrosis often lack fecal oxalate degraders, perhaps because of intensive drug therapy [60]. The antibiotic-induced deficiency of gut ODB has been considered one of the potential causes of intestinal oxalate absorption upregulation and hyperoxalemia [18, 48].

Currently, the medical community is pulling out all the stops to shift the paradigm from diagnosis and treatment to prediction and prevention. One of the primary goals of Predictive, Preventive and Personalized (PPPM / 3P) medicine is to promote innovative analytical approaches that enable targeted prevention and treatments tailored to the patient [28]. The gut microbiota is becoming the focus of advanced PPPM / 3P strategies through its key role in the prediction and prevention of many diseases including kidney stone disease [9, 14–16, 59].

The number of data on the efficacy of particular probiotic strains in the treatment of various diseases is growing day by day, but there is still not enough evidence-based data on probiotics'contribution to disease pathophysiology and their applicability in clinical care [63]. The effectiveness of personalized probiotic prescription based on patient symptoms has been reported to treat metabolic syndrome and hyperuricemia and to restore kidney function in gout. The authors suggested that patients with gout may be able to achieve the same results by eating yogurt or taking an over-the-counter probiotic supplement, but the personalized approach provided a higher likelihood of effectiveness [14–16].

Taking into account the difficulties in assessing ODB due to the wide range of their colonization density, which may reflect both diet and competing microorganisms, and considering the need for multiple assays to detect ODB colonization [54, 71], we hypothesized that the total ability of intestinal ODB to degrade oxalate, rather than the number of ODB, is a key determinant of oxalate homeostasis.

Therefore, in the present chapter, based on the results of our previous research and other published data, we present a new perspective on the translational relevance of assessing total fecal ODA in predicting the formation of nephrolithiasis and personalized prescribing of probiotics or synbiotics.

12.2   The Differential Sensitivity of Narrow-Profile Vs. Broad-Profile Oxalotrophs to Oxalate-Rich Diet and Antibiotic Violence: A Preclinical Study

The relevant animal model is the cornerstone of translational medicine. In our previously published study, we screened the widely used laboratory animal species (nonlinear rats, Wistar rats, BalbC mice, Chinchilla rabbits) for the presence of narrow-profile (obligate) and broad-profile (facultative) oxalotrophs in the intestine, their response to diet and antibiotic administration to find the appropriate species to study the role of gut microbiota in oxalate homeostasis [4]. Fecal biopsy of nonlinear rats (n = 12), Wistar rats (n = 12), BalbC mice (n = 12), and Chinchilla rabbits (n = 10) were inoculated into highly selective medium Oxalate Medium (the only source of energy is $Na_2C_2O_4$). To distinguish between obligate and facultative oxalotrophs, after cultivation on Oxalate Medium, all colonies were transferred to MRS agar, Bifidobacterium agar, and MPA media and cultured under anaerobic conditions. Cultures grown on these media after inoculation were classified as facultative oxalotrophs, and those grown only on Oxalate Medium were classified as obligate oxalotrophs (Fig. 12.1). Among all the experimental animals studied, only 100% of the Wistar rats were colonized with both narrow-profile (obligate) and broad-profile (facultative) oxalotrophs. Other animals had obligate oxalotrophs only, in particular, 58% of nonlinear rats, 42% of Balb C mice, and 7% of chinchilla rabbits. The Wistar rats had the greatest number of obligate oxalotrophs in the fecal biopsy (l g 6.12 ± 0.63 CFU/g), whereas in the other laboratory species studied obligate oxalotrophs ranged from l g 2.1 ± 0.4–2.97 ± 0.25 CFU/g. There was no difference in the

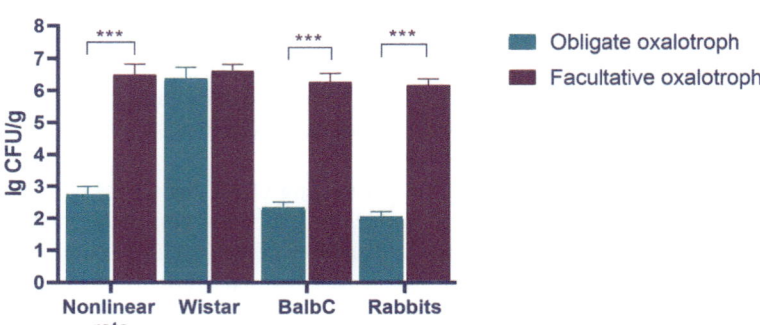

Fig. 12.1 The amount (M ± SD l g CFU/g) of oxalate-degrading bacteria in the fecal biopsy of different experimental animal species. (*** P < 0.001)

number of facultative oxalotrophs in nonlinear rats, Wistar rats, BalbC mice, Chinchilla rabbits [4].

The aforementioned results are in agreement with the study conducted by Niv Palgi et al. in which a significant difference has been found in the percentage of *O. formigenes* in three lines of laboratory rats [55].

Bearing in mind the importance of an oxalate-restricted diet in urolithiasis prevention [50], the following questions arise: (1) Is there a relationship between the consumption of oxalate-rich foods and the ODB number? (2) How does oxalate affect the mammalian gut microbiota in general? (3) How does an impaired gut microbiota affect the overall host oxalate metabolism since the majority of previous research have focused on the role of individual taxa in oxalate degradation? [35]. To try to answer these questions, we further investigated the impact of oxalate-enriched and oxalate-free diets on ODB numbers. The daily oral administration of 2 mg/kg sodium oxalate solution to experimental rats for 14 days was supported by a significant increase in the number of obligate oxalotrophs in fecal biopsy compared with a control group on a regular diet (1 g 4.21 ± 0.37 CFU/g *vs.* 1 g 2.97 ± 0.25 CFU/g). Conversely, the number of the facultative oxalotrophs was 1.5-fold lower compare with the control rats (p < 0.05) [4] (Fig. 12.2).

Unlike *O. formigenes*, the growth of *L. acidophilus* and *S. thermophilus* is inhibited by the presence of oxalate, even though these bacteria have been shown to effectively degrade oxalate *in vitro* [17]. Therefore, the growth of obligate oxalotrophs in the oxalate-rich diet cannot be considered a purely positive effect, because of the tendency to reduce the number of facultative oxalotrophs, which are predominantly anaerobic, sugar-fermenting bacteria.

An interesting comprehensive study on fecal microbiota using high-throughput sequencing has been done on the wild mammalian herbivore *Neotoma albigula* which consume a diet composed of nearly 100% *Opuntia* species cactus (1.5% dry

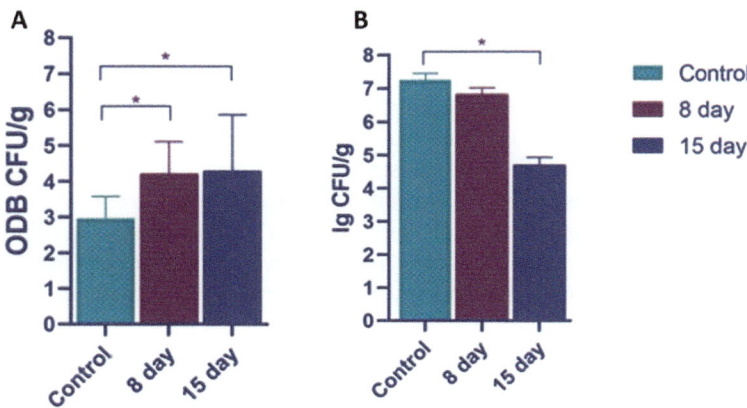

Fig. 12.2 The dynamics of changes in the number (M ± SD 1 g CFU/g) of obligate (**a**) and facultative (**b**) oxalotrophs in rat fecal biopsies of (n = 12) depending on the duration of sodium oxalate (2 mg/kg, *per os*) administration. (*– P < 0.05)

weight of oxalate content) and can degrade >90% of dietary oxalate most likely via microbial metabolism [49]. It was found a linear correlation between the amount of oxalate in diet and the rate of its degradation. Furthermore, feeding with an oxalate-enrich diet increases the abundance of certain taxa in *N. albigula* (the relative abundances of 117 operational taxonomic units exhibited a significant correlation with oxalate consumption). Some bacteria in this group, such as *O. formigines*, may engage in oxalate degradation directly. Others, such as *Oscillospira, Clostridium, Streptococcus*, and *Ruminococcus* are not directly involved in the oxalate degradation function and may benefit indirectly from oxalate degradation possibly via the process of acetogenesis or methanogenesis [49].

It is well-known that *O. formigenes* are characterized by high sensitivity to antibiotic quinolones, macrolides, tetracyclines, and metronidazole [41, 52]. However, there is scarce information on the effect of antibiotic therapy on the total ODB number and sensitivity of facultative *vs.* obligate oxalotrophs to antibiotics.

Recently, 2 models of dysbiotic disorders in male Wistar rats were used: co-administration of ampicillin (75 mg/kg) and metronidazole (50 mg/kg) for three days [25] and intramuscular injections of ceftriaxone (300 mg/kg) for seven days [70]. However, the long-term effects of antibiotic therapy on the ODB number have never before been evaluated. We assessed quantitative changes in obligate and facultative oxalotrophs in the fecal biopsy on the first, 14th and 56th days after antibiotic withdrawal. The ampicillin with metronidazole mixture as well ceftriaxone induced a rapid decrease in the number of obligate oxalotrophs starting from the first day after antibiotics withdrawal. Moreover, we did not observe a restoration of growth of obligate oxalotrophs even on the 56th day after antibiotic withdrawal. The facultative oxalotrophs were less sensitive to antibiotic treatment compared with obligate oxalotrophs. However, we still observed a decrease in the facultative oxalotrophs number on the 1st - 14th days after the withdrawal of ampicillin with a metronidazole mixture. Notably, despite a gradual increase in facultative oxalotrophs on day 56, their number was lower compared to the control group. After the first day of ceftriaxone withdrawal, there was an increase in the number of facultative oxalotrophs by almost 2 folds from l g 7.60 ± 0.15 CFU/g to l g 9.0 ± 0.23 CFU/g ($P < 0.01$). While by the 56th day after ceftriaxone withdrawal the number of facultative oxalotrophs was 2 folds lower than the control values (Fig. 12.3).

Recently, Nassal et al. have demonstrated the prevalence *of O. formigenes* in 50% of healthy subjects only [54]. It should be noted, that *O. formigenes* colonization was significantly and persistently suppressed in antibiotic-exposed subjects but remained stable in controls. The authors have shown no significant recolonization of *O. formigenes* by 6 months after therapy, which is consistent with persistent microbiome disturbance.

Moreover, the study provides evidence that antibiotic therapy may be one of the key triggers for the development of hyperoxaluria, since the presence of obligate oxalotrophs, e.g. *O. formigenes* is highly individualized, and its changes can substantially alter the harmony of the bacterial taxa consortium associated with urinary oxalate. In this context, we proposed that the total ability of intestinal ODB to degrade oxalate rather than their quantity is the key determinant of oxalate

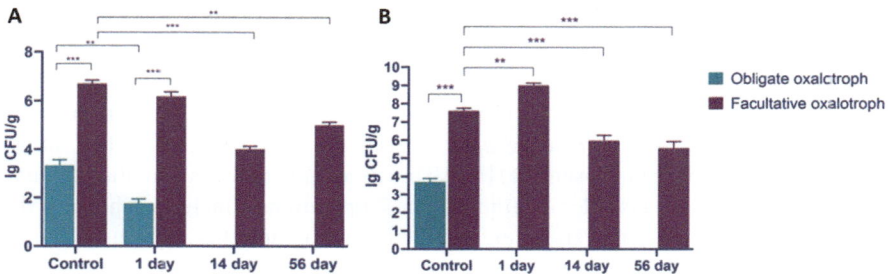

Fig. 12.3 Time-dependent changes in the number of oxalate-degrading bacteria in rat fecal biopsies after the withdrawal of the mixture of ampicillin and metronidazole (**a**) or ceftriaxone (**b**). (** $p < 0.01$; *** $p < 0.001$)

homeostasis. The identification of ODA may provide a personalized signature for kidney stone prevention in healthy adults.

12.3 Methods for Urine Oxalate Determination

The diagnosis of hyperoxaluria is based on the measurement of oxalate in the urine. Currently, the leading methods of oxalate quantification are various analytical approaches such as ion chromatography, high-performance liquid chromatography, gas chromatography, colorimetry, etc.

Colorimetric methods have the advantage of being simple, convenient, sensitive and do not require expensive equipment. The methods have been developed using metal-indicator complexes, which have clear advantages: easy of measurement in an aqueous medium, reduction of complex organic synthesis, and easy of preparation of various metal-indicator complexes using different metal ions [72]. Since oxalate has four oxygen atoms that can bind to metal ions, metal complexes must be effective chemosensors for the detection of oxalates in an aqueous solution. Studies have shown that metal complexes, including calcium, copper, nickel, or zinc can be used as good colorimetric or fluorescent chemosensors to detect oxalate.

In the *gas phase chromatography*, a filtered and degassed buffer (pH = 3) solution of 20 mm potassium phosphate and orthophosphate acid is used. Before analysis, the column is cleaned and balanced bypassing the mobile phase. The chromatographic peak exit time is 35 min at a flow rate of 07 mL/min The limit of oxalate detection is 21 ng, the linearity of the calibration graph is maintained up to 11 μg [2]. The disadvantages of this method include the duration and cost of analysis.

Capillary electrophoresis with indirect detection of ultraviolet absorption is also used to determine oxalates [76]. Electrophoresis is performed in a capillary coated with polyethylene amine. Samples diluted 1:10 and filtered, injected hydrodynamically under pressure. The analyte is identified by comparing the migration of analytes and standard solutions. Peak identity is confirmed by the method of additives. Quantitative determination is performed on the corrected peak area, which is defined as the ratio between the peak area of the analyte and the corresponding migration time and comparing this ratio with the internal standard (sodium sulfate).

The method of *ion chromatography* allows the detection of oxalate in an amount of 25 ng (5 μg in the sample) [30]. The disadvantages of this method include duration (about 1 h), complexity (more than 7 operations), including heating the sample and the cost of analysis (using liquid chromatography).

Permanganatometric titration is a simple and relatively inexpensive method for the quantification of oxalates but it requires a very long sample preparation. This method is based on the precipitation of calcium oxalate, the subsequent dissolution of this salt in sulfuric acid, and the determination of the amount of oxalate by titration with permanganate [77].

The described methods require painstaking preparation of samples, highly qualified specialists, and expensive equipment. Many of these methods have practical applications in medical practice for the quantification of oxalate in biological fluids such as urine and blood.

12.4   Methods for Total Fecal Bacteria ODA Measuring

Most of the existing studies are based on the quantitative determination of ODB in coprofiltrates. However, the quantification methods independently of them assess way (bacteriological, molecular, etc) could only estimate the strains' number but not their total fecal ODA [1, 31]. Total ODA is conservatively estimated as the difference between radioactive [14C]-oxalate consumed and [14C]-oxalate excreted in feces and urine [33]. Isolation of pure microbial culture is routinely used to estimate total ODA in fecal microbiota [29, 51]. Only one study has attempted to measure total ODA directly in fecal samples from patients with a jejunoileal bypass diluted in anaerobic dilution solution with [14C]-oxalate [7]. Therefore, there is a general lack of published data on non-radioactive methods for measuring the total ODA directly in the fecal biopsy to adaptation and further routine use in clinical practice. Moreover, there is a general lack of accepted methodology for the ODA determination .

We adapted the method of *redoxmetric titration with $KMnO_4$ to determine the bacteria ODA without isolation in pure cultures*, which was published [69]. The fecal samples were collected from the study individuals within 4 h of defecation, 0.01 g feces were cultured in highly selective media Oxalate Medium (g/L): K_2HPO_4–0.25, KH_2PO_4–0.25, $(NH_4)_2SO_4$–0.5, $MgSO_4 \cdot 7H_2O$ – 0.025, CH_3COON – 0.82, yeast extract – 1.0, resazurin – 0,001, Na_2CO_3–4, L-cysteine-HCl – 0,5, Trace element solution SL-10 – 1 mL (mix/L: HCl (25%; 7.7 M) – 10.00 mL, $FeCl_2 \times 4H_2O$ – 1.50 g, $ZnCl_2$–70.00 mg, $MnCl_2 \times 4H_2O$ – 100.00 mg, H_3BO_3–6.00 mg, $CoCl_2 \times 6H_2O$ – 190.00 mg, $CuCl_2 \times 2H_2O$ – 2.00 mg, $NiCl_2 \times 6H_2O$ – 24.00 mg, $Na_2MoO_4 \times 2H_2O$ – 36.00 mg; $Na_2C_2O_4$–5 g) and cultivated anaerobically at 37 °C for 48 h (test solution). In 48 h, a 10 mL aliquot of the test solution or 10 mL of Oxalate Medium (control) was centrifuged at 3000 g for 15 min, T_{room}. 10 mL of the supernatant was transferred to a 50 mL beaker. Calcium oxalate was precipitated from the samples by adding 10 mL of 0.4 M $Ca(NO_3)_2$. The precipitate formed was filtered through duplicate paper filters with a low filtration rate (80 g/m2). The filtrate was discarded. The precipitated calcium oxalate was dissolved with 25 mL of H_2SO_4 (1:4). About 10 mL of the acidified calcium oxalate solution was mixed with 20 mL of deionized water and heated to 80 °C prior to titration. About 10 mL of H_2SO_4 (1:4) solution was immediately added and titrated with $KMnO_4$ (0.02 N) solution until a pink color persisted for 30 seconds. The results were expressed as % of oxalate degradation per 0.01 g of feces.

We confirmed the accuracy and reliability of the method of redoximetric titration with a KMnO4 solution to evaluate the total ODA level in fecal microbiota incubated in an Oxalate Medium. This method is suitable for measuring the total ODA in the fecal biopsy, even at high oxalate concentrations and has been tested in both clinical and experimental settings [65, 69].

12.5 The Proof of Concept on the Significance of Total Fecal ODA *Vs.* ODB to Support Oxalate Homeostasis: A Preclinical Study

The term total ODA in fecal microbiota could be defined as the general ability of different strains of ODB to metabolize oxalate [64]. Several *in vitro* studies have been devoted to oxalate degradation abilities of *Oxalobacter formigenes, Bifidobacterium* and *Lactobacillus* strains in which the efficiency degrade oxalate from 15% to 98% has been reported [51, 53]. However, the effect of antibiotics on total ODA in fecal microbiota has never been evaluated before. Consequently, it is still unclear whether the use of antibiotics could affect not only the number of ODB but also their ODA and thus influence the oxalate concentrations in blood and urine. In our study on the ceftriaxone-treated Wistar rats, we found a decrease in the total fecal ODA (which was evaluated by the method of redoximetric titration) independently of fecal ODB level after ceftriaxone withdrawal compared to the vehicle group. These changes were associated with a significant increase in the urine oxalate excretion immediately after ceftriaxone withdrawal despite the increased ODB level. On the 57th day after ceftriaxone withdrawal, the number of ODB and the urine oxalate excretion did not differ compared with the vehicle group while ODA was statistically lower in the ceftriaxone-treated group. Moreover, plasma oxalate concentration in the ceftriaxone-treated rats was significantly higher compared to the vehicle-treated group [69]. These results also accord with our earlier studies, which reported a strong association between antibiotics administration, intestinal barrier dysfunction and hyperoxaluria formation [66]. Surprisingly, we observed a significant increase in the total ODB number with a simultaneous decrease in total fecal ODA after ceftriaxone exposure compared to the vehicle-treated group.

The observed result of an increase in ODB number after antibiotic exposure is in line with the recent papers in which the abundance and extraintestinal dissemination of *Enterococcus spp* and *Lactobacillus spp.* after ceftriaxone treatment [19], and a substantial increase in the relative abundance of *Enterococcus spp., Lactobacillus spp.* and *Bifidobacterium spp.* after the use of a combination of four antibiotics (bacitracin, meropenem, neomycin and vancomycin) [38] have been demonstrated. In this context, it is logical to assume a compensatory increase in ODB number with a lesser ability to degrade oxalate, which could explain a significant decrease in total fecal ODA simultaneously with a transient increase in the ODB number in the ceftriaxone-treated rats. Previously we reported sustained decreased levels of short-chain fatty acids (SCFAs) in rat feces long after ceftriaxone withdrawal with the most profound changes observed for propionic and butyric acids, which led to an increase in the relative amount of the acetic acid [34]. SCFAs are the main end-products of anaerobic fermentation of dietary fiber by large intestine microbiota, which provide the major source of energy for colonocytes, therefore, their levels may reflect the metabolic activity of intestinal microbiota [61].

These data are the first confirmation of the key role of total fecal ODA vs ODB to support oxalate homeostasis. However simultaneous studies on the ODB number and their total functional ability to degrade oxalate have never been conducted before, hence the obtained results cannot be directly compared with the results of previous reports.

12.6 The Proof of Concept on the Significance of Total Fecal ODA *Vs.* ODB to Support Oxalate Homeostasis: A Clinical Study

Bearing in mind the promising experimental results of total fecal ODA evaluation, we tried to reproduce the method of redoximetric titration in humans and assessed the role of ODA in fecal microbiota in oxalate homeostasis. To this end, we evaluated total fecal ODA, plasma oxalate concentration and daily urinary oxalate excretion in 24 healthy volunteers and 78 end-stage kidney disease (ESKD) patients. This cohort of patients was chosen because they have many causes for changes in the intestinal microbiota, such as uremia, malabsorption, increased inflammation and immunosuppression, pharmacological therapy, and dietary restrictions [27, 36, 62].

The results we obtained are partially reflected in several published works [67, 68]. Briefly, in healthy subjects, increased daily urinary oxalate excretion was directly associated with high fecal ODA ($r = 0.58$; $p < 0.0001$). The comparative analysis demonstrated a significantly high total fecal ODA in the volunteers with hyperoxaluria compared with the normooxaluric subjects 5.5 [2–20] vs. -1[-4—7] % ($p = 0.008$).]

Total fecal ODA in ESKD patients ranged from −23 to 24%/0.01 g of feces and was statistically lower compared with the healthy volunteers: 3.0 (−4.5–8) vs 4.5 (2–15.5) %/0.01 g of feces, p = 0.03. For further analysis, the patients were allocated into 2 groups according to total fecal ODA status: Group 1 included ESKD patients with positive total fecal ODA status (≥ 1% of oxalate degradation per 0.01 g of feces) and Group 2 consisted of the patients with negative total fecal ODA status (≤ −1% of oxalate degradation per 0.01 g of feces). Logically, high POx concentration and low UOx excretion were diagnosed in the patients with negative total ODA in fecal microbiota compared with the Group 1 patients: 5.5 [4.5–7.0] vs 3.3 [2.5–4.5] mg/L, p = 0.001 and 34.2 [24.4–39] vs 61.8 [51.6–70.3] mg/d, p = 0.002, respectively. Moreover, the ESKD patients with positive total fecal ODA status had significantly lower serum indoxyl sulfate [28.1 (15.3–41.2) vs 55.2 (35.1–82.7) μg/mL, p < 0.001] and POx [26.8 (24.1–35.7) vs 50.0 (42.2–78.2) μmol/L, p < 0.001] levels compared with the healthy participants. Total fecal ODA was inversely associated with serum indoxyl sulfate (r = −0.41, p < 0.001), atherogenic dyslipidemia (Fig. 12.4) and POx (r = −0.67, p < 0.001) concentrations in ESKD patients.

In addition, we observed lower concentrations of serum pro-inflammatory cytokines interleukin 6 and monocyte chemoattractant protein 1 in ESKD patients with positive ODA status compared with the negative one: 1.4 (0.1–4.9) vs 3.7 (0.7–9.8) pg/mL, p = 0.001 and 290.3 (246.1–331.2) vs 339.2 (321.7–407.2) pg/mL, p = 0.008, respectively [69].

In the present study, we analyzed total ODA in the normalized amount of feces which indicated total ODA function in fecal microbiota independently of strains and bacteria numbers. Together, the method of redoximetric titration with a KMnO4 solution to evaluate the total ODA level in fecal microbiota can be used in clinical

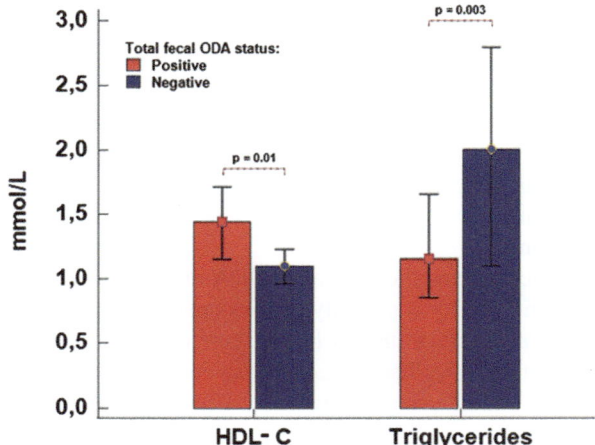

Fig. 12.4 The concentration of HDL-C and triglycerides in ESKD patients according to the total fecal ODA status

practice. The normal oxalate-degrading functioning ability of fecal microbiota can lead to a decrease in plasma oxalate concentration and an increase in urinary oxalate excretion. Moreover, we provided preliminary evidence that a decrease in total ODA in fecal microbiota was associated with atherogenic dyslipidemia, elevated serum indoxyl sulfate, and systemic inflammation in ESKD patients.

Therefore, we believe that our results might be useful for providing the groundwork for future research projects.

12.7   ODA in Fecal Microbiota Is a Key Point for Personalized Probiotic Therapy in Kidney Stone Disease Patients

Various pro-, pre- and synbiotic supplements have been proposed for restoring gut microbiota after antibiotic treatment and preventing urolithiasis [21, 26, 45, 58]. The overwhelming majority of the published experimental and *in vitro* studies have demonstrated a promising effect of probiotics and synbiotics on increasing ODB number and reducing urinary oxalate excretion [21, 45, 47]. However, little is known about the ODA of synbiotic *vs.* probiotic strains. We hypothesized that the ODA of a specific probiotic or synbiotic could explain the lack of its clinical efficacy and provide a personalized approach to their choice to reduce hyperoxaluria and prevent kidney stone disease.

As mentioned above, *O. formigenes* is the first and most studied oxalate destructor [8]. The use of *O. formigenes* as a probiotic is problematic due to the complexity of its cultivation and storage [46]. Currently, O. *formigenes* strains are being sought that are capable of surviving in the absence of oxalate and surviving for long periods of time in freeze-dried or yogurt-mixed forms, since *O. formigenes* requires oxalate in the culture medium which can be a barrier to large-scale production due to its cultivation and storage conditions [23].

The alternative bacterial species with high ODA look like reliable options. Several studies validate the ODA of different *Lactobacillus* species. In an early study, Turroni et al. have postulated some species of *Lactobacillus spp.* with pronounced oxalate-degrading features [73].It was established that all *L. acidophilus* and *L. gasseri* isolates were able to degrade oxalate because of the Frc and Oxc genes in their genome.

Mogna et al. have detected the efficiency ODA of different *Lactobacillus* strains and showed that their activity is not only species-specific but also strain-specific. In their study, *L. gasseri* were found to be the best oxalate converter that destruct 68.5% of ammonium oxalate. The *L. acidophilus* and *L. plantarum* srains degraded 54.2% and 40.3% oxalate respectively [51]. Chamberlain et al. calculated the percent-degradation of the oxalate substrate by *L. acidophilus* and *L. gasseri* using liquid scintillation and showed 100% degradation. by *L. acidophilus* [20]. Campieri et al. rated the best oxalate converters accordingly to the percentage of 10 mM ammonium oxalate degradation by HPLC. There were *L. acidophilus* − 11.8%, *B. infantis* - 5.3%, *S. thermophilus* - 2.3% and L. *brevis* − 0.9% of ammonium oxalate, respectively [17].

In the next study, we aimed to isolate bacteria from various foods and investigate their ODA. Bacteria capable of oxalate degradation were isolated from foods (spinach, tomato, sour cream, cream, grained cheese, sourdough) on MRS-agar medium with the addition of 20 mg/ml $Na_2C_2O_4$. As a result, 37 strains were isolated. To confirm the ability of oxalate degradation, the isolated strains were inoculated on Oxalate Medium. As a result, 23 strains retained the ability to grow. Subsequently, to identify stable strains, performed five-fold re-inoculation on Oxalate Medium and, as a result, only seven strains were isolated. All newly isolated strains were rod-shaped, Gram-positive, catalase-negative, and did not form gas on glucose medium. According to morphological-cultural and physiological-biochemical characteristics, the newly isolated strains were assigned to the genus *Lactobacillus*. For species identification, the spectrum of carbohydrate fermentation was investigated using ANAEROtest 23. All isolated *Lactobacillus* species were capable of fermenting arabinose, cellobiose, glucose, sucrose, maltose, galactose, lactose, fructose, trehalose, mannose, rhamnose and they did not use xylose and raffinose. No culture could produce indole and was unable to break down urea. As a result, the following species of the genus *Lactobacillus* were identified: *L. nagelii*, *L. rhamnosus*, *L. frumenti*, *L. plantarum* and *L. acidophilus*. In the next step, the ODA of the newly isolated strains was studied at the points of transition of the microbial population from the Lag-phase to the Log-phase and from the Log-phase to the stationary one. The most active oxalate destroyer on Oxalate Medium was *L. plantarum* C3, with 42% ODA; it was followed by *L. acidophilus* C5 and *L. nagelii* C2 strains with 38% and 35% ODA, respectively; *L. rhamnosus* K7 and *L. nagelii* C12 showed the worst results and degraded only 7% of sodium oxalate [5].

To move closely from bench to bedside we have analyzed the ODA of three widely used commercially available probiotics *vs.* synbiotic: (i) the manufactured by LLC "Biopharma Plasma", Ukraine, and consisted of 2 strains of *Lactobacillus spp.* (*L. fermentum* 2 × 109 CFU/g, *L. plantarum* 2 × 109); (ii) the manufactured by LLC "Biopharma Plasma", Ukraine, and consisted of *Bacillus subtilis* «UKMV-5020» 1 × 1010 CFU/g; (iii) the manufactured by *Sanofi* Aventis, Italy, and consisted of *Bacillus clausii* 2 × 109 CFU/g; (iv) the synbiotic was manufactured by LLC "Element of Health", Ukraine, and consisted of 4 strains of *Lactobacillus spp.* (*L. acidophilus* 5 × 108 CFU/g, *L. rhamnosus* 9 × 108 CFU/g, *L. plantarum* 2 × 107 CFU/g, *L. casei* 4 × 108 CFU/g), 2 strains of *Bifidobacterium spp.* (*B. bifidum* 5 × 108 CFU/g and *B. longum* 8 × 108 CFU/g), *Saccharomyces boulardii* 3 × 107 CFU/g, selenium 0.05 mg, oligofructose 40 mg and inulin 450 mg. The obtained ODA of examined probiotics and the synbiotic are presented in Table 12.1.

We found that both monospecies probiotics consisting of *Bacillus spp.* showed the lowest ODA rate compared to the probiotic containing *Lactobacillus spp.* *Bacillus subtilis* degraded only 13% oxalate in Oxalate Medium, while *Bacillus clausii* degraded 23% oxalate. The combination of 2 strains of *Lactobacillus spp.*

Table 12.1 ODA of commercially available probiotics and synbiotic in nutrient media with high oxalate content (5 g/l)

	Oxalate-degrading activity, %		
	MRS + $Na_2C_2O_4$	MPB + $Na_2C_2O_4$	Oxalate Medium
Probiotic 1	10	–	26
Probiotic 2	–	18	23
Probiotic 3	–	4	13
Synbiotic	42	–	69

(L. fermentum, L.plantarum) was able to degrade 26% oxalate in Oxalate Medium. The synbiotic consisting of *Lactobacillus spp. (L.acidophilus, L. rhamnosus, L. plantarum, L. casei), Bifidobacterium spp. (B. bifidum, B. longum), Saccharomyces boulardii,* selenium, oligofructose, and inulin, degraded 69% oxalate in Oxalate Medium. We established that the synbiotic had a high ODA compared to probiotics [65].

In our recent study, we found that the synbiotic was able to restore fecal ODA and numerous *in vitro* studies have addressed the beneficial effect of probiotics on ODA of gut microbiota and reducing hyperoxaluria [1, 51]. The clinical results are not so encouraging and require further large-scale studies [45, 46].

It should be noted that only a few studies have been conducted to investigate the synbiotics effect on human health [45, 47] and the only one addressed the effects of prebiotic and synbiotic on oxalate degradation *in vitro* [3, 21].

Moreover, there are limited data on the interaction between antibiotics and synbiotics [38]. Thus, it is not well understood how synbiotics alter antibiotic-induced oxalate homeostasis imbalance and whether the synbiotic supplementation changes the short- or long-term effects of antibiotics.

12.8   Conclusion

PPPM Strategies in the Field

Collectively, in the presented chapter for the first time, we demonstrated that total ODA in the fecal microbiota rather than the ODB number plays a pivotal role in oxalate homeostasis. These findings are crucial experimental justification for the implementation of predictive diagnostics based on each individual's ODA level and establishing a targeted prevention strategy by predicting the susceptibility for kidney stone disease formation. It is also expected to be utilized for personalized medical services according to the patient's ODA level including medical care and medical foods. However, it should be noted that certain limitations of the experimental protocol used preclude an unambiguous conclusion concerning the use of synbiotics for the treatment of hyperoxaluria or prevention of kidney stone disease recurrences in humans.

12.9   Expert recommendations in the Framework of 3 PM to the Practical Medicine

This study addresses several multidisciplinary aspects of synbiotic prescribing in view of PPPM/3P medicine. Future PPPM/3P-related studies should consider the proposals presented below:

- Total fecal ODA but not the ODB number should be thoroughly examined in the future to develop predictive diagnostics methods, targeted prevention and personalized treatment in kidney stone disease;
- The total fecal ODA determination could be an innovative tool for clearly delineating the need for the probiotic or synbiotic prescribing in a particular patient which would allow shifting the reactive medicine paradigm to PPPM/3P;
- To increase clinical efficacy and provide cost-effective targeted prevention, the choice of probiotic or synbiotic to reduce hyperoxaluria and prevent urolithiasis could be based on their specific ODA rate;
- Synbiotics might be useful for more targeted kidney stone disease prevention due to their beneficial effect on the total ODA in fecal microbiota even when used simultaneously with ceftriaxone;
- The probiotic or synbiotic prescribing in patients with hyperoxaluria or Ca oxalate-lithiasis makes sense as a person-tailored approach to providing conditions for the restoration of the intestine biocenosis and stimulation of ODB activity.

Acknowledgments The present study was supported: by the Ministry of Education and Science of Ukraine grants (19BF036-01, 22BF036-01) and the Grant of the President of Ukraine for Talented Youths # 2016.

References

1. Abratt VR, Reid SJ (2010) Oxalate-degrading bacteria of the human gut as probiotics in the management of kidney stone disease. Adv Appl Microbiol 72:63–87. https://doi.org/10.1016/S0065-2164(10)72003-7
2. Adams MA, Chen Z, Landman P, Colmer TD (1999) Simultaneous determination by capillary gas chromatography of organic acids, sugars, and sugar alcohols in plant tissue extracts as their trimethylsilyl derivatives. Anal Biochem 266:77–84. https://doi.org/10.1006/abio.1998.2906
3. Afkari R, Feizabadi M, Ansari-Moghadam A (2019) Simultaneous use of oxalate-degrading bacteria and herbal extract to reduce the urinary oxalate in a rat model: a new strategy. Int Braz J Urol 45(6):1249–1259. https://doi.org/10.1590/S1677-5538.IBJU.2019.0167
4. Akulenko I, Stetska V, Serhiychuk T et al (2018) Dependence of quantitative composition of oxalate-degrading bacteria in fecal biopsy of rats on the quantity of oxalate in diet. Visnyk Taras Shevchenko National University of Kyiv Biology 1(75):55–58
5. Akulenko I, Skovorodka M, Serhiichuk T et al (2020) The oxalate-degrading activity of lactobacillus spp. isolated from different sources as the potential probiotic modulators for oxalate homeostasis. J Microbiol Exp 8(3):118–123. https://doi.org/10.15406/jmen.2020.08.00295

6. Allison MJ, Cook HM (1981) Oxalate degradation by microbes of the large bowel of herbivores: the effect of dietary oxalate 212(4495):675–676. https://doi.org/10.1126/science.7221555
7. Allison MJ, Cook HM, Milne DB et al (1986) Oxalate degradation by gastrointestinal bacteria from humans. J Nutr 116:455–460. https://doi.org/10.1093/jn/116.3.455
8. Allison MJ, Dawson KA, Mayberry WR, Foss JG (1985) Oxalobacter formigenes gen. Nov., sp. nov.: oxalate-degrading anaerobes that inhabit the gastrointestinal tract. Arch Microbiol 141(1):1–7. https://doi.org/10.1007/BF00446731
9. Alvarenga L, Cardozo LFMF, Lindholm B et al (2020) Intestinal alkaline phosphatase modulation by food components: predictive, preventive, and personalized strategies for novel treatment options in chronic kidney disease. EPMA J 11(4):565–579. https://doi.org/10.1007/s13167-020-00228-9
10. Arafa A, Eshak ES, Iso H (2020) Oxalates, urinary stones and risk of cardiovascular diseases. Med Hypotheses 137:109570. https://doi.org/10.1016/j.mehy.2020.109570
11. Arvans D, Jung YC, Antonopoulos D et al (2017) Oxalobacter formigenes-derived bioactive factors stimulate oxalate Transport by intestinal epithelial cells. J Am Soc Nephrol 28(3):876–887. https://doi.org/10.1681/ASN.2016020132
12. Behnam JT, Williams EL, Brink S et al (2006) Reconstruction of human hepatocyte glyoxylate metabolic pathways in stably transformed Chinese-hamster ovary cells. Biochem J 394(2):409–416. https://doi.org/10.1042/BJ20051397
13. Brzica H, Breljak D, Burckhardt BC et al (2013) Oxalate: from the environment to kidney stones. Arch Hig Rada Toxikol 64(4):609–630. https://doi.org/10.2478/10004-1254-64-2013-2428
14. Bubnov R (2019) Probiotics effectively restore the function and structure of damaged kidney in gout. https://www.physiology.org/detail/news/2019/10/04/short-term-probiotics-regimen-may-help-treat-gout-kidney-disease?SSO=Y
15. Bubnov RV, Babenko LP, Lazarenko LM et al (2018) Specific properties of probiotic strains: relevance and benefits for the host. EPMA J 9(2):205–223. https://doi.org/10.1007/s13167-018-0132-z
16. Bubnov R, Babenko L, Lazarenko L et al (2019) Can tailored nanoceria act as a prebiotic? Report on improved lipid profile and gut microbiota in obese mice. EPMA J 10(4):317–335. https://doi.org/10.1007/s13167-019-00190-1
17. Campieri C, Campieri M, Bertuzzi V et al (2001) Reduction of oxaluria after an oral course of lactic acid bacteria at high concentration. Kidney Int 60(3):1097–1105. https://doi.org/10.1046/j.1523-1755.2001.0600031097.x
18. Chai W, Liebman M, Kynast-Gales S et al (2004) Oxalate absorption and endogenous oxalate synthesis from ascorbate in calcium oxalate stone formers and non-stone formers. Am J Kidney Dis 44(6):1060–1069. https://doi.org/10.1053/j.ajkd.2004.08.028
19. Chakraborty R, Lam V, Kommineni S et al (2018) Ceftriaxone administration disrupts intestinal homeostasis, mediating noninflammatory proliferation and dissemination of commensal enterococci. Infect Immun 86(12):e00674–e00618. https://doi.org/10.1128/IAI.00674-18
20. Chamberlain CA, HatchM GTJ (2019) Metabolomic profiling of oxalate-degrading probiotic lactobacillus acidophilus and lactobacillus gasseri. PLoS One 14(9):e0222393. https://doi.org/10.1371/journal.pone.0222393
21. Darilmaz ÖD, Sönmez Ş, Beyatli Y (2019) The effects of inulin as a prebiotic supplement and the synbiotic interactions of probiotics to improve oxalate degrading activity. Int J Food Sci Technol 54(1):121–131. https://doi.org/10.1111/ijfs.13912
22. Dominguez-Gutierrez PR, Kusmartsev S, Canales BK et al (2018) Calcium oxalate differentiates human monocytes into inflammatory M1 macrophages. Front Immunol 9:1863. https://doi.org/10.3389/fimmu.2018.01863
23. Ellis ML, Dowell AE, Li X et al (2016) Probiotic properties of Oxalobacter formigenes: an in vitro examination. Arch of Microbiol 198(10):1019–1026. https://doi.org/10.1007/s00203-016-1272-y

24. Ermer T, Eckardt KU, Aronson PS et al (2016) Oxalate, inflammasome, and progression of kidney disease. Curr Opin Nephrol Hypertens 25:363–371. https://doi.org/10.1097/MNH.0000000000000229
25. Ermolenko E, Ggromova L, Borscev Y et al (2013) Influence of different probiotic lactic acid bacteria on microbiota and metabolism of rats with Dysbiosis. Biosci Microbiota Food Health 32(2):41–49. https://doi.org/10.12938/bmfh.32.41
26. Ferraz RRN, Marques NC, Froeder L et al (2009) Effects of lactobacillus casei and Bifidobacterium breve on urinary oxalate excretion in nephrolithiasis patients. Urological Res 37:95e100
27. Georgieva R, Yocheva L, Tserovska L et al (2015) Antimicrobial activity and antibiotic susceptibility of lactobacillus and Bifidobacterium spp. intended for use as starter and probiotic cultures. Biotechnol Biotechnol Equip 29:84–91. https://doi.org/10.1080/13102818.2014.987450
28. Golubnitschaja O, Topolcan O, Kucera R et al (2020) 10th anniversary of the European Association for Predictive, preventive and personalised (3P) medicine - EPMA world congress supplement 2020. EPMA J 11:1–133. https://doi.org/10.1007/s13167-020-00206-1
29. Gomathi S, Sasikumar P, Anbazhagan K et al (2014) Screening of indigenous oxalate degrading lactic acid bacteria from human faeces and south Indian fermented foods: assessment of probiotic potential. ScientificWorldJournal 2014:648059. https://doi.org/10.1155/2014/648059
30. Harris AH, Freel RW, Hatch M (2004) Serum oxalate in human beings and rats as determined with the use of ion chromatography. J Lab Clin Med 144(1):45–52. https://doi.org/10.1016/j.lab.2004.04.008
31. Hatch M, Cornelius J, Allison M et al (2006) Oxalobacter sp. reduces urinary oxalate excretion by promoting enteric oxalate secretion. Kidney Int 69:691–698. https://doi.org/10.1038/sj.ki.5000162
32. Hatch M (2017) Gut microbiota and oxalate homeostasis. Ann Transl Med 5(2):36. https://doi.org/10.21037/atm.2016.12.70
33. Hodgkinson A (1977) Oxalic acid in biology and medicine. Academic, London
34. Holota Y, Dovbynchuk T, Kaji I et al (2019) The long-term consequences of antibiotic therapy: role of colonic short-chain fatty acids (SCFA) system and intestinal barrier integrity. PLoS One 14(8):e0220642. https://doi.org/10.1371/journal.pone.0220642
35. Hoppe B, Beck B, Gatter N et al (2007) Oxalobacter formigenes: a potential tool for the treatment of primary hyperoxaluria type. Kidney Int 70:1305–1311. https://doi.org/10.1038/sj.ki.5001707
36. Hu J, Zhong X, Yan J et al (2020) High-throughput sequencing analysis of intestinal flora changes in ESRD and CKD patients. BMC Nephrol 21:12. https://doi.org/10.1186/s12882-019-1668-4
37. Huang Y, Zhang YH, Chi ZP et al (2020) The handling of oxalate in the body and the origin of oxalate in calcium oxalate stones. Urol Int 104:167–176. https://doi.org/10.1159/000504417
38. Jačan A, Kashofer K, Zenz G et al (2020) Synergistic and antagonistic interactions between antibiotics and synbiotics in modifying the murine fecal microbiome. Eur J Nutr 59:1831–1844. https://doi.org/10.1007/s00394-019-02035-z
39. Jiang Z, Asplin JR, Evan AP et al (2006) Calcium oxalate urolithiasis in mice lacking anion transporter Slc26a6. Nat Genet 38(4):474–478. https://doi.org/10.1038/ng1762
40. Khan SR (2014) Reactive oxygen species, inflammation and calcium oxalate nephrolithiasis. Transl Androl Urol 3:256–276. https://doi.org/10.3978/j.issn.2223-4683.2014.06.04
41. Kharlamb V, Schelker J, Francois F et al (2011) Oral antibiotic treatment of helicobacter pylori leads to persistently reduced intestinal colonization rat es with Oxalobacter formigenes. J Endourol 25:1781–1785. https://doi.org/10.1089/end.2011.0243
42. Knauf F, Ko N, Jiang ZR et al (2011) Net intestinal transport of oxalate reflects passive absorption and SLC26A6-mediated secretion. J Am Soc Nephrol 22(12):22472255. https://doi.org/10.1681/asn.2011040433

43. Korol L, Stepanova N, Vasylchenko V et al (2021) Plasma oxalic acid as a trigger for oxidative processes in end-stage renal disease patients. J Nephrol Dialy 1:46–53. https://doi.org/10.31450/ukrjnd.1(69).2021.07
44. Lange JN, Wood KD, Wong H et al (2012) Sensitivity of human strains of Oxalobacter formigenes to commonly prescribed antibiotics. Urology 79(6):1286–1289. https://doi.org/10.1016/j.urology.2011.11.017
45. Li C, Niu Z, Zou M et al (2020) Probiotics, prebiotics, and synbiotics regulate the intestinal microbiota differentially and restore the relative abundance of specific gut microorganisms. J Dairy Sci 103(7):5816–5829. https://doi.org/10.3168/jds.2019-18003
46. Lieske J, Ellis M, Shaw K et al (2017) Probiotics for prevention of urinary stones. Analysis of commercial kidney stone probiotic supplements. Urology 85(3):517–521. https://doi.org/10.21037/atm.2016.11.86
47. Markowiak P, Śliżewska K (2017) Effects of probiotics, prebiotics, and Synbiotics on human health. Nutrients 9(9):1021. https://doi.org/10.3390/nu9091021
48. Massey LK, Liebman M, Kynast-Gales SA (2005) Ascorbate increases human oxaluria and kidney stone risk. J Nutr 135(7):1673–1677. https://doi.org/10.1093/jn/135.7.1673
49. Miller AW, Oakeson KF, Dale C et al (2016) Effect of dietary oxalate on the gut microbiota of the mammalian herbivore Neotoma albigula. Appl Environ Microbiol 82(9):2669–2675. https://doi.org/10.1128/AEM.00216-16
50. Mitchell T, Kumar P, Reddy T et al (2019) Dietary oxalate and kidney stone formation. Am J Physiol Renal Physiol 316(3):409–413. https://doi.org/10.1152/ajprenal.00373.2018
51. Mogna L, Pane M, Nicola S et al (2014) Screening of different probiotic strains for their in vitro ability to metabolise oxalates. Clin Gastroenterol 48:91–95. https://doi.org/10.1097/MCG.0000000000000228
52. Mufarrij PW, Knight J (2012) Sensitivity of human strains of Oxalobacter formigenes toc ommonly prescribed antibiotics. Urology 79:1286–1289. https://doi.org/10.1016/j.urology.2011.11.017
53. Murru N, Blaiotta G, Peruzy MF et al (2017) Screening of oxalate degrading lactic acid bacteria of food origin. Ital J Food Saf 13(6(2)):6345. https://doi.org/10.4081/ijfs.2017.6345
54. Nazzal L, Francois F, Henderson N et al (2021) Effect of antibiotic treatment on Oxalobacter formigenes colonization of the gut microbiome and urinary oxalate excretion. Sci Rep 11:16428. https://doi.org/10.1038/s41598-021-95992-7
55. Palgi N, Ronen Z, Pinshow B (2008) Oxalate balance in fat sand rats feeding on high and low calcium diets. J Comp Physiol B 178:617–622. https://doi.org/10.1007/s00360-008-0252-1
56. PeBenito A, Nazzal L, Wang C et al (2019) Comparative prevalence of Oxalobacter formigenes in three human populations. Sci Rep 9:574. https://doi.org/10.1038/s41598-018-36670-z
57. Perinpam M, Enders FT, Mara KC et al (2017) Plasma oxalate in relation to eGFR in patients with primary hyperoxaluria, enteric hyperoxaluria and urinary stone disease. Clin Biochem 50:1014–1019. https://doi.org/10.1016/j.clinbiochem.2017.07.017
58. Sanders ME, Guarner F, Guerrant R (2013) An update on the use and investigation of probiotics in health and disease. Gut 62:787–796. https://doi.org/10.1136/gutjnl-2012-302504
59. Seong E, Bose S, Han SY et al (2021) Positive influence of gut microbiota on the effects of Korean red ginseng in metabolic syndrome: a randomized, double-blind, placebo-controlled clinical trial. EPMA J 12(2):177–197. https://doi.org/10.1007/s13167-021-00243-4
60. Sidhu H, Hoppe B, Hesse A et al (1998) Absence of Oxalobacter formigenes in cystic fibrosis patients: a risk factor for hyperoxaluria. Lancet 352(9133):1026–1029. https://doi.org/10.1016/S0140-6736(98)03038-4
61. Sivaprakasam S, Bhutia YD, Yang S et al (2017) Short-chain fatty acid transporters: role in colonic homeostasis. Compr Physiol 8(1):299–314. https://doi.org/10.1002/cphy.c170014
62. Stadlbauer V, Horvath A, Ribitsch W et al (2017) Structural and functional differences in gut microbiome composition in patients undergoing haemodialysis or peritoneal dialysis. Sci Rep 7:15601. https://doi.org/10.1038/s41598-017-15650-9

63. Stanford J, Charlton K, Stefoska-Needham A et al (2020) The gut microbiota profile of adults with kidney disease and kidney stones: a systematic review of the literature. BMC Nephrol 21:215. https://doi.org/10.1186/s12882-020-01805-w
64. Stepanova N (2021c) Role of impaired oxalate homeostasis in cardiovascular disease in patients with end-stage renal disease: an opinion article. Front Pharmacol 12:692429. https://doi.org/10.3339/fphar.2021.692429
65. Stepanova N, Akulenko I, Serhiichuk T et al (2022) Synbiotic supplementation and oxalate homeostasis in rats: focus on microbiota oxalate-degrading activity. Urolithiasis 50:249–258. https://doi.org/10.1007/s00240-022-01312-7
66. Stepanova N, Tostanova G, Stashevska N et al (2017) The antibiotic prophylaxis affect on the colon oxalobacter formigenes colonization in patients with recurrent pyelonephritis and hyperoxaluria (pilot study). Nephrol Dial Transplant 32:iii432–iii433. https://doi.org/10.1093/ndt/gfx160.MP013
67. Stepanova N, Tolstanova G, Akulenko I et al (2020a) Oxalate-degrading activity in fecal microbiota associated with blood lipid profile in dialysis patients. Nephrol Dial Transplant 35(3):gfaa139.SO011. https://doi.org/10.1093/ndt/gfaa139.SO011
68. Stepanova N, Tolstanova G, Akulenko I et al (2021a) Altered total fecal oxalate-degrading activity is associated with atherosclerosis and systemic inflammation in end-stage renal disease patients. Atherosclerosis 331:E73–E74. https://doi.org/10.1016/j.atherosclerosis.2021.06.214
69. Stepanova N, Tolstanova G, Korol L et al (2021b) A potential role of fecal oxalate-degrading activity in oxalate homeostasis in end-stage renal disease patients; a descriptive pilot study. J Renal Inj Prev 10(3):e19. https://doi.org/10.34172/jrip.2021.1
70. Stetska VO, Holota YV, Gonchar SY et al (2019) Comparison of long-term effect of two dysbiosis models in Wistar rats. Мікробіологія і біотехнологія 2:6–15
71. Suryavanshi M, Bhute S, Jadhav S et al (2016) Hyperoxaluria leads to dysbiosis and drives selective enrichment of oxalate metabolizing bacterial species in recurrent kidney stone endures. Sci Rep 6:34712. https://doi.org/10.1038/srep34712
72. Tang L, Wu D, Huang Z et al (2016) A fluorescent sensor based on binaphthol-quinoline Schiff base for relay recognition of Zn2+ and oxalate in aqueous media. J Chem Sci 128(8):1337–1343. https://doi.org/10.1007/s12039-016-1124-y
73. Turroni S, Vitali B, Bendazzoli C et al (2007) Oxalate consumption by lactobacilli: evaluation of oxalyl-CoA decarboxylase and formyl-CoA trenasferase activity in lactobacillus acidophilus. J Appl Microbiol 103(5):1600–1609
74. Whittamore JM, Hatch M (2017) The role of intestinal oxalate transport in hyperoxaluria and the formation of kidney stones in animals and man. Urolithiasis 45(1):89–108. https://doi.org/10.1007/s00240-016-0952-z
75. Williams HE (1978) Oxalic acid and the Hyperoxaluric syndromes. Kidney Int 13:410–417. https://doi.org/10.1038/ki.1978.59
76. Zhao S, Yin D, Du H et al (2018) Determination of oxalate and citrate in urine by capillary electrophoresis using solid-phase extraction and capacitively coupled contactless conductivity based on an improved mini-cell. J Sep Sci 41(12):2623–2631. https://doi.org/10.1002/jssc.201701432
77. Zharovsky FG, Pylypenko AT, Pyatnitsky IV (1982) Analytical chemistry. High School 544

Chapter 13
In Vitro Study of Specific Properties of Probiotic Strains for Effective and Personalized Probiotic Therapy

Rostyslav V. Bubnov, Lidiia P. Babenko, Liudmyla M. Lazarenko, Victoria V. Mokrozub, and Mykola Spivak

Abstract Probiotics have tremendous potential to develop healthy diets, treatment and prevention. Investigation of *in vitro* cultural properties of health-promoting microorganisms like lactic acid bacteria (LAB) and Bifidobacteria is crucial to select probiotic strains for treatments based on gut microbiota modulation to justify individualized and personalized approach for nutrition and prevention of variety of diseases.

The studied strains of LAB and bifidobacteria did not form spores, were positively stained by Gram, grow on medium in a wide range of pH (1.0–9.0, optimum pH 5.5–6.5), were sensitive to wide range of antibiotics; and showed different resistance to gastric juice, bile and pancreatic enzymes.

The most resistant to antibiotics were *L. rhamnosus* LB-3 VK6 and *L. delbrueckii* LE VK8 strains. The most susceptible to gastric juice was *L. plantarum* LM VK7, which stopped its growth at 8% of gastric juice; *L. acidophilus* IMV B-7279, *B. animalis* VKL and *B. animalis* VKB strains were resistant even in the 100% concentration.

R. V. Bubnov (✉)
Zabolotny Institute of Microbiology and Virology, National Academy of Sciences of Ukraine, Kyiv, Ukraine

Clinical Hospital 'Pheophania' of State Affairs Department, Kyiv, Ukraine

L. P. Babenko · L. M. Lazarenko · V. V. Mokrozub
Zabolotny Institute of Microbiology and Virology, National Academy of Sciences of Ukraine, Kyiv, Ukraine

M. Spivak
Zabolotny Institute of Microbiology and Virology, National Academy of Sciences of Ukraine, Kyiv, Ukraine

PJSC «SPC Diaproph-Med», Kyiv, Ukraine

© The Author(s), under exclusive license to Springer Nature   355
Switzerland AG 2023
N. Boyko, O. Golubnitschaja (eds.), *Microbiome in 3P Medicine Strategies*,
Advances in Predictive, Preventive and Personalised Medicine 16,
https://doi.org/10.1007/978-3-031-19564-8_13

Strains *L. acidophilus* IMV B-7279, *L. casei* IMV B-7280, *B. animalis* VKL, *B. animalis* VKB, *L. rhamnosus* LB-3 VK6, *L. delbrueckii* LE VK8 and *L. delbrueckii* subsp. *bulgaricus* IMV B-7281 were resistant to pancreatic enzymes.

Adhesive properties of the strains according to AIM index were high in *L. casei* IMV B-7280, *B. animalis* VKL and *B. animalis* VKB; were moderate in *L. delbrueckii* subsp. *bulgaricus* IMV B-7281; and were low in *L. acidophilus* IMV B-7279, *L. rhamnosus* LB-3 VK6, *L. delbrueckii* LE VK8 and *L. plantarum* LM VK7.

Keywords Predictive preventive personalized medicine · Lactobacillus · Bifidobacterium · Probiotics · Gut microbiota · Antibiotics · Gastric juice · Bile · Pancreatic enzymes · Adhesive properties · Pili · Patient phenotype · Individualized medicine

13.1 Relevance of *In Vitro* Research to Support Strains Stratification for Effective Personalized Probiotic Interventions

Intestinal microbiota (mainly represented by *Firmicutes* and *Bacteroidetes*) impact on health and homeostasis [1]. The definition of a probiotic was determined in 2001 by the World Health Organization (WHO) and Food and Agriculture Organization of the United Nations (FAO) [2] and confirmed in 2014 by the experts of International Scientific Association for Probiotics and Prebiotics (ISAPP) [3]. Probiotic microorganisms have large potential as a part of healthy diets and treatment of immunity-related disease [3–12]. They are effective actors both in distant sites and in the gut [12] and have potential for use in nutrition and personalized medicine [13, 14].

Gut microbiota modification in metabolic syndrome and other chronic diseases [14, 15] is a leading task of microbiome studying and for clinical use of probiotics [16]. Excluding few aspects, evidence-supported knowledge on the role of probiotics in applicability to clinical care and disease pathophysiology is not yet sufficient [17]. The effects of probiotic microorganisms are considered evidence-based in cases of diarrhea associated with *Clostridium difficile* and respiratory tract infections [18]. *In vitro* and *in vivo* studies, including ecology of gut microbiota, mechanism of probiotics functioning, and metabolomic researches for strains screening are needed to implement personalized probiotic treatment [8, 19, 20]. Taking into account that clinical studies are complicated to conduct and design, *in vitro* studies [21] and the use of animal models [22] can still provide a high quality data. For now original postulates of Koch were adapted and requirements for different microorganism to be considered a probiotic were formulated:

- Commensal microorganism' strain is associated with the health of the host, which is regularly manifested in healthy hosts, but less common in patients with disease.
- Commensal microorganism' strain can be identified as pure culture and cultivated in the laboratory.

- Commensal microorganism' strain improves or alleviate the disease when introduced into a new host organism.
- Commensal microorganism' strain can be detected after its introduction into a host [23].

The principle mechanisms of the role of probiotics in several diseases pathogenesis is the ability of microorganisms to change the immune response, matabolize nutrients, regulate balance of energy and prevent pathogens colonization [1, 2]. Probiotics have antagonistic relations with pathogens due to the producing of a number of organic acids, lysozyme, bacteriocins, and hydrogen peroxide [2, 3, 24]. Probiotic microorganisms are also able to synthesize digestive enzymes and vitamins C, K, A, PP, E, B and others [25, 26], produce metabolites such as histamine and short-chain fatty acids [27]. Due to the presence of these substances, probiotic microorganisms are able to change microbiota and participate in the metabolism of bile, fatty acids, cholesterol, glucose, bilirubin, and choline; as well as influence on the metabolism of calcium and iron and have antitoxic and immunomodulatory properties [28].

Each strain of probiotic microorganism may have specific phenotype/genotype and have multiple mechanisms of action [10]. *In vitro* tests have demonstrated the strain-dependent immunomodulation properties of bifidobacteria [29, 30]. *In vitro* studies have several limitations but they should obviously be used for preliminary screening of bacterial cells' effects [31, 32], such as *in vitro* models using macrophage-like cells and cells isolated from the gut-associated lymphoid tissues (GALT) [33]. Biological properties such as parameters of the cell wall [30, 31] play an essential role in different aspects of immune response modulating and EPS-producing phenotype [32]. Such crosslinks between genotype and phenotype can warrant to stratify strains on their modulatory activity on innate immunity to justify an individual approach to prevention and nutrition.

13.2 Cultural-Morphological and Tinctorial Properties

To find the potential of lactic acid bacteria (LAB) and bifidobacteria as probiotics we determined their morphological and tinctorial properties by conventional research methods. A total of eight isolates were researched, as *L. casei* IMV B-7280 (B-7280 strain), *L. acidophilus* IMV B-7279 (B-7279 strain), *L. rhamnosus* LB-3 VK6 (VK6 strain), *L. delbruecki* subsp. *bulgaricus* IMV B-7281 (B-7281 strain), *L. delbrueckii* LE VK8 (VK8 strain), *L. plantarum* LM VK7 (VK7 strain), *B. animalis* VKB (VKB strain) and *B. animalis* VKL (VKL strain), which we obtained from the intestines of clinically healthy people. Freeze-dried LAB and bifidobacteria were used in our studies, therefore, we tested the viability of these bacteria by monitoring their growth on Man-Rogoza-Sharp (MRS) agar (MRSA) or bifidum agar (BA), respectively, for 24-48 h at 37 °C. Electron microscopy of bacterial samples was performed according to the generally accepted method [34] on an electron microscope JEM-1400 (Zabolotny Institute of Microbiology and Virology of the National Academy of Sciences of Ukraine), at 80 kV (Fig. 13.1).

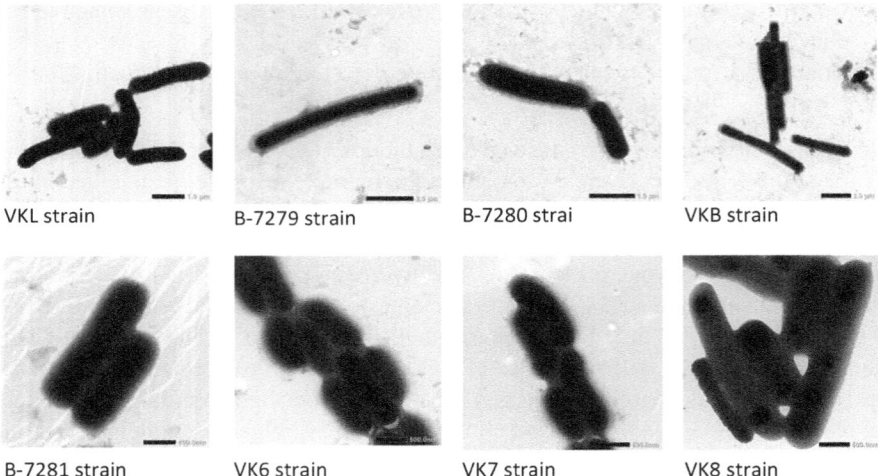

| VKL strain | B-7279 strain | B-7280 strai | VKB strain |

| B-7281 strain | VK6 strain | VK7 strain | VK8 strain |

Fig. 13.1 Subcellular imaging (electronic microscopy) of the strains of LAB and bifidobacteria. The scale is shown in the images

These tested LAB and bifidobacteria strains were Gram-positive and grew on nutrient media in the pH range 1.0–9.0 (optimal pH 5.5–6.5). They did not form a dispute, they were motionless. Bifidobacteria cells were polymorphic. Instead, LAB strains were typically rod-shaped. Figure 13.1 illustrates that the shape of bifidobacterial cells dependent on the stage of culture development. Thus, it ranged from rod-shaped to spindle-shaped, hairpin, amorphous, Y- or X-shaped cells.

The dynamics of growth of these LAB and bifidobacteria strains during incubation in MRS nutrient medium at 37 °C was also investigated. We determined the beginning of the stationary phase of their growth. The beginning of the stationary phase of their growth that was 7–12 h after, t did not depend on the species or genus of these probiotic cultures. The stationary growth phase of VKB, VKL and B-7279 strains was approximately 8 h after their inoculation into the MRS nutrient medium. The stationary growth phase of VK6 and B-7280 strains began at 10 h. However, the stationary growth phase of B-7281, VK8 and VK7 strains occurred only after 11–12 h, i.e. it turned out to be the longest.

13.3 Tolerance of Probiotic Bacteria to Gastric Juice, Bile Salts and Proteolytic Enzymes (Pancreatin)

The tested LAB and bifidobacteria strains were examined for gastric juice (of 1, 2, 5, 8, 10, 20, 30, 50, 75, and 100% in medium; for 2.5 h), bile (of 0.1, 0.25, 0.5, 0.75, 1, 2, 3, 4, and 5%; for 5 h) and proteolytic enzymes tolerance (of 0.1, 0.25, 0.5, 0.75, 1, 2, 3, 4, and 5%; for 15 h) in this research (Fig. 13.2). Daily probiotic cultures

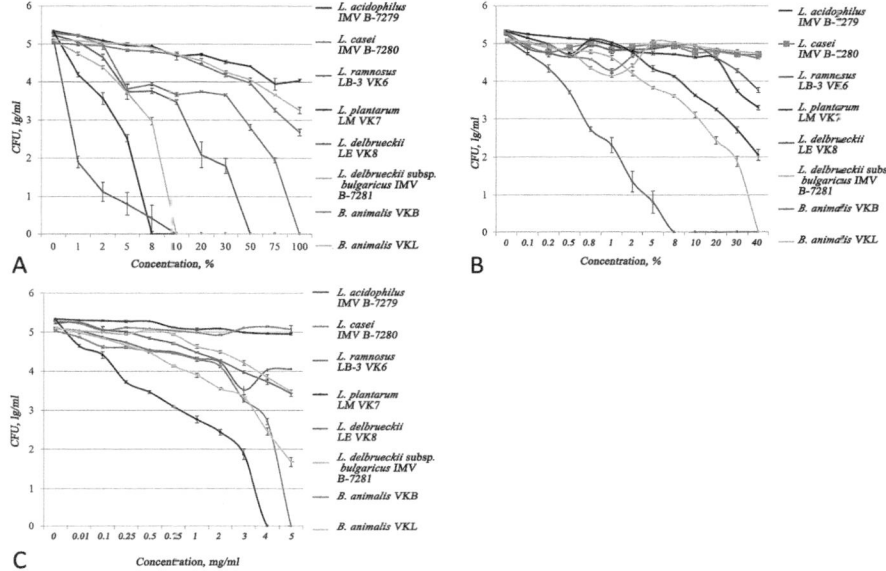

Fig. 13.2 Resistance of studied strains of LAB and bifidobacteria to gastric juice (**a**), bile acids (**b**), and proteolytic enzymes (**c**)

strains were grown in liquid media at 37 °C with these aggressive factors, and after this, bacteria were sown to MRSA and BA, respectively. The number of colony-forming units (CFU) was counted after cultivation at 37 °C for 24–48 h. Complex effect of these aggressive factors (gastric juice 2%, for 2.5 h; of bile salts 1%; for 5 h; proteolytic enzymes 1%; for 15 h) on the LAB and bifidobacteria was also evaluated. Thus, tested LAB and bifidobacteria strains were grown with the gradual addition [35, 36].

It was found that these probiotic cultures had different tolerance to these aggressive factors. The resistance to gastric juice even at 100% concentration were for three strains, as B-7279, VKL and VKB. The growth of VK6 strains was inhibited at a concentration of 50% gastric juice in MRS medium, and VK6 and B-7280 strains were characterized by moderate resistance to its effect. The gastric juice of 10% completely inhibited the growth of VK8 strain. Note that the growth of these bacteria was significantly inhibited by gastric juice at 1% concentration. In Fig. 13.2 illustrated that complete inhibition of growth of B-7281 strain was observed at 10% concentration of gastric juice in nutrient medium. The most sensitive was the VK7 strain, the growth of which was inhibited at a concentration of gastric juice of 8%.

The bile salts at a concentration of 1–40% had also different degrees of inhibition on the tested LAB and bifidobacteria strains in our study; their tolerances to bile salts are also shown in Fig. 13.2. The B-7279, B-7280, VK6, VKB, VKL strains were the most resistant to bile salts; their viability was not completely inhibited in the presence of this aggressive factor in concentrations up to 40%. According to our unpublished data, the growth of VKB and VKL strains began to be inhibited after the introduction of bile salts into the nutrient medium at concentrations of 50% and 75%, respectively. The B-7281 and VK7 strains resisted 40 or 10% bile salts, respectively. Medium bile salt resistance was found for B-7281 and VK7 strains. At the same time, complete inhibition of the VK8 strain occurred at a concentration of bile salts of only 8%. That is, this strain of LAB was the most sensitive to them. The viability of tested LAB and bifidobacteria with different proteolytic enzymes tolerance is also strain-dependent. Similar tolerance to proteolytic enzymes was observed among the seven strains. Thus, B-7279, B-7280, VKL, VKB, VK6, VK8, B-7281 strains had the highest survival in the MRS nutrient medium with proteolytic enzymes in concentrations up to 5% (inclusive). The growth of the VKB strain was inhibited at 5% concentration of proteolytic enzymes. But, VK7 and VKB strains were the most susceptible to them. Proteolytic enzymes at a concentration of 4% completely inhibited their growth.

Subsequent studies have investigated the complex effects of gastric juice, bile and proteolytic enzymes on tested LAB and bifidobacteria strains, as live probiotic bacteria must gradually resist these aggressive factors of the digestive human or animal tract (Fig. 13.3). Thus, only four tested strains were the most tolerant. In particular, the survival of B-7279, B-7280, VKB and VKL strains was over 90% (96.7; 95.7; 90.2 and 91.9%, respectively). The survival of VK6, VK8 and B-7281 strains was lower (50.0; 86.0; 65.6 and 79.2%, respectively). That is, they were moderately sensitive to the complex action of these aggressive factors. But the survival rate of VK7 strain was only 49.3%.

Probably more critical for the complete survival of live probiotic bacteria in the digestive tract is their resistance to bile salts and pancreatic enzymes, as B-7280 strain has moderate resistance to gastric juice (alone). But more research is needed to confirm this.

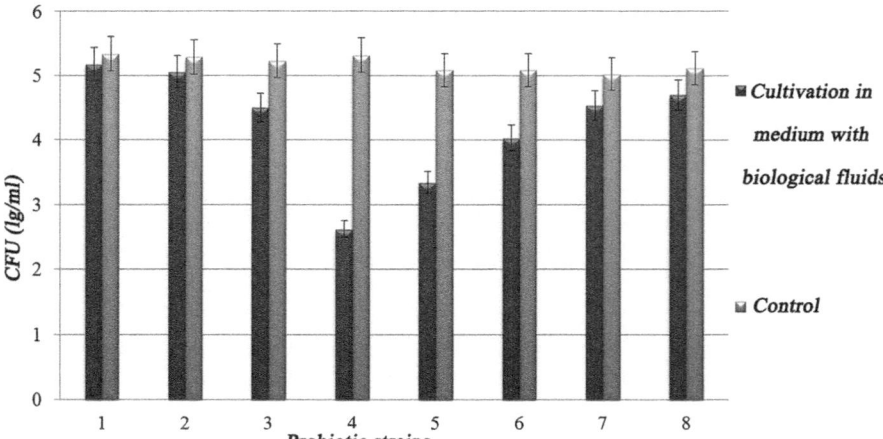

Fig. 13.3 Stability of probiotic strains to the complex effect of biological fluids in the gastrointestinal tract. (*1*). B-7279 strain, (*2*) B-7280 strain, (*3*) VK6 strain, (*4*) VK7 strain, (*5*) VK8 strain, (*6*) B-7281 strain, (*7*) VKB strain, and (*8*) VKL strain

13.4 Adhesive Properties of Probiotic Bacteria

For analysis of some cell surface characteristics, the adhesive properties of the tested LAB and bifidobacteria strains were measured (Table 13.1). Adhesive properties of these probiotic culture were assessed, as AAR (average adhesion rate; this is an average number of bacteria, which attached to the one epitheliocyte); AIM (index of adhesion of microorganisms; this is the average number of bacteria attached to one epitheliocyte); and PRE (the participation rate of epithelial cells; this is a percent of epithelial cells, having on its surface-adhered bacteria), AIM is calculated by the formula [37]. It was considered that bacteria had high adhesive activity at AIM >4.0; average adhesive activity at AIM 2.51–4.04; at AIM 1.75–2.5 were low adhesive activity, and at AIM index ≤1.75 there was no adhesive activity [29]. As shown in Table 13.2, the adhesive properties of these strains were significantly different (p < 0.05), and AIM index ranged from 2.14 to 7.81. Our data showed that the tested bacteria were distributed in this way by adhesive properties: B-7280 strain ≥ VKB strain ≥ VKL strain ≥ B-7281 strain ≥ B-7279 strain ≥ VK8 strain ≥ VK6 strain ≥ VK7 strain. Consequently, the adhesive properties of B-7280, VKB and VKL strains are higher than that of other tested probiotic cultures.

13.5 Antibiotic Resistant of Probiotic Bacteria

The tested LAB and bifidobacteria strains resistance to antibiotics (by disc-diffusion method [19, 22]), that are inhibitors of protein, or peptidoglycan, or nucleic acid biosynthesis, was studied to determine the possibility of their use together with antimicrobial drugs (Table 13.2). In our studies the standardized discs with antibiotics were used. According to the size of the zone of growth inhibition, we determined the degree of sensitivity of bacteria to antibiotics: over 20 mm as sensitive (*S*); 10–20 mm as medium resistant (*M*); and less than 10 mm as resistant (*R*) [31].

Table 13.1 Adhesive properties of LAB and bifidobacteria strains, M ± m

Strain	Parameter of adhesion		
	AAR, units	PRE, %	AIM, units
B-7279	2.3 ± 0.1	91.9 ± 2.2	2.5 ± 0.1
B-7281	2.0 ± 0.1	75.2 ± 4.6	2.6 ± 0.2
B-7280	6.8 ± 0.3	87.5 ± 3.3	7.8 ± 0.9
VKB	4.8 ± 0.4	93.8 ± 2.7	5.1 ± 0.5
VKL	4.0 ± 0.5	85.2 ± 4.5	4.7 ± 0.5
VK6	2.0 ± 0.2	88.0 ± 4.0	2.2 ± 0.3
VK7	1.9 ± 0.1	88.0 ± 5.1	2.1 ± 0.3
VK8	2.1 ± 0.2	92.0 ± 3.0	2.3 ± 0.2
B-7279	2.3 ± 0.1	91.8 ± 2.2	2.5 ± 0.1

Table 13.2 The resistance of investigated strains of LAB and bifidobacteria to antibiotics

Mechanism of effect	Group of antibiotics Generation/name	Strain of bacteria/antibiotic resistance (diameter of growth inhibition zone)							
Inhibition of cell wall synthesis		**VKL**	**VKB**	**B-7279**	**B-7280**	**B-7281**	**VK6**	**VK7**	**VK8**
Penicillins									
I	Penicillin	M	M	S	S	M	R	S	M
II	Oxacillin	R	R	R	R	R	R	M	R
	Ampicillin	S	S	S	S	M	R	M	M
	Carbenicillin	S	S	S	S	S	R	S	S
	Azlocillin	S	S	S	S	S	R	S	M
	Piperacillin	M	S	S	S	S	R	M	M
	Amoxicillin	S	S	S	S	S	R	S	M
	Amoxicillin/ clavula-nic acid	S	S	S	S	S	R	S	M
	Ticarcillin	M	S	S	S	S	R	S	M
	Ticarcillin/ cla-vulanic acid	M	S	S	S	S	M	S	S
Cephalosporins									
I	Cefazolin	R	M	M	M	M	R	M	R
	Cefalexin	M	M	M	M	R	M	M	R
II	Cefuroxime	M	R	M	M	R	R	M	M
	Cefaclor	M	S	M	M	M	M	R	M
III	Cefotaxime	R	S	S	M	S	R	S	R
	Ceftriaxone	M	M	S	M	S	R	M	M
	Cefoperazone	M	M	S	M	S	R	M	M
	Ceftazidime	R	M	S	R	R	R	R	R
	Cefamandole	M	R	S	S	S	S	R	M
	Ceftibuten	R	R	R	M	R	R	R	R
IV	Cefepime	R	M	M	M	R	R	R	R
Glycopeptides									
	Teicoplanin	R	R	R	R	R	R	R	M
	Vancomycin	R	R	R	R	R	R	R	M
Carbapenems									
	Meropenem	M	S	S	M	M	M	S	M
	Imipenem	S	S	S	M	M	M	M	M
Inhibition of cell wall synthesis		**VKL**	**VKB**	**B-7279**	**B-7280**	**B-7281**	**VK6**	**VK7**	**VK8**
Aminoglycosides									
I	Streptomycin	M	M	S	S	M	R	M	M
	Neomycin	S	M	S	M	R	M	M	M
	Kanamycin	R	R	S	M	R	R	R	M
II	Garamycin	S	S	S	M	M	M	S	M
	Tobramycin	R	S	S	M	M	M	S	M

(continued)

Table 13.2 (continued)

Mechanism of effect	Group of antibiotics Generation/name	Strain of bacteria/antibiotic resistance (diameter of growth inhibition zone)							
	Sisomycin	M	M	S	S	M	M	S	S
	Amikacin	M	M	S	M	R	M	R	R
III	Netilmicin	M	M	S	M	M	R	M	M
Macrolides									
I	Oleandomycin	S	S	S	S	M	R	S	M
	Erythromycin	M	S	S	S	M	R	S	S
II	Azithromycin	R	S	S	S	M	R	S	S
	Clarithromycin	S	R	S	S	S	R	S	S
	Roxithromycin	M	M	S	S	S	R	S	M
Lincosamides									
	Lincomycin	M	S	S	S	M	M	S	S
	Clindamycin	M	S	S	S	S	R	S	M
Tetracyclines									
I	Tetracycline	M	S	S	M	M	S	S	M
	Chlortetrazyklin	M	R	S	S	S	S	S	S
II	Doxycycline	M	S	S	M	S	R	S	M
Violation of the synthesis of respiratory enzyme and biosynthesis of cell membrane		**VKL**	**VKB**	**B-7279**	**B-7280**	**B-7281**	**VK6**	**VK7**	**VK8**
Nitrofurans									
I	Nitrofurantoin	R	M	S	R	S	R	M	M
	Furazolidone	R	M	S	R	M	R	R	R
	Fusidin	M	M	S	R	M	R	R	R
II	Nitroxoline (oxyquinolines)	S	M	S	S	S	M	S	S
	Rifampicin	M	S	S	S	S	R	S	S
	Levomycetin	S	S	S	S	S	R	M	M
Violation of synthesis of nucleic acids									
Fluoroquinolones									
I	Nalidixic acid	R	R	R	R	R	R	R	R
II	Ofloxacin	R	M	M	M	R	R	M	M
	Ciprofloxacin	R	R	S	M	R	R	M	M
	Norfloxacin	R	M	R	M	M	R	M	R
III	Levofloxacin	S	R	R	M	R	R	M	M
	Sparfloxacin	S	S	S	R	R	R	M	M
II	Pefloxacin	M	M	M	M	R	R	M	M

R resistant, *M* medium resistant, *S* sensitive

In particular, VK7, B-7279 and B-7280 strains were highly sensitive to antibiotics that inhibit protein biosynthesis. The VKB strain was resistant to chlortetracycline, clarithromycin and roxithromycin. To aminoglycosides was highly sensitive only to the B-7279 strain. To nitrofurans VKB or B-7279 strains were moderately resistant or highly sensitive, respectively. But nitrofurans did not inhibit the growth of B-7280 and VK6 strains. The B-7281 strain was sensitive to furadonin, and VK8 and VK7 strains were moderately resistant to furadonin and resistant to fucidin and furazolidone. To most antibiotics, that blocking the peptidoglycan biosynthesis, were sensitive B-7279 (except oxacillin), B-7280 and VKB (except oxacillin, penicillin, imipenem, meropenem) strains. VK8 strain was also predominantly sensitive to these antibiotics. But strain VK6 3 was resistant to most of them.

Importantly, vancomycin, oxacillin, and teicoplanin did not inhibit the growth of LAB and bifidobacteria strains we studied. As shown in Table 13.1, most of them were moderately resistant or resistant to cephalosporins, except for B-7279 and B-7281 strains. The tested probiotic cultures also had different sensitivity to antibiotics, inhibitors of nucleic acid synthesis. Only the VK6 strain was resistant to these antibiotics. At the same time, they moderately inhibited the growth of VK8, VK7 and B-7280 strains. Levofloxacin and sparfloxacin inhibited the growth of VKL strain, and sparfloxacin and ciprofloxacin inhibited the growth of B-7279 strain. However, the VKB strain was sensitive only to sparfloxacin. All of these tested probiotic cultures were resistant to nalidixic acid. Our results showed that the antibiotic resistance of LAB strains and bifidobacteria dependent on the strain and not on the genus and species of microorganisms.

Therefore, based on the results of our study, the strains with the best probiotic properties were selected. These LAB and bifidobacteria strains, as B-7279, B-7280, VKB and VKL, with the highest survival (over 90%) at complex effects of gastric juice, bile salts and proteolytic enzymes, proved their tolerance to digestive tract conditions. But the adhesive properties were high only of B-7280, VKB and VKL strains (their AIM index ranged from 4.72 to 7.81). Average adhesive activity was in the B-7281 strain with moderately tolerance (79.2%) to the complex of these aggressive factors. B-7279 strain and other tested probiotic cultures had low adhesive activity (their AIM index ranged from 2.14 to 2.45) and different tolerance to gastric juice, bile salts and proteolytic enzymes. Thus, strains B-7280, B-7281, VKB and VKL, isolated from the intestines of clinically healthy people, have some probiotic properties and are promising for use in probiotic preparations. But other comprehensive preclinical studies of their safety and probiotic properties using different models, as well as clinical trials, are needed.

13.6   Data Interpretation

Obtained data partially coincide with the data of other researchers [38–44]. Antibiotic resistant bacteria could be transmitted from animals to humans via food chain [41]. Almost all LAB isolated from farm animals and humans are susceptible to ampicillin, amikacin, different cephalosporins, gentamicin, imipenem, erythromycin, oxacillin, and penicillin.

Teuber et al. [45] showed that bifidobacteria and LAB have sensitivity to ampicillin, penicillin, cephalosporin, erythromycin, and tetracycline and resistance to vancomycin, gentamicin, and streptomycin. A matter of concern to use strains resistant to antibiotics is that antibiotic resistance is not a safety parameter due to the risk of plasmid transfer to pathogenic strains [35, 36, 46].

The strains studied by us meet such important criteria of selection as resistance to antibiotics according to probiotics guidelines like European Food Safety Authority (EFSA) [37].

As far as metabolism of cholesterol has strong crosslinks with circulation of bile acids and the potential associations between modulatory of immune response and hypocholesterolemic activity of probiotics [8], search for bile resistant strains might have a strong input on treatment patients with atherosclerosis, cholestasis, and other conditions. Some LAB and bifidobacteria strains had high hypocholesterolemic activity in different *in vivo* models of metabolic disorders and associated *in vitro* with degradation of bile acids [47].

Gram-positive LAB strains have bacterial pili for adhesion strengthening and protection against environmental stresses. However, their role in immune interaction is still largely unknown [48–50]. Pilus-associated SpaC pilin and gene cluster spaCBA makes possible exertion of both intimate and long distance contact with host tissue and provides mucus-binding [50]. Glycosylation as a sortase-dependent pili modification was shown to play a great role in the immunomodulation [50]. These findings altered the assumption about glycoconjugates's underappreciated role in interplay between bacteria and host. Studying of cell wall of probiotic bacteria using imaging and molecular methods is important to predict or evaluate adhesion and immunomodulatory properties of the strain [31, 51–53].

Gut microbiota is a key player in host energy homoeostasis' regulation and in the pathogenesis of metabolic syndrome and obesity for now is properly studied in clinical set due to number of unpredictable and uncontrollable factors in humans [54–56] such as genetic and environmental factors (drug assumption, diet, life style) [8, 56]. The essential task for probiotic and microbiome research *in clinico* is the search for reproducible in large population and reliable microbiome phenotype markers for longitudinal observation.

Immunomodulatory properties and anti-inflammatory properties [5, 7, 31] has been hypothesized to be likely correlate with clinically relevant effects, in particular the ability to demonstrate liver protective and hypocholesterolemic properties and modulate metabolic conditions [8]. All probiotic strains have primary native antiviral [57, 58], antibacterial [59] and antifungal properties [60–62], and have a high perspective to be alternative for antibiotics during treatment of various infections. Oxygen tolerance of probiotic microorganisms is another important parameter that should be studied in the near future, as far as for now there are only few studies in the field [63, 64]. LAB strains can potentiate hypoxia-inducible factor in the gut [65]. These results open new perspectives to propose individualized treatment to patients with Flammer syndrome [66]; manage hypoxia-associated conditions and stress [65]; use patient stratification on important gut hypoxia signaling marker, mesenteric ischemia in patients with atherosclerosis [67]. Potential of probiotic microorganisms for enhancement of safety and regenerative therapy efficacy [68], stem cells transplantation is difficult challenge aimed to develop biological preparations of a new generation. Finally, many of the mentioned properties of probiotic strains should be used in cancer case management as a part of supportive therapy [18] to facilitate, treatment-associated symptoms [69, 70].

This chapter includes concepts to the topic developed by the authors within previously performed studies [71–73].

References

1. Parekh PJ, Balart LA, Johnson DA (2015 Jun 18) The influence of the gut microbiome on obesity, metabolic syndrome and gastrointestinal disease. Clin Transl Gastroenterol 6:e91. https://doi.org/10.1038/ctg.2015.16
2. WHO/FAO scientific document. http://who.int/foodsafety/fs_management/en/probiotic_guidelines.pdf. Accessed 11 Feb 2018
3. Hill C, Guarner F, Reid G, Gibson GR, Merenstein DJ, Pot B et al (2014) Expert consensus document. The International Scientific Association for Probiotics and Prebiotics consensus statement on the scope and appropriate use of the term probiotic. Nat Rev Gastroenterol Hepatol 8:506–514. https://doi.org/10.1038/nrgastro.2014.66
4. Gibson GR, Hutkins R, Sanders ME, Prescott SL, Reimer RA, Salminen SJ et al (2017) Expert consensus document: the International Scientific Association for Probiotics and Prebiotics (ISAPP) consensus statement on the definition and scope of prebiotics. Nat Rev Gastroenterol Hepatol 14(8):491–502. https://doi.org/10.1038/nrgastro.2017.75
5. Bubnov RV, Spivak MY, Lazarenko LM, Bomba A, Boyko NV (2015) Probiotics and immunity: provisional role for personalized diets and disease prevention. EPMA J 6:14
6. Aron-Wisnewsky J, Clément K (2016 Mar) The gut microbiome, diet, and links to cardiometabolic and chronic disorders. Nat Rev Nephrol 12(3):169–181. https://doi.org/10.1038/nrneph.2015.191
7. Lazarenko LM, Babenko LP, Bubnov RV, Demchenko OM, Zotsenko VM, Boyko NV et al (2017) Imunobiotics are the novel biotech drugs with antibacterial and immunomodulatory properties. Mikrobiol Z 79(1):66–75
8. Bubnov RV, Babenko LP, Lazarenko LM, Mokrozub VV, Demchenko OA, Nechypurenko OV et al (2017) Comparative study of probiotic effects of Lactobacillus and bifidobacteria strains on cholesterol levels, liver morphology and the gut microbiota in obese mice. EPMA J 8(4):357–376. https://doi.org/10.1007/s13167-017-0117-3
9. Jobin C (2018 Jan 5) Precision medicine using microbiota. Science 359(6371):32–34. https://doi.org/10.1126/science.aar2946
10. Lebeer S, Bron PA, Marco ML, Van Pijkeren JP, O'Connell Motherway M, Hill C et al (2017) Identification of probiotic effector molecules: present state and future perspectives. Curr Opin Biotechnol 49:217–223. https://doi.org/10.1016/j.copbio.2017.10.007
11. Marchesi JR, Adams DH, Fava F, Hermes GD, Hirschfield GM, Hold G et al (2016) The gut microbiota and host health: a new clinical frontier. Gut 65(2):330–339. https://doi.org/10.1136/gutjnl2015-309990
12. Reid G, Abrahamsson T, Bailey M, Bindels LB, Bubnov R, Ganguli K et al (2017 Aug 24) How do probiotics and prebiotics function at distant sites? Benef Microbes 8(4):521–533. https://doi.org/10.3920/BM2016.0222

13. Golubnitschaja O, Baban B, Boniolo G, Wang W, Bubnov R, Kapalla M et al (2016) Medicine in the early twenty-first century: paradigm and anticipation—EPMA position paper 2016. EPMA J 7:23

14. Shapiro H, Suez J, Elinav E (2017) Personalized microbiome-based approaches to metabolic syndrome management and prevention. J Diabetes 9(3):226–236. https://doi.org/10.1111/1753-0407.12501.Review

15. Dao MC, Clément K (2018) Gut microbiota and obesity: concepts relevant to clinical care. Eur J Intern Med 48:18–24. https://doi.org/10.1016/j.ejim.2017.10.005

16. van den Nieuwboer M, Browne PD, Claassen E (2016) Patient needs and research priorities in probiotics: a quantitative KOL prioritization analysis with emphasis on infants and children. PharmaNutrition 4(1):19–28

17. Park S, Bae JH (2015) Probiotics for weight loss: a systematic review and meta-analysis. Nutr Res 35:566–575

18. Rondanelli M, Faliva MA, Perna S, Giacosa A, Peroni G, Castellazzi AM (2017) Using probiotics in clinical practice: where are we now? A review of existing meta-analyses. Gut Microbes 8(6):521–543. https://doi.org/10.1080/19490976.2017.1345414

19. Papadimitriou K, Zoumpopoulou G, Foligné B et al (2015) Discovering probiotic microorganisms: in vitro, in vivo, genetic and omics approaches. Front Microbiol 6:58. https://doi.org/10.3389/fmicb.2015.00058

20. Fijan S (2014) Microorganisms with claimed probiotic properties: an overview of recent literature. Int J Environ Res Public Health 11(5):4745–4767

21. Shah P, Fritz JV, Glaab E, Desai MS, Greenhalgh K, Frachet A et al (2016) A microfluidics-based in vitro model of the gastrointestinal human-microbe interface. Nat Commun 7:11535. https://doi.org/10.1038/ncomms11535

22. Nguyen TL, Vieira-Silva S, Liston A, Raes J (2015) How informative is the mouse for human gut microbiota research? Dis Model Mech 8(1):1–16. https://doi.org/10.1242/dmm.017400

23. Neville BA, Forster SC, Lawley TD (2017) Commensal Koch's postulates: establishing causation in human microbiota research. Curr Opin Microbiol 42:47–52. https://doi.org/10.1016/j.mib.2017.10.001

24. Qin J, Li R, Raes J, Arumugam M, Burgdorf KS, Manichanh C et al (2010) A human gut microbial gene catalogue established by metagenomic sequencing. Nature 464:59–65

25. D'Aimmo MR, Mattarelli P, Biavati B, Carlsson NG, Andlid T (2012) The potential of bifidobacteria as a source of natural folate. J Appl Microbiol 112(5):975–984. https://doi.org/10.1111/j.1365-2672.2012.05261.x

26. Lilly DM, Stillwell RH (1965) Growth promoting factors produced by probiotics. Science 147:747–748

27. Gao C, Ganesh BP, Shi Z, Shah RR, Fultz R, Major A et al (2017 Oct) Gut microbe-mediated suppression of inflammation-associated colon carcinogenesis by luminal histamine production. Am J Pathol 187(10):2323–2336. https://doi.org/10.1016/j.ajpath.2017.06.011

28. Bermudez-Brito M, Plaza-Díaz J, Muñoz-Quezada S, Gómez-Llorente C, Gil A (2012) Probiotic mechanisms of action. Ann Nutr Metab 61(2):160–174. https://doi.org/10.1159/000342079

29. Ruiz L, Delgado S, Ruas-Madiedo P, Sánchez B, Margolles A (2017) Bifidobacteria and their molecular communication with the immune system. Front Microbiol 8:2345. https://doi.org/10.3389/fmicb.2017.02345

30. Kobayashi H, Kanmani P, Ishizuka T, Miyazaki A, Soma J, Albarracin L et al (2017) Development of an in vitro immunobiotic evaluation system against rotavirus infection in bovine intestinal epitheliocytes. Benef Microbes 8:309–321. https://doi.org/10.3920/BM2016.0155

31. Mokrozub VV, Lazarenko LM, Sichel LM, Bubnov RV, Spivak MY (2015) The role of beneficial bacteria wall elasticity in regulating innate immune response. EPMA J 6:13

32. Hidalgo-Cantabrana C, Sánchez B, Milani C, Ventura M, Margolles A, Ruas-Madiedo P (2014) Genomic overview and biological functions of exopolysaccharide biosynthesis

in Bifidobacterium spp. Appl Environ Microbiol 80(1):9–18. https://doi.org/10.1128/AEM.02977-13
33. Hidalgo-Cantabrana C, Sánchez B, Álvarez-Martín P, López P, Martínez-Álvarez N, Delley M et al (2015) A single mutation in the gene responsible for the mucoid phenotype of Bifidobacterium animalis subsp. lactis confers surface and functional characteristics. Appl Environ Microbiol 81(23):7960–7968. https://doi.org/10.1128/AEM.02095-15
34. Wang T, Cai G, Qiu Y, Fei N, Zhang M, Pang X et al (2012) Structural segregation of gut microbiota between colorectal cancer patients and healthy volunteers. ISME J 6:320–329
35. European Food Safety Authority (EFSA) (2008) Technical guidance—update of the criteria used in the assessment of bacterial resistance to antibiotics of human or veterinary importance. EFSA J 732:1–15. https://doi.org/10.2903/j.efsa.2008.732
36. Ruiz L, Margolles A, Sánchez B (2013) Bile resistance mechanisms in Lactobacillus and Bifidobacterium. Front Microbiol 4:396. https://doi.org/10.3389/fmicb.2013.00396
37. Tanaka H, Doesburg K, Iwasaki T, Mierau I (1999) Screening of lactic acid bacteria for bile salt hydrolase activity. J Dairy Sci 82:2530–2535
38. Penders J, Stobberingh EE, Savelkoul PHM, Wolffs PFG (2013) The human microbiome as a reservoir of antimicrobial resistance. Front Microbiol 4:87. https://doi.org/10.3389/fmicb.2013.00087
39. D'Aimmo MR, Modesto M, Biavati B (2007) Antibiotic resistance of lactic acid bacteria and Bifidobacterium spp. isolated from dairy and pharmaceutical products. Int J Food Microbiol 115(1):35–42
40. Teuber M, Meile L, Schwarz F (1999) Acquired antibiotic resistance in lactic acid bacteria from food. Antonie Van Leeuwenhoek 76(1–4):115–137. Review
41. Singer RS, Finch R, Wegener HC, Bywater R, Walters J, Lipsitch M (2003 Jan) Antibiotic resistance—the interplay between antibiotic use in animals and human beings. Lancet Infect Dis 3(1):47–51
42. Gueimonde M, Sánchez B, de los Reyes-Gavilán CG, Margolles A (2013) Antibiotic resistance in probiotic bacteria. Front Microbiol 4:202. https://doi.org/10.3389/fmicb.2013.00202
43. Zheng M, Zhang R, Tian X, Zhou X, Pan X, Wong A (2017) Assessing the risk of probiotic dietary supplements in the context of antibiotic resistance. Front Microbiol 8:908
44. Sharma P, Tomar SK, Goswami P, Sangwan V, Singh R (2014) Antibiotic resistance among commercially available probiotics. Food Res Int 57:176–195
45. Tannock GW, Luchansky JB, Miller L, Connell H, Thode-Andersen S, Mercer AA et al (1994) Molecular characterization of a plasmid-borne (pGT633) erythromycin resistance determinant (ermGT) from Lactobacillus reuteri 100-63. Plasmid 31(1):60–71
46. EFSA NDA Panel (EFSA Panel on Dietetic Products, Nutrition and Allergies) (2016) General scientific guidance for stakeholders on health claim applications. EFSA J 14(1):4367 [38 pp.] https://doi.org/10.2903/j.efsa.2016.4367
47. Lebeer S, Vanderleyden J, De Keersmaecker SC (2008) Genes and molecules of lactobacilli supporting probiotic action. Mucosal adhesion properties of the probiotic Lactobacillus rhamnosus GG SpaCBA and SpaFED pilin subunits. Microbiol Mol Biol Rev 72(4):728–764
48. Lebeer S, Claes I, Tytgat HL, Verhoeven TL, Marien E, von Ossowski I et al (2012) Functional analysis of Lactobacillus rhamnosus GG pili in relation to adhesion and immunomodulatory interactions with intestinal epithelial cells. Appl Environ Microbiol 78(1):185–193. https://doi.org/10.1128/AEM.06192-11
49. von Ossowski I, Reunanen J, Satokari R, Vesterlund S, Kankainen M, Huhtinen H et al (2010) Mucosal adhesion properties of the probiotic Lactobacillus rhamnosus GG SpaCBA and SpaFED pilin subunits. Appl Environ Microbiol 76(7):2049–2057. https://doi.org/10.1128/AEM.01958-09
50. Tytgat HL, van Teijlingen NH, Sullan RM, Douillard FP, Rasinkangas P, Messing M et al (2016) Probiotic gut microbiota isolate interacts with dendritic cells via glycosylated heterotrimeric pili. PLoS One 11(3):e0151824. https://doi.org/10.1371/journal.pone.0151824. eCollection.2016

51. Burgain J, Gaiani C, Francius G, Revol-Junelles AM, Cailliez-Grimal C, Lebeer S et al (2013) In vitro interactions between probiotic bacteria and milk proteins probed by atomic force microscopy. Colloids Surf B Biointerfaces 104:153–162. https://doi.org/10._016/j. colsurfb.2012.11.032

52. Tytgat HL, Schoofs G, Vanderleyden J, Van Damme EJ, Wattiez R, Lebeer S et al (2016) Systematic exploration of the glycoproteome of the beneficial gut isolate Lactobacillus rhamnosus GG. J Mol Microbiol Biotechnol 26(5):345–358. https://doi.org/10.1159/000447091

53. Guerin J, Burgain J, Borges F, Bhandari B, Desobry S, Scher J et al (2017) Use of imaging techniques to identify efficient controlled release systems of Lactobacillus rhamnosus GG during *in vitro* digestion. Food Funct 8(4):1587–1598. https://doi.org/10.1039/c6fo01737a

54. Garcia SL, Buck M, McMahon KD, Grossart HP, Eiler A, Auxotrophy WF (2015) Intrapopulation complementary in the 'interactome' of a cultivated freshwater model community. Mol Ecol 24(17):4449–4459. https://doi.org/10.1111/mec.13319

55. Garcia SL, Stevens SLR, Crary B, Martinez-Garcia M, Stepanauskas R, Woyke T et al (2018) Contrasting patterns of genome-level diversity across distinct co-occurring bacterial populations. ISME J 12(3):745–755. https://doi.org/10.1038/s41396-017-0001-0

56. Compare D, Rocco A, Zamparelli MS, Nardone G (2016) The gut bacteria-driven obesity development. Dig Dis 34(3):221–229

57. Al Kassaa I, Hober D, Hamze M, Chihib NE, Drider D (2014) Antiviral potential of lactic acid bacteria and their bacteriocins. Probiotics Antimicrob Proteins 6(3–4):177–185. https://doi. org/10.1007/s12602-014-9162-6

58. Dillon SM, Frank DN, Wilson CC (2016) The gut microbiome and HIV1 pathogenesis: a two-way street. AIDS 30(18):2737–2751

59. Babenko LP, Lazarenko LM, Shynkarenko LM, Mokrozub VV, Pidgorskyi VS, Spivak MY (2012 Nov–Dec) The effect of lacto- and bifidobacteria compositions on the vaginal microflora in cases of intravaginal staphylococcosis. Mikrobiol Z 74(6):80–89

60. Jiang TT, Shao TY, Ang WXG, Kinder JM, Turner LH, Pham G et al (2017) Commensal fungi recapitulate the protective benefits of intestinal bacteria. Cell Host Microbe 22(6):809–816.e4. https://doi.org/10.1016/j.chom.2017.10.013

61. Ilavenil S, Park HS, Vijayakumar M, Arasu MV, Kim DH, Ravikumar S et al (2015) Probiotic potential of Lactobacillus strains with antifungal activity isolated from animal manure. Sci World J 2015:802570–802510. https://doi.org/10.1155/2015/802570

62. Hager CL, Ghannoum MA (2017) The mycobiome: role in health and disease, and as a potential probiotic target in gastrointestinal disease. Dig Liver Dis 49(11):1171–1176. https://doi. org/10.1016/j.dld.2017.08.025

63. Talwalkar A, Kailasapathy K (2003) Metabolic and biochemical responses of probiotic bacteria to oxygen. J Dairy Sci 86(8):2537–2546

64. Talwalkar A, Kailasapathy K (2004 Mar) The role of oxygen in the viability of probiotic bacteria with reference to L. acidophilus and Bifidobacterium spp. Curr Issues Intest Microbiol 5(1):1–8

65. Wang Y, Kirpich I, Liu Y, Ma Z, Barve S, McClain CJ et al (2011) Lactobacillus rhamnosus GG treatment potentiates intestinal hypoxia-inducible factor, promotes intestinal integrity and amelicrates alcohol-induced liver injury. Am J Pathol 179(6):2866–2875. https://doi. org/10.1016/j.ajpath.2011.08.039

66. Bubnov R, Polivka J Jr, Zubor P, Koniczka K, Golubnitschaja O (2017) Premetastatic niches in breast cancer: are they created by or prior to the tumour onset? "Flammer syndrome" relevance to address the question. EPMA J 8:141–157. https://doi.org/10.1007/s13167-017-0092-8

67. Bubnov RV (2011) Ultrasonography diagnostic capability for mesenteric vascular disorders. Gut 60(Suppl 3):A104

68. Taur Y, Jenq RR, Perales MA, Littmann ER, Morjaria S, Ling L et al (2014) The effects of intestinal tract bacterial diversity on mortality following allogeneic hematopoietic stem cell transplantation. Blood 124(7):1174–1182

69. Grech G, Zhan X, Yoo BC, Bubnov R, Hagan S, Danesi R et al (2015) Position paper in cancer: current overview and future perspectives. EPMA J 6(1):9. https://doi.org/10.1186/s13167-015-0030-6

70. York A (2018) Microbiome: gut microbiota sways response to cancer immunotherapy. Nat Rev Microbiol 16(3):121. https://doi.org/10.1038/nrmicro.2018.12

71. Bubnov RV, Babenko LP, Lazarenko LM, Mokrozub VV, Spivak MY (2018) Specific properties of probiotic strains: relevance and benefits for the host. EPMA J 9(2):205–223. https://doi.org/10.1007/s13167-018-0132-z

72. Bubnov R, Babenko L, Lazarenko L, Kryvtsova M, Shcherbakov O, Zholobak N, Golubnitschaja O, Spivak M (2019) Can tailored nanoceria act as a prebiotic? Report on improved lipid profile and gut microbiota in obese mice. EPMA J 10(4):317–335. https://doi.org/10.1007/s13167-019-00190-1

73. Golubnitschaja O (ed) (2019) Flammer syndrome – from phenotype to associated pathologies, prediction, prevention and personalisation, vol 11. isbn 978-3-030-13549-2 isbn 978-3-030-13550-8 (eBook). https://doi.org/10.1007/978-3-030-13550-8

Chapter 14
Probiotic Concepts of Predictive, Preventive, and Personalized Medical Approach for Obesity: Lactic Acid Bacteria and Bifidobacteria Probiotic Strains Improve Glycemic and Inflammation Profiles

Liudmyla Lazarenko, Oleksandra Melnykova, Lidiia Babenko, Rostyslav Bubnov, Tetyana Beregova, Tetyana Falalyeyeva, and Mykola Spivak

Abstract The use of probiotics demonstrate efficacy against obesity and metabolic syndrome (MetS). Detection of effective probiotic strains for hyperglycemia and immunity correction is important task. Current study was evaluating an influence of *Lactobacillus casei* IMV B-7280 separately and composition *L. casei* IMV B-7280 / *Bifidobacterium animalis* VKB / *B. animalis* VKL on the levels of blood glucose and immunity in obese mice. Obesity was induced by fat-enriched diet (FED) in male BALB/c mice. Obese mice were transferred to standard diet and received *per os* probiotic strains daily during 10 days. We measured tumor necrosis factor-alpha (TNF-alpha) in blood serum using enzyme-linked immunosorbent assay and func-

L. Lazarenko · L. Babenko · R. Bubnov (✉)
Zabolotny Institute of Microbiology and Virology, National Academy of Sciences of Ukraine, Kyiv, Ukraine

O. Melnykova
Zabolotny Institute of Microbiology and Virology, National Academy of Sciences of Ukraine, Kyiv, Ukraine

ESC "Institute of Biology and Medicine" of Taras Shevchenko National University of Kyiv, Kyiv, Ukraine

T. Beregova · T. Falalyeyeva
ESC "Institute of Biology and Medicine" of Taras Shevchenko National University of Kyiv, Kyiv, Ukraine

M. Spivak
Zabolotny Institute of Microbiology and Virology, National Academy of Sciences of Ukraine, Kyiv, Ukraine

PJSC «SPC Diaproph-Med», Kyiv, Ukraine

© The Author(s), under exclusive license to Springer Nature Switzerland AG 2023
N. Boyko, O. Golubnitschaja (eds.), *Microbiome in 3P Medicine Strategies*, Advances in Predictive, Preventive and Personalised Medicine 16, https://doi.org/10.1007/978-3-031-19564-8_14

tional activity of peritoneal exudate macrophages (PEMs). Glucose levels in blood were defined with glucometer.

We ascertained that all probiotic strains induced reducing mice weight and visceral fat, normalization of TNF-alpha production and functional activity of PEMs. Treatment with *L. casei* IMV B-7280 was associated with decreasing blood glucose levels. No normalization of glucose and TNF-alpha levels was observed in obese mice, transferred to standard diet without probiotic treatment; although we revealed decreasing their weight and visceral fat and partial recover of functional activity of PEMs. In conclusion, probiotic strain *L. casei* IMV B-7280 (separately) and composition *L. casei* IMV B-7280 / *B. animalis* VKB / *B. animalis* VKL can re-equilibrate metabolic and inflammation indices in mouse obesity model. *L. casei* IMV B-7280 alone was more efficient in decreasing glucose levels than composition of strains.

Keywords Metabolic syndrome · Diabetes mellitus · Obesity · Lactobacilli · Bifidobacteria · Glucose · Macrophages · Inflammation · Mouse model · Predictive preventive personalized medicine

Abbreviations

FED	Fat-enriched diet
LAB	Lactic acid bacteria
FR	Functional reserve
HDL	High density lipoprotein
IL	Interleukin
LPS	Lipopolysaccharides
TNF-α	Tumor necrosis factor-α
IFN-γ	Interferon-γ
MetS	Metabolic syndrome
NBT	Nitro-blue tetrazolium
PEMs	Peritoneal exudate macrophages
PN	Phagocytic number
PBS	Phosphate buffered saline
PI	Phagocytic index
RBA	Respiratory burst activity
SCFA	Short-chain fatty acid
DM	Diabetes mellitus
T2DM	Type 2 diabetes mellitus
FBG	Fasting blood glucose

14.1   Introduction

Obesity is a major risk factor for the development of type 2 diabetes mellitus (T2DM) and being more than doubled during the last 30 years, is considered to be the most important health problem globally [1–5].

Metabolic syndrome (MetS) is a common condition, associated with profound violation of metabolic processes in the human body, and include development of diabetes mellitus (DM), cardiovascular (coronary heart disease, atherosclerosis, arterial hypertension) and hepatobiliary diseases (gallstone disease, bile dyskinesia, chronic cholecystitis), pathology of the musculoskeletal system, as well as several types of cancer and premature death [1].

The **diagnosis of "metabolic syndrome"** can be made if at least three of the following five criteria [2] are met: visceral adiposity; dyslipidemia; high density lipoprotein (HDL) cholesterol ≤ 40 mg/dL; hypertension of 130/85 mmHg or more; and *type 2 diabetes mellitus* (T2DM).

Abdominal obesity most often occurs as a result of increased energy consumption with food, uncontrolled use of nutritional supplements, low levels of physical activity and genetic predisposition. The complex relation of obesity/overweight and DM as well as their associations with the **microbiome** has been demonstrated [2–5]. Microbiome is highly variable and a very important player mutually related to DM as well as obesity, e.g., weight gain may be also considered as a complication of insulin treatment [5]. Dietary intervention was found to modulate the gut microbiota and improve glucose control in individuals with T2DM [6].

The role of **immunity,** the development of inflammatory processes in diabetes type 1 and 2 and all aspects of metabolic disorders of obesity, have been extensively discussed and different immunity patterns switch on during both diabetes type 1 and 2 during the modification of gut microbiota have been hypothesized [7–10]. One of the potential proinflammatory pathways in development of obesity and T2DM may occur through activation as the innate so the adaptive immune response via toll-like receptors (TLRs), which can recognize antigens like lipopolysaccharides (LPS) and initiate the production of proinflammatory cytokines and TNF-α [9]. Thus, in **type 1 diabetes mellitus**, a greater immune response may be a result of the lowered expression of adhesion proteins within the intestinal epithelium that favours destruction of pancreatic β cells by CD8+ T-lymphocytes, and increased expression of interleukin(IL)-17, related to autoimmunity [8]. The development of T2DM association with increased levels of LPS has been observed in some clinical trials [9]. The cytokine-induced inflammatory or "acute-phase response" hypothesis claims that T2DM and the MetS are associated with proinflammatory cytokine IL-6 and acute-phase reactants such as C-reactive protein [10]. The gut microbiota benefits humans via short-chain fatty acid (SCFA) production from carbohydrate fermentation, and deficiency in SCFA production is associated with T2DM. Targeted restoration of these SCFA producers may present a novel ecological approach for managing T2DM [11].

On the other hand, it has been proven that both the high-fat as well as carbohydrate-rich diet are independent factors causing the reduction of the number of Bacteroidetes and the increase of the amount of Firmicutes [12]. The relative raise of Firmicutes in response to a fat-dominant diet has been described in both mice and human observations [12].

The pathophysiological mechanisms of obesity can involve a number of neuro-immunodecrine disorders [13, 14]. The development of chronic systemic **inflammation**, dysfunction of specific hypothalamic neurons, and activation of the **hypothalamic-pituitary-adrenal axis** during obesity cause disturbances in the control of body weight in the lipostat system. In obese patients, with the low level of adiponectin, products of leptin and proinflammatory cytokines increase adipocytes of adipose tissue, which in turn activates the production of IL-1β, IL-6, tumor necrosis factor-α (TNF-α), interferon-γ (IFN- γ) and chemokines by the cells of the immune system. Proinflammatory cytokines affect the activity of adipocytes, lipolysis, insulin reaction, glucose metabolism and adipokine production, which results in fat deposition and systemic chronic inflammation [13, 14]. Activation of proinflammatory cytokines is considered to be the main factor in the development of resistance to insulin and hyperlipidemia [14].

Clinical data taken from several meta-analyses of randomized controlled trials (RCTs) [15–19] confirm that probiotics may have beneficial effects for diabetes, especially for T2DM patients. Consuming probiotics may improve glucose metabolism by a modest degree, with a potentially greater effect when the duration of intervention is ≥8 weeks, or multiple species of probiotics are consumed [15]. The findings by Horvath et al. [18] explain the beneficial effects of probiotics on immune function via increased serum neopterin levels and the production of reactive oxygen species by neutrophils to improve liver function in cirrhosis. A meta-analysis of RCTs by Jun et al. [19] stated that application of probiotics may decrease the lipid profile indices, blood pressure, and fasting blood glucose (FBG) in patients with T2DM and can be a new method for lipid profiles and blood pressure management in T2DM [19]. It has been concluded in the study by Wang et al. [17] that probiotics may have beneficial effects on the reduction of glucose, insulin and HbA1c for diabetes, especially for T2DM patients.

However, the limitations of completed RCTs [15–19] have been noted to be as follows [17, 19]: differences in the inclusion criteria and exclusion criteria for patients; different patients with previous disease and treatments were unavailable; most trials with low quality and low Jadad score; pooled data used for analysis, and unavailable individual patients' data. It was suggested that improvement of quality of clinical trials urgently needed and it would require to consider all known potential biases, update the list periodically, and score the trial by multiplying (not adding) the component scores with regard to risks of potential bias to obtain an overall score, etc. [20].

Probiotics have a great potential to alleviate many aspects of MetS and associated immunity [21–24]. Thus, recently we have studied efficacy of use of probiotic for pillar conditions of MetS, namely obesity, dyslipidemia and liver dysfunction, and also for modifying gut microbiota [24] and against glutamate-induced and immune-mediated obesity [12] and the potential for individualized use according to strain-specific probiotic properties has been recently studied [25]. We have found that the original probiotic strains of lactic acid bacteria (LAB) and bifidobacteria (*Lactobacillus casei* IMV B-7280, *Bifidobacterium animalis* VKB and *B. animalis* VKL and others), having hypocholesterolemic, immunomodulatory and anti-inflammatory properties [25] also demonstrate high levels of antagonistic activity against pathogenic and opportunistic bacteria [26], the resistance to aggressive conditions of the gastrointestinal tract, high or moderate adhesion to the cells of the epithelium of the gastrointestinal mucosa [25] and immunomodulatory properties associated with bacterial wall structure [27]. It has been proved on different models of experimental obesity in animals that probiotics improved the metabolism and modified the intestinal microbiota. The most effective were *L. casei* IMV B-7280 and composition *L. casei* IMV B-7280 / *B. animalis* VKB / *B. animalis* VKL [14, 24].

The probiotic strain *L. casei* IMV B-7280 in recent studies demonstrated most pronounced beneficial effect for most applications.

Despite the number of human and animal studies have been conducted, many aspects of using probiotics for MetS have not been finalized up to date. Individualized use according to the probiotic properties and host's phenotype and aiming to pick a proper strain in particular clinical case. We believe that focused studies can provide new data on mechanics of probiotic activity for MetS treatment and add a value to the paradigm of personalized / individualized use of probiotics. According to our hypothesis the lack of personaliztion in the studies is a main limitation supporting existence of gap in the evidence.

Therefore, we have set this study in order to finalize the puzzle to study effects of mentioned strains against conditions that constitute MetS.

We have supposed the model with use as a malnutrition leading trigger of MetS development. In this regard the diet that was given to mice was largely unhealthy and corresponded rather to "industrial trans-fatty acids-enriched" diet vs "fat-enriched diet". It was an intention to model obesity and hyperglycemia affecting gut microbiota and even overall metabolic syndrome.

The aim of the study was to evaluate the effect of probiotic strain *L. casei* IMV B-7280 (separately) and composition *L. casei* IMV B-7280 / *B. animalis* VKB / *B. animalis* VKL on the level of glucose in blood and also inflammatory response (inflammation) on the experimental obesity in mice, induced by a fat-enriched diet (FED).

14.2   Study Design

14.2.1   Animals

Experimental study was carried out on BALB/c male mice at the age of 6–8 weeks (17–24 g). Throughout the experiment, animals are kept under standard vivarium conditions, in plastic cells in a separate room at a constant air temperature

(20–22 °C). They received a full-fledged meal and had free access to auto-squirrels. The keeping of animals and all manipulations against them was carried out in accordance with the Law of Ukraine No. 3447-IV "On the Protection of Animals from Cruel Treatment"and the "European Convention for the Protection of Vertebrate Animals Used for Experimental and Scientific Purposes of September 20, 1985", Strasbourg, 1986), and according to the "General Ethical Animal Experiments" (First National Congress on Bioethics, 2001) and "Code of Practice for the Housing and Care of Animals Used in Scientific Procedures" [28].

The whole group was conducted under anesthetics of pentobarbital sodium, and all efforts were made to minimize animal suffering.

The randomization method was used as follows: all the animals enrolled to the experiment were put into one large cell, and then they were consequently randomly one by one animal taken from the large cell and allocated for each group.

14.2.2 Ethics

The keeping of animals and all manipulations on them have been carried out in accordance with the Law of Ukraine No. 3447-IV "On the Protection of Animals from Cruel Treatment"and the "European Convention for the Protection of Vertebrate Animals Used for Experimental and Scientific Purposes of September 20, 1985", Strasbourg, 1986), and according to the "General Ethical Animal Experiments" (First National Congress on Bioethics, 2001) and "Code of Practice for the Housing and Care of Animals Used in Scientific Procedures" [28].

The experiments have been conducted under using of anesthetics (pentobarbital sodium), and all accessible efforts have been made to minimize animal suffering.

This study has been approved by the ethics committee of institutional review board and Special Academic Council on Doctoral Thesis of D.K. Zabolotny Institute of Microbiology and Virology of the National Academy of Sciences of Ukraine (protocol N 7 issued 03.07.2018).

No human subjects were included to the study.

14.2.3 Bacteria and Culture Conditions

We used original probiotic strains of LAB and bifidobacteria – *L. casei* IMV B-7280, which is deposited in the Depository of Microorganisms of the D.K. Zabolotny Institute of Microbiology and Virology of National Academy of Sciences of Ukraine, as well as *B. animalis* VKL and *B. animalis* VKB. The studies were conducted using bacteria freeze dried in Cuddon Freeze Dryer FD1500 (New Zealand). Prior to each experimental study, the activity of these freeze-dried probiotic bacteria was checked by control of their growth on a Man-Rogosa-Sharpe Agar or Bifido Agar medium (Merck, Germany) at 37 °C during 24–48 hours.

14.2.4   Diet

The composition of **FED** consisted of 30% fats, 40% proteins and 30% carbohydrates and included following ingredients: margarine – 1200 g; liver sausage – 1800 g; wheat porridge – 750 g; milk whey – 750 ml. FED did not contain vitamin, mineral supplements.

Regular diet was the standard mixed feed balanced by amino acid composition, mineral substances and vitamins; did not contain preservatives, hormones, nitrates. **Regular diet** included ingredients: grains, wheat flour, wheat bran, sunflower meal, soybean meal, dry milk, vegetable oil, yeast, fish meal, vitamin-mineral complex.

14.2.5   Design of the Experiment

The experimental model of FED-induced obesity in mice, previously developed in [24], was used in current study.

Mice were randomly allocated into five different groups of 6 animals each: intact mice (controls) (n = 6); and mice that received FED during 35 days (n = 24), and thereafter were divided to: obese mice that continued to receive FED (n = 6); obese mice, transferred to a regular diet, and did not receive probiotic bacteria (n = 6); obese mice, transferred to a regular diet, that received *L. casei* IMV B- (n = 6); obese mice, transferred to a regular diet, that received the probiotic composition *L. casei* IMV B-7280 / *B. animalis* VKB / *B. animalis* VKL (n = 6).

All groups of mice, except the controls, received FED.

Mice with obesity were injected with *L. casei* IMV B-7280 (separately) or composition *L. casei* IMV B-7280 / *B. animalis* VKB / *B. animalis* VKL (in a ratio of 2: 1: 1) orally at a dose of 5×10^6 cells / animal every day during 10 days. To reproduce the rational therapeutic scheme, these mice were transferred to a regular diet on day 35. Probiotics were given from day 35 to day 45.

Mice were weighed weekly. The blood samples were collected from all animals including control animals that continued to receive standard diet. The blood was taken from the caudal vein of mice to determine the level of glucose prior to the beginning of the FED and at 7, 14, 21, 28, 35, 45 and 52 days of observation from the beginning of FED. Serum was received from the blood, and was stored at −20 °C before the experimental studies were performed. At 52 day mice were killed by cervical dislocation after complete anesthesia, visceral fat was removed from the abdominal cavity and weighed.

14.2.6   Glucose Measurements

To determine the blood glucose content, the Glucometer FreeStyle Optium NEO (Abbot Diabetes Care Ltd. (England)) was used.

14.2.7   Cytokines Production Measurements

The concentration of TNF-α in serum was determined using the enzyme-linked immunosorbent assay (ELISA) using the immunoassay system Thermo Fisher Scientific Inc. (Bender MedSystems GmbH, Austria) according to manufacturer's recommendations and expressed in pg / ml.

14.2.8   Macrophages Harvest

On the 52 day after the probiotic bacteria injection, a fluid was collected from the peritoneal cavity in mice of all groups that were killed by cervical dislocation after complete anesthesia. The cooled culture medium 199 was injected into the abdominal cavity, and after the soft massage of the abdomen, the abdominal fluid was collected by aspiration. Peritoneal exudate cells were centrifuged three times for 10 min at $400 \times g$.

The precipitate was then resuspended in a cold culture medium RPMI-1640, which was added to 10% fetal calf serum, 50 µg / ml gentamicin and 1 M Hepes. The cell suspension was placed in a 35 mm culture plate and incubated for 1 hour at 37 °C in a CO_2 incubator. Nonadherent cells have been removed by gently washing three times with warm phosphate buffered saline (PBS). After that, the adherent cells – the peritoneal exudate macrophages (PEMs) were detached from the surface by jetting chilled RPMI-1640 into the cells and counted using Goryaev's chamber. A suspension of PEMs was prepared in RPMI-1640 with 10% fetal calf serum, 50 µg / ml gentamicin and 1 M Hepes in a quantity of 5×10^7 cells / ml.

14.2.9   Respiratory Burst and Phagocytic Activity Measurements

Respiratory burst activity (RBA) of PEMs was evaluated using a spontaneous and stimulated nitro-blue tetrazolium (NBT) reduction test [29].

The principle of this method is the restoration of NBT in the PEMs cytoplasm to formazan under the influence of the superoxide anion formed during cell activation. Formazan has the appearance of dark blue granules, the number of which varies depending on the severity of the oxygen explosion. The advantage was given to the cytochemical record of the results, based on the microscopic examination and counting of the percent NBT-positive (formazan-containing) PEMs. Briefly, in a spontaneous NBT test, 100 µl suspension of PEMs (5×10^7 cells / ml), 100 µl of 0.1% NBT (Sigma, USA) in PBS and 100 µl of PBS were applied to the substrate and incubated at 37 °C for 2 hours in an atmosphere of 5% CO_2. In the stimulated NBT test, 100 µl suspension of PBS by daily culture of *Staphylococcus aureus* strain 8325–4 (inactivated by heating at 57 °C during 30 min) at a concentration of 5×10^9 cells / ml was introduced into the PEMs. After cultivation, PEMs were washed twice with PBS, carefully dried, fixed with methanol and stained with 1% safranin solution. In the field of view of the microscope, 100 or more cells were counted, the percent of NBT-positive PEMs was determined. The difference between the indices of the spontaneous and stimulated NBT test was calculated by the PEMs functional reserve (FR).

The phagocytic activity of PEMs was determined using latex beads (1.1 µm, Sigma, USA) microscopically [2]. Briefly, a 100 µl suspension of PEMs (5×10^7 cells / ml) and a suspension of latex particles in PBS was applied to the substrate and incubated at 37 °C for 2 hours in an atmosphere of 5% CO_2. Then PEMs was washed twice with PBS, carefully dried, fixed with methanol and stained with Giemsa. A total of 100 or more PEMs were calculated and determined the *Phagocytic index* (PI) as the number of phagocytic cells (in %) and *phagocytic number* (PN) – the average number of latex particles phagocyted by a single PEM (presented in units).

14.2.10   Statistics

Statistical analysis of data was carried out by the software package "Statistica 8.0". The comparison between control and experimental groups was checked using the Student's *t* test for independent samples and also non-parametric Kolmogorov-Smirnov test.

The data was represented as M ± Std.Dev. The difference between the indicators was considered statistically significant at $P < 0.05$.

14.3   Achievements

At day 7 from the beginning of FED in mice, observed the following changes in behavioral reactions: mice became sluggish, their response to external stimuli decreased, and the body temperature increased. At the 14th and 21st day, we detected a tendency to increase their weight, however, the difference compared with the values for intact mice (controls) was insignificant (Fig. 14.1). Weight of mice that consumed FED exceeded normal values ($P < 0.05$) on average by 3–4 g, starting from

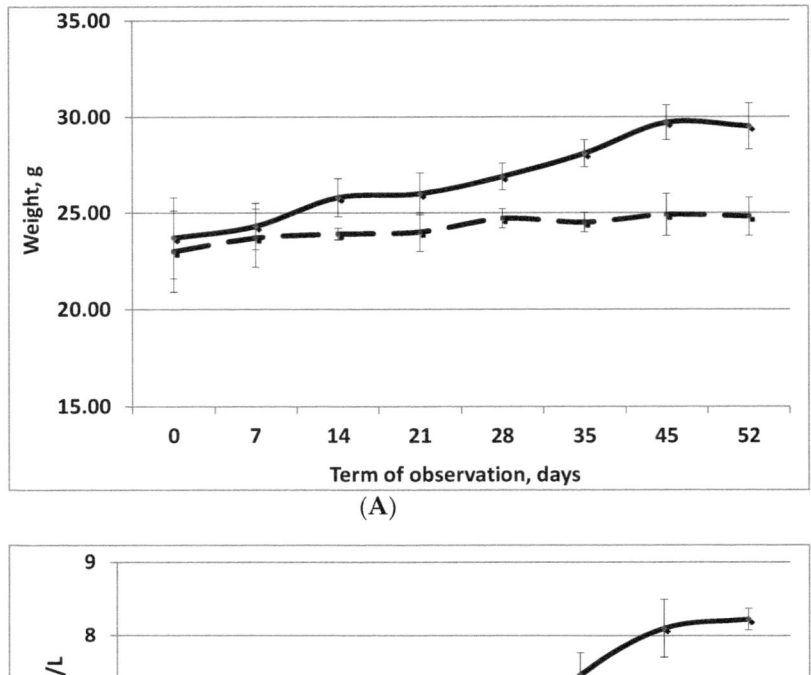

(A)

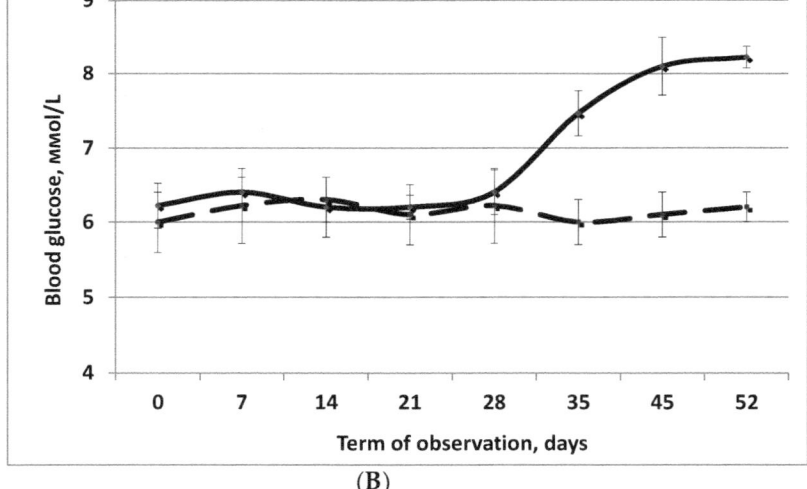

(B)

Fig. 14.1 Changes in the weight (**a**) and glucose (**b**) levels in the blood of mice receiving FED or a regular diet (n = 6 in each group, M ± Std.Dev). 1 – intact mice receiving regular diet (dotted line); 2 – mice receiving FED (solid line)

the 28th day from the beginning of FED and during the entire subsequent follow-up period. We have found that the consumption of FED by mice led to the development of alimentary visceral obesity, confirmed by significant accumulation of visceral fat in the abdominal cavity, and more than 18-fold increased mass (1206.0 ± 437.4 mg / animal, in control of 64.8 ± 2.2 mg / animal; $P < 0.05$) after 52 days. Along with this, the glucose level in blood of obese mice, was observed to be higher than normal at 35, 45 and 52 days of observation and was measured as 7.46 ± 0.30; 8.10 ± 0.40 and 8.23 ± 0.15 mmol / l accordingly (P < 0.05), which may indicate the development of mild hyperglycemia. Thus, an increase in the weight of mice as a result of FED consumption was preceded by an increase in blood glucose levels (Fig. 14.1).

In the case of injection of *L. casei* IMV B-7280 (separately) or *L. casei* IMV B-7280 / *B. animalis* VKB / *B. animalis* VKL, the weight of obese mice decreased to the level of control indices, however, there was no probable difference from mice, which continued to consume FED (Table 14.1). After administration of *L. casei* IMV B-7280 (separately) or probiotic composition to obese mice, the weight of the visceral fat was reduced at the 52nd day accordingly to 366.8 ± 17.8 and 343.0 ± 89.5 mg / animal relative to the mice, who continued to consume FED (P < 0.05), but remained higher than control (P < 0.05).

Note that similar results were obtained when weighing of the visceral fat in mice with obesity, translated into a regular diet, which did not receive probiotic bacteria (Table 14.1). The weight of the visceral fat in these mice was also higher than in controls (358.0 ± 65.4 mg / animal, P < 0.05), although it decreased compared with the mice that continued consuming FED (P < 0.05).

The blood glucose levels were measured as normal only in obese mice injected with *L. casei* IMV B-7280 (Fig. 14.2). Under the influence of *L. casei* IMV B-7280 / *B. animalis* VKB / *B. animalis* VKL composition, the blood glucose level in obese mice did not recover to the control level, although there was a tendency to decreasing this parameter. Hyperglycemia has been retained in obese mice transferred to a regular diet that did not receive probiotic bacteria. Consequently, the transfer of obese mice to the regular diet did not have a significant effect on the level of blood glucose.

Table 14.1 Weight of obese mice received *L. casei* IMV B-7280 (separately) and composition *L. casei* IMV B-7280 / *B. animalis* VKB / *B. animalis* VKL

A group of animals	Weight of mice, g / period of observation from the beginning of FED			
	day 0	day 35	day 45	day 52
Intact mice (controls)	23.5 ± 0.1	24.5 ± 0.9	24.9 ± 0.8	24.8 ± 1.0
Obese mice that continued to receive FED	22.0 ± 0.9	30.5 ± 1.1•*	29.7 ± 1.9•*	29.5 ± 1.2•*
Obese mice, transferred to a regular diet, and did not receive probiotic bacteria	22.7 ± 1.1	30.9 ± 1.1•*	24.0 ± 1.1	24.0 ± 2.0
Obese mice, transferred to a regular diet, that received *L. casei* IMV B-7280	21.5 ± 0.8	28.9 ± 0.8•*	25.3 ± 1.8	25.4 ± 2.8
Obese mice, transferred to an regular diet, that received the probiotic composition	22.9 ± 1.5	28.5 ± 0.9•*	25.3 ± 1.0	25.0 ± 1.9

Notes: • - P < 0.05 vs indices of intact mice; * - P < 0.05 vs parameters before FED (day 0)

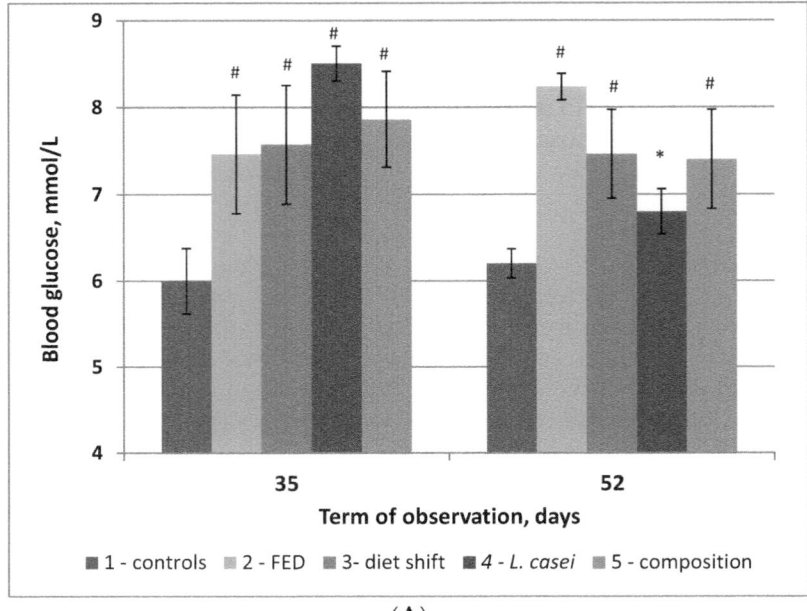

(A)

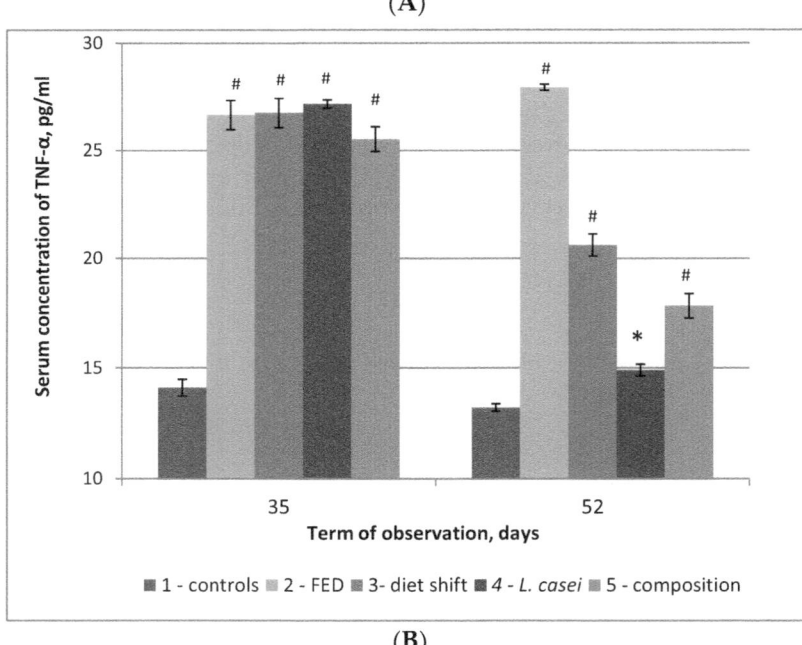

(B)

Fig. 14.2 The level of glucose (**a**) in the blood of obese mice and the levels of serum TNF-α (**b**) under conditions of administration of probiotic strains (5×10^6 cells / animal) (n = 6 in each group, M ± Std.Dev): 1 - intact mice (controls); 2 – obese mice, that continued to receive FED (FED); 3 – obese mice, transferred to a regular diet that did not receive probiotic bacteria (diet shift); 4 – obese mice, transferred to a regular diet that received *L. casei* IMV B-7280 (*L. casei*); 5 – obese mice, transferred to an regular diet that received *L. casei* IMV B-7280 / *B. animalis* VKB / *B. animalis* VKL composition (composition). # P < 0.05 – vs indices of intact mice; * P < 0.05 - vs indices of obese mice that continued to consume FED

The hyperglycemia in mice with obesity was associated also with the development of *inflammation* at the systemic level: a slight increase in serum TNF-α level was observed for 35 and 52 days of observation compared to controls (Fig. 14.2).

The probiotic strain *L. casei* IMV B-7280 (separate) and the *L. casei* IMV B-7280 / *B. animalis* VKB / *B. animalis* VKL composition demonstrated an antiinflammatory effect in obese mice. So, under conditions of their therapeutic treatment, mice with obesity observed normalization of the serum TNF-α levels (Fig. 14.2). Instead, in obese mice transferred to the regular diet, without treatment with probiotic bacteria, there was observed a decreasing the levels of TNF-α in serum, but the difference compared to obese mice that continued to consume FED and controls was insignificant.

Probably, a manifestation of the systemic inflammation development in FED-induced obesity in mice was the observed hyperactivation of oxygen-dependent bactericidal activity of PEMs against the background of violations of their ability to absorb latex particles and the depletion of reserve capabilities of intracellular systems, which may predispose chronic inflammation and increased sensitivity to the development of infectious and other pathologies.

As it is shown in Table 14.2, the indices of the spontaneous NBT test of PEMs in obese mice increased significantly compared to controls. At the same time, PEMs of these mice did not respond on additional stimulation *in vitro*, no difference was found between the indices of the spontaneous and stimulated NBT test, there was no FR of the PEMs detected. It should be noted that the hyperproduction of oxidants

Table 14.2 PEMs activity in obese mice that received *L. casei* IMV B-7280 (separately) and *L. casei* IMV B-7280 / *B. animalis* VKB / *B. animalis* VKL composition

A group of animals	Indices of activity of PEMs				
	PI, %	PN, units	Spontaneous NST test, %	Induced NST test, %	Functional reserve of PEMs, units
Intact mice (controls)	31.5 ± 2.6	4.0 ± 1.0	42.0 ± 4.3	59.7 ± 1.5	9.9 ± 0.3
Obese mice that continued to receive FED	20.0 ± 1.2•	1.5 ± 0.6•	65.0 ± 4.0•	59.0 ± 8.0	0
Obese mice, transferred to a regular diet, and did not receive probiotic bacteria	30.0 ± 2.1*	3.8 ± 0.9*	60.0 ± 8.0•	57.0 ± 10.0	0
Obese mice, transferred to a regular diet, that received *L. casei* IMV B-7280	38.0 ± 4.2*	3.2 ± 0.1*	34.0 ± 7.0*	43.0 ± 5.0*	9.0 ± 3.0*
Obese mice, transferred to a regular diet, that received the probiotic composition	28.0 ± 3.4*	3.8 ± 1.0*	32.0 ± 2.0*	42.0 ± 9.1*	10.0 ± 2.0*

Notes: • - $P < 0.05$ vs indices of intact mice; * - $P < 0.05$ vs indices of obese mice that continued to consume FED

by activated phagocytic cells, in particular due to the development of inflammation, can lead to an increase in the level of intracellular autooxic processes and inhibition of their functional activity [3]. This might explain a decreasing phagocytic activity of the PEMs in FED-induced obesity in mice according to PI (by 36.5%) and to PN (by 62.5%) compared with controls. After administration of *L. casei* IMV B-7280 (separately) or *L. casei* IMV B-7280 / *B. animalis* VKB / *B. animalis* VKL composition, the functional activity of the PEMs was restored to the control values (Table 14.2). Instead, in obese mice that were transferred to regular diet and did not receive probiotic bacteria, yet there was a detected the normalization of phagocytic activity of the PEMs, but the NBT test values remained significantly higher than in the control. There was also absent the FR of the PEMs of these mice.

14.4 Data Interpretation

Thus, the results obtained demonstrate the potentially preventive effect of *L. casei* IMV B-7280 (separately) and *L. casei* IMV B-7280 / *B. animalis* VKB / *B. animalis* VKL composition on FED-induced obesity in mice. Supplementing the usual diet with probiotic therapy led to a decrease in the weight of obese mice and visceral fat mass against the background of normalization of the production of TNF-α and the functional activity of PEMs. There was a recovery to control values of glucose content in blood of obese mice only after injection of *L. casei* IMV B-7280.

It has been proven that probiotics based on strains of LAB, representatives of commensal intestinal microbiota, can reduce body weight and improve obesity biomarkers such as hyperglycemia, dyslipidemia and inflammation [30].

According to the literature data, chronic hyperglycemia can trigger lipid peroxidation, decrease the stability of the antioxidant system, and damage the vascular endothelium [31, 32], resulting in increased levels of proinflammatory cytokine, like TNF-α and other markers of systemic inflammation in the blood, as well as the functional activity of macrophages [33–37]. Since the level of TNF-α excretion is characterized by a clear positive correlation with the degree of obesity and insulin resistance, the correction of products of this proinflammatory cytokine is considered by many researchers as a promising therapeutic approach to the treatment of obesity and insulin resistance [33]. This is explained by the fact that increased intestinal *permeability* due to the disturbance of the intestinal microbiota spectrum enhance penetration of LPS of gram-negative bacteria into the blood, liver and adipose tissue, where they are recognized by the Toll- and NOD-like receptors expressed on macrophages and dendritic cells, activating the innate immune response and initiate the development of inflammation, being involved in the pathophysiology of obesity and the formation of insulin resistance. Increasing the LPS content in the blood for obesity correlated with the activation of products of TNF-α and other proinflammatory cytokines [34]. LPS triggers a strong proinflammatory reaction in colonic macrophages and secretion of proinflammatory cytokines that include TNF-α, interferon-γ, and IL-1β, whereas antiinflammatory cytokines IL-10 and IL-17 decrease and restore the intestinal permeability [35].

Commensal microbiota was reported to induce hyporesponsiveness to LPS via the production of IL-10 [36]. A randomized controlled clinical trial by Mohamacshahi et al. [37] showed that consumption of probiotic yogurt for 8 weeks improved lipid profile in type 2 diabetic patients induced a significant decrease in HbA1c accompanied by reduction levels of proinflammatory cytokine TNF-α.

On the other hand, in current study the normalization of blood glucose and TNF-α production was not observed in obese mice transferred to a regular diet, and not treated with probiotic bacteria, although their weight loss was detected, the weight of the visceral fat reduced, and the partial restoration of the functional activity of PEMs.

We believe that the change in the spectrum of gastrointestinal microbiota plays a key role in the implementation of the anti-inflammatory action of *L. casei* IMV B-7280 (separately) and *L. casei* IMV B-7280 / *B. animalis* VKB / *B. animalis* VKL composition for FED-induced obesity in mice. As is well-known, its imbalance, arising from the consumption of a HFD and more often found in the increase in the number of Firmicutes, Proteobacteria and in the reduction of LAB, is also one of the causes of systemic chronic inflammation [38].

Previously, we have found [24] that in the intestinal contents of obese mice receiving *L. casei* IMV B-7280 (separately) or *L. casei* IMV B-7280 / *B. animalis* VKB / *B. animalis* VKL composition, increased the amount of LAB, bifidobacteria and gram-negative bacteria (*coliform* bacteria), although there was no normalization of the total number of aerobic and optionally anaerobic bacteria and the number of gram-positive cocci. This is combined with the normalization of lipid metabolism (the level of general and free serum cholesterol decreased and improving the morphological structure of the liver (decreased or disappeared distinct degenerative changes such as necrosis, fatty degeneration of hepatocytes and multiple hemorrhages). The current data might support the idea of multifactorial mechanism of preventive action of these probiotic strains of bacteria in FED-induced obesity in mice, being in agreement with broad literature data.

Number of studies demonstrates that other probiotic strains of LAB and bifidobacteria had a strain-dependent therapeutic effect on FED-induced obesity in mice of different genetic lines [39–43]. Thus, the effectiveness of the strains, such as *L. acidophilus* AD031, *B. bifidum* BGN4 and *B. longum* BORI, was reduced by the following: weight of mice (*B. longum* BORI), triglycerides (*B. bifidum* BGN4 and *L. acidophilus* AD031) and total cholesterol (*B. longum* BORI), as well as morphological manifestations of liver steatosis [39]. Introduction of *L. acidophilus* NS1 to obese mice was due to a decrease in the level of total cholesterol and cholesterol of low density lipoprotein [40], and *L. acidophilus* NCDC caused a modification of the intestinal microbiota composition [41]. When *L. plantarum* K21 was used, the level of leptin, cholesterol and triglycerides in blood plasma of obese mice decreased the intensity of liver damage decreased and the number of Lactobacillus spp. and Bifidobacterium spp. in the gastrointestinal tract increased [42]. Probiotic strains

L. curvatus HY7601 and *L. plantarum* KY1032 prevented fat accumulation in obese mice, reduced their weight, as well as insulin, leptin, and total cholesterol levels in blood plasma, modified intestinal microbiota and suppressed the expression of genes of proinflammatory molecules (TNF-α, IL-6, IL-1β, etc.) in adipose tissue [43].

It has been shown that *Bifidobacterium* spp. improved homeostasis of glucose in mice, reduced the levels of accumulation of fat and reduced insulin secretion in FED-induced obesity model. Under the influence of *B. breve* B-3, the weight of obese mice and the accumulation of visceral fat decreased with a decrease in total cholesterol, glucose and insulin in serum [44]. Following the introduction of *C. pseudocatenulatum* CECT 7765 to obese mice, changes in expression of key genes involved in regulation of energy metabolism and lipid transport were detected [45].

We have established that probiotic strain *L. casei* IMV B-728 (separate) and the composition of *L. casei* IMV B-7280 / *B. animalis* VKB / *B. animalis* VKL were effective also for glutamate-induced and immune-mediated obesity in Wistar rats [14]. Probiotics prevented growth retardation, reduced body mass index and protected from excessive accumulation of visceral fat. The mechanisms of influence of these probiotic bacteria also included normalization of lipid and carbohydrate metabolism, the antioxidant system and mitigation of inflammation processes. The more effective treatment option for obesity induced by sodium glutamate had the *L. casei* IMV B-7280 / *B. animalis* VKB / *B. animalis* VKL composition [14], while in FED-induced obesity in mice – *L. casei* IMV B-7280 (separately), which is probably due to the pathophysiological features of the development of these diverse types of obesity.

14.5   Future Outlooks

Recent findings can initiate reconsidering strategies for probiotic therapies.

Thus, according to *"Developmental Origins of Health and Disease"*(DOHaD) hypothesis [46] infant microbiota alterations due to colonisation from obese mothers may lead to increased risk of childhood obesity and diabetes [47]. The maternal gut microbiome is the source of the majority of transmitted strains shaping the early microbial diversity in the infant gut [48]. This provides novel insights for earliest prevention of MetS via targetting mother's gut microbiome.

Dead bacteria might also improve diabetic profiles. Thus, the study by Kikuchi et al. [49] indicates that sterilized bifidobacteria suppressed fat accumulation, improved insulin resistance, and lowered blood glucose levels in mice on a high-fat diet.

Current model of obesity may corresponds to MetS model according to the definition [2] and consequently this study completes an agenda on the research strategy for metabolic syndrome treatment with probiotics since studied strains: reduce obesity and visceral fat [14, 24], improve dyslipidemia [24], and re-equilibrate glycemic and inflammatory profiles in obesity according to the results of current study.

The *"single strain" concept* could be preliminary considered, since the probiotic strain *L. casei* IMB B-7280 has been demonstrated to be most effective in decreasing glucose and serum TNF-α levels than composition of the probiotic strains.

Therefore, it is promising to develop the probiotics of multifactorial actions that can be used for personalized treatment of patients with MetS and DM and for prevention of obesity.

On the other hand, individualized approach should be carefully considered for personalized dietary and probiotic therapy for obese patients, as well as to set preventive programs, taking into account patient's individual profiles: age, sex, spectrum of intestinal microbiota, immune system status, nutritional habits, etc.

Further research with clearly formulated endpoints should aim the study the effectiveness vs currently established efficacy for the identification of new effective compositions of probiotic strains, the selection of optimal treatment schemes and rational dosage.

Conflicts of Interest Declare conflicts of interest or state "The authors declare no conflict of interest."

Ethics This study has been approved by the ethics committee of institutional review board and Special Academic Council on Doctoral Thesis of D.K. Zabolotny Institute of Microbiology and Virology of the National Academy of Sciences of Ukraine (protocol N 7 issued 03.07.2018). No human subjects were included to the study.

Funding This research received financial support from the State Fund for Fundamental Research of Ukraine via the framework of the project Φ64/28–2015.

References

1. WHO/FAO scientific document. http://who.int/foodsafety/fs_management/en/probiotic_guidelines.pdf. Accessed 7 August 2018
2. Harsch IA, Konturek PC (2018) The Role of Gut Microbiota in Obesity and Type 2 and Type 1 Diabetes Mellitus: New Insights into "Old" Diseases. Med Sci (Basel) 6(2):E32. https://doi.org/10.3390/medsci6020032
3. Musso G, Gambino R, Cassader M (2010) Obesity, diabetes, and gut microbiota: the hygiene hypothesis expanded? Diabetes Care 33(10):2277–2284. https://doi.org/10.2337/dc10-0556
4. Patterson E, Ryan PM, Cryan JF, Dinan TG, Ross RP, Fitzgerald GF, Stanton C (2016) Gut microbiota, obesity and diabetes. Postgrad Med J 92(1087):286–300. https://doi.org/10.1136/postgradmedj-2015-133285
5. Chobot A, Górowska-Kowolik K, Sokołowska M, Jarosz-Chobot P (2018) Obesity and diabetes-not only a simple link between two epidemics. Diabetes Metab Res Rev 21:e3042. https://doi.org/10.1002/dmrr.3042
6. Houghton D, Hardy T, Stewart C, Errington L, Day CP, Trenell MI, Avery L (2018) Systematic review assessing the effectiveness of dietary intervention on gut microbiota in adults with type 2 diabetes. Diabetologia 61:1700–1711. https://doi.org/10.1007/s00125-018-4632-0

7. Vaarala O, Atkinson MA, Neu J (2008) The "perfect storm" for type 1 diabetes: the complex interplay between intestinal microbiota, gut permeability, and mucosal immunity. Diabetes 57(10):2555–2562. https://doi.org/10.2337/db08-0331

8. Gomes AC, Bueno AA, de Souza RG, Mota JF (2014) Gut microbiota, probiotics and diabetes. Nutr J 17(13):60. https://doi.org/10.1186/1475-2891-13-60

9. Creely SJ, McTernan PG, Kusminski CM, Fisher FM, Da Silva NF, Khanolkar M, Evans M, Harte AL, Kumar S (2007) Lipopolysaccharide activates an innate immune system response in human adipose tissue in obesity and type 2 diabetes. Am J Physiol Endocrinol Metab 292:E740–E747

10. Fernández-Real JM, Pickup JC (2008) Innate immunity, insulin resistance and type 2 diabetes. Trends Endocrinol Metab 19(1):10–16. https://doi.org/10.1016/j.tem.2007.10.004

11. Zhao L, Zhang F, Ding X, Wu G, Lam YY, Wang X, Fu H, Xue X, Lu C et al (2018) Gut bacteria selectively promoted by dietary fibers alleviate type 2 diabetes. Science 359:1151–1156. https://doi.org/10.1126/science.aao5774

12. Zhang M, Yang XJ (2016) Effects of a high fat diet on intestinal microbiota and gastrointestinal diseases. World J Gastroenterol 22(40):8905–8909. https://doi.org/10.3748/wjg.v22.i40.8905

13. Guijarro A, Laviano A, Meguid MM (2006) Hypothalamic integration of immune function and metabolism. Prog Brain Res 153:367–405. https://doi.org/10.1016/S0079-6123(06)53022-5

14. Savcheniuk OA, Virchenko OV, Falalyeyeva TM, Beregova Tetyana V, Babenko LP, Lazarenko LM, Demchenko OM, Bubnov RV, Spivak MY (2014) The efficacy of probiotics for monosodium glutamate-induced obesity: dietology concerns and opportunities for prevention. EPMA Journal 5:2. https://doi.org/10.1186/1878-5085-5-2

15. Zhang Q, Wu Y, Fei X (2016) Effect of probiotics on glucose metabolism in patients with type 2 diabetes mellitus: a meta-analysis of randomized controlled trials. Medicina (Kaunas) 52(1):28–34. https://doi.org/10.1016/j.medici.2015.11.008

16. Zhang Q, Yucheng W, Fei X (2015) Effect of probiotics on glucose metabolism in patients with type 2 diabetes mellitus: a meta-analysis of randomized controlled trials. Medicina 52(1):28–34

17. Wang X, Juan QF, He YW, Zhuang L, Fang YY, Wang YH (2017) Multiple effects of probiotics on different types of diabetes: a systematic review and meta-analysis of randomized, placebo-controlled trials. J Pediatr Endocrinol Metab 30(6):611–622. https://doi.org/10.1515/jpem-2016-0230

18. Horvath A, Leber B, Schmerboeck B, Tawdrous M, Zettel G, Hartl A, Madl T, Stryeck S, Fuchs D, Lemesch S, Douschan P, Krones E, Spindelboeck W, Durchschein F, Rainer F, Zollner G, Stauber RE, Fickert P, Stiegler P, Stadlbauer V (2016) Randomised clinical trial: the effects of a multispecies probiotic vs. placebo on innate immune function, bacterial translocation and gut permeability in patients with cirrhosis. Aliment Pharmacol Ther 44(9):926–935. https://doi.org/10.1111/apt.13788

19. He J, Zhang F, Han Y (2017) Effect of probiotics on lipid profiles and blood pressure in patients with type 2 diabetes: A meta-analysis of RCTs. Chen. W. Medicine 96(51):e9166. https://doi.org/10.1097/MD.0000000000009166

20. Berger VW, Alperson SY (2009) A general framework for the evaluation of clinical trial quality. Rev Recent Clin Trials 4(2):79–88

21. Hill C, Guarner F, Reid G, Gibson GR, Merenstein DJ, Pot B, Morelli L, Canani RB, Flint HJ, Salminen S, Calder PC (2014) Sanders ME Expert consensus document. The International Scientific Association for Probiotics and Prebiotics consensus statement on the scope and appropriate use of the term probiotic. Nat Rev Gastroenterol Hepatol 11(8):506–514

22. Bubnov RV, Spivak MY, Lazarenko LM, Bomba A, Boyko NV (2015) Probiotics and immunity: provisional role for personalized diets and disease prevention. EPMA J 6:14. https://doi.org/10.1186/s13167-015-0036-0

23. Reid G, Abrahamsson T, Bailey M, Bindels LB, Bubnov R, Ganguli K, Martoni C, O'Neill C, Savignac HM, Stanton C, Ship N, Surette M, Tuohy K, van Hemert S (2017) How do probiot-

ics and prebiotics function at distant sites? Benef Microbes 20:1–14. https://doi.org/10.3920/BM2016.0222

24. Bubnov RV, Babenko LP, Lazarenko LM, Mokrozub VV, Demchenko OA, Nechypurenko OV, Spivak MY (2017) Comparative study of probiotic effects of lactobacillus and Bifidobacteria strains on cholesterol levels, liver morphology and the gut microbiota in obese mice. EPMA J 8(4):357–376. https://doi.org/10.1007/s13167-017-0117-3

25. Bubnov RV, Babenko LP, Lazarenko LM, Mokrozub VV, Spivak MY (2018) Specific properties of probiotic strains: relevance and benefits for the host. EPMA J 9(2):205–223. https://doi.org/10.1007/s13167-018-0132-z

26. Lazarenko LM, Babenko LP, Bubnov RV, Demchenko OM, Zotsenko VM, Boyko NV et al (2017) Imunobiotics are the novel biotech drugs with antibacterial and immunomodulatory properties. Mikrobiol Z 79(1):66–75. https://doi.org/10.15407/microbiolj79.01

27. Mokrozub VV, Lazarenko LM, Sichel LM, Bubnov RV, Spivak MY (2015) The role of beneficial bacteria wall elasticity in regulating innate immune response. EPMA J 6:13. https://doi.org/10.1186/s13167-015-0035-1

28. Home Office Animals (Scientific Procedures) Act (1986) Code of practice for the housing and Care of Animals Used in scientific procedures. Available online: http://www.official-documents.gov.uk/document/hc8889/hc01/0107/0107.pdf. accessed 28 July 2018

29. Dubaniewicz A, Hoppe A (2004) The spontaneous and stimulated nitroblue tetrazolium (NBT) tests in mononuclear cells of patients with tuberculosis. Rocz Akad Med Bialymst 49:252–255

30. Aggarwal J, Swami G, Kumar M (2013) Probiotics and their effects on metabolic diseases: an update. J Clin Diagn Res 7(1):173–177. https://doi.org/10.7860/JCDR/2012/5004.2701

31. Góralczyk K, Szymańska J, Szot K, Fisz J, Rość D (2016) Low-level laser irradiation effect on endothelial cells under conditions of hyperglycemia. Lasers Med Sci 31(5):825–831. https://doi.org/10.1007/s10103-016-1880-4

32. Macharia M, Kengne AP, Blackhurst DM, Erasmus RT, Matsha TE (2014) The impact of chronic untreated hyperglycaemia on the long-term stability of paraoxonase 1 (PON1) and antioxidant status in human sera. J Clin Pathol 67(1):55–59. https://doi.org/10.1136/jclinpath-2013-201646

33. Tzanavari T, Giannogonas P, Karalis KP (2010) TNF-alpha and obesity. Curr Dir Autoimmun 11:145–156. https://doi.org/10.1159/000289203

34. Dragano NR, Haddad-Tovolli R, Velloso LA (2017) Leptin, Neuroinflammation and obesity. Front Horm Res 48:84–96. https://doi.org/10.1159/000452908

35. Hiippala K, Jouhten H, Ronkainen A, Hartikainen A, Kainulainen V, Jalanka J, Satokari R (2018) The potential of gut commensals in reinforcing intestinal barrier function and alleviating inflammation. Nutrients 10(8):E988. https://doi.org/10.3390/nu10080988

36. Ueda Y, Kayama H, Jeon SG, Kusu T, Isaka Y, Rakugi H, Yamamoto M, Takeda K (2010) Commensal microbiota induce LPS hyporesponsiveness in colonic macrophages via the production of IL-10. Int Immunol 22(12):953–962. https://doi.org/10.1093/intimm/dxq449

37. Mohamadshahi M, Veissi M, Haidari F, Javid AZ, Mohammadi F, Shirbeigi E (2014) Effects of probiotic yogurt consumption on lipid profile in type 2 diabetic patients: a randomized controlled clinical trial. J Res Med Sci 19(6):531–536

38. Boulangé CL, Neves AL, Chilloux J, Nicholson JK, Dumas ME (2016) Impact of the gut microbiota on inflammation, obesity, and metabolic disease. Genome Med. 8(1):42. https://doi.org/10.1186/s13073-016-0303-2

39. Li Z, Jin H, Oh SY, Ji GE (2016) Anti-obese effects of two lactobacilli and two Bifidobacteria on ICR mice fed on a high fat diet. Biochem Biophys Res Commun 480(2):222–227. https://doi.org/10.1016/j.bbrc.2016.10.031

40. Song M, Park S, Lee H, Min B, Jung S, Park S, Kim E, Oh S (2015) Effect of lactobacillus acidophilus NS1 on plasma cholesterol levels in diet-induced obese mice. J Dairy Sci 98(3):1492–1501. https://doi.org/10.3168/jds.2014-8586

41. Arora T, Anastasovska J, Gibson G, Tuohy K, Sharma RK, Bell J, Frost G (2012) Effect of lactobacillus acidophilus NCDC 13 supplementation on the progression of obesity in diet-induced obese mice. Br J Nutr 108(8):1382–1389. https://doi.org/10.1017/S0007114511006957

42. Wu CC, Weng WL, Lai WL, Tsai HP, Liu WH, Lee MH, Tsai YC (2015) Effect of lactobacillus plantarum strain K21 on high-fat diet-fed obese mice. Evid Based Complement Alternat Med 2015:391767. https://doi.org/10.1155/2015/391767

43. Park DY, Ahn YT, Park SH, Huh CS, Yoo SR, Yu R, Sung MK, McGregor RA, Choi MS (2013) Supplementation of lactobacillus curvatus HY7601 and lactobacillus plantarum KY1032 in diet-induced obese mice is associated with gut microbial changes and reduction in obesity. PLoS One 8(3):e59470. https://doi.org/10.1371/journal.pone.0059470

44. Kondo S, Xiao JZ, Satoh T, Odamaki T, Takahashi S, Sugahara H, Yaeshima T, Iwatsuki K, Kamei A, Abe K (2010) Antiobesity effects of Bifidobacterium breve strain B-3 supplementation in a mouse model with high-fat diet-induced obesity. Biosci Biotechnol Biochem 74(8):1656–1661

45. Moya-Pérez A, Romo-Vaquero M, Tomás-Barberán F, Sanz Y, García-Conesa MT (2014) Hepatic molecular responses to Bifidobacterium pseudocatenulatum CECT 7765 in a mouse model of diet-induced obesity. Nutr Metab Cardiovasc Dis 24(1):57–64. https://doi.org/10.1016/j.numecd.2013.04.011

46. Wadhwa PD, Buss C, Entringer S, Swanson JM (2009) Developmental origins of health and disease: brief history of the approach and current focus on epigenetic mechanisms. Semin Reprod Med 27(5):358–368. https://doi.org/10.1055/s-0029-1237424

47. Soderborg TK, Borengasser SJ, Barbour LA, Friedman JE (2016) Microbial transmission from mothers with obesity or diabetes to infants: an innovative opportunity to interrupt a vicious cycle. Diabetologia 59(5):895–906. https://doi.org/10.1007/s00125-016-3880-0

48. Ferretti P, Pasolli E, Tett A, Asnicar F, Gorfer V, Fedi S, Armanini F, t al. (2018) Mother-to-infant microbial transmission from different body sites shapes the developing infant gut microbiome. Cell Host Microbe 24(1):133–145.e5. https://doi.org/10.1016/j.chom.2018.06.005

49. Kikuchi K, Ben Othman M, Sakamoto K (2018) Sterilized bifidobacteria suppressed fat accumulation and blood glucose level. Biochem Biophys Res Commun 501(4):1041–1047. https://doi.org/10.1016/j.bbrc.2018.05.105

Chapter 15
Oral Microbiome and Innate Immunity in Health and Disease: Building a Predictive, Preventive and Personalized Therapeutic Approach

Jack C. Yu, Hesam Khodadadi, Évila Lopes Salles, Sahar Emami Naeini, Edie Threlkeld, Bidhan Bhandari, Mohamed Meghil, P. Lei Wang, and Babak Baban

Abstract An average person carries 1 to 2 kg of microbes in the alimentary track, including the oral cavity. There are more bacteria in a person's mouth than the total human population in the entire world. Oral health is critical to the general systemic health of an individual. The harmonious co-existence between more than 1000 bacterial species and the host's immune system underpins sustained, long-term homeostasis, the sine qua non of oral health. In a similar manner, global oral health is essential for general population health of the world. Since our last review of this

J. C. Yu (✉)
Department of Surgery, Medical College of Georgia, Augusta University, Augusta, GA, USA
e-mail: jyu@augusta.edu

H. Khodadadi · É. L. Salles · S. E. Naeini · P. L. Wang · B. Baban
Department of Oral Biology and Diagnostic Sciences, Dental College of Georgia, Augusta University, Augusta, GA, USA

Center for Excellence in Research, Scholarship and Innovation (CERSI), Dental College of Georgia, Augusta University, Augusta, GA, USA

E. Threlkeld
Medical College of Georgia, Augusta University, Augusta, GA, USA

B. Bhandari
The Graduate School, Augusta University, Augusta, GA, USA

M. Meghil
Department of Oral Biology and Diagnostic Sciences, Dental College of Georgia, Augusta University, Augusta, GA, USA

Center for Excellence in Research, Scholarship and Innovation (CERSI), Dental College of Georgia, Augusta University, Augusta, GA, USA

Department of Periodontics, Dental College of Georgia, Augusta University, Augusta, GA, USA

N. Boyko, O. Golubnitschaja (eds.), *Microbiome in 3P Medicine Strategies*, Advances in Predictive, Preventive and Personalised Medicine 16, https://doi.org/10.1007/978-3-031-19564-8_15

subject in 2019, while significant clinical advances continue, the disparity, lack of prevention, insufficient care, and political unrest have persisted or significantly deteriorated. This review focuses on the following important questions:

1. What is oral microbiome? How to detect, characterize, compare, report, and interpret the results?
2. How does oral microbiome affect and respond to local and systemic innate immunity?
3. What is the role of oral microbiome in the pathogenesis of diseases of the mouth?
4. What are the impacts of oral health or the lack of it at the systemic level?
5. Why is oral health important at the population level?
6. How can the healthcare providers restore and sustain harmonious co-existence between host and oral microbiome?

Keywords Microbiome · Oral · Health · Prevention · Personnalized · Predictive

15.1 Introduction

The microbial community of the oral cavity has four domains if one considers virus a domain, the other three being *Archaea*, *Bacteria*, and *Eukarya* [1]. While the colonization of the fetal gut starts during the second trimester measured by 16 S rRNA sequencing, some 8 weeks after the first development of gut mucosal immunity, the oral colonization occurs after birth. The oral ecology evolves with the host throughout the life of the individual from neonate to senescence, affecting and responding to the actions and reactions of the host immune system [2]. Both oral health and oral diseases, such as caries and periodontitis, are emergent phenomena—the results of numerous interactions of complex adaptive systems (CAS). Oral health requires symbiosis and eubiosis between the host and the oral microbiota, characterized by sustained and sustainable harmonious homeostasis. The disruption of this dynamic steady state results in phase transition that is oral diseases. Unless effective treatments can achieve a new steady state, the dental and/or periodontal health will continue to deteriorate due to the functional demand (mastication) and the unfavorable host-microbes and microbes-microbes interactions [3]. Once the pathological process advances to pulpal or furcation involvement, the loss of the dental-alveolar unit becomes inevitable. With this eventual loss of the tooth, the local ecology changes. While it prevents the local, regional, and systemic spreads of the infection, the edentulous alveolar segment, devoid of enamel, dentine, cementum, periodontal pockets, perturbs the integrity of the masticatory system. The return to a new steady state, due to altered microbial community, more sustainable host-microbe interactions and resolution of the oral infection, is incomplete and comes with the heavy price of some masticatory functional loss. The disruption of the dental arch integrity changes the dental occlusion, often with super eruption of the opposing tooth and drifting/ tilting of the tooth behind the gap so created (Fig. 15.1) [4].

Fig. 15.1 With the loss of right mandibular first molar (Tooth # 30), the tooth behind it (# 31) tilts into the gap with signs of supra-eruption of the opposing maxillary first molar

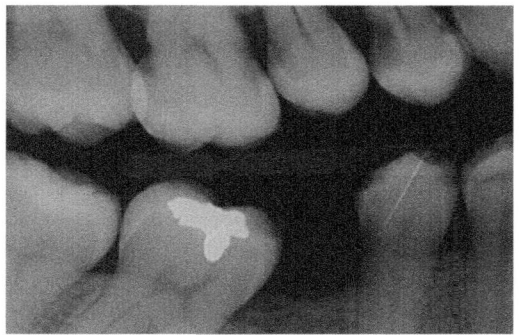

There are several unique features of the oral cavity, which houses the majority of the masticatory system. The deciduous dentition has 20 teeth and the permanent dentition, second and final, has 32 teeth. The dental formula for each quadrant of the deciduous dentition is $I_2C_1M_2$, denoting 2 incisors 1 canine and 2 molars. It is $I_2C_1P_2M_3$ for the permanent dentition. Each tooth has a specialized function. Firmly embedded in the alveolar process of the maxilla and mandible, these teeth must perform the function of initial breakdown of food by incising and chewing. In no other parts of the human body is there a protrusion from bone, which is sterile, through the integumentary surface with communication to the external environment, which contains many microbes. The enamel covering of each permanent tooth, 1.5 to 2.5 mm in thickness with a Young's modulus of 90 GPa in compression, undergoes attrition from the moment it comes into occlusion without any ability for further synthesis [5]. This complete lack of turnover allows for continual microbial build up and evolution, forming complex biofilms.

This review, following our initial publication in 2019, highlights again the importance or oral health to the general systemic health of the individual and expand this to the population level. The oral microbiome, like the major histocompatibility complex (MHC), is unique to each individual. Personalized data can allow healthcare providers to predict and thus prevent pathological conditions [2]. The present review adds a discussion of oral mycome and oral virome, which is of particular interest and importance in light of the recent pandemic by SARS-CoV-2 [6]. It has six sections to cover the following important topics:

1. The oral microbiome: What is oral microbiome? How to detect, characterize, compare, report, and interpret the results?
2. The host-oral microbe interaction: How does oral microbiome affect and respond to local and systemic innate immunity?
3. Oral microbiome and oral diseases: What is the role of oral microbiome in the pathogenesis of diseases of the mouth?
4. Oral health and systemic health: What are the impacts of oral health or the lack of it at the systemic level?
5. Oral health and population health: Why is oral health important at the population level?
6. The role of the healthcare providers: How can the healthcare providers restore and sustain harmonious co-existence between host and oral microbiome?

15.1.1   Oral Microbiome

The total genomic contents of all oral microbes is the oral microbiome. The microbes themselves make up the microbiota. On a surface of approximately 47 × 47 cm, slightly larger than an average tray from the cafeteria, the oral cavity of 2200 cm² contains a trillion bacteria belonging to 7 phyla and more than 1000 taxa. In addition to bacteria, the other oral microbes include viruses, Achaea (mostly methanogens), protozoans, and fungi [7]. The most numerous and least well-studied microbial agents making up the oral microbiome are the viruses. They outnumber bacteria 100 to 1. For 1 mg of dental plaque, there are 10^{10} viral particles, each measuring 10 to 100 nm. By mass, bacteria are the most abundant oral microbes, with each mg of dental plaque having 10^8 colony forming units (CFU). The detection and identification of bacteria no longer depends on methods initially developed by Julius Richard Petri for Robert Koch more than 140 years ago. Such methods require growing the bacteria from the specimen and separating them into axenic cultures (pure culture through limiting dilution) followed by identifications. These methods are woefully inadequate because of the very large numbers of bacterial species present in varying density and with vastly different growth rates in culture [8]. The genomic approach has vastly improved the identification of oral bacteria. Taking advantage of the constant, conserved region of 16 s ribosomal RNA gene, which is only present in bacteria, the PCR primer amplifies these bacterial sequences. The individual species identification is possible by sequencing the variable regions [9] (Fig. 15.2).

Using this type of methods, the human oral microbiome has 619 taxa in 13 phyla, as follows: *Actinobacteria, Bacteroidetes, Chlamydiae, Chloroflexi, Euryarchaeota, Firmicutes, Fusobacteria, Proteobacteria, Spirochaetes, SR1, Synergistetes, Tenericutes,* and *TM7*[10]. The five phyla making up the core component are *Firmicutes, Proteobacteria, Bacteroidetes, Fusobacteria, and Actinobacteria*. With the advances in culture-independent methods and large number of taxa, there is an increasing need to describe the diversity within group and between groups with precision. Alpha diversity is the measure of "richness" (total count of different

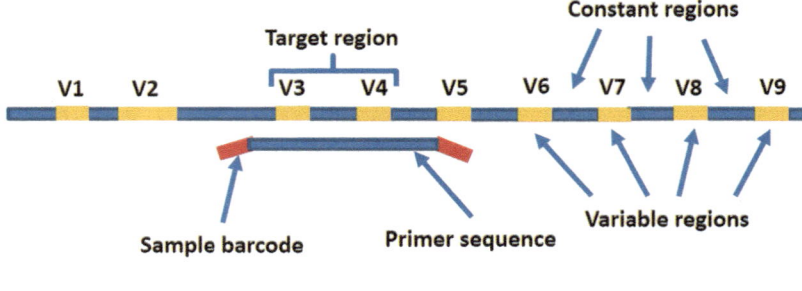

Fig. 15.2 The identification of bacterial species using 16S rRNA sequencing

species) based on counting of how many species (or operational taxonomic unit, OUT) are in different areas within the local niche. The comparison of diversity between two local niches, or regions, is the beta diversity. The diversity of all regions under study is the gamma diversity [10].

About 70% of the initial neonatal oral microbiota are maternal in origin with 65% derived from maternal oral microbiota through vertical transmission. Perinatal factors such as mode of delivery, microbiome of meconium and amniotic fluid, and maternal antibiotic therapy will alter the neonatal microbiome [11]. Oral swab taken from the neonate immediately upon presentation of the head during vaginal delivery without the use of topical 10% povidone iodine disinfectant show the best diversity with an abundant presence of *Lactobacilli*, a non-motile, catalase-negative, anaerobic homolactic fermenter [12]. These very early colonizers are from maternal vaginal flora. The *Lactobacilli* predominance continues in babies fed breast milk, followed by *Bifidobacteria* and *Streptococci* (mostly *Streptococcus sanguis* and *Streptococcus mitis*). Within 2 to 5 days after birth, the oral microbiome of the neonate changes, becoming similar to oral microbiome of the mother. Maternal transfer of immune cells (80,000,000/day) and non-nutritive soluble factors (TNF-alpha, type I IFN, immunoglobulins, mostly IgA) through her milk assist the developing neonatal immune system and shape the infant oral microbiome, which in turn guide the acquisition and maintenance of the gut microbiome [13]. With the eruption of deciduous lower central incisors by age 6 months, the infant oral microbiome changes, preceded by the presence of all three types of innate lymphoid cells (ILC) at 4 months of age. The non-renewing, hard enamel surface is ideal for biofilm formation. The first layer is the pellicle, which forms in minutes, consisting of organic maters, mostly glycoproteins, condensed from the saliva with calcium phosphates. Attached to the pellicle are complex adhesive molecules (adhesins) from bacteria and saliva. They serve as anchors for the early colonizers including many streptococcal species such as *S. mitis*, *S. oralis*, and *S. sanguis*, and *S. mutan*. These are facultative anaerobes with complex nutritional needs, which are well suited for the supra-gingival microenvironment. Often, the early colonizers will co-aggregate among themselves, facilitating adhesion to the tooth surface. On top of this group, come *Actinomyces*, *Capnocytophaga*, *Veillonella*, *Eikenella*, and *Prevotella*. Of them, *Capnocytophaga*, *Prevotella*, *Veillonella*, *Eikenella* and are Gram-negative, with the first two belong to the phylum *Bacteriodota,* and the second two the phylum *Pseudomonadota*. *Actinomyces*, on the other hand, are Gram-positive endospore forming, producers of short organic acids with 1 to 4 carbons including formic acid, acetic acid, lactic acid, and succinic acid. This sequence repeats each time after effective tooth brushing [14]. Given sufficient time, the adhesive bacteria, the larger *Fusobacteria* start to build on the early colonizers and provide the substratum for the late colonizers such as *Treponema* and *Porphyromonas*, which are more pathogenic. Because of this "bridging" role, these bacteria are the bridging colonizers. *Fusobacterium nucleatum* is an obligatory anaerobe non-spore-forming producer of short chain fatty acid such as butyrate. These short chain fatty acids serve as important nutrient sources for other bacteria of the microbiota and are alter the host innate response. As this bacterial colony expands the microenvironment on and within it changes, especially with the production of extracellular matrix, known as EPS, or extracellular polymeric substrates. The center will experience hypoxia with progressive decrease in O_2 content. *Treponema denticola*, *Porphyromonas gigivalis*, and *Tannerella forsythia*, all Gram-negative obligatory anaerobes, make up the Red Complex capable of producing periodontitis (Fig. 15.2). Due to the low O_2 content of the gingival crevice (50% of the atmospheric level), the bacteria of Red Complex inhabit these niches. Other anaerobes found in the gingival sulci are *Bacteroides melaninogenicus*, and members of clostridia, and peptostreptococci.

In addition to bacteria, the oral microbiota contains members from the Domain Eukarya (fungi) and viruses. With normal oral microbiota, the types of fungi are those that can resist bacterial attacks by producing antibacterial compounds. Fungi are eukaryotic and much larger than bacteria, measuring in 10^{-5} m and kept in dynamic balance with bacteria [15]. 100 species of fungi live in the oral cavity belonging to 4 major groups: *Candida*, *Aspergillus*, *Cryptococcus*, and *Fusarium*, with *Candida albicans* found in 40% of the normal healthy individuals. When bacterial populations change drastically, due to antibiotic use or host immune suppression, fungi overgrowth occur, causing opportunistic infections. These mycotic infections can progress rapidly and cause tissue necrosis by blocking the cutaneous capillaries due to their large size (Fig. 15.3).

On the other end of the length scale are the viruses, measured in 10^{-8} m. Just as fungi must defend themselves against smaller and more numerous bacteria (by synthesizing antibiotics), bacteria must defend against viral predation (using CRISPR, or clustered regularly interspaced short palindromic repeats). CRISPR allows bacteria to excise viral genetic materials from their own genomes. By estimates, there are 10^{15} virus, or virus-like, particles in and on the human body. There are several ways to classify viruses, by RNA or DNA, single or double stranded, enveloped or non-enveloped nuclocapsids, type of intermediate and final hosts. Within the oral cavity, the virome consists of two major groups, depending on the target: eukaryotic

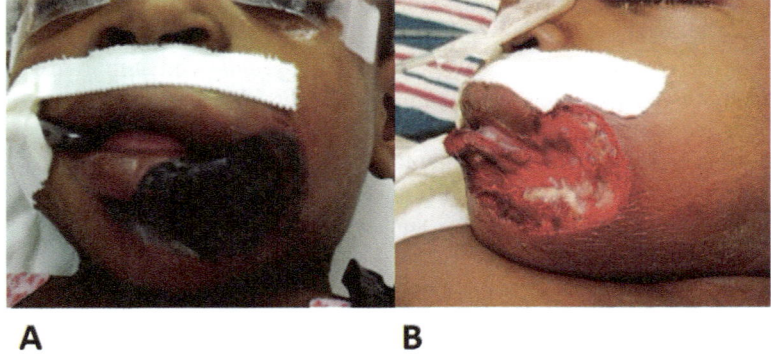

A   **B**

Fig. 15.3 (**a**): The clinical appearance of a rapidly enlarging cutaneous necrosis due to mycotic infection in an immune compromised infant. (**b**) The patient after surgical debridement of all the affected area

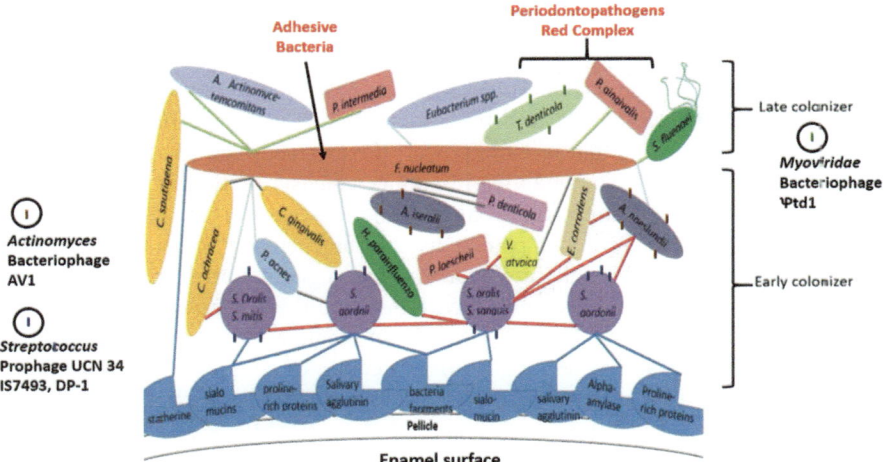

Fig. 15.4 Modified from Yu et al. 2019 [3] showing the microbial community on enamel surface. The colored lines represent adhesion-cognate receptors responsible for co-aggregation. For the bacterial genus, starting from the bottom: *S Streptococcus*, *P Propionibacterium* (acnes), *C Capnocytophaga*, *H Hemophilus*, *P Prevotella (loescheii, denticola, and intermedia)*, *V Veillonella*, *E Eikenella*, *A Actinomyces (isralii and naslundii)*, *T Treponema*, *P Porphyromonas (gingivalis)*. Five common bacteriophage species are included

viruses and the more abundant prokaryotic viruses (bacteriophage). Eight of the more than 100 herpesviruses (DNA eukaryotic virus, from Family *Herviviricetes*, enveloped DNA virus 180–250 nm in diameter) use human as host. There are three important types of herpesviruses of significant medical relevance: Alphaherpesviruses (herpes simplex viruses, HSV), Betaherpesvirus (Cytomegalovirus, CMV), and Gammaherpesvirus (Epstein-Barr virus, EBV). While HSV has neuro-tropism, CMV and EBV attack oral epithelium, fibroblasts, and lymphocytes. Both are present in the gingival crevicular fluids with EBV showing a prevalence of 66% in periodontal pockets 5 mm or deeper and CMV having an increased odds ratio of 51.4 (95% confidence interval 5.4–486.5) for aggressive periodontitis when *P. gingivalis* is also present. The bacteriophages attack prokaryotes and thus regulate oral bacterial population and participate in DNA transfer. In turn, bacterial species select the bacteriophages to "accept": DP-1, UCN34, IS7493 for streptococcus, AV-1 for actinomyces, P-7 and Lambda for enterococcus, and φtd1 for *T. denticola* [16] (Fig. 15.4). In addition to resident viruses, oropharynx is also the principal portal for viral entry. Many air-borne viruses gain access through the mouth. One such example is SARS-CoV-2, a member of the Family *Coronaviridae*, which binds to the Angiotensin Converting Enzyme II Receptor. The diversity of the oral microbiome influenced the infection severity, and duration of CVID-19, with eubiotic, high diversity microbiome associated with less symptoms and shorter duration [17].

15.2   Innate Immunity and Oral Microbiome

To defend against pathogenic microbes, the body has two different kinds of immune systems: a rapidly deployable first responders known as innate immunity, and a second more slower one that is more powerful, sustained, and specific, known as adaptive immunity [3]. The adaptive immunity consists of B- and T-lymphocyte families capable of detecting and producing antibodies and cytotoxic cells, which target specifically the invading pathogens that are causing infection and tissue destruction. Both B and T cells use hypervariable regions in their genes to recombine to produce proteins that can bind with high affinities to novel antigen targets once properly presented [3, 18, 19]. The cells of innate immunity do not have such ability to recombine and generate effector molecules or cells with high specificity. Because of this, they can mobilize and respond to pathogen invasions (through receptors, such as Toll-Like Receptor, or TLR for PAMP- pathogen associated molecular pattern) and tissue damage (through receptors for DAMP- damage associated molecular pattern) much faster than the adaptive immune system. Inflammation is the first innate immune response and activates within seconds to minutes. One critical requirement for innate immunity, due to the less specific targeting and rapidity of activation, is that it does not respond inappropriately by attacking commensals or self [3, 18].

The relationship between innate immunity and microbiome is highly complex and with a long history of co-evolution. In human, it is a well-tuned defense system, which establishes and maintains an immunologic truce between the host and over 100 trillion of microorganisms residing in and on the human body. The intricate symbiosis between host immune system and microbiome starts *in utero*. The immune system is immature at birth. It develops further during the neonatal period, maturing in late childhood and adolescence, continuously evolving through adulthood and senescence, changing and changed by the microbiome [3, 20, 21].

The task of guarding the oral aperture, the gate to the alimentary canal, is particularly challenging. Too much (tissue destruction) is as bad as too little (pathogen invasion). The interplay between innate immunity and oral microbiome is critical in striking this delicate balance. In addition to a trillion bacteria, the oral cavity, from the lips to the oropharynx, has both solids and liquids, soft and hard tissues including, teeth, saliva, hard and soft palates, floor of the mouth, oral mucosa, tongue, attached gingiva, and gingival sulcus. These diverse bio-structures have high variability biochemically, microbiologically, and biophysically. The core components of cellular and humoral innate immunity are neutrophils, macrophages, dendritic cells, natural killer cells, innate lymphoid cells, and many cytokines and chemokines [3, 20, 21]. With time, there emerges an "immunobiography" -- the most important element for achieving diverse yet appropriate interactions between innate immunity and microbiome [22]. That is oral homeostasis. The immunobiography is the "immunologic history" of the individual, consisting of type, dose, intensity and temporal sequence of all prior pathogenic challenges in the context of that person's general health, habits, and lifestyle. These interplays between oral microbiome and innate immunity "educate" the system, generating the immunobiography, affecting the general health of individuals locally and systemically [22].

No education is successful without memory. A significant consequence of the interaction between innate immunity and oral microbiome is "innate memory", also known as "trained memory" [23, 24]. Innate memory confers more intense and faster responses to a second challenge by a previous pathogen. Most importantly, microbiome plays a central role in determining whether innate memory (trained memory) should stimulate or suppress the immune response. At its core, the binary decision is either to tolerate or attack. In the oral cavity, like many other tissues with microbial contacts, the direction and magnitude of immune responses are heavily dependent on immunobiography, the strength of homeostasis, and the composition of oral microbiome [22–24].

The relationship between innate immunity and oral microbiome is highly variable within an individual over time, depending on the general health, and prior exposure and responses to pathogens, which is the "immunobiography" of that person, using innate memory. Because of this historical dependence and that innate memory is unique to each patient, the symbiotic synergism between oral microbiome and innate immunity of the individual lends itself to personalized approach, to predict, prevent and even treat diseases. The composition of oral microbiome and the components of innate immune system are constantly co-evolving, shaped by individual lifestyle, immunobiography, epigenetic and genetic factors. As a result, they are unique to each person, providing a viable platform to apply predictive, preventive and personalized medicine (PPPM) in treating oral diseases, maintaining harmonious homeostasis, prolonging healthspan [3, 22–24].

15.2.1 Oral Microbiome and Diseases of the Mouth

Excluding trauma and cancer, oral diseases are either dental or periodontal. The prevalence of dental caries in children 2 to 8 years of age is 37%, and for adults 20 to 64 years of age, the prevalence is 26% in the U. S. Regarding periodontal diseases, the U. S. CDC reports that 47% of adults 30 years or older have some form of periodontal diseases. The prevalence of periodontal diseases increases to 70% by age 65 years and older. Reflecting and in agreement with all these, 17% of the adults 65 years or older are edentulous. [CDC 22019] For dental caries to develop there must be bacteria and sugar. Other factors such as salivary composition and quantity, genetic factor (amylogenesis imperfecta and dentinogensis imperfecta) influence cariogenicity, but the two key elements are sugar and bacteria, specifically, but not limited to, *Streptococcus mutan* and *S. sobrinus* [25, 26] (Fig. 15.5).

Many bacterial species can ferment sugar and produce short chain organic acids such as formic acid, lactic acid, acetic acid, butyric acid, and succinic acid. The list of acid-producing organisms has been growing and now include *Bifidobacterium*, *Actinomyces* and *Propionibacterium* species, and *Scardovia wiggsiae*. The pKa of these short chain organic acids ranges from 3.75 (formic acid) to 4.75 (acetic acid),

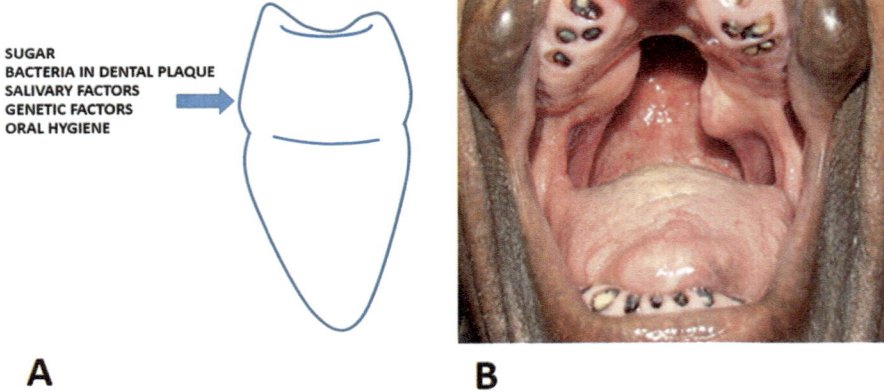

A   **B**

Fig. 15.5 (**a**) Five key factors affecting the development of dental caries: sugar, bacteria, salivary and genetic variables, and oral hygiene. Proper oral hygiene by brushing (2 minutes twice a day) and flossing (once daily) with reduction and control in sugar intake will greatly improve dental health. (**b**) An individual with acquired xerostomia (dry mouth due to orofacial clefts) has sustained insufficient salivary quantity, resulting in extensive and rampant dental caries of all teeth down to the gingival level

easily reaching below the critical pH of 5.5 for dissolution of enamel. Thus, dental caries is preventable if one can keep salivary pH above this critical level by oral hygiene or diet with limited fermentable sugars [27]. However, once the repeated and sustained decalcification results in an enamel defect, typically as a cone oriented to the direction of enamel prisms. The presence of such incipient carious lesions creates niches for more bacterial colonies that are no longer amenable to removal by brushing. The progression of acid dissolution beyond the DEJ (dentino-enamel junction) creates a second cone-shaped area, forming the classic diagnostic stacked triangular radiolucent appearance in bitewings or periapical radiographs [28]. Since odontoblastic processes within the dentinal tubules innervate dentine, transmitting signals to nociceptive nerve endings in the pulp, the patient will begin experiencing temperature sensitivity and pain. Because dentine is more soluble at pH produced by the organic acids, the rate of progression of the carious lesion increases. As the destruction approaches the dental pulp, host innate response produces inflammation with hyperemia, increase in capillary permeability, and extra-vascular migration of neutrophils, macrophages, and other cellular effectors, and chemical mediators. There is a rapid increase of intra-pulpal pressure. This is pulpitis. Following the closure of the apical foramen, which happens early in adolescence, there is no room to accommodate the swelling, setting up a vicious cycle of inflammation and ischemia. The initial reversible pulpitis frequently deteriorates to irreversible pulpitis causing severe, unrelenting odontalgia, paradoxically relieved with eventual pulpal necrosis and severe destruction of the tooth structures (Fig. 15.6).

If bacterial infection spreads beyond the root apex, peri-apical abscess can form, with destruction of the alveolar bone. In advanced cases, this progresses to purulent

Fig. 15.6 Dental radiograph showing advanced caries of the right maxillary first premolar (Tooth #5) with severe destruction of the distal occlusal part of the crown

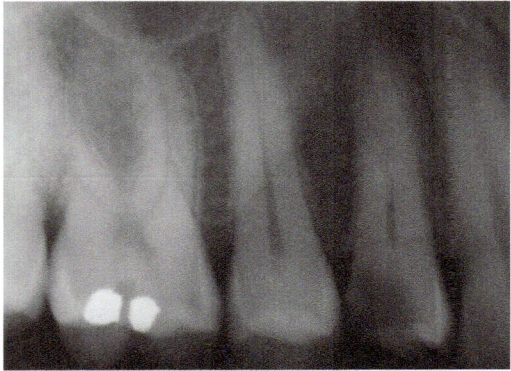

Fig. 15.7 A peri-apical abscess with destruction of the alveolar bone. forming a fistula to the right cheek. The treatment includes resection of the cutaneous opening, the entire fistula, and bone surrounding the abscess at the apex of the roots

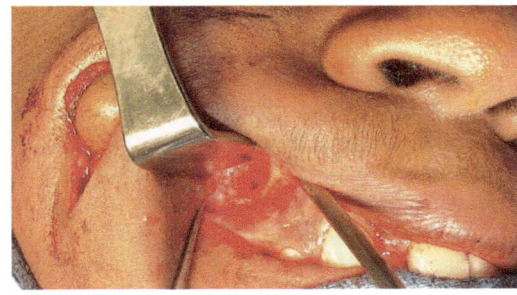

oro-cutaneous fistula (Fig. 15.7). Serious life-threatening infections of odontogenic origin, such as Ludwig's angina and Lemieere's disease, have a reported mortality rate of 5.8%. If the infection spreads to the posterior mediastinum, 2 out of 3 patients may die [29].

Though biofilm-induced like caries, the development of periodontal disease follows a different path. The host-microbiome dysbiosis is much more relevant. Six in ten adults older than 65 years of age in the U. S. have periodontal disease. One in twelve adults have advanced periodontal disease serious enough to affect their systemic health. Periodontal diseases are very costly. Data from 2018 indicate total cost (aggregate + indirect) to be $150 billion for the U. S. and €156 billion in Europe [30]. Periodontitis begins with the loss of host-oral microbiome homeostasis, by whatever the cause. Innate immune system, through ligand binding of receptors for PAMP and DAMP such as TLR, NLR (Nod-Like Receptor), and RLR (RIG-1-Like Receptor), mount an intense reaction against some members of the oral microbiome, especially those of the Red Complex: *Treponema denticola*, *Porphyromonas gigivalis*, and *Tannerella forsythia*. Of them, *P. gingivalis* is particularly problematic in that it can incite more inflammation and further destruction of the periodontium (tissue supporting the tooth, such as periodontal ligament, alveolar bone, attached gingiva, etc.). For this, *P. gingivalis* is the keystone pathogen. The tissue destruction from inflammation provide nutrients for newly pathogenic bacteria

Fig. 15.8 A cone-beam
CT coronal section of
severe advanced
periodontitis with
extensive bone loss and
erosion into the dental pulp

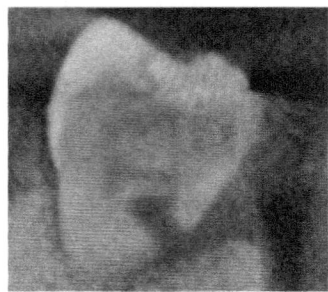

formed from previously non-pathogenic bacteria (thus pathobionts). The spillage of
sequestered intracellular parts from destroyed tissue such ATP, fragments of nucleic
acids, subcellular parts are potent alarmins, which together with bacterial components, increases the innate immune response with outpouring of classic inflammatory mediators (prostaglandins, leukotrienes), chemokines (TNF-alpha, interferons),
interleukins (IL-2, IL-6, IL-17, etc.) and activated neutrophils, macrophage, ILCs,
NKs. However, such intense responses do not rid the tissue of biofilms made of
polymicrobial pathogens. Like any prolonged battle, the loser is the battlefield. The
depth of periodontal pockets increases, further lowering O_2 tension promoting
obligatory anaerobic pathogens, while making it physically impossible for their
elimination by brushing [31] (Fig. 15.8).

Bone resorption becomes inevitable as the battle rages and with formation of
osteoclasts and release of RANKL (Receptor Activator of Nuclear Factor Kappa-B
Ligand, a member of the TNF superfamily). As the alveolar crest recedes, bone support becomes compromised. Even normal masticatory forces exceed the limit and
increase trauma to the remaining periodontium (secondary occlusal traumatism)
[32]. The futile efforts by the innate immune system fail to remove the pathogen but
succeed in activating the more powerful adaptive immune system. Most of the time,
the host-pathobiont duel ends with the exfoliating of the involved tooth or teeth.
Edentulous alveolus, like belated peace after the cities have been utterly destroyed,
allows for de-escalation on both sides, and a new homeostasis.

15.3   Systemic Health and Oral Microbiome

That oral health has significant impact on systemic health is indisputable [33, 34].
This section will review how oral microbiome and oral health affect systemic health
from the following perspectives: neurological, cardiovascular, reproductive medicine, immunological, metabolic, and gastrointestinal. Importantly, we will critically
examine the data supporting these conjectures.

Recent studies have shown a clear link between certain oral microbes
(*Pasteurellacae* and *Lautropia mirabilis*) and Alzheimer disease, which has a prevalence of 5.5 million in the U. S., costing $ 215 billion in 2010 [35]. One of the major

difficulties regarding measuring cognitive function in the aging population is the perpetual problem of having to set thresholds to convert a continuous distribution (cognitive function) to discrete groups (normal versus disease). On both ends of this continuous spectrum, normal aging and Alzheimer dementia, the picture is very clear. It is the middle of the continuum that poses the diagnostic dilemma. At what point does decline related to normal aging become mild cognitive Impairment? The value of such diagnostic classification resides in our ability to predict and, more importantly through intervention, to prevent future disease development. The current understanding is that there is a feedforward cycle involving cognitive impairment, oral microbiome, and poor oral hygiene, with chronic inflammation at the center (Fig. 15 9). However, reports of association between tooth loss and worsening cognitive impairment may be due to both having a common cause: aging. On the other hand, the isolation of oral bacteria such as *P. gingivalis, F. nucleatum*, and *P. intermedia* in the brain of patients died of Alzheimer's dementia lends more credence to this theory of oral microbiome-induced chronic systemic inflammation being a common cause of multiple pathologies. Unfortunately, in a well-designed and successfully executed study, with sufficient power, failed to demonstrate a difference in the oral microbiome between those with mild cognitive impairment and those without. Nor was there a difference in the serum levels of two inflammatory markers [36]. It is entirely possible that the systemic antibiotic therapies employed did not achieve sufficient changes in Shannon diversity or beta diversity.

Medical and dental literatures abound with articles showing associations between cardiovascular diseases and oral microbiome. The pathological mechanisms of

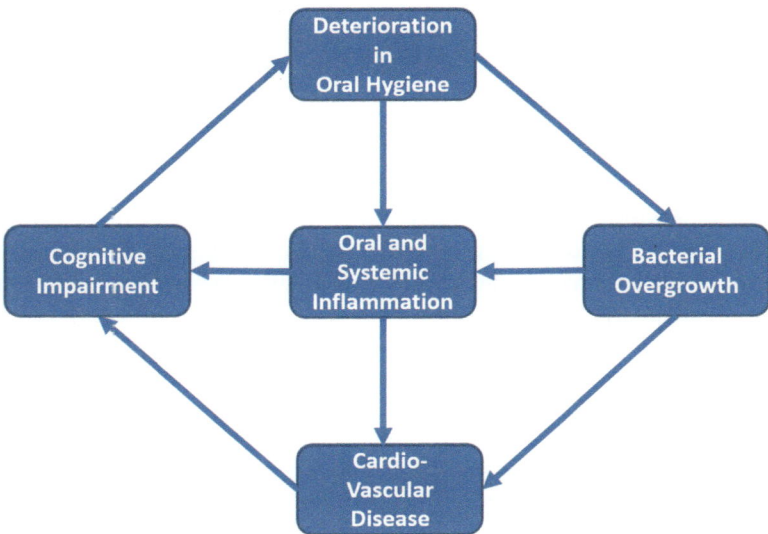

Fig. 15.9 The feedforward cycle linking cognitive impairment and atherosclerotic cardiovascular diseases and oral microbiome, with elevated inflammation at the center of the complex web of events

developing atherosclerotic cardiovascular diseases (ASCVD) include systemic endothelial dysfunction due to oral microbes, their products, or excessive immune response to them. In the early stages, patients with periodontitis have increase stiffness of their arteries (measurable by increase in pulse wave velocity—PWV), decrease in flow-mediated dilation (FMD), and a thickening of the carotid intima-media (cIMT) with elevated arterial calcification scores. Within the oral cavity, periodontal health requires normal oral microbiome *and* a competent, tolerant innate immune system capable of discriminating commensal from pathogenic microbes and self from non-self. Failure of such oral homeostasis is periodontitis, a non-communicable infectious disease due to dysbiotic oral microbiome (Fig. 15.8). Since bacteremia originating from oral flora occur several times daily, it is logical to consider inflammation due to this shedding of microbes into the blood stream from dysbiotic oral microbiome contributes to atherosclerosis. From an evolutionary perspective, injuries inevitably occur with struggles for existence and with that entry of microbes into blood stream. The endothelial surface and circulating innate immune cells such as neutrophils must develop defensive responses once they come in contact with these microbes and microbial products, binding to them with TLRs- one of the most important proinflammatory receptor. Endothelium can concentrate myeloperoxidase (MPO) produced by neutrophils upon detection of pathogens in the blood, causing arterial contraction due to break down of nitric oxide (NO) by MPO. There is also breakdown of the glycocalyx layer above the surface of the endothelium with opening of the intercellular junctions to allow macrophage to enter the subendothelial space, phagocytizing host and microbial debris, forming foamy macrophages. If sustained and repeated, the vascular inflammation progresses to formation of atheromatous plagues, some containing members of the Red Complex such as *P. gingivalis*. The inevitable consequence is narrowing of the vascular lumen, increasing in blood flow velocity, resulting in higher shear stress, which exacerbates and accelerate atheromatous plaque formation. This is a dangerous feedforward cycle. Many epidemiological investigations confirm the link between oral bacteria and atherosclerosis. Bartova et al. reported in 2014 a 25–50% increase in ASCVD in periodontal patients when compared to control population [37]. Case-control studies also showed clearly that the antibody titers against oral microbial species associated with periodontitis are higher in patients with ASCVD. However, large prospective clinical trials using antibiotics specifically against periodontal pathogens in more than 16,000 patients with coronary artery diseases and more recent trials showed no long-term improvements [38]. Again, this may simply reflect the difficulty of altering oral microbiome with systemic antibiotic therapy.

From the perspective of reproductive medicine, oral microbiome plays an enormous role, with oral pathogens such as *Fusobacterium nucleatum, Porphyromonas gingivalis, Filifactor alocis,* and *Campylobacter rectus*, having been well documented to cause premature labors [39]. National Health and Nutrition Examination Survey from a decade ago has documented the tremendous high prevalence of periodontitis, 46% or 65 million people, (age 30 to 65 years). Intrauterine inflammation, measured by amniotic fluid interleukin 6 levels ≥ 11.3 ng/mL, is present in 20% of the preterm labors. Premature labor and low birth weight are the principal cause of infant mortality [40]. Several correlational studies have found periodontal diseases increased odds ratio of preterm labor. In addition to preterm labor, preeclampsia (spilling protein in the urine with new onset hypertension by 20 weeks of gestation) is a dreaded complication that occurs in 2% - 8% of all pregnancies. The breakdown of NO by MPO, as outlined, may well be part of the pathogenesis. Perhaps the best supporting evidence is from the identification of periodontitis keystone pathogens within the placenta of preterm labors. As with many other health issues related to oral microbiome, the prevention is simple and effective: better education of the population on proper oral hygiene [41].

15.4   Oral Microbiome and Diseases: Impact of COVID-19 on Oral Microbiome

After the gut, the oral microbiota is the largest one in the human body [42] and contributes significantly to the training/development of the immune system and the maintenance of immune homeostasis [43].

The most updated data shows that on March 25th, 2022, the COVID-19 pandemic infected more than 480 million people, resulting in over 6.1 million deaths around the world [44]. Many laboratories been exploring the influence of the oral microbiota and SARS-CoV-2 infection, and vice-versa, on the susceptibility and severity of COVID-19 disease [43, 45, 46].

The interaction of the SARS-CoV-2 virus with lung and oral micro-flora may cause dysbiosis, an imbalance between the types of organisms presents in the host, due to changes in the immune cells' responses and cytokines profile [46]. Soffritti and colleagues (2021) showed that COVID-19 patients had lower alpha-diversity and other indices, meaning that those patients had lower species richness and evenness [6]. On the other hand, the oral microbiome metabolites can modulate the host's immune response to several infectious diseases, influencing the disease outcome [47].

Poor oral hygiene habits can lead to the accumulation of many periodontal microorganisms, causing infected, and inflamed gingiva (gingivitis) that may be related to higher risks of complications from COVID-19, including death. Patients with gingivitis and other periodontal disease had higher levels of indicators of inflammation in the blood, which may explain the higher rates of complications [48].

A better understanding of oral microbiota and its influence on lung infectious diseases is crucial to the application of the PPPM (Personalized, predictive, and preventive medicine) concept in health care. The concept of PPPM enables the health care system to predict a patient's predisposition to diseases before onset, provide preventive measures and create a personalized treatment [49].

COVID-19 patients usually receive antimicrobial therapy to prevent co-infections [45, 50, 51]. This is an example in which the PPPM concept could be applied. The use of the right antimicrobial drugs for each COVID-19 patient according to its microecological changes could be decisive to the COVID-19 onset. The same concept can be applied to oral health. Personalized oral care would help to reduce the incidence of periodontal diseases and their destructive effects related to lung infectious diseases [43, 45, 48].

15.5   Practice Guidelines for the Maintenance of Oral Health

Oral microbiome is critically important in oral and systemic health. For the modest efforts required, the reward in improvement and maintenance of individual's wellbeing is enormous. All healthcare providers can and should advise their patients to make these small investments in their personal health. There are several simple and effective guidelines and recommendations for the maintenance of oral health in children, adolescents, and seniors, including those with special health care needs [52, 53]. As shown in Fig. 15.10, risk factors for oral disease can be categorized into two main categories of modifiable (e.g., sugar consumption, smoking, vaping, drinking alcohol, etc.) and non-modifiable (e.g., gender, age, etc.) [54, 55].

Although we cannot change the impact of non-modifiable risk factors, however, by eliminating the modifiable risk factors, we can improve the outcome of oral health. As a consequence of global population aging and increase in the number of vulnerable group of people for oral diseases, practicing these guidelines and recommendations will help to prevent the predictive, unwanted oral diseases in a personalized fashion based on each person's unique situation. It is our hope that this prevention will reduce the emergent need for treatment of diseases related to poor dental care in the future [3].

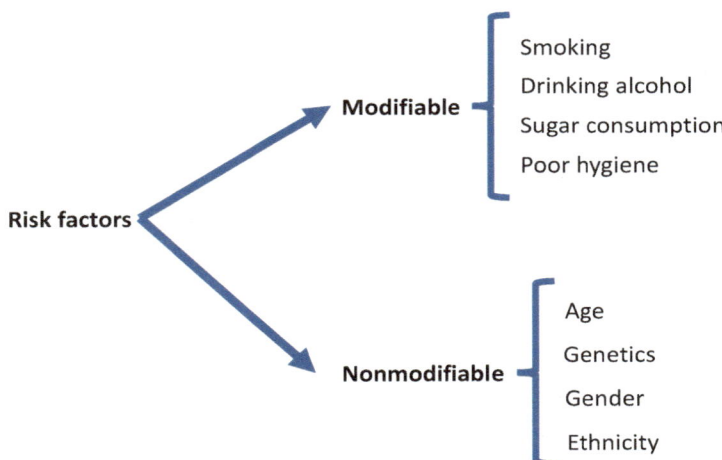

Fig. 15.10 Schematic risk factors for oral disease

Followings are the summary of the acceptable methods and recommendations to maintain the oral health in general population:

- Annual dental checkup by visiting dentist.
- Stop smoking and vaping.
- Reduce alcoholic beverage consumption.
- Reduce amount of sugar consumption, avoid high fructose corn syrup.
- Fluoride therapy by using fluoride toothpaste and drinking fluoridated water.
- Good oral hygiene routine by brushing and flossing teeth thoroughly two times per day for at least two minutes.
- Individuals from vulnerable groups, such as those with obesity, diabetes, hypertension, need to pay more attention to control their underlying disease and dental hygiene to reduce the oral diseases possibility [53].

References

1. Deo PN, Deshmukh R (2019) Oral microbiome: unveiling the fundamentals. J Oral Maxillofac Pathol 23:122–128
2. Gomez A, Nelson KE (2017) The Oral microbiome of children: development, disease, and implications beyond Oral health. Microb Ecol 73:492–503
3. Yu JC, Khodadadi H, Baban B (2019) Innate immunity and oral microbiome: a personalized, predictive, and preventive approach to the management of oral diseases. EPMA J 10:43–50
4. Nakamura S, Donatelli D, Rosenberg E (2021) Posterior bite collapse and diagnostic grading for periodontitis. Int J Periodontics Restorative Dent 41:61–69
5. Kailasam V, Rangarajan H, Easwaran HN, Muthu MS (2021) Proximal enamel thickness of the permanent teeth: A systematic review and meta-analysis. Am J Orthod Dentofac Orthop 160:793–804.e3
6. Soffritti I et al (2021) Oral microbiome Dysbiosis is associated with symptoms severity and local immune/inflammatory response in COVID-19 patients: a cross-sectional study. Front Microbiol 12:1–15
7. Willis JR, Gabaldón T (2020) The human oral microbiome in health and disease: from sequences to ecosystems. Microorganisms 8:1–28
8. Baker JL, Bor B, Agnello M, Shi W, He X (2017) Ecology of the Oral microbiome: beyond bacteria. Trends Microbiol 25:362–374
9. Yuan C, Lei J, Cole J, Sun Y (2015) Reconstructing 16S rRNA genes in metagenomic data. Bioinformatics 31:i35–i43
10. Moon JH, Lee JH (2016) Probing the diversity of healthy oral microbiome with bioinformatics approaches. BMB Rep 49:662–670
11. Sedghi L, DiMassa V, Harrington A, Lynch SV, Kapila YL (2000) The oral microbiome: role of key organisms and complex networks in oral health and disease. Periodontol 2021(87):107–131
12. Li H et al (2019) Impact of maternal intrapartum antibiotics on the initial oral microbiome of neonates. Pediatr Neonatol 60:654–661
13. Baban B, Malik A, Bhatia J, Yu JC (2018) Presence and profile of innate lymphoid cells in human breast milk. JAMA Pediatr 172:594–596
14. Ihara Y et al (2019) Identification of initial colonizing bacteria in dental plaques. Clin Sci Epidemiol 4:1–11
15. Lof M, Janus MM, Krom BP (2017) Metabolic interactions between bacteria and fungi in commensal oral biofilms. J Fungi 3

16. Martínez A, Kuraji R, Kapila YL (2000) The human oral virome: shedding light on the dark matter. Periodontol 2021(87):282–298
17. Haran JP et al (2021) Inflammation-type dysbiosis of the oral microbiome associates with the duration of COVID-19 symptoms and long COVID. JCI Insight 6:1–14
18. Netea MG, Schlitzer A, Placek K, Joosten LAB, Schultze JL (2019) Innate and adaptive immune memory: an evolutionary continuum in the Host's response to pathogens. Cell Host Microbe 25:13–26
19. Bauer J, Nelde A, Bilich T, Walz JS (2019) Antigen targets for the development of immunotherapies in leukemia. Int J Mol Sci 20
20. Cullender TC et al (2013) Innate and adaptive immunity interact to quench microbiome flagellar motility in the gut. Cell Host Microbe 14:571–581
21. Zenobia C, Herpoldt KL, Freire M (2021) Is the oral microbiome a source to enhance mucosal immunity against infectious diseases? Npj Vaccines 6:1–12
22. Baban B et al (2021) Inflammaging and Cannabinoids. Ageing Res Rev 72:101487
23. Boraschi D, Italiani P (2018) Innate immune memory: time for adopting a correct terminology. Front Immunol 9:1–4
24. McCoy KD, Burkhard R, Geuking MB (2019) The microbiome and immune memory formation. Immunol Cell Biol 97:625–635
25. Karpiński TM, Szkaradkiewicz AK (2013) Microbiology of dental caries. Rev J Biol Earth Sci 3:21–24
26. Mattos-Graner RO, Klein MI, Smith DJ (2014) Lessons learned from clinical studies: roles of Mutans streptococci in the pathogenesis of dental caries. Curr Oral Heal Reports 1:70–78
27. Siska Ella Natassa SP, Ilyas S (2019) Dental radiography effectivity of probiotic and non-probiotic milk consumption on salivary pH and streptococcus mutans count. J Dent Med Sci 18:67–72
28. Joen Iannucci LJH (2021) Dental radiography- principles and techniques
29. Bali R, Sharma P, Gaba S, Kaur A, Ghanghas P (2015) A review of complications of odontogenic infections. Natl J Maxillofac Surg 6:136
30. Botelho J et al (2022) Economic burden of periodontitis in the United States and Europe: an updated estimation. J Periodontol 93:373–379
31. Hajishengallis G (2014) Immunomicrobial pathogenesis of periodontitis: keystones, pathobionts, and host response. Trends Immunol 35:3–11
32. Fan J, Caton JG (2018) Occlusal trauma and excessive occlusal forces: narrative review, case definitions, and diagnostic considerations. J Periodontol 89:S214–S222
33. Lee YH et al (2021) Progress in oral microbiome related to oral and systemic diseases: an update. Diagnostics 11
34. Thomas C et al (2021) Oral microbiota: a major player in the diagnosis of systemic diseases. Diagnostics 11:1–29
35. Guo H et al (2021) Profiling the oral microbiomes in patients with Alzheimer's disease. Oral Dis:1–15. https://doi.org/10.1111/odi.14110
36. Yang I et al (2021) The oral microbiome and inflammation in mild cognitive impairment. Exp Gerontol 147:111273
37. Bartova J et al (2014) Periodontitis as a risk factor of atherosclerosis. J Immunol Res 2014:1–9
38. Seinost G et al (2020) Periodontal treatment and vascular inflammation in patients with advanced peripheral arterial disease: a randomized controlled trial. Atherosclerosis 313:60–69
39. Terzic M et al (2021) Periodontal pathogens and preterm birth: current knowledge and further interventions. Pathogens 10:1–13
40. Leaños-Miranda A et al (2021) Interleukin-6 in amniotic fluid: a reliable marker for adverse outcomes in women in preterm labor and intact membranes. Fetal Diagn Ther 48:313–320
41. Gare J et al (2021) Zperiodontal conditions and pathogens associated with pre-eclampsia: a scoping review. Int J Environ Res Public Health 18
42. Dewhirst FE et al (2010) The human oral microbiome. J Bacteriol 192:5002–5017

43. Wu Y et al (2021) Altered oral and gut microbiota and its association with SARS-CoV-2 viral load in COVID-19 patients during hospitalization. Npj Biofilms Microbiomes 7
44. WHO. No Title. https://covid19.who.int/
45. Bao L et al (2020) Oral microbiome and SARS-CoV-2: beware of lung co-infection. Front Microbiol 11:1–13
46. Iebba V et al (2021) Profiling of Oral microbiota and cytokines in COVID-19 patients Front Microbiol 12:1–13
47. Mammen MJ, Scannapieco FA, Sethi S (2000) Oral-lung microbiome interactions in lung diseases. Periodontol 2020(83):234–241
48. Marouf N et al (2021) Association between periodontitis and severity of COVID-19 infection: a case–control study. J Clin Periodontol 48:483–491
49. Golubnitschaja O, Kinkorova J, Costigliola V (2014) Predictive, preventive and personalised medicine as the hardcore of 'horizon 2020': EPMA position paper. EPMA J 5:1–29
50. Rawson TM et al (2020) Bacterial and fungal co-infection in individuals with coronavirus: a rapid review to support COVID-19 antimicrobial prescribing Timothy. Clin Infect Dis 71:2459–2468
51. Verweij PE et al (2020) Diagnosing COVID-19-associated pulmonary aspergillosis. The Lancet Microbe 1:e53–e55
52. Krishnan K, Chen T, Paster BJ (2017) A practical guide to the oral microbiome and its relation to health and disease. Oral Dis 23:276–286
53. CDC. Centers for Disease Control and Prevention. https://www.cdc.gov/oralhealth/basics/adult-oral-health/tips.html. Accessed on 28 March 2022 at 4:20 pm
54. Sheiham A (2005) Oral health, general health and quality of life. Bull World Health Organ 83:644
55. Kinane DF, Stathopoulou PG, Papapanou PN (2017) Periodontal diseases. Nat Rev Dis Prim 3:17038

Index

© The Editor(s) (if applicable) and The Author(s), under exclusive license to
Springer Nature Switzerland AG 2023
N. Boyko, O. Golubnitschaja (eds.), *Microbiome in 3P Medicine Strategies*,
Advances in Predictive, Preventive and Personalised Medicine 16,
https://doi.org/10.1007/978-3-031-19564-8

Printed by Printforce, the Netherlands